Android 编程宝典

秦建平　编著

北京航空航天大学出版社

内容提要

本书讲解 Android 手机平台开发从入门到精通的相关知识，全书内容共分为 3 篇。

第 1 篇是 Android 基础编程，主要介绍 Android 平台编程的基础知识，包括活动、意图、广播接受者、用户界面布局、常用控件、菜单、通知、闹钟服务、事件处理、数据存储、内容提供者以及 Android异步处理机制等相关内容。基础编程这一篇所涉及的知识点贯穿于 Android 客户端开发工作的始终，是进行 Android 客户端开发的一条重要线索。

第 2 篇是 Android 高级编程，主要介绍 Android 平台编程的高级知识，包括服务、基于位置的服务、网络编程、多媒体、传感器、图形和图像、硬件接口以及 Android 的桌面组件等相关内容。

第 3 篇是 Android 实战应用。此篇介绍了一个基于 B/S 架构的电子订餐系统，包括 Android 客户端的开发以及服务端的开发等相关知识，服务端采用了完整的 JAVA EE 分层架构，整个应用具有良好的可扩展性和示范性。

本书并不局限于枯燥的理论介绍，而是采用实例的方式来讲授知识点，以便读者可以更好地阅读以及进行相关知识点的理解和发散。在内容上，涉及当前移动互联网领域一些拥有大量用户数的客户端应用的一些特色功能的原理介绍以及代码实现。

如果读者对 Java 语法比较熟悉，并且有一定的事件驱动的程序编程经验，那么阅读本书就可以很快掌握 Android 应用开发。本书适合想全面学习 Android 开发的人员阅读，对经常使用 Android平台做开发的人员，更是一本不可多得的案头必备参考书。

图书在版编目(CIP)数据

Android 编程宝典 / 秦建平编著. --北京 ：北京航空航天大学出版社，2013.3

ISBN 978-7-5124-1080-0

Ⅰ. ①A… Ⅱ. ①秦… Ⅲ. ①移动终端—应用程序—程序设计 Ⅳ. ①TN929.53

中国版本图书馆 CIP 数据核字(2013)第 040372 号

Android 编程宝典

秦建平 编著

责任编辑 陈 旭

*

北京航空航天大学出版社出版发行

北京市海淀区学院路 37 号(邮编 100191) http://www.buaapress.com.cn

发行部电话：(010)82317024 传真：(010)82328026

读者信箱：emsbook@gmail.com 邮购电话：(010)82316936

涿州市新华印刷有限公司印装 各地书店经销

*

开本：710×1 000 1/16 印张：35.25 字数：751 千字

2013 年 3 月第 1 版 2013 年 3 月第 1 次印刷 印数：3 000 册

ISBN 978-7-5124-1080-0 定价：79.00 元

前 言

Android 是一种基于 Linux 的开放源代码的操作系统，主要用于便携设备，如智能手机和平板电脑。目前尚未有统一的中文名称，中国大陆地区较多人使用“安卓”或“安致”。Android 操作系统最初由 Andy Rubin 开发，主要支持手机。2005 年由 Google 收购注资，并组建开放手机联盟开发改良。随后逐渐扩展到平板电脑及其他领域上。第一部 Android 智能手机发布于 2008 年 10 月。2011 年第一季度，Android在全球的市场份额首次超过塞班系统，跃居全球第一。2012 年 11 月的数据显示，Android 占据全球智能手机操作系统市场 76％的份额，中国市场占有率为 90％。

在智能手机上开发各种应用，被认为是继 PC 时代、互联网时代后的第 3 波 IT 技术浪潮。基于手机开发各种互联网应用，可以说是未来 10～20 年的基本技术趋势。任何希望在 IT 技术界有所建树、有所发明创新的人不可忽视 Android 开发平台，需要尽早学习，进入这个领域。

笔者结合自己的 Android 平台开发经验和心得体会，花费了一年多的时间写作本书。希望各位读者能在本书的引领下跨入 Android 平台开发的大门，并成为一名开发高手。本书最大的特色就是结合大量的说明插图，全面、系统、深入地介绍了 Android 平台的开发技术，并以大量实例贯穿于全书的讲解之中，最后还详细介绍了结合服务端与客户端的一个完整的实战应用。学习完本书后，读者应该可以具备独立进行编程开发的能力。

本书特点

1. 大量教学插图，读书学习不再枯燥乏味

本书最大的特点就是通篇采用图片讲解，将传统的文字讲解转换为各种形式的图形图表，最大限度地提升读者的阅读兴趣，让读者在潜移默化中掌握 Android 平台的开发精髓。

2. 讲解由浅入深，循序渐进，适合各个层次的读者阅读

本书从 Android 平台编程基础开始讲解，逐步深入到 Android 平台的高级应用，内容梯度从易到难，讲解由浅入深，循序渐进，适合各个层次的读者阅读。

3. 贯穿大量的开发实例和技巧,迅速提升开发水平

本书在讲解知识点时贯穿了大量的典型实例,并给出了大量的开发技巧,以便读者更好地理解各种概念和开发技术,体验实际编程,迅速提高开发水平。

本书内容及体系结构

❑ **第 1 篇 Android 基础编程(第 1～10 章)**

本篇主要内容包括:初识 Android、活动、意图和广播接收者、用户界面、常用控件、菜单、通知以及闹钟服务、Android 事件处理、数据存储、内容提供者以及 Android 异步处理机制。通过本篇的学习,读者可以掌握 Android 编程的基础知识。

❑ **第 2 篇 Android 高级编程(第 11～18 章)**

本篇主要内容包括:服务、LBS、网络编程、多媒体、传感器、Android 图形和图像、Android 硬件接口以及 Android 桌面组件。通过本篇的学习,读者可以掌握 Android编程的高级技术。

❑ **第 3 篇 Android 实战应用(第 19 章)**

本篇主要内容包括:介绍了一个基于 B/S 结构的电子订餐系统,包括客户端的开发以及服务端的开发,是一个 Android＋Servlet＋JDBC＋JSP 整合的应用。本应用服务端采用了完整的 JAVAEE 应用架构,应用架构采用了具有高度可扩展性的控制器层(Servlet)＋ 视图层(JSP)＋数据访问层(DAO)的分层架构。Android 客户端通过网络与服务器的控制器组件(Servlet)交互,整个应用具有极好的可扩展性和示范性。

本书读者对象

❑ Android 平台初学者;
❑ 想全面学习 Android 平台开发技术的人员;
❑ Android 平台专业开发人员;
❑ 利用 Java 语言做 Android 开发的工程技术人员;
❑ Android 平台的开发爱好者;
❑ 大中专院校的学生;
❑ 社会培训班学员。

致谢

秦建平是本书主要的编著者,另外,魏春、张伟、水淼、高智雷、张利峰、关玉琴、王海兴、张昆、朱虹颖、李克、赵凤艳、李东博、向旭宇、秦姣华、刘桂珍等也参与了本书的资料收集和部分内容的编撰工作。在此对他们的工作表示感谢。

本书在编写过程中,编者虽然未敢稍有疏虞,但纰漏和不尽如人意之处在所难免,诚请读者提出意见或建议,以便修订并使之更臻完善。

作　者
2012 年 12 月

目　录

第1篇　Android 基础编程

第 2 篇 Android 高级编程

第 1 篇　Android 基础编程

第 1 章 初识 Android

Android 一词的本义指“机器人”，同时也是 Google 于 2007 年 11 月 5 日宣布的基于 Linux 平台的开源手机操作系统的名称，该平台由操作系统、中间件、用户界面和应用软件组成。它采用软件堆层（Software Stack，又名软件叠层）的架构，主要分为 3 部分。底层以 Linux 内核工作为基础，由 C 语言开发，只提供基本功能。中间层包括函数库 Library 和虚拟机 Virtual Machine，由 C＋＋开发。最上层是各种应用软件，包括通话程序、短信程序等，应用软件则由各公司自行开发，以 Java 作为编写程序的一部分。

Google 通过与软、硬件开发商、设备制造商、电信运营商等其他有关各方结成深层次的合作伙伴关系，希望借助建立标准化、开放式的移动电话软件平台，在移动产业内形成一个开放式的生态系统。全球为数众多的移动电话用户正在使用各种基于 Android 的电话。Google 的目标是让（移动通信）不依赖于设备甚至平台。出于这个目的，Android 将补充而不会替代 Google 长期以来奉行的移动发展战略：通过与全球各地的手机制造商和移动运营商结成合作伙伴，开发既有用又有吸引力的移动服务，并推广这些产品。

本章将会介绍如下知识：

- 为什么要开发 Android 应用程序；
- 搭建 Android 开发环境；
- 编写 Hello World；
- Android 编程基础。

1.1 为什么要开发 Android 应用程序

事实上真正的问题应该是“为什么不呢？”，你想要你写的应用程序被世界上各个国家的用户所使用吗？你想要在开发和测试完成之后马上发布你的应用程序吗？你喜欢在开放平台进行开发吗？如果你对这些问题的答案都是肯定的，那么你就应该去开发 Android 应用。

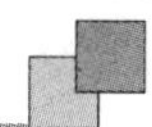

1. Google Play

作为一名开发者，你有机会让你的应用程序被世界上的大多数 Android 用户所使用。有了 Google Play(前身是 Android Market)，Android 用户就没必要通过互联网搜索去发现自己想要安装的应用了，直接去 Google Play 上面安装自己喜欢的任何应用程序(apps)即可。由于 Google Play 客户端一般都会预装在 Android 终端手机里面，所以用户只需要打开 Google Play 的客户端，在搜索栏中输入自己想要的应用程序的名称，就可以搜索到自己想要的任何应用程序，然后安装这些应用。

由于 Google 提供了非常方便的 API(Application Programming Interface)，所以开发者在相对较短的时间内很容易去开发一些满足用户需求的应用程序，在给应用程序正式签名过后，开发者就可以在 Google Play 上面去发布应用程序。

2. 开放平台

Android 操作系统作为一个开放平台，意味着 Android 不会与某一个硬件生产商或者是硬件提供商相绑定。所以读者就可以想象出来，为什么 Android 能够在这么短的时间内拥有这么大的市场份额，因为所有的硬件生产商和提供商都可以制造和销售它们的 Android 设备。https://android.googlesource.com 是 Android 操作系统的源码地址，允许用户去查看和进行修改。开源的系统允许手机生产商去定制用户界面，还可以添加其他内置的应用。

3. 兼容性

Android 操作系统可以在很多不同屏幕尺寸大小和不同分辨率的设备上运行，Android 也提供了一些工具帮助开发者去开发兼容性好的应用程序。Google 允许开发者的应用程序只在兼容的设备上运行，举个例子，如果开发者的应用程序需要使用前置摄像头，那么只有带前置摄像头的 Android 设备在打开 Google Play 客户端的时候才会看到开发者发布的应用程序，这一项功能叫做特征监测。为了让 Android 设备具备兼容性，终端厂商必须去遵循一些硬件的指导方针。这些指导方针包括以下几个方面但不限制于以下几个方面：

- ❑ 摄像头。
- ❑ 指南针。
- ❑ GPS 定位。
- ❑ 蓝牙。

读者可以通过访问 http://source.android.com/compatibility/overview.html 来查看什么样的设备配置会被认为是具备兼容性的。设备的兼容性保证了应用程序可以在任意 Android 设备上面运行。

4. 聚合能力

聚合能力是指可以将一项或者多项服务组合起来去创建一个应用程序。开发者可以将摄像头应用与定位服务结合起来创建应用，举个例子，在拍照的时候，可以将拍照地点的位置信息显示在图片上。使用 Android 系统本身提供的 API，可以很方便地将两项或者多项 Android 系统本身提供的内置功能用于制作开发者自己的应用程序。比如，开发者可以使用 Google Maps API 将所有的联系人显示在地图上面。

1.2 搭建 Android 开发环境

用于开发 Android 应用的所有的软件都是免费的。在接下来的一节中，读者将会发现开发 Android 应用所用到的所有构建块包括开发工具、Android 框架以及 Android 源码都是免费的。当然读者开发所用到的计算机不是免费的，但是开发 Android 应用所要准备的开发环境都是免费的。在本节中，将会介绍一些必要的步骤去正确搭建 Android 的开发环境。

1.2.1 Android 源码

读者应该意识到 Android 源代码是开源的，这就意味着读者不仅仅可以免费使用这些源码，而且还可以对这些源代码进行修改。读者可以去下载 Android 的源代码，然后创建一个新的 Android 版本出来，这也是没有问题的。读者可以通过 http://source.android.com 这个链接下载 Android 最新版本的源代码。

1.2.2 Android 整体架构

图 1-1 显示了 Android 系统主要的组件。

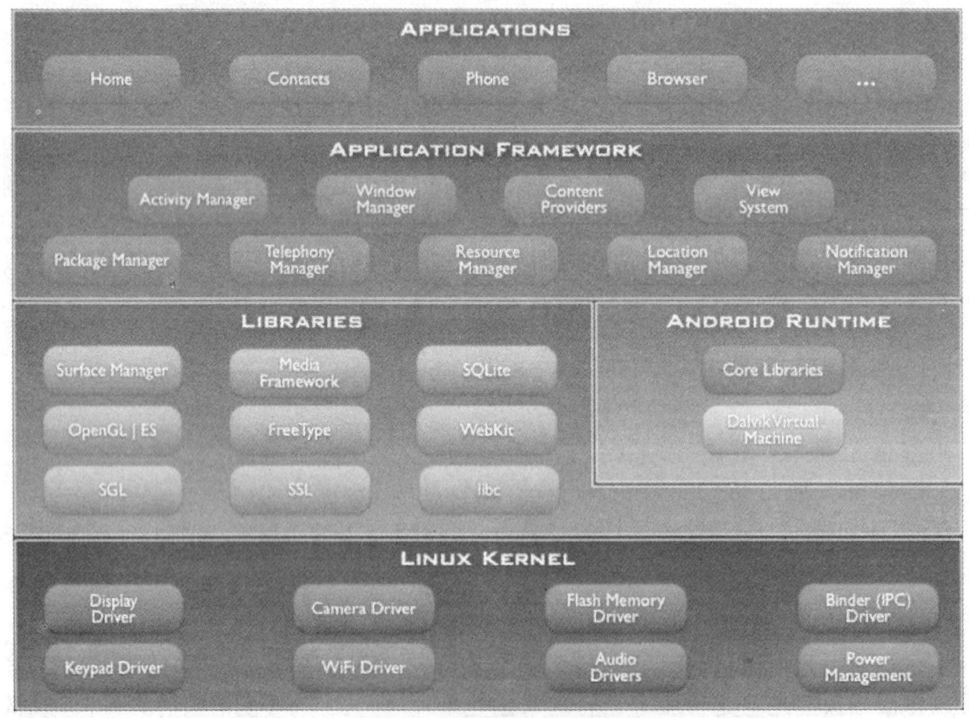

图 1-1 Android 系统架构

1. Linux 2.6 内核

Android 操作系统是创建于开源的 Linux 2.6 内核之上。Linux 2.6 内核包含的功能有以下几个(但不仅仅局限于这几个):

- ❑ 安全模型:Linux 内核处理应用程序与系统之间的安全。
- ❑ 内存管理:Linux 内核处理内存管理,使得开发者不需要去进行内存管理。
- ❑ 进程管理:Linux 内核处理进程管理,为进程分配所需要的资源。
- ❑ 网络栈:Linux 内核处理网络通信。
- ❑ 驱动模型:硬件生产商可以将他们的驱动集成于内核之中。

2. Android 框架

Linux 2.6 内核上是 Android 框架。Android 框架提供了很多功能,这些功能大多数都是从一些开源项目当中提取出来的。这些功能主要包括以下几个:

- ❑ The Android run time:由 Java 核心库和 Dalvik 虚拟机组成。
- ❑ Open GL:绘制 2D 以及 3D 图形。
- ❑ WebKit:开源 Web 浏览器引擎提供了显示 Web 内容的功能以及简化了页面加载。
- ❑ SQLite:用于嵌入式设备的开源关系数据库引擎。
- ❑ Media Framework:播放和录制音视频。

3. 应用程序框架

通过上面介绍的 Linux 2.6 内核以及 Android 框架,读者可能会想:这些东西都很好很强大,但是站在开发者的角度如何去用它们呢?事实上很简单。所有这些开源的框架都可以通过 Android 提供的应用程序框架来使用。开发者不用担心 Android 应用程序框架如何与下面的 Android 框架如何进行交互,只需要把 Android 提供的应用程序框架当作工具、当作类库来使用就可以了。Android 开发团队通过向开发者暴露接口,已经提供了很多功能完善的类库。这些类库主要包括以下几个(但不仅仅局限于):

- ❑ Activity Manager:用于管理 Activity 的生命周期。
- ❑ Telephony Manager:提供了对电话服务的访问。
- ❑ View System:处理用户界面的视图和布局。
- ❑ Location manager:对设备的地理位置信息进行定位。

提示:有的时候,开发者开发的应用程序需要去访问 Android 系统里面的资源,比如:要去访问内置 Settings 应用程序的图标。通过查看 Android 源代码,可以去浏览很多资源文件,当然开发者可以下载这些资源文件以用于自己开发的应用程序。

1.2.3 搭建 Android 开发环境

“工欲善其事,必先利其器”。所以,在开始开发 Android 应用程序之前,有以下一些开发工具和 SDK 需要去安装和配置:

- ❑ Java JDK:为 Android SDK 奠定基础。
- ❑ Android SDK:开发者可以通过 Android SDK 去访问 Android 提供的类库,从而编写 Android 应用程序。
- ❑ Eclipse 集成开发环境:把 Java、Android SDK 以及 ADT 都结合起来的集成开发环境,为开发者编写 Android 应用程序提供了相应的工具。
- ❑ Android ADT:Eclipse 平台下用来开发 Android 应用程序的插件。用 ADT 可以进行 AVD(Android 模拟器)的管理以及最新版本 SDK 的下载。

1.2.4 下载和安装 JDK

安装 JDK 的具体过程如下:

(1) 在浏览器中输入地址:http://www.oracle.com/technetwork/java/javase/downloads/index.html,输入地址之后,Java SE 的下载页面将会出现在 Web 浏览器当中,如图 1-2 所示。

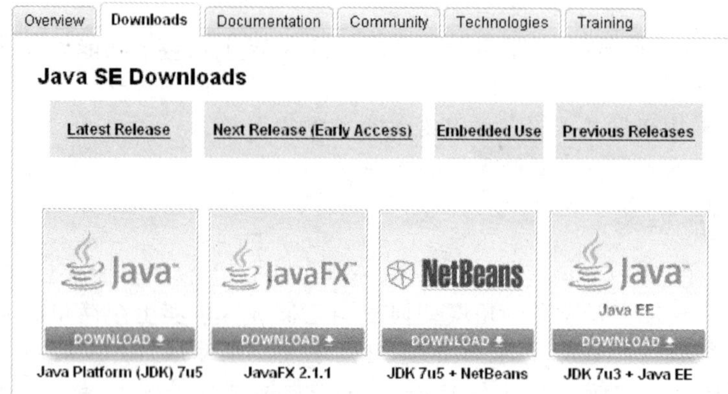

图 1-2 JDK 下载首页

(2) 单击图 1-2 的最左边 JDK 的下载链接,这个页面才是 JDK 的真正的下载链接页面,根据开发者计算机的操作系统的差异性,读者选择适合自己计算机操作系统的 JDK 下载。

(3) 确认好计算机的操作系统之后,单击相应的下载链接进行下载。

(4) JDK 下载完成之后,双击下载的文件进行安装。由于安装步骤比较简单,在这里笔者就不一一罗列出来,请读者自行进行安装。

(5) JDK 安装完成之后,需要配置 JAVA_HOME 的环境变量,如果不配置 JAVA_HOME 的环境变量,后面在安装 Eclipse 的时候会报还没有安装 JDK 的错误。具体步骤为:右击"我的电脑",选择"属性"命令,弹出"系统属性"对话框。选择"高级"选项卡,单击"环境变量"按钮,在弹出的"环境变量"对话框中的下面一栏的系统变量中,单击"新建"按钮,如图 1-3 所示。

(6) 在"变量名"文本框中输入 JAVA_HOME,并在"变量值"文本框中输入 JDK

安装的目录路径，笔者 JDK 安装的目录路径为：C:\jdk1.6.0_05，如图 1-4 所示。输入完成之后，单击“确定”按钮，至此 JDK 安装完毕。

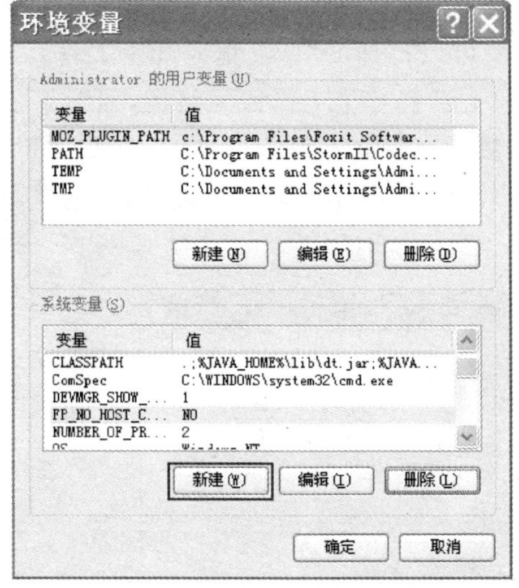

图 1-3　弹出新建系统变量对话框

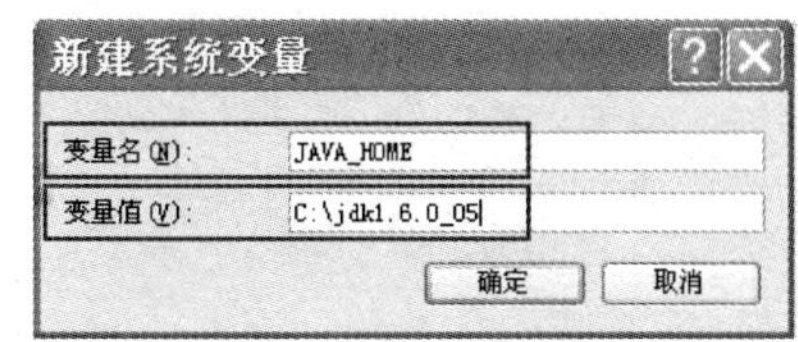

图 1-4　新建 JAVA_HOME 环境变量

提示：如果读者感兴趣，还可以顺便将安装的 JDK 所在目录下面的 bin 子目录的路径添加至系统变量 path 当中。这样，就可以通过运行 cmd 命令进入到 Windows 操作系统的命令行界面，在命令行界面通过输入 javac 或 java 等其他一些常用命令，执行一些具体的操作。

1.2.5　下载和安装 Eclipse 和 ADT

1. 下载安装 Eclipse

打开网址 http://www.eclipse.org/downloads/进入到 Eclipse 的下载页面，读者可以选择下载 Eclipse 的 IDE 版本或者 Eclipse For JAVA EE 开发者的版本都是没有问题的，如图 1-5 所示。

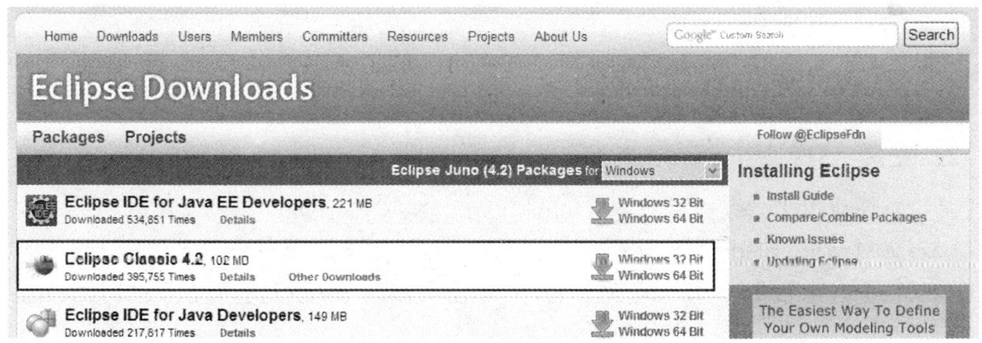

图 1-5　Eclipse 下载页面

提示：如果是 32 位的机器，就进入到 Windows 32 位机器的 Eclipse 下载页面；如果是 64 位的机器，就进入到 Windows 64 位机器的 Eclipse 下载页面。

下载完成之后，由于 Eclipse 是采用解压缩的方式进行安装，所以解压下载得到的压缩包之后，双击 eclipse. exe 可执行文件，就可以打开 Eclipse 集成开发工具。打开集成开发工具时，会弹出一个对话框，让用户去选择一个工作区间路径，如图 1－6 所示。选择工作区间所存放的路径之后，以后读者在建立工程的时候，工作区间的路径也就是所建的工程的存放路径。用户也可以通过单击弹出框中的复选框，选择默认的工作区间。

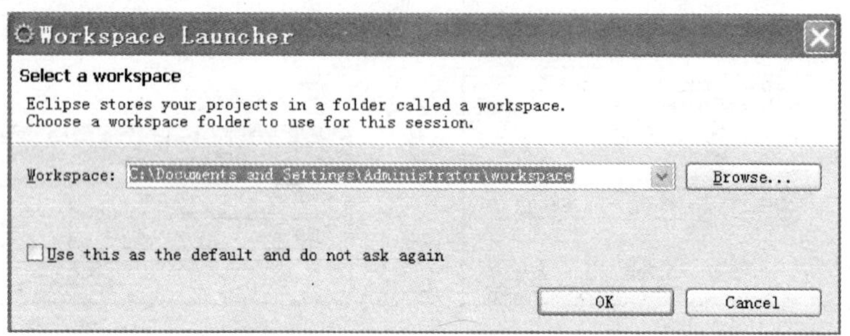

图 1－6　选择工作区间对话框

如果读者打算同时开发多个应用程序，笔者建议给每一个项目分别指定一个独立的工作区间来进行存放。如果是多个项目同时存在同一个工作区间，那么组织管理这些项目就会显得比较混乱。如果是将每一个工程都存放于独立的工作区间，这样比较有利于进行项目开发和维护。当 Eclipse 完成加载的时候，读者就会看到以下的 Eclipse 欢迎界面，如图 1－7 所示。

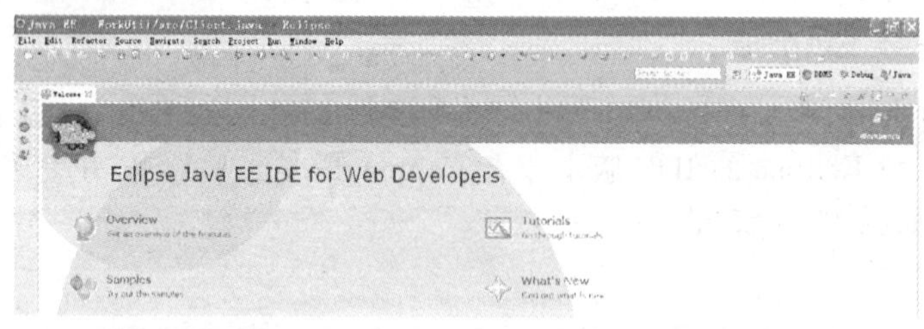

图 1－7　Eclipse 欢迎界面

2. 安装 ADT

(1) 启动 Eclipse，然后选择 Help | Install New Software 命令，弹出 Available Software 对话框，如图 1－8 所示。

(2) 在 Available Software 对话框中，单击右侧的 Add 按钮，弹出 Add repository 对话框。

(3) 在 Name 文本框中输入命名的远程站点的名称(例如：ADT)，在 Location 文本

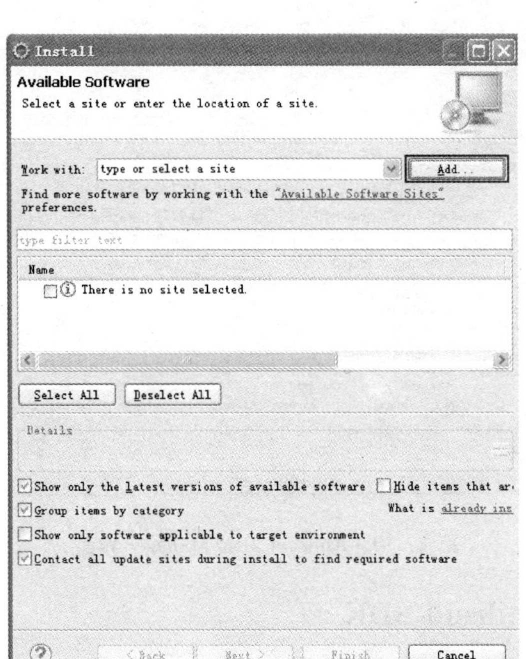

图 1-8 Available Software 弹出框

框输入 https://dl-ssl.google.com/android/eclipse，如果访问该插件出现问题，可以将 URL 当中的 https 改成 http，单击 OK 按钮，如图 1-9 所示。

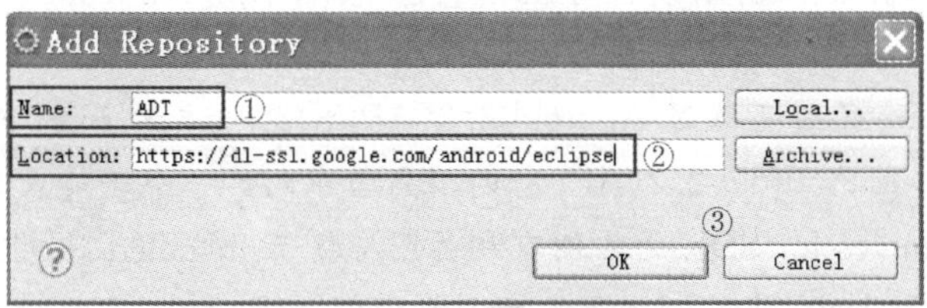

图 1-9 Add Repository 对话框

（4）回到 Available Software 界面，应该看到列表当中的 Developer Tools 以及 NDK plugins 选项（笔者下载的 Eclipse 为 4.2 的版本，如果是之前的比较老一点的 Eclipse 版本安装 ADT 插件的时候，可能看不到 NDK plugins 选项）。同时选择 Developer Tools 和 NDK plugins 这两个选项之后，会同时将 Android DDMS、Android Development Toos、Android Hierarchy Viewer、Android Traceview、Tracer for OpenGL ES 以及 Android Native Development Tools 选项同时选中。单击 Next 按钮阅读并接受许可协议，同时安装以上列出插件，然后单击 Finish 按钮。

（5）重新启动 Eclipse，装有 ADT 插件的 Eclipse 提示 SDK 没有安装，单击对话框中的 Open SDK Manager 按钮，弹出 Android SDK Manager 界面，如图 1-10 所示。

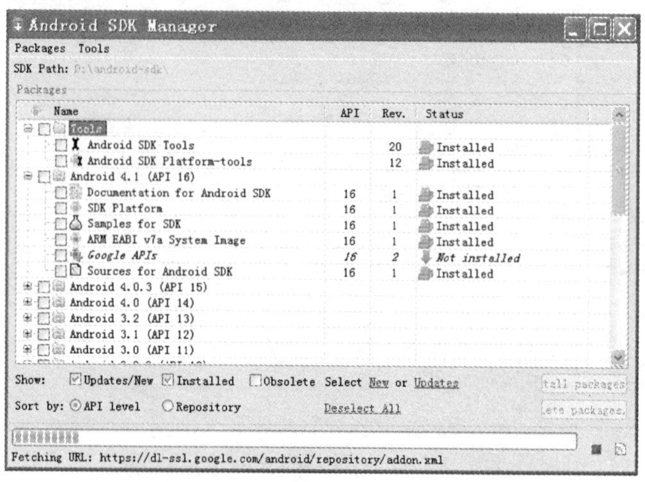

图 1-10 Android SDK Manager 界面

1.2.6 安装 Android SDK

在装有插件的 Eclipse 的工具栏当中，有一个“机器人”按钮可以打开 Android SDK Manager，顾名思义，Android SDK Manager 就是对 SDK 进行管理的工具。通过该工具可以查看哪些版本的 SDK 已经安装，哪些版本的 SDK 暂时没有安装，最新的 SDK 版本是多少，开发者可以选择相应的 SDK 的版本进行下载。由于本书实例使用的 SDK 2.2 的版本，所以读者可以和笔者一样，下载 Android SDK 2.2(API 8)的版本。

说明：由于笔者已经下载过 Android 2.2(API 8)的版本的 SDK Platform 和 Samples for SDK，所以在 Android 2.2(API 8)的节点当中的第一个子节点 SDK Platform 和第二个子节点 Samples for SDK 的右侧有相关图标提示 Installed，表明这两项已经安装过。如果是通过 Android SDK Manager 下载的 SDK 版本，默认是存放在 C:\Documents and Settings\Administrator\Local Settings\Application Data\Android\android-sdk 目录下面，在此目录的子目录 platforms 当中，不同的 SDK 版本分别用不同的文件夹保存起来，例如：文件夹 android-8 存放的是 Android 2.2(API 8)的 SDK Platform，android-16 存放的是 Android 4.1(API 16)的 SDK Platform。

除了通过 Android SDK Manager 下载 SDK 以外，还可以通过其他的方式去下载 Android 的 SDK，读者如果感兴趣，可以去网上搜索一些资料查阅，这里笔者不做深入探讨。下载过后，可以通过指定 Android SDK 的存放目录来告诉 Eclipse Android SDK 的存放路径。读者可以通过选中 Window|Preferences|Android，在选中 Android 选项之后的右侧 SDK Location 一栏中，可以选择 SDK 的保存路径，如图 1-11 所示。

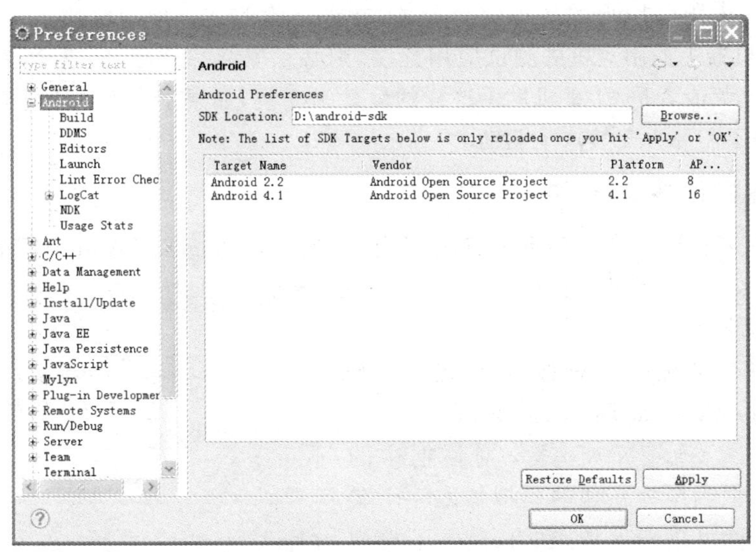

图 1-11 Android SDK 路径设定

1.2.7 管理模拟器

1. 模拟器介绍

AVD 所指的就是 Android 虚拟设备(Android Virtual Device),也就是读者可能常听到的模拟器。Google 为开发者提供模拟器的目的不仅仅是方便开发者开发应用程序,模拟器还可以帮助开发者去测试各种各样的屏幕大小,各种不同分辨率的 Android 机器。测试应用程序对各屏幕尺寸大小以及各屏幕分辨率的兼容性的时候,开发者没有必要把每一种屏幕尺寸,每一种分辨率的真机都拿来进行测试,使用模拟器就可以完成相应的工作。但是模拟器也会有一些限制,模拟器不能去模拟一些硬件设备,比如加速器,蓝牙等。但是不用担心,大多数应用程序用模拟器就可以完成应用程序的开发与测试。

如果读者要去开发一个跟蓝牙有关的应用程序,则必须使用真机来进行调试和测试,因为模拟器无法去模拟一些硬件设备。如果读者的计算机配置很好,速度很快,那么打开模拟器的速度就会非常快,反之,如果计算机配置稍微差一点的话,打开模拟器的速度就会很慢,所以建议在这种情况下,读者在开发应用程序的时候,尽量用真机去进行调试和测试,因为用真机调试应用程序的速度会比较快。

之前安装的 ADT 当中有一个 DDMS 的插件,在 Eclipse 当中 DDMS 是以一种视图的方式存在的,读者可以通过选择 Window|Open Perspective|Other|DDMS 来打开 DDMS 视图。DDMS 允许开发者使用真机来调试应用程序。假设读者要去开发一款实时跟踪用户位置的应用,当然只可以通过真机才能测试出来应用程序到底

能不能实时去跟踪用户的地理信息位置，并将正确的结果展示给用户。如果要在Windows机器上面用真机去测试应用程序，则必须安装一个驱动，这个驱动读者可以通过豌豆荚或者是91手机助手这样的软件来安装，如果是在MAC或者是Linux的机器，则不需要去安装这个驱动。

2. 新建模拟器

（1）在装有插件的Eclipse的工具栏当中，有一个Open the Android Virtual Device Manager按钮可以打开Android AVD的管理对话框。

（2）单击右侧的new按钮，弹出Create new Android Virtual Device对话框。

（3）在Create New AVD对话框中的Name输入框当中为新建的模拟器取一个名称，这个名称可以随便取，在Target下拉列表框当中，选择模拟器使用的SDK版本，在这里可以选择Android 2.2的SDK版本，在SD Card组合框中的Size输入框当中输入开发者希望模拟器模拟出来的SD Card的空间大小是多大，在这里可以设置为100 MiB的大小，在Skin栏Built-in的下拉列表框当中选择模拟器的屏幕分辨率，在这里默认选择HVGA（320×480）所对应的分辨率，如图1-12所示。单击Create AVD按钮，成功创建模拟器。

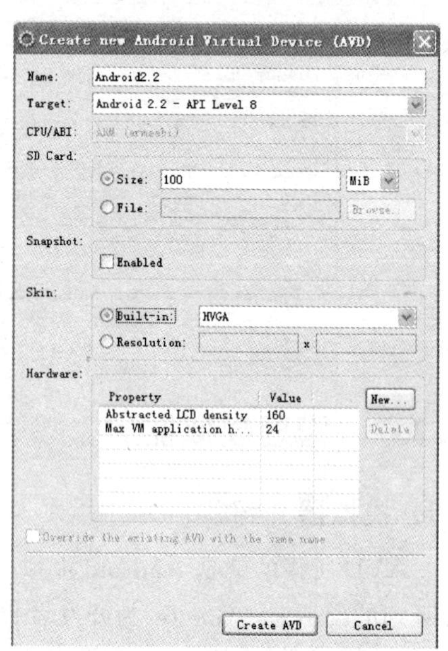

图1-12　Create AVD界面

3. 启动模拟器

建立好模拟器之后，通过Android Virtual Device Manager，选中想要启动的模拟器，单击右侧的Start按钮，模拟器就启动起来了。模拟器启动之后，读者就可以编写各种各样的Android应用程序，然后在模拟器上运行、调试以及测试了。

1.3　编写Hello World

笔者相信读者对于Hello World是有着深厚感情的，如果选择了程序员这条道路，无论学习哪种语言，学习哪种平台，刚开始接触到的第一个实例程序，一般都是所谓的Hello World程序。对于Android平台，笔者也不想打破这种规律，所以给读者介绍的第一个实例程序也是在界面上输出Hello World的字样。

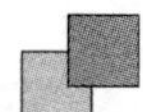

1.3.1 新建 Hello World 工程

下面笔者通过新建 Hello World 工程，为读者演示一下如何通过应用程序向导搭建一个 Hello World 工程。

(1) 启动 Eclipse，选择 File|New|Project 命令，在弹出的 New Project（新建工程）对话框中，选择 Android 节点下面的 Android Application Project，（如果没有出现 Android 节点，说明 Eclipse 没有安装 ADT 插件），单击 Next 按钮，如图 1-13 所示。

(2) 在弹出的 New Android App 对话框中，在 Applicaton Name 文本框中输入应用程序的名称（当这个应用程序安装在模拟器或者真机的时候，Application Name 文本框中输入的名称将会显示在应用程序 Launcher 里面），在 Project Name 文本框中输入项目的名称，在 Package Name 文本框中输入应用程序的包名（输入包名的时候，要遵循 JAVA 语言当中一贯的包名的命名规则，读者要确保所输入的包名在所有应用程序当中是独一无二的，所以应该使用标准的域名形式的包名），在 Build SDK 下拉列表框中选择应用程序编译使用的 SDK 版本，在 Minimum Required SDK 下拉列表框中选择程序最低兼容的 SDK 版本，单击 Next 按钮，如图 1-14 所示。

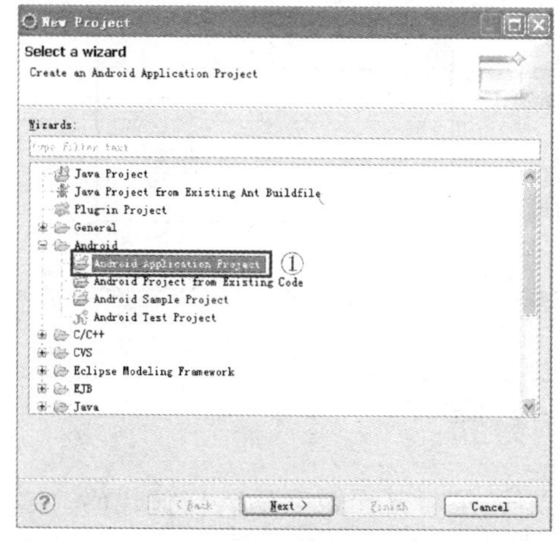

图 1-13 新建工程

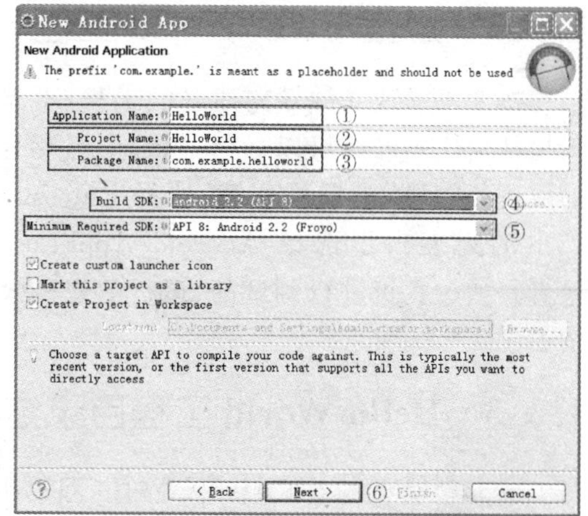

图 1-14 New Android App 对话框

提示：Android 操作系统版本是向后兼容的。如果读者选择的 Build SDK 版本号是 1.6，那么用这个 SDK 版本编译出来的应用程序可以在 Android 版本 1.6 以及以上的版本的机器上运行。使用旧一点的 Framework 的好处就是用户群体会更广，但是也有一些缺点，例如如果使用 1.6 版本的 Framework，就不能访问有关蓝牙的 API，因为有关蓝牙的 API 直到 2.0 之后才引入进来。

Minimum Required SDK 的选择是用来设置应用程序要想正常能够运行需要的 Android 设备的最低版本号是多少。如果要建立的应用程序不是对所有 Android 的版本都正常兼容，例如用到了有关蓝牙的 API，则应该是设置 Minimum Required SDK 为 Android 2.0。

(3) 在弹出的下一个界面当中，是有关程序图标属性的相关设置，由于是 Android SDK 4.1 引入的新功能，在此暂时略过，读者只需按照默认的设置单击 Next 按钮即可。

(4) 在弹出的 Create Activity 界面当中，选中 Create Activity 复选框，选中 Blank Activity(空白的活动)，单击 Next 按钮。

(5) 弹出的新界面是要进行 Activity(后面会介绍 Activity 到底是什么)相关属性的设置，需要在 Activity Name 文本框当中输入主入口活动的名称，在 Layout Name 文本框当中输入主入口活动使用的布局文件的名称，在 Navigation Type 下拉列表当中选择主入口活动使用的导航类型，在 Hierarchy Parent 选择框中选择要继承的相关类，在 Title 输入框当中输入主入口 Activity 显示的标题，可以按照默认给出的设置，单击 Finish 按钮。

1.3.2 运行 Hello World 工程

按照之前建立的导向建立的工程，不需要新增任何代码已经可以运行了，选中想要运行的应用程序，单击鼠标右键，选择 Run As，在弹出的列表当中选择 Android Application。笔者是通过真机进行调试的，运行的真机效果图如图 1-15 所示。

图 1-15 Hello World 真机运行效果图

1.3.3 Hello World 工程目录结构分析

以上创建了第一个 Hello World 的应用程序，笔者甚至没有输入任何代码。这都是因为安装的 ADT 插件给开发者提供了用来快速开发应用程序的工具，但是通过应用程序向导创建的应用程序仅仅只是一个模板，如果需要开发更多功能的应用程序，开发者必须在此模版基础上做很多修改。读者以后开发应用程序，都要在这些目录之间不停的切换，不停地为这些目录添加新的文件或者修改旧的文件。所以读者必须明白这些目录文件是哪些，以及它们的作用是怎么样的。接下来就来分析通过模板新建的 Hello World 应用程序的整个工程的目录结构。

展开 Hello World 项目之后，工程下面有十多个子目录。这些子目录不仅仅是 Hello World 这个应用程序才有的，只要是通过 Eclipse 创建 Android 应用程序的创建向导创建出来的应用程序都会包含以上所列举的所有子目录。除了这些默认生成

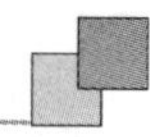

的子目录之外，还有一些自动生成的文件，例如 AndroidManifest. xml 以及 project. properties。AndroidManifest. xml 文件会告诉操作系统这个应用程序是由哪些组件所组成的，project. properties 文件则帮助开发者进行一些 Android 属性的管理，例如可以指定应用程序构建的目标系统 SDK 的版本等，下面逐一分析各个子目录。

1. src

这个子目录是用来存放 Java 源代码的目录，应用程序当中所有的 Java 源文件都存放在 src 这个子目录当中。通过选中 src 子目录，然后单击目录左边的箭头从而展开 src 目录，读者将会看见工程当中的默认包：com. example. helloworld。展开默认包，将会看见 MainActivity. java 这个 Java 文件。在真正开发一个应程序过程当中，开发者通常要根据功能的不同将不同的 Java 源文件放在不同的默认包的子包当中。例如，如果程序当中有一个类专门负责使用 XML 的数据格式向服务端发送 http 请求，可能还有一个用来表示现实生活中客户这么一个具体模型的 Customer 类。开发者应该要将这两个不同功能的类放在不同的默认包的子包当中。可以将跟 http 请求相关的类，放入 com. example. helloworld. http 子包当中，可以将与模型相关的类放入 com. example. helloworld. models 子包当中。

2. Android version

笔者跳过了 gen 目录，在提到 res 目录的时候，笔者会提及 gen 目录的作用。在使用应用程序向导创建 Hello World 的时候，笔者选的 Build SDK 版本为 Android 2.2(API 8)，所以读者看到的截图当中所用到的框架是 Android 2.2 的框架。展开这个节点，里面包含了一个应用程序编译时所要用到的 android. jar 库文件，读者还会看到这个库文件的实际硬盘存储位置。

3. Android Dependencies

顾名思义，这个目录里面存放的是一些通过应用程序向导建立的应用程序所依赖的一些其他库文件。通过展开这个节点，里面默认包含两个 jar 文件，分别是 annotations. jar 和 android-support-v4. jar。通过引入 annotations. jar，读者在编写 Android 应用程序的时候，可以使用 annotations. jar 所提供的与注解相关的一些功能，而 android-support-v4. jar 提供了一些在旧版本的 SDK 当中无法使用的一些 API，以及一些在旧版本 SDK 不曾包含的工具类。引入 android-support-v4. jar 的目的就是通过为开发者提供更多种类的 API 从而降低开发者开发应用程序的难度，缩短应用程序开发的时间。

4. assets

通过应用程序向导建立的工程的 assets 目录默认是空的。这个目录是用来存放一些程序当中需要用到的一些文件，举个例子，如果读者将要开发一个在不联网环境下，带有字典查询功能的应用程序，读者可以将数据文件，一般都是 XML 格式或者是以 SQLite 数据库的格式存放在 assests 目录下面。存放在 assets 目录里面的文件打包后会原封不动地保存在 APK 包中，不会被编译成二进制。Android API 提供了 Asset Manager 对 assets 目录的文件进行管理。

Android 把 assets 目录也作为一种资源管理的方式，在 assets 目录当中可以放置任意类型的文件。但是相对于 res 目录用资源 ID 来进行资源管理的方式，assets 资源管理的方式略显沉闷和复杂。开发者如果想要读取 assets 下面的文件，必须先把文件转换为字节流，然后再对获取到的字节流进行进一步的处理。

5．bin

工程编译之后的一些文件将会存放在 bin 目录当中，比如最重要的 apk 安装包文件，编译之后就是存放在 bin 目录当中，当然也包括一些其他文件，比如 Android-Manifest. xml 项目清单文件，再比如 Dalvik 虚拟机能够运行的 classes. dex 文件等。

6．libs

libs 目录当中是存放一些应用程序运行时所需要的除 Android Frameworks 之外的库文件，例如上面曾经有提到过的 android-support-v4. jar 这个文件。如果程序当中还引入了一些其他的第三方的库文件，例如进行网络操作时要用到的跟 Http-Client 相关的一些第三方 jar 包；以及做分享功能时，所要用到的一些跟 OAuth 鉴权机制有关的一些第三方 jar 包。除了要把这些 jar 包作为 Android Dependencies 引入作为依赖库之外，还需要将这些 jar 包放入 libs 目录中。

7．res

res 目录当中包含了多种多样的应用程序所要使用到的资源文件。字符串和图片是最典型的资源文件，为了避免在程序当中进行“硬编码”，程序当中需要字符串的时候，开发者可以将所需要的字符串放到 res 目录下面 values 子目录下面的 strings. xml 文件当中进行管理，然后在代码当中进行引用。开发者还需要根据不同的设备配置，为不同的设备配置提供不同的资源文件放入不同的设备配置的文件夹当中。在运行的时候，Android 操作系统将会根据应用程序运行的设备的一些参数配置从而决定使用哪一套资源文件。举例来说，展开 res 节点的时候，res 目录当中包含 drawable-hdpi，drawable-ldpi，drawable-mdpi 等子目录，如果运行应用程序的设备是高分辨率的机器，那么 Android 会选择使用 drawable-hdpi 当中存放的图片资源，如果运行应用程序的设备是低分辨率的机器，那么 Android 会选择使用 drawable-ldpi 当中存放的图片资源，如果运行应用程序的设备是中等分辨率的机器，那么 Android 会选择使用 drawable-mdpi 当中存放的图片。这种机制给开发者提供了一种实现多分辨率适配，多语言设配，多屏幕尺寸适配的解决方案。

一般来讲，放在 res 目录当中的资源文件，都会有一个唯一的资源 ID 与之相对应。编译的时候，ADT 插件会将这些唯一的资源 ID 与 gen 目录下面的 R. java 文件

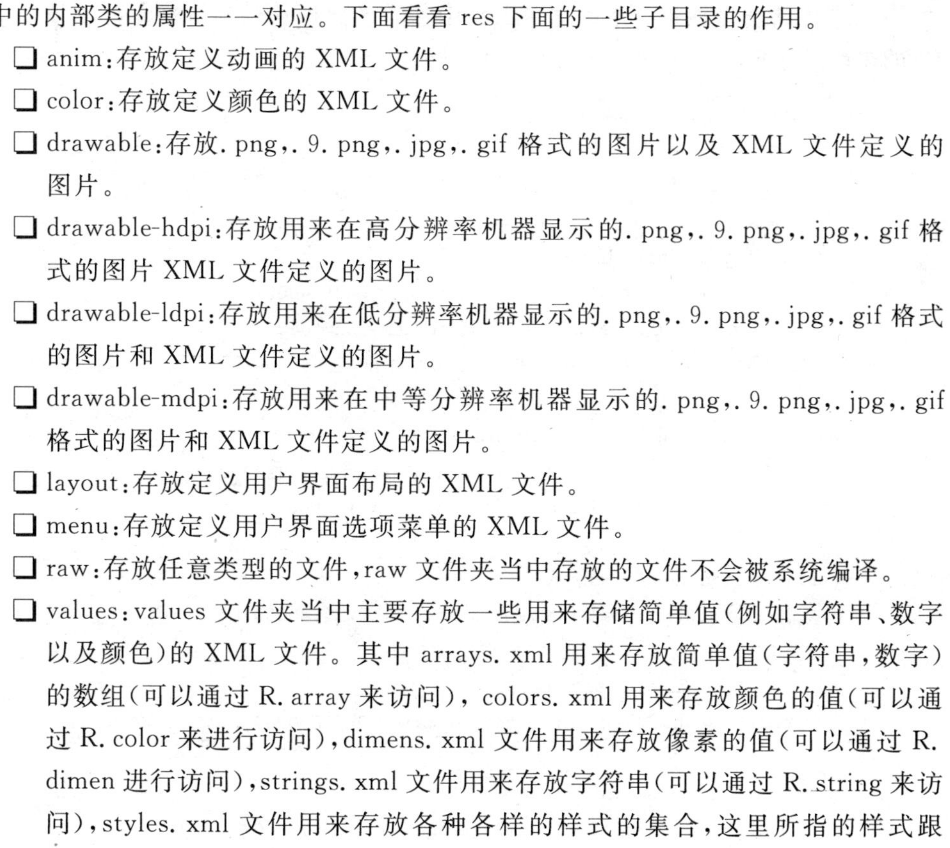

当中的内部类的属性一一对应。下面看看 res 下面的一些子目录的作用。

- ❑ anim：存放定义动画的 XML 文件。
- ❑ color：存放定义颜色的 XML 文件。
- ❑ drawable：存放.png，.9.png，.jpg，.gif 格式的图片以及 XML 文件定义的图片。
- ❑ drawable-hdpi：存放用来在高分辨率机器显示的.png，.9.png，.jpg，.gif 格式的图片 XML 文件定义的图片。
- ❑ drawable-ldpi：存放用来在低分辨率机器显示的.png，.9.png，.jpg，.gif 格式的图片和 XML 文件定义的图片。
- ❑ drawable-mdpi：存放用来在中等分辨率机器显示的.png，.9.png，.jpg，.gif 格式的图片和 XML 文件定义的图片。
- ❑ layout：存放定义用户界面布局的 XML 文件。
- ❑ menu：存放定义用户界面选项菜单的 XML 文件。
- ❑ raw：存放任意类型的文件，raw 文件夹当中存放的文件不会被系统编译。
- ❑ values：values 文件夹当中主要存放一些用来存储简单值（例如字符串、数字以及颜色）的 XML 文件。其中 arrays.xml 用来存放简单值（字符串，数字）的数组（可以通过 R.array 来访问），colors.xml 用来存放颜色的值（可以通过 R.color 来进行访问），dimens.xml 文件用来存放像素的值（可以通过 R.dimen 进行访问），strings.xml 文件用来存放字符串（可以通过 R.string 来访问），styles.xml 文件用来存放各种各样的样式的集合，这里所指的样式跟 CSS 的概念很相似，可以通过 R.style 来进行访问。

8. gen

根据向导创建好应用程序之后，在第一次编译程序之前，gen 目录是不存在的。一旦第一次编译完成之后，ADT 插件就帮开发者自动生成了 gen 目录，并将编译产生的一些文件放入到 gen 目录当中。gen 目录当中包含的 Java 文件都是 ADT 生成的，其中最重要的一个文件就是 R.java 文件。笔者之所以在介绍 gen 目录之前先介绍 res 目录，就是因为 R.java 文件当中包含的内部类以及内部类的属性都是由 res 目录当中包含的资源文件所生成的。由于开发者在应用程序开发过程当中，需要经常跟 res 目录下面的那些资源文件打交道，而开发者使用的语言又是 Java 语言，所以 R.java 文件相当于是 Java 源代码与 res 目录资源文件当中的一个桥梁，Java 源代码通过引用 R.java 文件当中内部类的属性，就可以找到想要找到的资源，例如通过在代码当中引用 R.menu.activity_main 就可以找到所对应的 menu 子目录下面 activity_menu.xml 这个选项菜单定义文件。

双击 R.java 类文件，读者可以发现 R 类文件当中嵌套了其他的一些内部类，这些内部类的名称与 res 目录下面的子目录的名称是保持一致的。在每一个嵌套的子类当中，读者将会发现每一个嵌套子类当中拥有的属性名称与资源文件当中定义的ID的名称是一致的。R.java类文件源代码如下所示：

```
package com.example.helloworld;
public final class R {
    public static final class attr {
    }
    public static final class dimen {
        public static final int padding_large = 0x7f040002;
        public static final int padding_medium = 0x7f040001;
        public static final int padding_small = 0x7f040000;
    }
    public static final class drawable {
        public static final int ic_action_search = 0x7f020000;
        public static final int ic_launcher = 0x7f020001;
    }
    public static final class id {
        public static final int menu_settings = 0x7f080000;
    }
    public static final class layout {
        public static final int activity_main = 0x7f030000;
    }
    public static final class menu {
        public static final int activity_main = 0x7f070000;
    }
    public static final class string {
        public static final int app_name = 0x7f050000;
        public static final int hello_world = 0x7f050001;
        public static final int menu_settings = 0x7f050002;
        public static final int title_activity_main = 0x7f050003;
    }
    public static final class style {
        public static final int AppTheme = 0x7f060000;
    }
}
```

在编译应用程序的时候，ADT会帮助开发者生成R.java文件，开发者不用去关心ADT生成R.java文件的细节，也不能手动去编辑生成的R.java文件。如果手动编辑了的话，应用程序在编译的时候可能会通不过。如果实在不小心编辑了R.java文件，应该手动删除这个文件，然后重新让Eclipse去编译。

现在来解释为什么R.java文件当中的内部类当中的属性与res目录下面含有资源ID的资源是一一对应的，展开res目录下面的menu子目录，读者会发现activi-

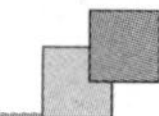

ty_main. xml 文件的存在，打开 activity_main. xml 文件，源代码如下所示：

```
<menu xmlns:android = "http://schemas.android.com/apk/res/android">
    <item android:id = "@ + id/menu_settings"
        android:title = "@string/menu_settings"
        android:orderInCategory = "100" />
</menu>
```

可以发现这个文件当中有一句代码：android:id＝"@＋id/menu_settings"，这是在资源文件当中定义资源 ID 的方式，在 R. java 文件当中的 id 嵌套类当中，包含一个静态常量 menu_settings，所以资源文件当中的这个 item 与 R. java 文件当中的 id 嵌套类当中的静态变量 menu_settings 是一一对应的。

1.3.4　AndroidManifest. xml 文件分析

AndroidManifest. xml 文件又叫做项目清单文件，存储在项目的根目录下面，每一个应用程序都必须要有一个项目清单文件。每一个应用程序的项目清单文件为 Android 系统提供了很多必要的信息，如果没有这些必要的信息，程序当中的代码就不会被正确的执行。项目清单文件提供如下信息：

- ❑ 应用程序包名，每一个应用程序的包名都必须是唯一的，这是 Android 系统和 Google Play 用来区分应用程序的唯一方式。
- ❑ 应用程序组件信息注册，所有活动以及后台服务的注册，部分广播接受者的注册（部分广播接受者在程序当中用代码注册）。
- ❑ 应用程序运行所需权限声明。
- ❑ 应用程序运行所需最低的 Android SDK 版本号。

除此之外，项目清单文件当中还声明了应用程序的版本信息，应用程序的版本信息跟 Android SDK 的版本的概念很类似。在项目开发早期，就应该确定好应用程序的版本定义策略。应用程序的版本信息包括版本号以及版本名称。

1. 版本号

版本号用来区别同一应用程序的不同版本信息，Google Play 也是根据应用程序的版本号信息来提示用户是否需要升级应用程序。开发者需要保证下一个正常发布的版本号要比前一个正常发布版本的版本号要大，虽然这不是 Google Play 强制规定的，但是这条规则已经成为所有开发者都去遵循的最佳实践。第一次发布应用程序的时候，可以将程序当中的版本号信息设置为 1，以后每发布一个版本都要去增加程序当中的版本号信息，当然每次增加的幅度有大有小，增加的幅度代表了这一次的应用程序版本与前一个版本的差异程度。幅度越大，代表差异程度越大，增强的功能越多。

在更新程序代码的同时，如果忘记了去更新程序当中的版本号，会造成同一个版本下面会有不同的代码基线。设想一下，当开发者第一次发布一个应用程序的时候，将版本号信息设置为 1，用户通过 Google Play 安装了开发者发布的第一个版本，通过多次的使用，用户发现了一些程序当中的缺陷，并将这些缺陷通过一些方式反馈给开发者，开发者收到反馈过后，进行了相应的修复，但是由于忘记去更改应用程序的版本号信息，当再次将应用程序上传到 Google Play 里面的时候，由于 Google Play 会将版本号进行比较，发现没有变化，则不会通知安装了此应用程序的用户去更新此应用程序，虽然开发者修改了代码，也上传到了 Google Play，但是用户并没有得到相应的更新下载，这是开发者和用户都不愿意去看到的。

2. 版本名称

版本名称是用来代表版本号并向用户进行展示的一串字符串信息，比如 2.1.4，就可以作为一个版本名称来使用。一般来讲版本名称都是要紧跟之前提到的版本号，然后中间会用顿号隔开的一串字符串，版本名称也可能是这样的字符串，比如 2.1。Android 系统会将版本名称展现给用户（版本号不会展示给用户），除此之外，Android 系统不会在其他地方使用清单文件当中的版本名称的信息。版本名称在不同的应用程序当中也可能是以其他的形式出现，只要是能够相对或绝对地成为版本的唯一标示就可以了。比如 Foursquare 这个应用程序，用的就是与日期相关的一个版本名称，比如 2010-06-28，这样的版本名称清楚地表明了应用程序版本发布的日期，对于开发者和用户来说都是非常有意义的名称。应用程序的版本名称完全是由开发者来决定的，只不过在制定版本名称策略的时候，一定是要对开发者和用户来讲都是有意义的名称。

3. 用户权限

假设读者开发的应用程序需要通过访问互联网从而获取网络数据，Android 系统默认情况下是禁止应用程序对互联网进行访问的。所以，如果要实现这个功能，必须在项目清单文件当中显示地声明应用程序需要拥有访问互联网的权限。下面列举了一些常用的权限：

- ❑ android. permission. INTERNET：互联网访问权限。
- ❑ android. permission. WRITE_EXTERNAL_STORAGE：向 SD Card 写入数据权限。
- ❑ android. permission. CAMERA：使用摄像头权限。
- ❑ android. permission. ACCESS_FINE_LOCATION：使用 GPS 定位权限。
- ❑ android. permission. READ_PHONE_STATE：读取电话状态权限。

除以上常用权限之外，还有其他权限，在这里笔者就不一一列举了，读者可以查阅 android 的开发者网站（http://developer. android. com）获取相关资料信息。添加上述常用权限信息过后，项目清单文件源码如下：

```
<manifest xmlns:android = "http://schemas.android.com/apk/res/android"
    package = "com.example.helloworld"
    android:versionCode = "1"
    android:versionName = "1.0" >
    <!-- 包名,应用程序版本号,应用程序版本名称 -->
    <!-- 声明程序运行的设备的 SDK 最低版本号为 8,编译使用的 SDK 版本为 15 -->
    <uses-sdk
        android:minSdkVersion = "8"
        android:targetSdkVersion = "15" />
    <!-- 声明应用程序所要使用到的一些权限 -->
    <uses-permission android:name = "android.permission.INTERNET" />
    <!-- 互联网访问权限 -->
    <uses-permission android:name = "android.permission.WRITE_EXTERNAL_STORAGE" />
    <!-- 往外部存储写入数据权限 -->
    <uses-permission android:name = "android.permission.CAMERA" />
    <!-- 使用摄像头权限 -->
    <uses-permission android:name = "android.permission.ACCESS_FINE_LOCATION" />
    <!-- 使用 GPS 定位权限 -->
    <uses-permission android:name = "android.permission.READ_PHONE_STATE" />
    <!-- 读取电话状态权限 -->
    <application
        android:icon = "@drawable/ic_launcher"
        android:label = "@string/app_name"
        android:theme = "@style/AppTheme" >
        <activity
            android:name = ".MainActivity"
            android:label = "@string/title_activity_main" >
            <intent-filter>
                <action android:name = "android.intent.action.MAIN" />
                <category android:name = "android.intent.category.LAUNCHER" />
            </intent-filter>
        </activity>
    </application>
</manifest>
```

1.4　Android 编程基础

在介绍 Android 的详细知识之前,笔者先给读者介绍一些 Android 的编程基础,使得读者能够对 Android 应用程序开发当中的一些知识有一个大致了解。

1. Java 语言

Android 的框架层代码都是使用 Java 语言编写的,不过 Android 运行时使用的虚拟机是专门为嵌入式设备优化过的 Dalvik 虚拟机，不是传统的 JVM 虚拟机。编写 Android 应用程序的开发者大部分都使用 Java 语言进行开发，除了使用 Java 语言之外，开发者还可以使用其他一些语言，例如 C# 和 PHP 等语言来进行 Android 应用程序的开发，已经有一些开源的项目在做这方面的支持工作。如果读者对 Java 语言不是很熟悉，需要先去熟悉一下 Java 语言，熟悉 Java 语言之后，再来学习 Android 应用程序的开发，会达到事半功倍的效果。当然遇到问题的时候，也可以多使用 Google、bing 等相关的一些搜索引擎来进行相关问题的搜索，由于大部分的开发者都乐于去分享自己的知识，所以现在从搜索引擎上面寻找问题的答案是最有效率的方法之一。除了使用搜索引擎之外，当然也可以直接去一些程序员的问答网站，例如 http://www.stackoverflow.com，进行相关问题的搜索，通常情况下搜索到的答案都比较靠谱，且花费的时间会比较少。

2. 活动(Activity)

Android 应用程序是由一个或多个活动所组成的。在之前建立的 Hello World 工程当中，笔者只创建了一个叫做 MainActivity 的活动。一般来讲，应用程序当中至少会有一个所谓的主入口活动，也就是当用户单击 Launcher 当中的应用程序图标时，应用程序启动起来第一个显示的界面就叫做主入口活动。活动事实上就是一个用户界面，每一个用户界面都是一个活动。在后面的章节当中，笔者会对活动进行深入的探讨和分析。

3. 意图(Intent)

意图组成了运行 Android 系统的核心消息机制，通过 Intent，可以进行消息的传递，除了进行消息的传递之外，通过意图，还可以将原本互不相干的应用程序相关联起来。比如在一个日程管理的程序当中，如果要进行相关日程的分享，应用程序开发者可以在这个日程管理程序当中，调用邮件或者短信等其他应用程序，而这些应用程序的调用，就可以通过发送相关的意图来实现。每一个应用程序都可以用来广播意图，也都可以作为意图的接受者。这也是笔者觉得 Android 系统最棒的功能。意图分为两种，一种是隐士意图，一种是显示意图。

对于隐式意图，在程序当中，需要指定将要进行的动作是什么，这个动作可以是查看联系人，编辑联系人，拨打电话，打开浏览器等一系列动作，除了指定动作之外，还需要指定意图将要操作的数据，这个数据可以是一个联系人的信息，可以是一个 URL 地址等信息。Android 系统会根据项目清单文件当中的各个组件的<intent-filter></intent-filter>节点的信息，去匹配最适合的组件来相应发送的意图。如果同时有多个应用程序当中的组件都能够正确响应被广播的意图，Android 系统则会提供一个列表让用户去选择，到底使用哪个应用程序来响应发送的意图，而且还可以设置默认方式，下次广播相同意图的时候，则会直接使用默认的应用程序来响应发送的意图。

对于显示意图，需要在程序当中直接指定接受意图的 Java 类，以及传递过去的信息。

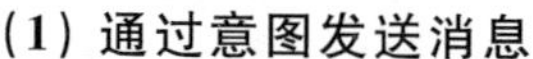

(1) 通过意图发送消息

当应用程序广播意图的时候，实际上就是应用程序发送了一个消息告诉 Android 系统应用程序要做什么。可能会启动同一个应用程序当中的组件，也可能会启动其他应用程序当中的组件。

(2) 注册意图过滤器(Intent Filter)

Android 系统接收到隐式意图的时候，会根据项目清单当中的<intent-filter></intent-filter>节点信息进行匹配。所以如果开发者想要自己的某一个组件，比如活动，可以对其他的一些意图进行响应，则需要在项目清单文件当中注册相关的意图过滤器信息，也就是相应的<intent-filter></intent-filter>节点信息，从而对相关意图进行监听。如果有多个<intent-filter></intent-filter>节点所对应的组件节点匹配的话，则会出现一个应用程序列表供用户选择。例如，笔者打开本机的文件管理应用，随便选择一首音乐，然后单击发送按钮，则会弹出来一系列应用程序供笔者去选择，应该要使用哪一个应用程序去发送。

注意：如果 Android 系统无法找到任何一个应用程序的组件与广播的意图相匹配，那么广播意图的应用程序则会因为没有捕获到的运行时异常而异常退出。在开发当中，最好使用隐式意图的方式，让 Android 系统根据意图的一些信息，从而去找到与之相匹配的应用程序，尽量避免使用显式意图的方式。

4. 广播接收者(Broadcast Receiver)

所谓的广播接收者，事实上就是一种观察者模式的实现。在观察者模式当中，有两种角色，第一种角色叫做发布者，第二种角色叫做订阅者。如果订阅者对发布者进行了订阅，那么当发布者有相关更新的时候，就会发送消息通知订阅者：发布者已经进行了相关更新。广播接收者可以对其感兴趣的事件进行订阅，订阅一般有两种方式，一种是在 JAVA 文件当中，通过代码的方式进行订阅，一种是在项目清单文件当中，通过 XML 的方式进行订阅。订阅过后，当相关事件发生的时候，所有对此事件进行订阅的广播接收者，都会回调广播接收者的 onReceive 方法。在这里事件一般是以意图广播的形式发生的。

5. 服务(Service)

服务跟活动不太一样，活动都有用户界面，可以供用户进行交互，而服务没有用户界面，服务仅仅是在应用程序后台运行的一些代码段，一般是长时间运行的代码段。最典型的例子就是音乐播放器，当用户看不到音乐播放器的用户界面时，用户还可以听到播放的声音。做音乐播放器时，就要使用到服务的相关知识。服务又分为两种，绑定的服务和未绑定的服务，绑定的服务可以通过 AIDL 提供相关具体的 API 接口供其他应用程序使用，而非绑定的服务就是指服务的声明周期不会跟启动它们的活动相绑定。

6. 视图(View)和用户界面控件(Widget)

视图指的就是一个基本的UI元素，实际上也就是屏幕上一块矩形界面，视图要负责对那一块的矩形截面进行绘制并且对事件进行处理。下面是一些视图的例子：

❑ ContextMenu：上下文菜单。

❑ Menu：选项菜单。

❑ View：视图。

用户界面控件是一些封装好了的更加复杂的用户界面元素，例如复选框。下面列举了一些常用的用户界面控件：

❑ Button：按钮。

❑ CheckBox：复选框。

❑ DatePicker：日期选择器。

❑ Gallery：相册。

❑ FrameLayout：框架布局。

❑ ImageView：图片视图。

❑ RelativeLayout：相对视图。

❑ PopupWindow：弹出窗口。

读者可以去Android的开发者网站去寻找更多的用户界面控件，在这里，笔者就不一一详细列出。

7. 异步调用

Android系统当中的线程模型是这样的，UI线程叫做主线程，但是由于客户端程序要去执行比较耗时的操作，比如从网络上获取数据。对于这种比较耗时的操作，如果也放在主线程当中去操作的话，势必会造成主线程的阻塞，使得程序在下载网络数据的时候，程序的用户界面无法与用户进行交互，无法响应用户的操作。如果用户的操作在一定时间内无法得到响应，Android系统规定这一时间为5 s，如果5 s之内，应用程序还未对用户的交互做出任何响应，Android系统则会弹出Application not responding(ANR)的错误提示框。当ANR错误发生时，用户可以选择强行关闭应用程序或者是继续等待。

所以开发者在开发应用程序的时候，必须将一些比较耗时的操作，一般是指网络操作，放入其余的线程当中运行。Android提供了AsyncTask这个类处理异步的操作。AsyncTask类允许程序在同一时间执行多个操作而不用去关心线程本身。AsyncTask不仅仅允许开发者启动新的线程，而且还会返回结果给相应的启动AsyncTask的活动。AsyncTask为异步调用提供了一个比较简洁的编程模型。

8. Android应用程序生命周期

默认情况下，每一个Android应用程序都是在自己独立的进程空间当中运行，每一个进程事实上都是Dalvik虚拟机的一个实例。Android应用程序本身无法对自己

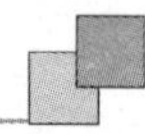

的生命周期进行管理，每一个应用程序的内存与进程管理都是交由 Android 运行时去管理。应用程序组件必须时刻监听应用程序状态的改变，从而对应用程序状态的改变进行相应的响应，特别是在内存不足的情况下，需要做好相关信息的保存。

由于手机设备内存有限，在一些特殊情况下，可能会导致系统内存不足。为了保证系统在内存不足的情况下，仍能够及时响应用户的操作，与用户进行交互，Android 系统运行时会去杀死一些优先级比较低的进程，把释放出来的内存资源供其他优先级比较高的进程使用。

Android 系统运行时为了释放系统资源而去杀死应用程序的顺序是按照应用程序进程优先级来排的，而一个应用程序的进程优先级是由一个应用程序当中所有组件的优先级的最大值决定的。应用程序的优先级有时候也是相互依赖的，如果一个应用程序会依赖第二个程序提供的服务或者内容提供者，那么第二个应用程序至少跟第一个应用程序的进程优先级是相等的。下面按照优先级从高到低，对应用程序进程的各个优先级逐一说明：

(1) 活动进程：所谓的活动进程，就是包含一些正在与用户进行交互的组件的应用程序所在的进程。活动进程的优先级最高。看一个应用程序是不是活动进程，主要看应用程序当中是否包含以下场景的组件：

- ❑ 正在与用户正在交互的活动；
- ❑ 正在执行 onReceive 方法的广播接收者；
- ❑ 正在执行 onStart、onCreate 或者 onDestory 方法的服务。

(2) 可见进程：含有可见活动的应用程序所在的进程叫做可见进程。所谓的可见活动是指，虽没有在前台与用户直接交互，但是仍然对用户可见的活动。读者可以想象一下下面的场景，一个原本在前台与用户交互的活动，被一个对话框样式的活动所覆盖了，那么此时与用户直接交互的活动就是那个对话框样式的活动，而原本在前台与用户进行交互的活动虽然没有与用户直接交互，但是仍然可见，这样的活动叫做可见活动，含有可见活动的进程就叫做可见进程。

(3) 启动后的服务进程：拥有启动了的服务的应用程序所在的进程叫做启动后的服务进程。服务需要在没有可视化用户界面的情况下持续执行。因为服务不需要与用户进行交互，所以启动后的服务进程的优先级比可视进程的优先级稍微低一点。但是启动后的服务进程仍然被认为是属于前台进程的范畴，当 Android 运行时需要为活动进程以及可见进程释放资源的时候才会杀死启动后的服务进程。

(4) 后台进程：应用程序当中包含的都是一些非可见的活动或者是一些未启动的服务的组件的应用程序所在进程叫做后台进程。后台进程的优先级比较低，很容易被 Android 运行时杀死。

(5) 空进程：为了提高整体的性能，当应用程序的生命周期结束的时候，Android 运行时会把应用程序保存在内存当中，相当于做了一个缓存。当下次应用程序再次启动的时候，就可以减少应用程序的启动时间。这些作为缓存的进程一般叫做空进程。

第2章 活动(Activity)

活动(Activity)就是一个包含用户界面的窗口,能够与用户进行交互是活动最主要的功能,一个应用程序会包含一个或多个活动。活动从出现在屏幕的那一刻到最终销毁的那一刻,整个过程叫活动的生命周期。了解活动的生命周期是开发 Android 应用程序比较关键的一点。在本章当中,笔者将和读者一起对活动做进一步的认识和理解。本章将会介绍如下知识:

- ❑ 创建活动;
- ❑ 理解活动生命周期;
- ❑ 活动的启动模式;
- ❑ 活动运用样式和主题;
- ❑ 隐藏活动标题栏;
- ❑ 弹出对话框;
- ❑ 弹出进度条对话框。

2.1 创建活动

为了创建应用程序的用户界面,开发者需要继承 Activity 这个基类,通过一些视图来提供用户交互。开发者需要为应用程序当中展示的每一个用户界面都创建一个新的活动。一般来讲,应用程序当中都会有一个主界面用来处理应用程序的主要功能,有一些其他辅助的功能是由另外一些用户界面来提供的,用户会从主要功能的界面进入到辅助功能的界面,这就涉及了活动的跳转。

大多数的活动都会完整地占据整个用户设备屏幕,也有一些活动展示的样式比较例外,可以是半透明的、悬浮的或者是对话框形式的。

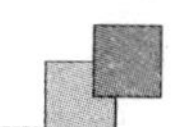

2.1.1　新建活动

首先创建工程 chapter2_1，按照应用程序向导创建出来的应用程序，一般会自动建立一个叫 MainActivity 的主入口活动，这个活动将作为应用程序最开始启动的时候展示给用户的界面。至于为什么 MainActivity 能够成为应用程序启动时展现给用户的界面，笔者将会在介绍意图以及意图过滤器相关内容的时候，为读者介绍。接下来笔者将在 chapter2_1 中新建一个活动，并将此活动作为应用程序启动的时候展现给用户的活动。

(1) 右击应用程序的包名，在弹出的快捷菜单中选择 New | Class 命令，弹出 New Java Class 对话框，如图 2－1 所示。

(2) 在 Name 文本框中，输入要建立的活动的类名 FirstActivity。单击 Superclass 输入框右侧的 Browse 按钮，弹出 Superclass Selection 对话框，如图 2－2 所示。

注意：输入类名的时候，一般要遵循一些约定俗成的 Java 命名法则，类名的每一个单词的首字母都要大写。

(3) 在 Choose a type 文本中输入活动要继承的基类 android. app. activity，单击 OK 按钮，Superclass selection 对话框关闭，回到 New Java Class 对话框，单击 Finish 按钮。这样，新活动 FirstActivity 就建立好了。

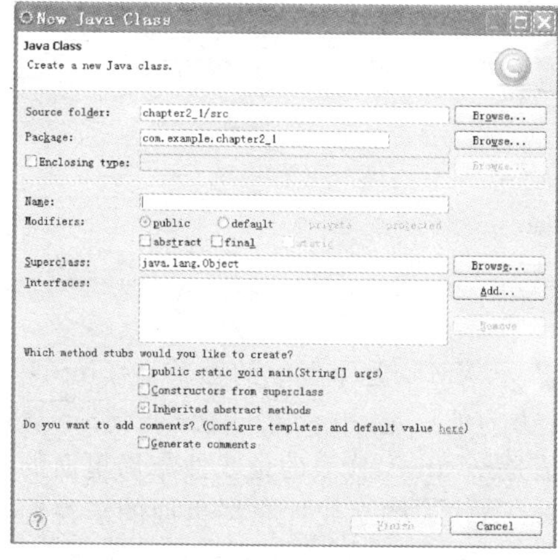
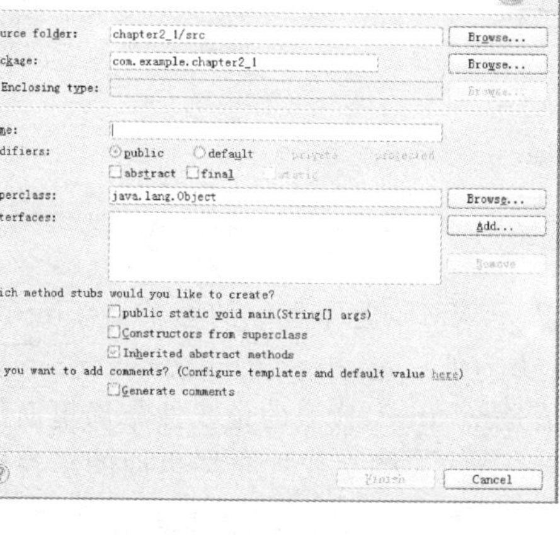

图 2－1　新建 Java 类对话框

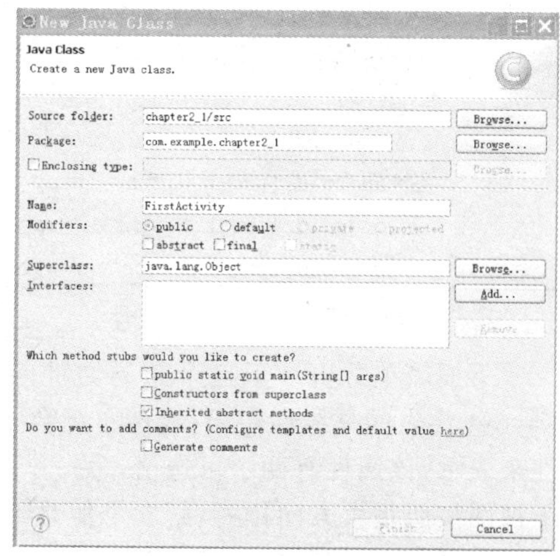

图 2－2　父类选择对话框

2.1.2　新建用户界面

由于每一个活动都要拥有一个用户界面，所以接下来笔者要为 FirstActivity 建立一个用户界面，用户界面的详细内容将会在后面章节当中介绍。简单起见，在 FirstActivity 的用户界面上只放置一个 Button 按钮。

(1) 单击工程根目录下面的 res 子目录，选中 res 子目录下面的 layout 子目录，右击 new | other 命令，在弹出的 New 对话框当中，选中 Android 节点下面的 Android

XML File 选项，单击 Next 按钮，则弹出 New Android XML File 对话框。

(2) 在 File 文本框中输入 FirstActivity 用户界面布局文件的名称：first_layout.xml，单击 Finish 按钮。

(3) 双击新建的 first_layout.xml 布局文件。Eclipse 提供两种视图来查看布局文件：

- ❑ 一种是图形界面视图，这种视图支持开发者以可视化的方式往界面上拖动各种各样的控件。
- ❑ 一种是代码视图，这种视图可以查看布局文件的源代码，允许开发者通过编写源代码的方式来控制用户界面的布局。

提示：由于 Eclipse 提供的这种图形界面的拖动控件的功能比较弱，所以 Android 应用程序的开发者一般使用的还是相对比较麻烦的通过手写源代码的方式来控制用户界面布局，这也是 Android 应用程序开发过程中比较麻烦的一点。所以，笔者在这里也是通过手写源代码的方式来控制用户界面布局的。

first_layout.xml 布局文件源代码如下：

```
<? xml version = "1.0" encoding = "utf-8"? >
<LinearLayout xmlns:android = "http://schemas.android.com/apk/res/android"
    android:layout_width = "match_parent"
    android:layout_height = "match_parent"
    android:orientation = "vertical" >
    <!-- Button 的定义 -->
    <Button
        android:id = "@ + id/myButton"
        android:layout_width = "wrap_content"
        android:layout_height = "wrap_content"
        android:text = "@string/my_button_str" />
</LinearLayout>
```

下面简单分析一下这个布局文件的源代码。XML 文件当中的根节点 LinearLayout 代表的是线性布局，线性布局会有一个方向的设定，用 android:orientation 属性来表示，该属性有两个可选值，第一个可选值是 vertical，代表的是线性布局里面的子视图按照垂直方向排列，第二个值是 horizontal，代表的是线性布局里面的子视图按照水平方向排列。

在 LinearLayout 节点中，有两个比较重要的属性，android:layout_width 以及 android:layout_height。

- ❑ android:layout_width 属性用来设置视图的宽度，有几个可选值分别是：fill_parent、match_parent 以及 wrap_content。
  - ➢ fill_parent 代表的意思是指视图与父视图的宽度一致。
  - ➢ match_parent 代表的意思与 fill_parent 一致，只不过这个值只可以在 android SDK 2.2 以上的版本当中才可以使用，2.2 之前的 SDK 版本没有这个属性。

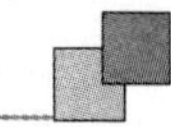

- wrap_content 的意思是指内容的宽度有多大，视图的宽度就有多大。
- android:layout_height 属性用来设置视图的高度，可选值以及可选值表示的意思与 android:layout_width 一致。

在 LinearLayout 节点中，包含一个子节点 Button。Button 节点用来代表一个按钮，第一个属性 android:id="@+id/myButton"，表明在编译的时候，会在 gen 目录下面的 R.java 文件的内部类 id 当中添加一个叫做 myButton 的属性，这样就可以直接在代码当中通过 R.id.myButton 来获取 id 为 myButton 的按钮控件的引用了。

除了 android:layout_width 以及 android:layout_height 属性之外，还有一个用来设置按钮上面显示文字的属性 android:text，这个属性引用的值是@sring/my_button_str。表明引用的是 values/strings.xml 文件当中的字符串名称为 my_button_str 所代表的值。values/strings.xml 文件源代码如下：

```
<resources>
    <string name = "app_name">chapter2_1</string>
    <string name = "hello_world">Hello world! </string>
    <string name = "menu_settings">Settings</string>
    <string name = "title_activity_main">MainActivity</string>
  <string name = "my_button_str">My First Button</string>   <!-- Button 显示文字 -->
</resources>
```

布局文件建好了之后，需要在 FirstActivity 当中进行引用。

(4) 打开 FirstActivity.java 文件。首先需要去重写父类的 onCreate 方法，这个方法是活动在创建的时候回调的方法。覆盖 onCreate 方法之后，FirstActivity 源文件代码如下：

```
public class FirstActivity extends Activity
{
    @Override
    protected void onCreate(Bundle savedInstanceState)
    {
        super.onCreate(savedInstanceState);
        setContentView(R.layout.first_layout);     //设置布局文件 first_layout.xml
                                                   //为活动的 UI 界面
    }
}
```

在这段代码当中，最重要的一句代码就是 setContentView(R.layout.first_layout)，首先解释一下这里的 R 类，也就是前面章节介绍 gen 目录时候的那个 R.java 文件，R.layout.first_layout 实际上引用的是 res 目录 layout 子目录下面的 first_layout.xml 文件，而通过调用父类的 setContentView(R.layout.first_layout)，实际上就是指定了这个活动的 UI 界面的定义文件就是 first_layout.xml。

每新建一个活动，都需要在项目清单文件当中进行相关声明，FirstActivity 在 AndroidManifest.xml 当中的声明代码如下：

```
<!-- FirstActivity 声明 -->
<activity
    android:name = ".FirstActivity"
    android:label = "@string/title_activity_main" >
</activity>
```

<activity>节点要放在<application>节点中，作为<application>的子节点存在。

其中，android:name 属性指定了活动的名称，这个属性指定的名称应该是一个完整的类名（比如，com.example.project.ExtracurricularActivity），然而作为完整类名的缩写（比如.FirstActivity），实际上代表的就是 manifest 节点当中 package 属性指定的包名下面的 FirstActivity 类。这个属性是必填项，没有默认值。android:label 属性指定的是活动的标题，如果这个属性没有设置，那么就会引用 application 节点中的 android:label 属性。

提示：当设置 android:label 属性的时候，最好通过引用 values 子目录下面的 strings.xml 中的字符串，这样比较容易实现字符串资源的多语言支持。

把活动作为应用程序的主入口活动，需要在<activity>节点中添加如下代码：

```
<intent-filter>
            <action android:name = "android.intent.action.MAIN" />
            <category android:name = "android.intent.category.LAUNCHER" />
</intent-filter>
```

至于具体 intent-filter 节点以及节点下面的子节点，笔者将在后面章节中介绍。修改过后的项目清单文件源码如下：

```
<manifest xmlns:android = "http://schemas.android.com/apk/res/android"
    package = "com.example.chapter2_1"
    android:versionCode = "1"
    android:versionName = "1.0" >
    <uses-sdk
        android:minSdkVersion = "8"
        android:targetSdkVersion = "15" />
    <application
        android:icon = "@drawable/ic_launcher"
        android:label = "@string/app_name"
        android:theme = "@style/AppTheme" >
        <!-- MainActivity 声明 -->
        <activity
            android:name = ".MainActivity"
```

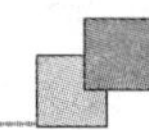

```
            android:label = "@string/title_activity_main" >
        </activity>
        <!-- FirstActivity 声明 -->
        <activity
            android:name = ".FirstActivity"
            android:label = "@string/title_activity_main" >
            <intent-filter>
                <action android:name = "android.intent.action.MAIN" />
                <category android:name = "android.intent.category.LAUNCHER" />
            </intent-filter>
        </activity>
    </application>
</manifest>
```

(5) 右击工程,在弹出的快捷菜单中选择 Run As | Android Application 命令,运行结果如图 2－3 所示。

图 2－3　chpater2_1 运行效果图

2.2　理解活动生命周期

Android 是通过一个活动栈的方式来管理活动的。当一个活动启动的时候,它会被通过压栈的方式放入到栈顶从而成为运行中的活动,而前一个运行的活动则不会来到前台直到当前的活动退出。当回退按钮被按下的时候,实际上就是对当前活动进行出栈的处理。

2.2.1　活动生命周期

一个活动有以下 4 种状态:

- 如果一个活动处于活动栈的顶部,也就是正在运行与用户交互当中,这时候的活动处于运行状态。
- 如果一个活动已经失去了焦点但是仍然有部分可见(可能有一个对话框样式的活动或者是透明的活动覆盖在了当前的这个活动之上),这时候的活动处于暂停状态。处于暂停状态的活动也是活着的,此时的活动仍然保留了所有的状态信息,但是在系统内存不足的情况下会被 Android 运行时销毁。
- 如果一个活动完全被其他活动所覆盖,这时候的活动处于停止状态,虽然处于停止状态的活动仍保留了所有的状态信息,但是对用户来说,处于停止状态的活动已经不可见了,无论系统在什么地方需要内存的时候,都可能将处于停止状态的活动销毁。
- 当活动处于暂停或者停止状态的时候,Android 运行时可能因为系统内存不足而去调用活动的 finish 方法,或者是直接销毁活动所处的应用程序的进程。活动由于内存不足而被系统销毁之后,下次重新启动的时候,一般来讲都要去恢复被 Android 运行时销毁之前的那个状态。

图 2－4 描述的是一个活动重要状态转变的路径图。其中，矩形部分代表的是一些活动状态改变的时候，被系统回调的一些方法，而椭圆形部分代表的是活动的一些状态。

2.2.2 LogCat 的使用

理解活动生命周期的最好的方式就是在活动的每一个回调函数里面加上日志，观察每一个回调函数被调用的时间和顺序。由于需要在应用程序里面添加相关日志，所以笔者先介绍一下 LogCat 如何使用。

Android 日志系统提供了记录和查看系统调试信息的功能。日志都是从各种软件和一些系统的缓冲区中记录下来的，缓冲区可以通过 logCat 命令来查看和使用。

提示：在使用 logCat 命令之前，需要将 android SDK 中 platform-tools 子目录路径加到系统变量 path 当中，添加和编辑系统变量的步骤笔者在前面章节当中已经有所介绍，在这里笔者就不重复介绍了。

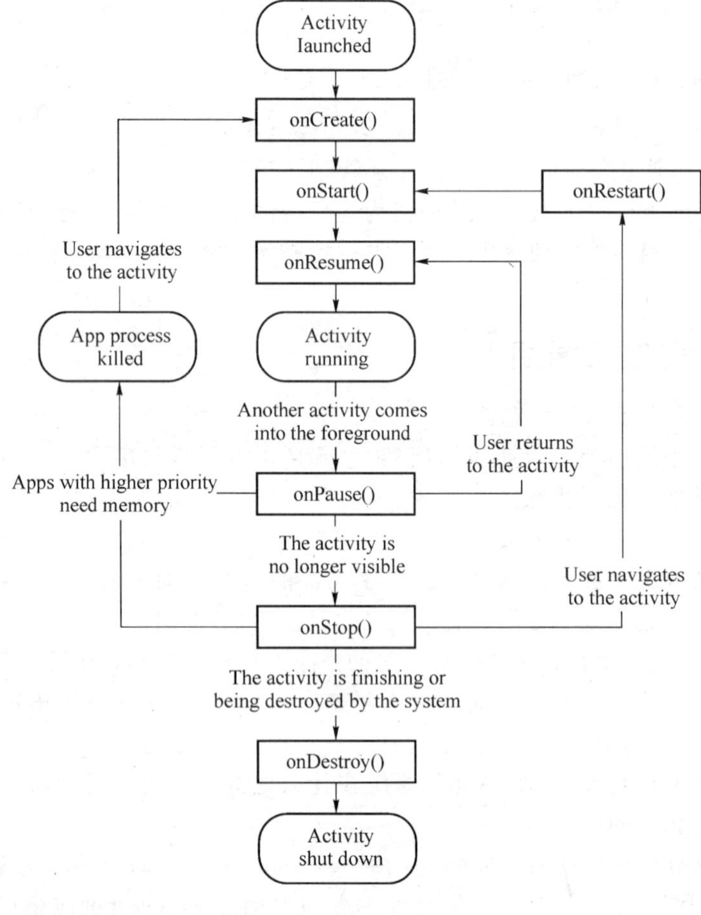

图 2－4　活动生命周期

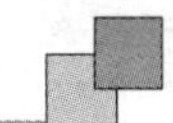

1. logCat 命令

用 logCat 命令来查看系统日志缓冲区的内容：[adb] logcat [<option>] ... [<filter-spec>] ...，可以通过两种方式来使用 logCat 命令。

- ❑ 直接进入计算机的命令行界面，输入 adb logcat 命令，查看日志信息。
- ❑ 直接进入计算机的命令行界面，输入 adb shell 命令，再输入 logCat 命令，查看日志信息。

2. DDMS LogCat 工具

除了使用命令行界面来查看日志信息之外，还可以通过在 Eclipse 中选择 Window|Show View|Other|Android|LogCat 命令，查看日志信息，如图 2－5 所示。

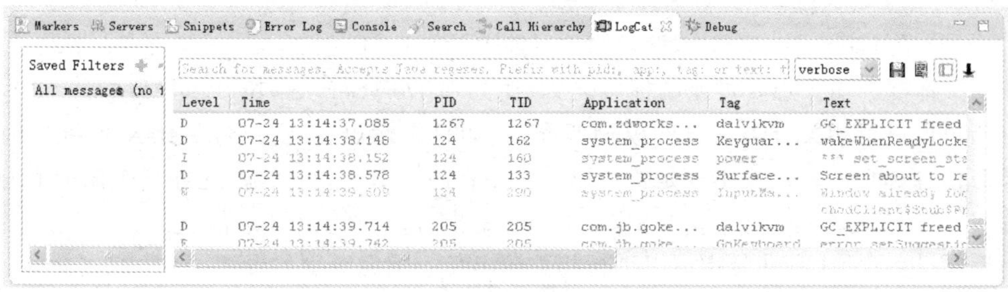

图 2－5　LogCat 界面

3. Log 的级别和过滤

Android Log 信息都有一个标签(TAG)和它的输出级别(Level)。Android Log 信息的级别由低到高分为以下几种级别：

- ❑ V(Verbose)；
- ❑ D(Debug)；
- ❑ I(Info)；
- ❑ W(Warn)；
- ❑ E(Error)；
- ❑ F(Fatal)；
- ❑ S(Silent)。

当使用 adb logcat 命令输出日志信息时，在命令行输出出来的日志的前两列信息中，读者就可以看到 logCat 的标签列表和输出级别(比如：I/ActivityManager(585)：Starting activity：Intent { action＝android. intent. action... })，那么这条 Log 信息的优先级就是 I，标签就是 ActivityManager。

提示：为了让 Log 输出达到可管理的级别，开发者可以用过滤器来限制 Log 输出，过滤器可以帮助只显示开发者感兴趣的标签和输出级别的组合，而隐藏其他的信息。

在指定了输出标签和级别之后，包含输出标签以及输出级别在指定的输出级别之上的所有日志信息都会输出出来。可以在一个过滤表达式当中提供多个(标签:输出级别)的声明，这些声明之间用空白符间隔。

举例来说：想要输出标签为 ActivityManager 且输出级别在 Info 以上(包括 Info 级别)的日志，标签为 MyApp 且输出级别在 Debug 以上(包括 Debug 级别)的信息，或者其他任意标签输出级别在 Silent(包括 Silent)以上的日志信息，则过滤表达式为 adb logcat ActivityManager:I MyApp:D *:S。

下面介绍如何使用 DDMS 的 LogCat 工具来进行日志信息的过滤。

(1) 在图 2-5 当中，单击视图左侧的蓝色"+"号按钮，弹出 Logcat Message Filter Settings 对话框。

(2) 在 Filter Name 文本框中输入过滤器的名称，比如 BatteryService；在 by Log Tag 文本框中输入想要过滤的标签名称，比如 BatteryService；by Log Message 文本框指可以进行相关日志信息的模糊匹配，比如只想输出含有 update start 字符串的日志信息；而 by PID 文本框是指可以对应用程序的 ID 进行过滤，比如只想输出 PID 为"176"的日志信息；by Application Name 是指可以对输出日志信息的应用程序进行匹配；by Log Level 是指可以进行相关输出日志信息级别的设置，如图 2-6 所示。

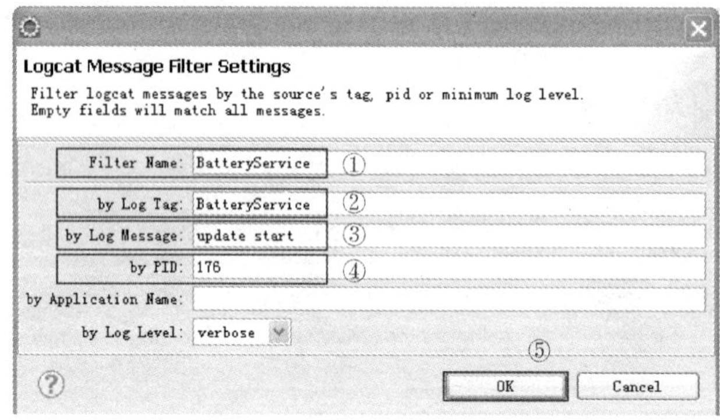

图 2-6 Logcat 日志信息过滤设置

(3) 上述过滤条件组合在一起，实际上就是要输出标签为 BatteryService，含有字符串 update start，应用程序的 PID 为"176"，且日志信息级别在 verbose(包括 verbose)以上的所有日志信息。

(4) 过滤过后的日志信息如图 2-7 所示。

下面通过重写 Activity 基类当中重要的事件回调方法，从而观察活动状态变化的时候，回调方法调用的时间以及顺序是怎样的。新建工程 chapter2_2，MainActivity.java 源代码如下：

```
public class MainActivity extends Activity
```

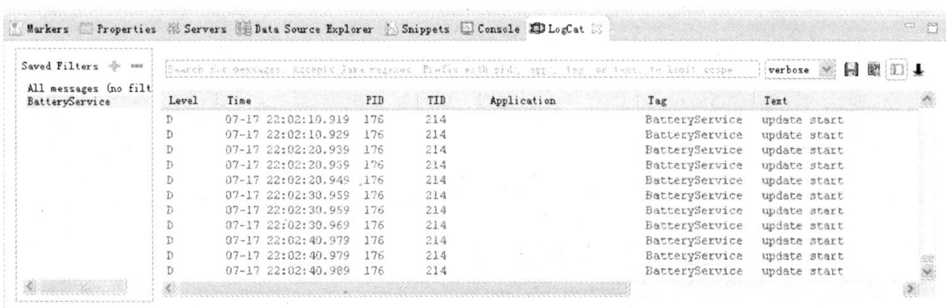

图 2-7　BatteryService 过滤器 Logcat 界面

```
{
    private static final String TAG = "MainActivity";//Logcat 日志信息的标签
    @Override
    public void onCreate(Bundle savedInstanceState)
    {//活动第一次启动的时候会调用 onCreate()方法
        super.onCreate(savedInstanceState);
        requestWindowFeature(Window.FEATURE_NO_TITLE);//隐藏活动标题栏
        setContentView(R.layout.activity_main);
        Log.d(TAG, "onCreate()方法被调用");
    }
    @Override
    protected void onRestart()
    {//当之前处于停止状态的互动重新获得用户焦点的时候调用 onRestart()方法
    super.onRestart();
    Log.d(TAG, "onRestart()方法被调用");
    }
    @Override
    protected void onStart()
    {//onStart()方法在 onCreate()和 onResume 之间被调用
    super.onStart();
    Log.d(TAG, "onStart()方法被调用");
    }
    @Override
    protected void onResume()
    {//调用 onResume()方法过后,活动正式进入运行状态
    super.onResume();
    Log.d(TAG, "onResume()方法被调用");
    }
    @Override
    protected void onPause()
    {//当活动失去焦点,但仍然部分可见的时候 onPause()方法被调用
```

```
    super.onPause();
    Log.d(TAG, "onPause()方法被调用");
    }
    @Override
    protected void onStop()
    {//当活动完全不可见,onStop()方法被调用
    super.onStop();
    Log.d(TAG, "onStop()方法被调用");
    }
    @Override
    protected void onDestroy()
    {//当活动被 finish 的时候调用 onDestory()方法
    super.onDestroy();
    Log.d(TAG, "onDestory()方法被调用");
    }
}
```

为了观察日志信息的输出，首先建立一个用来观察 chpater2_2 应用程序当中标签为 MainActivity 的日志信息的过滤器，通过 DDMS 视图的 LogCat 工具类创建这个过滤器的步骤笔者在这里就不重复介绍了。

(1) 右击 chapter2_2 工程，在弹出的快捷菜单中选择 Run As|Android Application 命令，观察日志信息的输出，如图 2-8 所示。

Markers Properties Servers Data Source Explorer Snippets Console LogCat

Level	Time	PID	TID	Application	Tag	Text
D	07-17 23:16:54.989	14383	14383	com.example.chapte...	MainActivity	onCreate()方法被调用
D	07-17 23:16:54.989	14383	14383	com.example.chapte...	MainActivity	onStart()方法被调用
D	07-17 23:16:54.989	14383	14383	com.example.chapte...	MainActivity	onResume()方法被调用

图 2-8　MainActivity 启动日志信息输出

(2) 在活动运行的时候，如果用户按下 Back 按键，观察日志信息输出，如图 2-9 所示。

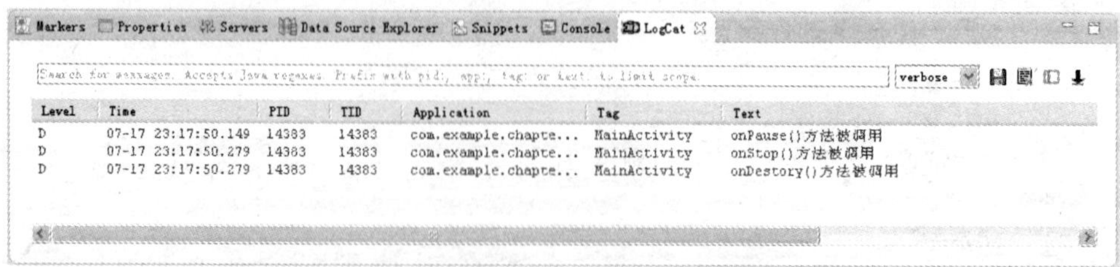

图 2-9　MainActivity 按下 Back 按键日志输出

（3）当活动运行的时候，如果按下 Home 键，观察日志信息输出，如图 2－10 所示。

Level	Time	PID	TID	Application	Tag	Text
D	07-17 23:18:30.869	14383	14383	com.example.chapte...	MainActivity	onPause()方法被调用
D	07-17 23:18:31.259	14383	14383	com.example.chapte...	MainActivity	onStop()方法被调用

图 2－10　MainActivity 按下 Home 按键日志输出

（4）按住 Home 键过后，重新单击应用程序图标，观察日志信息输出，如图 2－11 所示。

Level	Time	PID	TID	Application	Tag	Text
D	07-17 23:19:05.829	14383	14383	com.example.chapte...	MainActivity	onRestart()方法被调用
D	07-17 23:19:05.829	14383	14383	com.example.chapte...	MainActivity	onStart()方法被调用
D	07-17 23:19:05.829	14383	14383	com.example.chapte...	MainActivity	onResume()方法被调用

图 2－11　按住 Home 键过后重新启动 MainActivity 日志输出

从以上的例子当中可以看出，当活动第一次创建的时候，onCreate 方法将会被调用，紧接着 onStart 方法，onResume 方法依次被调用，onResume 方法调用过后，活动就会处于运行状态。当活动失去焦点，但是部分可见的时候，onPause 方法会被调用，当活动完全不可见的时候，onStop 方法会被调用，当用户按下回退按钮的时候，onPause 方法，onStop 方法和 onDestroy 方法会被依次调用，活动将会被销毁，当原先处于停止状态的活动重新启动的时候，onRestart 方法会被调用，紧接着 onStart 和 onResume 方法会被调用。

当活动处于暂停状态或者是停止状态的时候，活动所在应用程序进程有可能会因为进程优先级的关系而被 Android 运行时销毁，在这种特殊情况下，活动必须在应用程序被销毁之前之前保存一些状态信息，从而在再次启动的时候，能够把这些状态信息给读取到。无论是在用户按下 Back 按钮，还是按下 Home 按钮，onPause 方法和 onStop 方法都会被调用。

在 Android 应用程序开发的时候，在上述回调函数里面添加代码一般遵循以下规则：

- ❑ 使用 onCreate 方法来做一些创建和初始化活动当中将要使用的对象。
- ❑ 使用 onResume 方法来启动一些服务，注册一些广播接收者。
- ❑ 使用 onPause 方法来停止一些服务，注销一些广播接收者。
- ❑ 使用 onDestory 方法在活动销毁之前释放一些活动当中使用过的资源。

2.3 活动的启动模式(android:launchMode)

笔者在前面章节当中已经做过介绍,在新建活动的时候,需要在项目清单文件当中,对活动进行配置。其中<activity>这个节点当中有一个 android:launchMode 的属性用来配置对应的 Activity 的启动模式。目前有以下 4 种启动模式:standard、singleTop、singleTask 及 singleInstance。

默认情况下,Activity 的启动模式为 standard 模式。

新建工程 chapter2_3,修改主入口布局文件,一个 TextView 控件用来显示所在 Activity 的 HashCode,添加一个按钮,用来触发跳转页面的操作,笔者在程序当中设置的跳转是从当前活动跳转到 MainActivity,通过观察刚启动的时候 TextView 展示的 HashCode 与跳转过来之后的 HashCode 是否一致,来判断在 standard 模式下,Activity 是否会创建多次。

修改过后的 AndroidManifest.xml 文件源代码如下:

```
<RelativeLayout xmlns:android = "http://schemas.android.com/apk/res/android"
    xmlns:tools = "http://schemas.android.com/tools"
    android:layout_width = "match_parent"
    android:layout_height = "match_parent" >
<!-- 显示当前对象的 HashCode -->
    <TextView
        android:id = "@ + id/textView1"
        android:layout_width = "wrap_content"
        android:layout_height = "wrap_content"
        android:layout_alignParentTop = "true"
        android:layout_centerHorizontal = "true"
        android:layout_marginTop = "16dp"
        android:padding = "@dimen/padding_medium"
        android:text = "@string/hello_world"
        tools:context = ".MainActivity" />
    <!-- 单击按钮,触发跳转操作 -->
    <Button
        android:id = "@ + id/button1"
        android:layout_width = "wrap_content"
        android:layout_height = "wrap_content"
        android:layout_alignParentLeft = "true"
        android:layout_alignParentRight = "true"
        android:layout_below = "@ + id/textView1"
        android:text = "@string/button_str" />
</RelativeLayout>
```

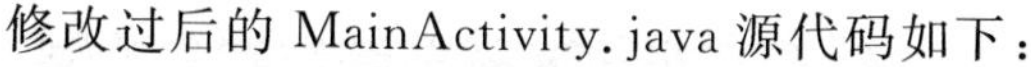

修改过后的 MainActivity.java 源代码如下：

```
public class MainActivity extends Activity implements OnClickListener
{
    TextView textView;
    Button myButton;
    @Override
public void onCreate(Bundle savedInstanceState)
{
        super.onCreate(savedInstanceState);
        setContentView(R.layout.activity_main);
        myButton = (Button)findViewById(R.id.button1);
        myButton.setOnClickListener(this);        //设置当前对象为按钮的事件监听器
        textView = (TextView)findViewById(R.id.textView1);
        textView.setText(this + "");              //显示当前对象 HashCode
    }
    @Override
    public void onClick(View v)
    {
        if(R.id.button1 == v.getId())
        {
            Intent intent = new Intent(this, MainActivity.class);
                                                  //从当前 Activity 跳转到 MainActivity
            startActivity(intent);                //跳转
        }
    }
}
```

笔者来解释一下代码。代码当中通过调用 findViewById()方法来获取布局文件当中 id 为 button1 的按钮的引用，再通过调用按钮的 setOnClickListenner()方法来为按钮设置单击事件的监听器，由于本例当中的 MainActivity 实现了 View.OnClickListenner 这个接口，所以可以把当前类的引用，也就是 this 作为按钮的单击事件监听器。

当按钮被单击的时候，会自动回调 MainActivity 当中的 onClick()方法，笔者通过使用意图(Intent)做了一次界面跳转的操作，Intent 的构造函数有两个参数，第一个参数用 this 表示跳转的当前活动，而第二个参数表示 MainActivity 是想要跳转到的界面，最后调用 startActivity(intent)来完成这次跳转(意图后面细讲)。程序启动的时候效果如图 2-12 所示，单击按钮之后的效果如图 2-13 所示。

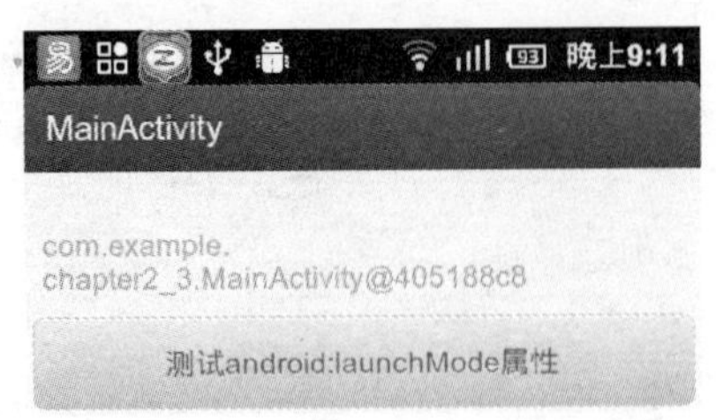

图 2-12 chapter2_3 运行效果图

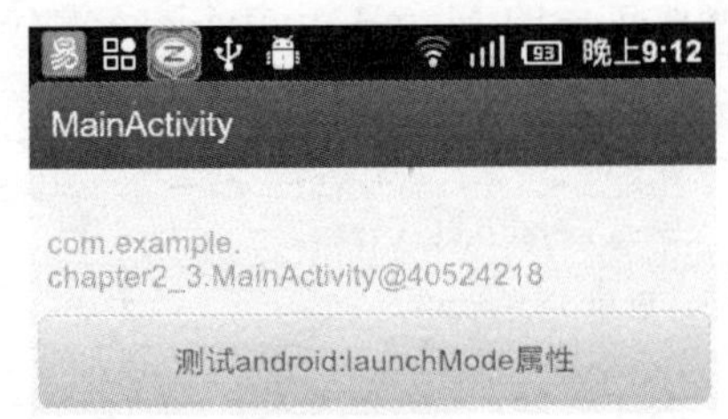

图 2-13 单击按钮之后的效果图

读者会发现，两次打印出来的 MainActivity 的 HashCode 是不一样的，说明刚启动时的那个 Activity 与单击一次按钮之后出现的那个活动不是同一个对象。如果在项目清单文件当中为 MainActivity 配置 android:launchMode 这个属性，并将它的值设置为 singleTop，则运行效果如图 2-14 所示，单击按钮过后，效果如图 2-15 所示。

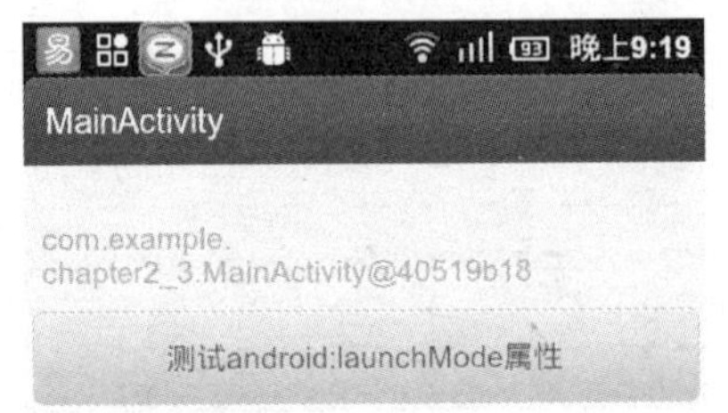

图 2-14 运行效果

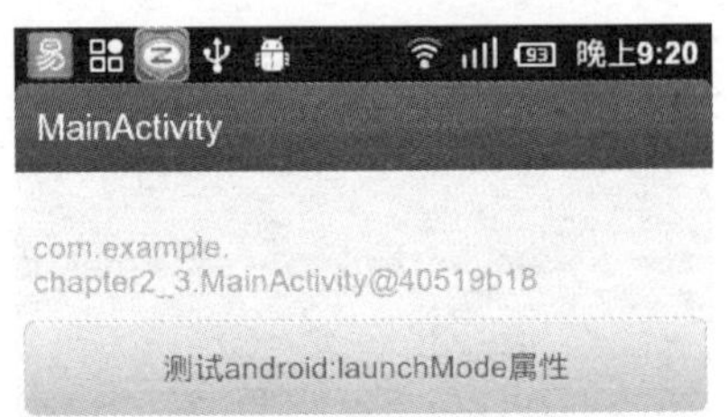

图 2-15 单击 singleTop 按钮的效果

由此可以得出结论：singleTop 和 standard 模式都会将 Intent 发送给新的 Activity 实例，不同的是，如果创建 Intent 的时候栈顶有要创建的 singleTop 模式下的 Activity 实例，则将 Intent 发送给该实例，不会再创建 Activity 的新实例，会调用 Activity 的 onNewIntent()方法。在 singleTop 模式下，如果在栈顶存在 Intent 中那个目标 Activity 的实例，就不会创建新的实例，而直接使用栈顶的对象，对于资源有限的移动设备来说，也是有实际意义的。

android:launchMode 为 singleTask 的 Activity 只能创建一个实例，当发送一个 Intent，目标 Activity 为 singletTask 模式时，系统会检查栈里面是否已经有该 Activity 的实例，如果有就直接将 Intent 发送给它，不会再创建 Activity 的新实例。由此可知，singleTask 模式的 Activity 在栈中只有一个实例，可以被重复使用。

android:launchMode 为 singleInstance 的 Activity 会在一个单独的 Task(可以理解成一个单独的应用)栈中，而且这个栈当中只能包含这一个活动，其他的活动不能存在于这个 Task 栈中，如果它启动了一个新的 Activity，不管新的 activity 的 launch mode 如何，新的 activity 都将会在别的 Task 里运行。

2.4　活动运用样式和主题

默认情况下，活动会全屏显示。当然，开发者也可以在活动上应用一个对话框的主题或者一些其他的主题。如果应用的是对话框的主题，那么当活动运行的时候，就会显示成一个浮动的对话框。这种浮动的对话框可以用来提示一些用户正在执行的操作，由于活动是以弹出的对话框的形式展示给用户的，所以对话框形式的活动会很容易获取到用户的焦点。

接着使用前面章节的工程 chapter2_2，如果需要给 MainActivity 活动应用对话框的主题，只需要在项目清单文件当中的相应的<activity>节点添加 android:theme 这样一个属性即可，修改过后的项目清单文件源代码如下：

```
<manifest xmlns:android = "http://schemas.android.com/apk/res/android"
    package = "com.example.chapter2_2"
    android:versionCode = "1" android:versionName = "1.0" >
    <uses-sdk   android:minSdkVersion = "8" android:targetSdkVersion = "15" />
    <application
        android:icon = "@drawable/ic_launcher"
        android:label = "@string/app_name" android:theme = "@style/AppTheme" >
      <activity
            android:name = ".MainActivity"  android:
            label = "@string/title_activity_main"
            android:theme = "@android:style/Theme.
            Dialog" >      <!-- 为活动应用对话框主
            题 -->
            <intent-filter>
                <action android:name = "android.in-
tent.action.MAIN" />
                <category android:name = "android.
intent.category.LAUNCHER" />
            </intent-filter>
        </activity>
    </application>
</manifest>
```

图 2-16　对话框样式活动

选中 chapter2_2 工程，右击选择 Run As|Android Application 命令，运行效果如图 2-16 所示。

2.5 隐藏活动标题栏

除了可以给活动应用主题和样式以外,还可以隐藏活动的标题栏,因为在有的应用场景下活动是不需要标题栏的。接着使用前面章节的工程 chapter2_2,笔者将通过代码方式隐藏掉 MainActivity 的标题栏,修改过后的 onCreate 方法源代码如下:

```
@Override
public void onCreate(Bundle savedInstanceState)
{//活动第一次启动的时候会调用 onCreate()方法
        super.onCreate(savedInstanceState);
        requestWindowFeature(Window.FEATURE_NO_TI-
TLE);//隐藏活动标题栏
        setContentView(R.layout.activity_main);
        Log.d(TAG, "onCreate()方法被调用");
}
```

图 2-17 无标题栏活动

同样是选中 chapter2_2 工程,右击 Run As|Android Application 命令,运行效果如图 2-17 所示。

2.6 弹出对话框

应用程序为了从用户那里得到一个确认信息,就需要显示一个对话框窗口,让用户去做选择。这时候在程序当中就需要去重写 Activity 基类当中的 onCreateDialog()方法。新建工程 chapter2_4,笔者将要在主入口活动的布局文件当中添加一个按钮,当用户单击这个按钮的时候,去显示一个对话框。

(1) 修改 res 目录下 layout 子目录下 activity_main.xml 文件,源代码如下:

```
<RelativeLayout xmlns:android="http://schemas.android.com/apk/res/android"
    xmlns:tools="http://schemas.android.com/tools"
    android:layout_width="match_parent"
    android:layout_height="match_parent" >
    <TextView
        android:id="@+id/textView1"
        android:layout_width="wrap_content"
        android:layout_height="wrap_content"
        android:layout_centerHorizontal="true"
        android:layout_centerVertical="true"
        android:padding="@dimen/padding_medium"
```

```
        android:text = "@string/hello_world"
        tools:context = ".MainActivity" />
    <!-- 用户单击此按钮,弹出对话框 -->
    <Button
        android:id = "@ + id/button1"
        android:layout_width = "wrap_content"
        android:layout_height = "wrap_content"
        android:layout_above = "@ + id/textView1"
        android:layout_centerHorizontal = "true"
        android:text = "@string/button_str" />
</RelativeLayout>
```

(2) 修改 MainActivity.java 文件,源代码如下:

```
public class MainActivity extends Activity implements View.OnClickListener
{
    Button myButton;
    CharSequence[] items = {"Tecent", "Sina", "Sohu"};
    boolean[] itemsChecked = new boolean[items.length];
    @Override
    public void onCreate(Bundle savedInstanceState)
    {
        super.onCreate(savedInstanceState);
        setContentView(R.layout.activity_main);
        myButton = (Button)findViewById(R.id.button1);    //找到 ID 为 button1 的按钮
        myButton.setOnClickListener(this);                //为按钮设置当前活动为单
                                                          //击事件的监听器
    }
    @Override
    public void onClick(View v)
    {
        showDialog(0);
    }
    @Override
    protected Dialog onCreateDialog(int id)
    {
        switch(id)
        {
            case 0://对话框 id 为 0 时
            return new AlertDialog.Builder(this).setIcon(R.drawable.ic_launcher).
setTitle("This is a dialog with some simple text")
                                        .setPositiveButton("OK", new DialogInterface.On-
ClickListener()
```

```
                        {
                        @Override
                          public void onClick(DialogInterface dialog, int
which)
                        {
                          Toast.makeText(getBaseContext(), "OK Clicked!",
Toast.LENGTH_SHORT).show();
                        }
                    })
                    .setNegativeButton("Cancel", new DialogInterface.On-
ClickListener()
                      {
                          @Override
                           public void onClick(DialogInterface dialog, int
which)
                          {
                                Toast.makeText(getBaseContext(), "Cancel
Clicked", Toast.LENGTH_SHORT).show();
                            }
                          }).setMultiChoiceItems(items, itemsChecked, new
DialogInterface.OnMultiChoiceClickListener()
                        {
                          @Override
                           public void onClick(DialogInterface dialog, int
which, boolean isChecked) {
                            Toast.makeText(getBaseContext(), items[which]
 + (isChecked ? " checked!" : " unchecked!"), Toast.LENGTH_SHORT).show();
                          }
                          }).show();
            }
            return null;
        }
    }
```

笔者解释一下上述代码。onCreateDialog()这个方法实际上是用来创建对话框的一个回调函数。当程序当中调用 showDialog()这个方法的时候，Activity 就会自动去调用 onCreaterDialog()这个方法。showDialog()和 onCreateDialog()这两个方法都是 Activity 这个基类当中已经定义的方法。showDialog()这个方法会接受一个整型的参数，用来标识每一个要显示的对话框。在 onCreateDialog()方法当中，笔者使用了一个 switch-case 的处理语句，用来针对每一个不同的整型参数，显示不同的对话框。

可以通过 AlertDialog 类当中的 Builder 这个内部类的构造函数来创建对话框，在上述代码当中，设置了很多对话框的属性，包括图标、标题以及按钮。代码通过调用 setPositiveButton()和 setNegativeButton()这两个方法，设置了对话框当中的两个按钮，一个按钮名称叫 OK，一个按钮名称叫 Cancel。通过调用 setMultiChoiceItems()方法，设置了可以让用户去选择的一个复选框的列表，并在用户选择列表当中的每一项的时候，给予用户相应的“吐西”提示。选中工程 chapter2_4，右击选择 Run As|Android Application 命令，运行效果如图 2-18 所示，单击 Click Me 按钮之后，运行效果如图 2-19 所示。

图 2-18　chapter2_4 运行效果

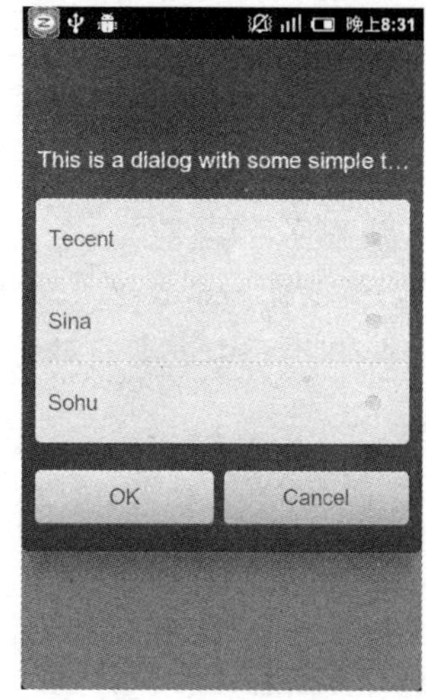

图 2-19　单击“Click Me”按钮

上下文对象(Context)

在开发 Android 应用程序的时候，开发者经常要遇到 Context 这个类以及它的实例。Context 类的实例一般用来提供对应用程序的引用。举例来说，在下面的代码片段当中，Toast 类当中的静态方法 makeTest()当中的第一个参数就需要一个 Context 类的实例。代码片段如下：

```
.setNegativeButton("Cancel", new DialogInterface.OnClickListener()
    {
    @Override
    public void onClick(DialogInterface dialog, int which)
        {
            Toast.makeText(getBaseContext(), "Cancel Clicked", Toast.LENGTH_SHORT).
show();
        }
    }
```

由于上面代码片段当中的 Toast.makeText()方法是在内部类当中使用的，所以此方法的第一个参数不能接受 this 这个参数，而可以通过 getBaseContext()方法的调用返回一个 Context 类的对象。在动态创建布局对象的时候，也需要用到 Context 类的对象。例如：

TextView tv = new TextView(this);

由于 TextView 的构造函数需要接受一个 Context 类的对象，而 Activity 类本身就是 Context 类的一个子类，所以可以将 this 作为 TextView 构造函数的参数。

2.7 弹出进度条对话框

应用程序有的时候需要去做一些比较耗时的操作，比如下载一些文件，而作为用户来讲，很想知道当前所等待的操作到底现在处于一种什么样的进度，所以这个时候，如果应用程序可以弹出一个显示当前进度的对话框的话，用户体验就会比较好。

下面新建工程 chapter2_5，为主入口活动的界面布局文件添加一个按钮，用户单击此按钮时，弹出一个进度条对话框，修改后的 activity_main.xml 文件源代码如下：

```
<RelativeLayout xmlns:android="http://schemas.android.com/apk/res/android"
    xmlns:tools="http://schemas.android.com/tools"
    android:layout_width="match_parent"
    android:layout_height="match_parent" >
   <TextView
        android:id="@+id/textView1"
        android:layout_width="wrap_content"
        android:layout_height="wrap_content"
        android:layout_centerHorizontal="true"
        android:layout_centerVertical="true"
        android:padding="@dimen/padding_medium"
        android:text="@string/hello_world"
        tools:context=".MainActivity" />
<!-- 用户单击此按钮时，弹出进度条对话框 -->
    <Button
        android:id="@+id/button1"
        android:layout_width="wrap_content"
        android:layout_height="wrap_content"
        android:layout_above="@+id/textView1"
        android:layout_centerHorizontal="true"
        android:text="@string/button_str" />
</RelativeLayout>
```

修改过后的 MainActivity.java 文件源代码如下：

```
public class MainActivity extends Activity implements OnClickListener
{
    Button myButton;
    @Override
    public void onCreate(Bundle savedInstanceState)
    {
        super.onCreate(savedInstanceState);
        setContentView(R.layout.activity_main);
        myButton = (Button)findViewById(R.id.button1);//找到布局文件中按钮的引用
    }
    //按钮单击回调 onClick 方法
    @Override
    public void onClick(View v)
    {
    final ProgressDialog progressDlg = ProgressDialog.show(this, "Do someting", "
Please hold on.....", true);
    new Thread(new Runnable()
      {        //开启新线程,模拟正在执行的异步操作
            @Override
            public void run()
            {
                SystemClock.sleep(5 * 1000);//子线程休眠 5 s 过后,隐藏对话框
                progressDlg.dismiss();
            }
        }).start();
    }
}
```

笔者解释一下上述代码。通过调用 ProgressDialog 的静态方法 show()来创建一个进度条对话框,这个静态方法有 4 个参数,第 1 个参数是一个 Context 类的对象,可以直接传入 this;第 2 个参数是进度条对话框的标题,可以传入一个字符串内容;第 3 个参数是进度条对话框的内容,也可以传入一个字符串内容;第 4 个参数是一个 bolean 值,当值为 false 的时候,滚动条的当前值自动在最小到最大值之间来回移动,形成这样一个动画效果,这个只是告诉用户"操作还没有结束",但不能提示工作进度到哪个阶段,当值为 true 的时候,则可以显示程序当中设置的进度。对于以下代码:

```
new Thread(new Runnable()
      {        //开启新线程,模拟正在执行的异步操作
            @Override
            public void run()
            {
```

```
                SystemClock.sleep(5 * 1000);//子线程休眠 5 s 过后,隐藏对话框
                progressDlg.dismiss();
            }
        }).start();
```

弹出来的进度条对话框是模态对话框,所以这个对话框在没消失之前,会一直阻塞住用户界面。为了模拟程序当中将要进行一个比较耗时的操作,所以笔者通过新建一个子线程,假设这些耗时的操作都是在这个子线程当中完成的,代码当中先让子线程所在的线程休眠 5 s,5 s 过后,会调用 dismiss()方法使进度条对话框消失。

一般来讲,在弹出的进度条对话框当中,需要给予用户明确的提示,现在的进度到哪里了。所以下面笔者来介绍在 Android 应用程序开发当中比较常见的一种进度条对话框。打开 chapter2_4 工程当中的 activity_main. xml 文件,在此布局文件当中再次添加一个按钮,当用户单击这个按钮的时候,弹出一个文件下载进度的对话框。修改过后的 activity_main. xml 文件源代码如下:

```
<LinearLayout xmlns:android = "http://schemas.android.com/apk/res/android"
    xmlns:tools = "http://schemas.android.com/tools"
    android:orientation = "vertical"
    android:layout_width = "match_parent"
    android:layout_height = "match_parent" >
   <TextView
        android:id = "@ + id/textView1"
        android:layout_width = "wrap_content"
        android:layout_height = "wrap_content"
        android:padding = "@dimen/padding_medium"
        android:text = "@string/hello_world"
        tools:context = ".MainActivity" />
<!-- 用户单击此按钮时,弹出进度条对话框 -->
    <Button
        android:id = "@ + id/button1"
        android:layout_width = "wrap_content"
        android:layout_height = "wrap_content"
        android:text = "@string/button_str" />
      <!-- 用户单击此按钮时,弹出文件下载进度条对话框 -->
      <Button
        android:id = "@ + id/file_download_button"
        android:layout_width = "wrap_content"
        android:layout_height = "wrap_content"
        android:text = "@string/file_down_button_str" />
</LinearLayout>
```

修改过后的 MainActivity. java 源代码如下:

```
public class MainActivity extends Activity implements OnClickListener
{
    Button myButton, fileDownloadButton;;
    @Override
    public void onCreate(Bundle savedInstanceState)
    {
        super.onCreate(savedInstanceState);
        setContentView(R.layout.activity_main);
        myButton = (Button)findViewById(R.id.button1);  //找到布局文件当中的按钮
                                                        //的引用
        //找到布局文件当中的 ID 为 file_download_button 的按钮的引用
        fileDownloadButton = (Button)findViewById(R.id.file_download_button);
        myButton.setOnClickListener(this);
        fileDownloadButton.setOnClickListener(this);
    }
    //按钮单击回调 onClick 方法
    @Override
    public void onClick(View v)
    {
        int viewId = v.getId();
        switch (viewId)
        {
        case R.id.button1:               //弹出普通进度条对话框
            final ProgressDialog progressDlg = ProgressDialog.show(this, "Do someting", "Please hold on.....", true);
          new Thread(new Runnable()
              {//开启新线程,模拟正在执行的异步操作
                  @Override
                  public void run()
                  {
                      SystemClock.sleep(5 * 1000);  //子线程休眠 5s 过后,隐藏对
                                                    //话框
                      progressDlg.dismiss();
                  }
              }).start();
          break;
      case R.id.file_download_button://弹出文件下载进度对话框
          final ProgressDialog fileDownloadDlg = new ProgressDialog(this);
          fileDownloadDlg.setIcon(R.drawable.ic_launcher);//设置图标
          fileDownloadDlg.setTitle("Downloading files.....");//设置标题
          fileDownloadDlg.setProgressStyle(ProgressDialog.STYLE_HORIZONTAL);
          fileDownloadDlg.setButton(DialogInterface.BUTTON_POSITIVE, "OK", new Dialog-
```

```
Interface.OnClickListener()
            {
                @Override
                public void onClick(DialogInterface dialog, int which)
                {
                   Toast.makeText(getBaseContext(), "OK Clicked!", Toast.LENGTH_SHORT).
show();
                }
            });
            fileDownloadDlg.setButton(DialogInterface.BUTTON_NEGATIVE, "Cancel", new Dia-
logInterface.OnClickListener()
            {
                @Override
                public void onClick(DialogInterface dialog, int which)
                {
                      Toast.makeText(getBaseContext(), "Cancel Clicked!" , Toast.LENGTH_
SHORT).show();
                }
            });
            fileDownloadDlg.show();
            fileDownloadDlg.setProgress(0);//设置进度条对话框当前进度
            new Thread(new Runnable()
          {
                @Override
                public void run()
                {
                     for(int i = 0; i < 10; i++)
                     {//10 次循环,每次进度增加 10
                          SystemClock.sleep(1000);
                          fileDownloadDlg.incrementProgressBy(100 / 10);//每休眠 1s,进度增
                                                                      //加 10%
                   }
                }
            }).start();
            break;
          }
       }
    }
```

MainActivity.java 文件当中,最核心的代码就是下面一段:

```
new Thread(new Runnable()
           {
```

```
            @Override
            public void run()
            {
              for(int i = 0; i < 10; i++)
              {//10 次循环,每次进度增加 10
                  SystemClock.sleep(1000);
                  fileDownloadDlg.incrementProgressBy(100 / 10);//每休眠 1s,进
                                                                //度增加 10%
              }
            }
        }).start();
```

这一段代码也是重新启动了一个子线程,同样是模拟一个比较耗时的操作,在 run()方法体当中,有一个 for 循环,总共会执行 10 次。每执行一次,子线程休眠 1 s,随后通过调用 incrementProgressBy()这个方法来增加进度条显示的进度(每次循环显示进度增加 10)。在新建进度条对话框的时候,由于没有指定进度条的最大值和最小值,所以会使用进度条对话框的默认最大值和最小值,也就是 100 和 0,所以 10 次循环执行过后,进度条的值会达到最大值,也就是 100。

右击工程 chapter2_5,在弹出的快捷菜单当中选择 Run As|Android Application 命令,单击第一个按钮运行效果如图 2-20 所示,单击第二个按钮的运行效果如图 2-21 所示。

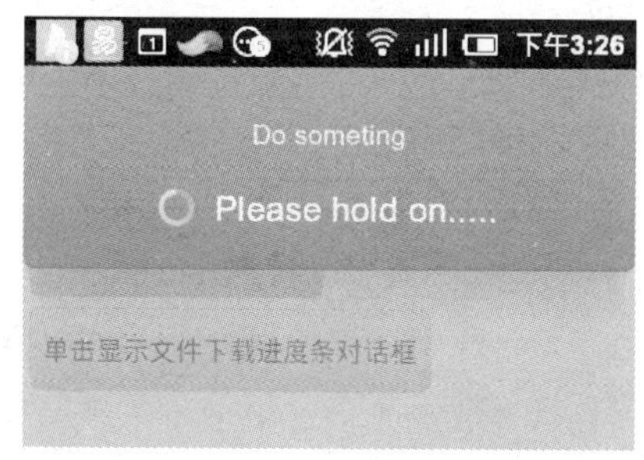

图 2-20　单击第一个按钮的运行效果

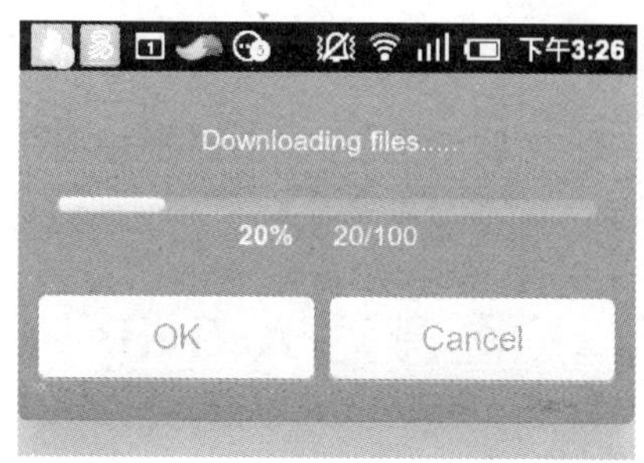

图 2-21　单击第二个按钮的运行效果

第 3 章 意图和广播接收者

在前面章节当中，笔者已经跟读者简单地介绍了意图(Intent)以及广播接收者(Broadcast Receiver)的概念。意图表明的是，程序想要做什么。意图组成了运行Android系统的核心消息机制，通过Intent可以携带数据，从而进行消息的传递。除了携带数据以外，意图还可以激活组件。Android应用程序的3大组件Activity、Broadcast Receiver和Service都是靠Intent对象来激活的。意图分为显示意图和隐式意图，使用隐式意图的时候，需要使用到意图的匹配规则。广播接收者可以对感兴趣的系统事件和自定义事件进行监听，当相应事件发生的时候，注册监听的广播接收者的onReceive()方法就会被回调。本章当中，笔者将会介绍以下内容：

- ❑ 意图激活 Activity；
- ❑ 隐式意图；
- ❑ 广播接收者。

3.1 意图激活 Activity

3.1.1 Activity 之间的跳转

一个Android应用程序一般来讲都会含有多个活动，之前的章节只介绍了如何创建一个新的活动，但是读者在读完了前面的章节之后，并不知道如何在这些活动之间进行跳转。意图(Intent)对象可以完成活动之间的跳转，并且在跳转的同时携带数据，传递消息。下面笔者就通过实例来介绍，如何利用Intent完成Activity之间的跳转。

(1) 新建工程 chapter3_1，通过应用程序向导创建的应用程序本身默认含有MainActivity，编辑MainActivity的布局文件activity_main.xml，源代码如下：

```
<RelativeLayout xmlns:android="http://schemas.android.com/apk/res/android"
```

```
    xmlns:tools = "http://schemas.android.com/tools"
    android:layout_width = "match_parent"
    android:layout_height = "match_parent" >
<!-- 触发跳转的按钮 -->
    <Button
        android:id = "@ + id/triggerButton"
        android:layout_width = "wrap_content"
        android:layout_height = "wrap_content"
        android:layout_centerHorizontal = "true"
        android:layout_centerVertical = "true"
        android:padding = "@dimen/padding_medium"
        android:text = "@string/hello_world"
        tools:context = ".MainActivity" />
</RelativeLayout>
```

(2) 新建活动 SecondActivity.java，展开工程 res 子目录，在 layout 子目录下面新建 second_layout.xml 布局文件，编辑 second_layout.xml，源代码如下：

```
<RelativeLayout xmlns:android = "http://schemas.android.com/apk/res/android"
    xmlns:tools = "http://schemas.android.com/tools"
    android:layout_width = "match_parent"
    android:layout_height = "match_parent" >
    <Button
        android:id = "@ + id/second_button"
        android:layout_width = "wrap_content"
        android:layout_height = "wrap_content"
        android:layout_centerHorizontal = "true"
        android:layout_centerVertical = "true"
        android:padding = "@dimen/padding_medium"
        android:text = "@string/second_screen_text"
        tools:context = ".MainActivity" />
</RelativeLayout>
```

(3) 由于活动需要在 AndroidManifest.xml 文件当中进行声明，编辑 AndroidManifest.xml 文件，源代码如下：

```
<manifest xmlns:android = "http://schemas.android.com/apk/res/android"
    package = "com.example.chapter3_1"
    android:versionCode = "1"
    android:versionName = "1.0" >
    <uses-sdk  android:minSdkVersion = "8"   android:targetSdkVersion = "15" />
    <application  android:icon = "@drawable/ic_launcher"    android:label = "@
string/app_name"      android:theme = "@style/AppTheme" >
```

```
        <!-- MainActivity 声明 -->
        <activity  android:name=".MainActivity"  android:label="@string/title
_activity_main" >
            <intent-filter>
                <action android:name="android.intent.action.MAIN" />
                <category android:name="android.intent.category.LAUNCHER" />
            </intent-filter>
        </activity>
        <!-- SecondActivity 声明 -->
        <activity  android:name=".SecondActivity"  android:label="@string/
second_activity_title" />
    </application>
</manifest>
```

(4) 编辑 SecondActivity.java,源代码如下:

```
public class SecondActivity extends Activity
{
        @Override
        protected void onCreate(Bundle savedInstanceState)
        {
            super.onCreate(savedInstanceState);
            setContentView(R.layout.second_layout);//指定 layout/second_layout.
                                                   //xml 文件为布局文件
        }
}
```

(5) 编辑 MainActivity.java,源代码如下:

```
public class MainActivity extends Activity implements OnClickListener
{
    Button button;//触发跳转的按钮
    @Override
public void onCreate(Bundle savedInstanceState)
{
        super.onCreate(savedInstanceState);
        setContentView(R.layout.activity_main);
        button = (Button)findViewById(R.id.triggerButton);
                                        //找到 Id 为 triggerButton 的按钮控件
        button.setOnClickListener(this);
}
@Override
public void onClick(View v)
{
```

```
        if(R.id.triggerButton == v.getId())
        {//ID 为 triggerButton 的按钮单击事件处理
            Intent intent = new Intent(this, SecondActivity.class);
            startActivity(intent);            //跳转到 SecondActivity
            }
        }
    @Override
    public boolean onCreateOptionsMenu(Menu menu)
    {
        getMenuInflater().inflate(R.menu.activity_main, menu);
        return true;
    }
}
```

笔者主要解释一下完成跳转的两句代码：

```
Intent intent = new Intent(this, SecondActivity.class);
startActivity(intent);                //跳转到 SecondActivity
```

Intent 类当中含有一个构造函数：

public Intent(Context packageContext, Class<? >cls)

这个构造函数需要两个参数，第一个参数是 Context 的对象，由于 Activity 本身继承自 Context 类，所以可以将当前对象的引用，也就是 this 作为第一个参数；第二个参数需要一个 Class 的对象，Java 当中任何一个类都有且仅有一个 Class 对象，可以通过类名.class 来引用相应的 Class 对象。

这实际上也就是显示意图的使用方式，所谓的"显示意图"，就是明确指出了要激活的组件，例如上述代码当中的 SecondActivity.class，实际上明确指出了意图要激活 SecondActivity 这个 Activity，然后通过调用 startActivity(intent)来完成 Activity 的跳转。

(6) 右击工程 chapter3_1，在弹出的快捷菜单中，选择 Run As|Android Application 命令，运行效果如图 3-1 所示，单击 Hello world 按钮，跳转到 SecondActivity，如图 3-2 所示。

Hello world!

图 3-1　chapter3_1 运行效果图

第二个界面

图 3-2　Activity 跳转效果

3.1.2　Intent 传递数据

Activity 在跳转的过程当中，一般还需要完成数据的传递。下面介绍如何通过 Intent 对象完成数据的传递。编辑 MainActivity.java 文件的 onClick 回调方法，源代码如下：

```
Intent intent = new Intent(this, SecondActivity.class);
intent.putExtra("username", "freedie.qin");
startActivity(intent);                    //跳转到 SecondActivity
```

第二句代码实际上表明的就是为 Intent 对象添加一个键值对的数据，其中键为 username，值为 freedie. qin，有了发送数据的逻辑之后，就需要编写接收数据的逻辑，编辑 SecondActivity. java 文件，源代码如下：

```
public class SecondActivity extends Activity
{
        Button button;
        @Override
        protected void onCreate(Bundle savedInstanceState)
        {
            super.onCreate(savedInstanceState);
            setContentView(R.layout.second_layout);//指定 layout/second_
                                                   //layout.xml 文件为布局文件
            button = (Button)findViewById(R.id.second_button);
            Intent intent = getIntent();        //获取传递过来的 Intent 对象
            button.setText(intent.getStringExtra("username"));  //获取键为
                                                                //username 的值
        }
}
```

其中 onCreate()函数中最后两句代码就是获取 Intent 对象当中键为 username 的值的操作。发送数据时，还可以通过 Bundle 对象来编写代码，如下所示：

```
Intent intent = new Intent(this, SecondActivity.class);
Bundle bundle = new Bundle();
bundle.putString("username", "freedie.qin");
intent.putExtras(bundle);
startActivity(intent);              //跳转到 SecondActivity
```

接收数据的代码除了以上提到的写法之外，还可以通过以下方式来获取传过来的数据：

```
Intent intent = getIntent();
//获取传递过来的 Intent 对象
button.setText(intent.getExtras().getString("
username"));//获取键为 username 的值
```

其中 intent. getExtras()方法得到的就是传递过来的 Bundle 对象。单击 Hello world 按钮过后，运行效果如图 3－3 所示。

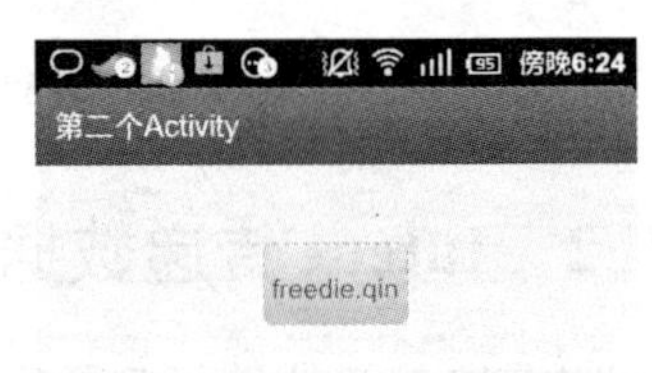

图 3－3　利用 Intent 传递数据

3.1.3　跳转至其他活动并获取结果

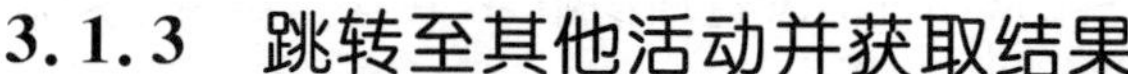

当前活动通过调用 startActivity()方法跳转到其他活动，当其他活动结束的时候，之前的活动无法从跳转到的活动获取到任何数据。在这里笔者习惯称之前的活动为父活动，而后面跳转到的活动为子活动。如果在父活动当中仅仅通过 startActivity()方法跳转到其他子活动，那么当其他子活动结束的时候，子活动无法传递相关数据给父活动。在开发 Android 应用程序的时候，有的应用场景，父活动不仅需要跳转到子活动，还需要在子活动结束的时候，返回相关数据给父活动。

接下来笔者继续在之前的 chapter3_1 工程上做相关修改，使得当子活动 SecondActivity 结束之后，返回相关数据给父活动，并且在父活动 MainActivity 界面上显示出来。

编辑 MainActivty.java 文件，修改 onClick 方法，源代码如下：

```
@Override
  public void onClick(View v)
  {
  if(R.id.triggerButton == v.getId())
  {//ID 为 triggerButton 的按钮单击事件处理
        Intent intent = new Intent(this, SecondActivity.class);
        Bundle bundle = new Bundle();
        bundle.putString("username", "freedie.qin");
        intent.putExtras(bundle);
        //startActivity(intent);              //跳转到 SecondActivity
      startActivityForResult(intent, 0);      //跳转到 SecondActivity 并需要
                                              //SecondActivity 返回相关结果
      }
  }
```

这段代码片段当中，最重要的一句代码就是：

```
startActivityForResult(intent, 0);  //跳转到 SecondActivity 并需要 SecondActivity 返
                                    //回相关结果
```

startActivityForResult(Intent intent, int requestCode)函数含有两个参数：

- ❑ intent 这个 Intent 类型参数，与之前调用 startActivity()方法传入的 Intent 类型参数的作用是一样的，用来在活动之间传递数据。
- ❑ requestCode 这个参数是一个整型的请求码，用来标明父活动的一个请求码，由于跳转到某一个子活动的活动往往不止一个父活动，所以通过这个返回码可以标示跳转到同一个子活动的不同的父活动，在这里笔者传入的请求码为 0。

由于 MainActivity 需要接收从 SecondActivity 返回的结果，在这里需要重新添加一个回调方法，源代码如下：

```
@Override
protected void onActivityResult(int requestCode, int resultCode, Intent data)
{
if(0 == requestCode && RESULT_OK  == resultCode)
  {
     button.setText(data.getStringExtra("result"));//将 SecondActivity 返回结果设置
                                                   //为按钮文字
     }
}
```

当 SecondActivity 通过调用 setResult()方法返回到父活动 MainActivity 的时候，onActivityForResult()这个回调方法就会被自动调用，onActivityForResult(int requestCode, int resultCode, Intent data)函数含有 3 个参数：

- ❑ requestCode：这个整型参数是之前在 startActivityForResult()方法当中指定的标明父活动的请求码。
- ❑ resultCode：这个整型参数是在子活动当中通过调用 setResult()方法传入的一个返回码，由于同一个父活动，可能要依据情况的不同而跳转到不同的子活动，而这个返回码就是用来标示这些不同的子活动的。
- ❑ data：这个 Intent 类型的参数是父活动当中用来接收从子活动当中传过来的数据的。

编辑 SecondActivity.java 文件，源代码如下：

```
public class SecondActivity extends Activity implements OnClickListener
{
        Button button;
        @Override
        protected void onCreate(Bundle savedInstanceState)
        {
             super.onCreate(savedInstanceState);
             setContentView(R.layout.second_layout);  //指定 layout/second_layout.
                                                      //xml 文件为布局文件
        button = (Button)findViewById(R.id.second_button);
        button.setOnClickListener(this);
        Intent intent = getIntent();                  //获取传递过来的 Intent 对象
        button.setText(intent.getExtras().getString("username"));
                                                      //获取键为 username 的值
    }
    @Override
```

```
    public void onClick(View v)
    {
        if(R.id.second_button == v.getId())
        {
            Intent result = new Intent();
            result.putExtra("result", "android 编程宝典");
            setResult(RESULT_OK, result);          //使活动返回到之前的调用者(调用
                                                   //startActivityForResult 的活动)
            finish();  //结束当前 Activity
        }
    }
}
```

笔者对 SecondActivity 界面当中的按钮的单击事件进行了处理。在调用 finish()方法之前(结束当前 SecondActivity 活动),笔者调用了 setResult(int resultCode, Intent data)这个方法,这个函数需要两个参数:

- resultCode:这个整型参数用来标示子活动返回给父活动的一个返回码,因为在同一个应用程序当中,同一个父活动可能会因为条件的不同,跳转到不同的子活动,用这个返回码来标示不同的子活动,一般来讲这个参数传入的比较多的是 Activity 基类当中已经定义好的的两个整型常量,RESULT_CANCELED 和 RESULT_OK,当然有的时候需要根据不同的情况,返回不同的整型参数。在这里笔者传入的实参是 RESULT_OK。
- data:第二个参数是一个 Intent 类型的对象,这里 Intent 对象的作用是用来传递数据,笔者在这里通过 Intent 对象向父活动返回一个简单的键值对。调用完 setResult()方法以及 finish()方法之后,会将相关数据结果返回给父活动。

单击 MainActivity 界面按钮跳转到 SecondActivity 界面,效果图如图 3-4 所示,单击 SecondActivity 界面上按钮,效果图如图 3-5 所示。

图 3-4　startActivityForResult 效果图　　　**图 3-5　单击 SecondActivity 按钮效果图**

3.2　隐式意图

在之前的章节中,笔者已经指出意图分为显示意图和隐式意图两种,显示意图的使用已经介绍过,而在本节当中,笔者将和读者一起来探讨一下 Android 当中提供的

隐式意图该如何使用。因为意图是在一系列数据上面执行的操作，Android 运行时会根据每一个应用程序当中的项目清单文件注册的各应用程序组件的<Intent-Filter>节点来决定要启动当前设备上的哪一个应用程序的组件。

3.2.1 意图过滤器

意图过滤器(Intent Filter)描述了一个组件愿意接受什么样的 Intent 对象，Android 将其抽象为 android. content. IntentFilter 类。在 AndroidManifest. xml 配置文件当中可以通过<intent-filter>节点为一个 Activity 指定其 Intent Filter，以便告诉系统该 Activity 可以响应什么样的 Intent。在项目清单当中注册广播接收者的时候，意图过滤器还用来表明广播接收者将会对哪些发生的事件或者动作进行监听。

在定义<intent-filter>组件节点的时候，有以下子节点需要声明：

(1) action：可以使用<action>节点当中的 android:name 属性来指定声明的组件可以对什么名称的动作的意图进行处理和服务。每一个意图过滤器至少应该包含一个<action>节点，否则任何 Intent 请求都不能和该<intent-filter>匹配。其中 android:name 属性必须使用独一无二的字符串来对可以处理和服务的意图的动作进行描述，其命名一般遵循 java 的包名命名规范。

(2) category：可以使用<category>节点当中的 android:name 属性来指定声明的组件可以对什么种类的意图进行处理和服务。每一个意图过滤器都可以包含多个<category>节点，在进行意图解析的时候，Intent 对象当中的种类必须能够在配置的 category 节点当中找到，只要有一个种类没有在配置的 category 节点当中找到，则相应的组件无法给与响应。开发者可以使用自定义的字符串来指定组件所能处理的意图的种类，也可以使用下面列举的几种比较常用的 Android 框架提供的标准的种类类型：

- ❑ BROWSABLE：如果将 Category 设置为 android. intent. category. BROWSABLE，则表明所声明的 Activity 可以在浏览器当中通过发送意图而激活。一般来讲，如果在浏览器当中要想激活某个组件，在设置意图 Category 的时候，都要添加 android. intent. category. BROWSABLE。
- ❑ DEFAULT：如果将 Category 设置为 android. intent. category. DEFAULT，则表明所声明的 Activity 可以对没有设置种类的 Intent 对象进行处理和服务。
- ❑ GADGET：如果将 Category 设置为 android. intent. category. GADGET，则表明所声明的 Activity 拥有嵌入在其他的 Activity 当中运行的能力。
- ❑ HOME：如果将 Category 设置为 android. intent. category. HOME，并且没有指定 action 节点，那么用户按下 Home 键的时候就会看到所声明的 Activity。
- ❑ LAUNCHER：如果将 Category 设置为 android. intent. category. LAUNCHER，那么所声明的 Activity 将会作为应用程序的主入口活动，并且出现在应

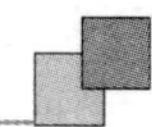

用程序启动器中。

(3) data：可以使用<data>节点对组件可以处理和服务的数据进行声明，<data>节点的使用语法如下：

```
<data android:host = "string"
      android:mimeType = "string"
      android:path = "string"
      android:pathPattern = "string"
      android:pathPrefix = "string"
      android:port = "string"
      android:scheme = "string"/>
```

<data>节点作为对可操作的数据进行描述，这个描述可以仅仅是一个数据类型(android:mimeType 属性)，也可以仅仅是一个 URI，也可以既包括数据类型也包括 URI，URI 是通过下面每一部分独立的属性组成起来的：scheme://host:port/path or pathPrefix or pathPattern。这里需要注意的是 scheme 而不是 schema，也许读者会记得 xmlns:android="http://schemas.android.com/apk/res/android"这段声明，就会想起其中的 schema，但这里的 schema 不是。虽然在写 AndroidManifest.xml 的时候，有智能提示，但是希望大家还是能注意到。

上面那句最后的 path or pathPrefix or pathPattern 是指后面的 path 验证可以使用 data 属性中的 android:path，android:pathPrefix 或 android:pathPattern，可以通过添加任意个 data 标签，由于是 or，因此只要其中任意一个 data 匹配，系统就会选择相应的 Activity 启动，当然，如果别的 Activity 也有相同的 data 标签，系统就会给用户弹出一个 Chooser Dialog。

上面提到的这些属性都是可选的，但是事实上它们又是相互依靠的。如果没有声明 android:scheme 这个属性，其余属性将会被忽略。如果没有声明 android:host 这个属性，android:port 属性以及所有跟 android:path 相关的属性都将会被忽略。可以把上面提到的属性都放在同一个<data>节点当中：

```
<intent-filter>
    <data android:scheme = "something"   android:host = "project.example.com"/>
</intent-filter>
```

也可以把每一个属性都放在一个<data>节点当中：

```
<intent-filter>
                <data android:scheme = "something"  />
                <data android:host = "project.example.com"/>
</intent-filter>
```

上面这两种声明<data>节点的方式是等价的。下面对各个属性进行相应的解释：

- android:host:URI 的主机名,例如 www. baidu. com,如果 android:scheme 属性没有指定,那么 android:host 属性的指定也没有任何意义。
- android:mimeType:MIME 的媒体类型,例如 image/jpeg 或者 audio/mpeg4-generic。可以通过指定(*)来标示声明组件可以处理任意 MIME 类型。比如:当开发者使用 Intent. setType("text/plain"),那么系统将会匹配所有注册 android:mimeType="text/plain"的 Activity。这里需要十分注意的是 Intent. setType(),Intent. setData()以及 Intent. setDataAndType()三个方法。其中 setType()调用后设置 mimeType,然后将 data 设置为 null;setData()调用后设置 data,然后将 mimeType 设置为 null;setDataAndType()调用后才会同时设置 data 与 mimeType。
  另外需要注意的是,如果开发者在 data 标签,既设置了 mimeType 又设置了 scheme,那么在发送 Intent 的时候,需要同时设置匹配的 data 和 mimeType,即需要调用 setDataAndType()方法,系统才能匹配到这个 Activity(即便 mimeType 设置为 */* 也是如此)。
- android:path:用来匹配完整的路径,如:http://example. com/blog/abc. html,这里将 path 设置为/blog/abc. html 才能够进行匹配。
- android:pathPrefix:用来匹配路径的开头部分,拿上面的 URI 来说,这里将 android:pathPrefix 设置为/blog 就能进行匹配了。
- android:pathPattern:用表达式来匹配整个路径,这里需要说明下匹配符号与转义。
  - 匹配符号:
    - "*"用来匹配 0 次或更多,如:"a*"可以匹配"a"、"aa"、"aaa"...。
    - "."用来匹配任意字符,如:"."可以匹配"a"、"b","c"...。

    因此".*"就是用来匹配任意字符 0 次或更多,如:".*html"可以匹配"abchtml"、"chtml","html","sdf. html"...。
  - 转义:

    因为当读取 Xml 的时候,"\"是被当作转义字符的(当它被用作 pathPattern 转义之前),因此这里需要两次转义,读取 Xml 是一次,在 pathPattern 中使用又是一次。如:"*"这个字符就应该写成"*","\"这个字符就应该写成"\\\\"。
- android:port:URI 的端口,只有在同时指定了 android:scheme 和 android:host 属性的情况下,android:port 属性的指定才会有意义。
- android:scheme:URL 的协议,只有指定了 android:scheme 属性,URI 的其他属性才有意义。URI 的格式,不包括尾随的冒号。

例子一:

如果想要匹配 http 以". pdf"结尾的路径,使得别的程序想要打开网络 pdf 的时

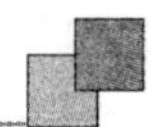

候,用户可以选择我们的程序进行下载查看。我们可以将 scheme 设置为“http”, pathPattern 设置为“. *\. pdf”,整个 intent-filter 设置为:

```
<intent-filter>
        <action android:name = "android.intent.action.VIEW"></action>
        <category android:name = "android.intent.category.DEFAULT"></category>
        <data android:scheme = "http" android:pathPattern = ". *\\.pdf"></data>
</intent-filter>
```

如果只想处理某个站点的 pdf,那么在 data 标签里增加 android:host=“yoursite. com”则只会匹配 http://yoursite. com/xxx/xxx. pdf,但这不会匹配 www. yoursite. com,如果也想匹配这个站点的话,就需要再添加一个 data 标签,除了 android:host 改为 www. yousite. com,其他都一样。

例子二:

如果我们做的是一个 IM 应用,或是类似于微博之类的应用,如何让别人通过 Intent 调用分享出现在选择框里呢?

只用注册 android. intent. action. SEND 与 mimeType 为 text/plain 或 */* 就可以了,整个 intent-filter 设置为:

```
<intent-filter>
      <action android:name = "android.intent.action.SEND"></action>
      <category android:name = "android.intent.category.DEFAULT"></category>
      <data mimeType = " */* "/>
</intent-filter>
```

设置 Category 的原因是,创建的 Intent 的实例默认 category 就包含了 Intent. CATEGORY_DEFAULT,google 这样做的原因是为了让这个 Intent 始终有一个 Category。

例子三:

如果我们做的是一个音乐播放软件,当文件浏览器打开某音乐文件的时候,如何让别人通过 Intent 使我们的应用出现在选择框里?

这类似于文件关联,其实做起来跟上面一样,也很简单,只用注册 android. intent. action. VIEW 与 mimeType 为 audio/* 就可以了,整个 intent-filter 设置为:

```
<intent-filter>
      <action android:name = "android.intent.action.VIEW"></action>
      <category android:name = "android.intent.category.DEFAULT"></category>
      <data mimeType = "audio/* "/>
</intent-filter>
```

3.2.2 Android 隐式意图的解析

当程序员使用 startActivity(intent)来启动另外一个 Activity 的时候,如果直接指定了 Intent 对象的 Component 属性，那么 Activity Manager 将试图启动其 Component 属性指定的 Activity。否则 Android 将通过 Intent 的其他属性从安装在系统中的所有 Activity 当中查找与之最匹配的一个启动，如果没有找到合适的 Activity，应用程序会得到一个系统抛出的异常，如果找到了多个与之匹配的 Activity，Android 将会弹出一个选择的列表对话框供用户去选择要启动哪个 Activity。

决定采用哪个 Activity 组件来处理和服务的过程叫做意图的解析。意图解析的目的是找到最符合条件的 Activity 组件，意图解析的步骤大致是这样的：

(1) 根据各个应用的项目清单文件，找到所有含有<Intent-Filter>子节点的组件的集合。

(2) 迭代这个组件的集合，删除 Action 匹配失败的 Intent Filter。如果 Intent 请求的 Action 和<intent-filter>中某一条<action>匹配，那么该 Intent 就通过了这条<intent-filter>的动作测试。如果 Intent 请求或<intent-filter>没有说明具体的 Action 类型，那么会出现下面两种情况：

- 如果<intent-filter>中没有包含任何 Action 类型，那么无论什么 Intent 请求都无法和这条<intent-filter>匹配。
- 反之，如果 Intent 请求中没有设定 Action 类型，那么只要<intent-filter>中包含有 Action 类型，这个 Intent 请求就将顺利地通过<intent-filter>的行为测试。

(3) 迭代这个组件的集合，删除 Category 匹配失败的 Intent Filter。一个 Intent 要通过 Category 测试，那么该 Intent 对象中的每个 Category 都必须和<intent-filter>中的某一个<category>元素所匹配。

理论上说，一个 Intent 对象如果没有指定 Category 的话，它应该能通过任意的 Category 测试。有一个例外：Android 把所有传给 startActivity()的隐式 Intent 看做至少有一个 Category："android. intent. category. DEFAULT"。因此，想要接收隐式 Intent 的 Activity 必须在<intent-filter>中加入"android. intent. category. DEFAULT"。(android. intent. action. Main)和(android. intent. category. LAUNCHER)的<intent-filter>例外。它们不需要将<category>设置为"android. itent. category. DEFAULT"。

(4) 迭代这个组件的集合，删除 Data 匹配失败的 Intent Filter。每个<data>元素指定了一个 URI 和一个数据类型。URI 每个部分都是不同的属性，scheme、host、port 以及 path。例如：在下面的 URI 当中：content://com. example. project:200/folder/subfolder/etc，scheme 为 content，host 为 com. example. project，port 为 200，path 为 folders/subfolder/etc。host 和 port 一起组成了 Uri authority。如果 host 没有指定，则 port 被忽略。要使一个 authority 有意义，必须指定一个 scheme。要使一个 path 有意义，必须指定一个 scheme 和 authority。当 Intent 对象中的 Uri 和<intent-filter>相比较时，只有在 scheme，authority 以及 path 都匹配的时候，且 Intent 对象的 type 与<intent-filter>的 mimeType 也匹配时，才能叫 Data 匹配成功。

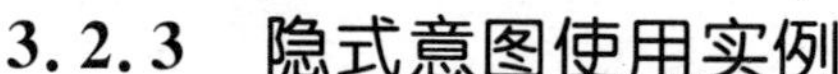

3.2.3 隐式意图使用实例

上面章节当中笔者已经详细介绍了 Android 如何对程序当中使用的隐式意图进行解析，下面笔者将通过代码实例的方式与之前所介绍的理论知识做一次结合。

（1）新建工程 chapter3_2，编辑主入口活动的界面布局文件 activity_main. xml，将主入口活动界面的 TextView 控件替换成按钮控件，当此按钮被用户按下时，发送相应的隐式视图来激活其他 Activity 组件。修改过后的 activity_main. xml 布局文件源代码如下：

```
<RelativeLayout xmlns:android = "http://schemas.android.com/apk/res/android"
    xmlns:tools = "http://schemas.android.com/tools"
    android:layout_width = "match_parent"
    android:layout_height = "match_parent" >
    <Button
        android:id = "@ + id/implicit_intent_test"
        android:layout_width = "wrap_content"
        android:layout_height = "wrap_content"
        android:layout_centerHorizontal = "true"
        android:layout_centerVertical = "true"
        android:padding = "@dimen/padding_medium"
        android:text = "@string/hello_world"
        tools:context = ".MainActivity" />
</RelativeLayout>
```

（2）在 com. example. chapter3_2 包中，新建活动 ImplicitIntentTestActivity，在 AndroidManifest. xml 文件当中进行相应的配置，配置的<activity>节点源码如下：

```
<!-- ImplicitIntentTestActivity 活动配置信息 -->
<activity android:name = ".ImplicitIntentTestActivity"
    android:label = "@string/title_activity_main">
    <intent-filter>
        <action android:name = "com.example.chapter3_2.test"/>
    </intent-filter>
</activity>
```

（3）在 layout 目录下面，新建布局文件 implicit_intent_activity. xml，编辑 implicit_intent_activity. xml 文件，源代码如下：

```
<RelativeLayout xmlns:android = "http://schemas.android.com/apk/res/android"
    xmlns:tools = "http://schemas.android.com/tools"
    android:layout_width = "match_parent"
    android:layout_height = "match_parent" >
    <TextView
```

```
        android:layout_width = "wrap_content"
        android:layout_height = "wrap_content"
        android:layout_centerHorizontal = "true"
        android:layout_centerVertical = "true"
        android:padding = "@dimen/padding_medium"
        android:text = "@string/implicit_intent_text"
        tools:context = ".MainActivity" />
</RelativeLayout>
```

(4) 编辑 ImplicitIntentTestActivity.java 文件,源代码如下:

```
public class ImplicitIntentTestActivity extends Activity
{
    @Override
    protected void onCreate(Bundle savedInstanceState)
    {
        super.onCreate(savedInstanceState);
        setContentView(R.layout.implicit_intent_activity);//设置 implicit_intent
                                          //_activity.xml 文件为活动的布局文件
    }
}
```

(5) 编辑 MainActivity.java 文件,源代码如下:

```
public class MainActivity extends Activity implements OnClickListener
{
    private static final String TAG = "MainActivity";
    Button sendIntentBtn;
    @Override
    public void onCreate(Bundle savedInstanceState)
    {
        super.onCreate(savedInstanceState);
        setContentView(R.layout.activity_main);
        sendIntentBtn = (Button)findViewById(R.id.implicit_intent_test);
                                    //找到 id 为 implicit_intent_test 的按钮控件
        sendIntentBtn.setOnClickListener(this);  //设置当前类为按钮的单击事件监听器
    }
    @Override
    public void onClick(View v)
    {
    if(R.id.implicit_intent_test == v.getId())
    {//当 id 为 implicit_intent_test 的按钮被单击的时候
        Intent intent = new Intent();
        intent.setAction("com.example.chapter3_2.test");//隐式意图的 action 属
```

```
                                        //性,与清单文件的<action>节点相对应
            try
            {
                startActivity(intent); //通过此方法发送的隐式视图的 Category 默认
                                       //为"android.intent.category.DEFAULT"
            }
            catch (Exception e)
            {//没有找到匹配的 Activity 的时候,会抛出异常
            Log.e(TAG, "No Component matches,exception type:\n" + e.getClass());
          }
        }
      }
      @Override
      public boolean onCreateOptionsMenu(Menu menu)
      {
          getMenuInflater().inflate(R.menu.activity_main, menu);
          return true;
      }
  }
```

当 id 为 implicit_intent_test 的按钮被用户单击的时候,代码当中的 onClick(View v)函数就会被回调。在 onClick(View v)函数当中,笔者通过调用 Intent 类的默认的构造函数,声明了一个隐式的 Intent 对象,并通过调用 Intent 对象的 setAction(String str)函数为隐式 Intent 对象的 action 属性赋值,Intent 对象的 action 属性与项目清单文件当中的<intent-filter>节点下面的<action>子节点是一一对应的。接着通过调用 startActivity(intent)方法来启动与声明的隐式 Intent 对象最匹配的 Activity 组件,如果找不到匹配的 Activity 组件,则会抛出一个异常,所以笔者在调用 startActivity(intent)函数的时候,对可能抛出的异常进行了捕获,如果抛出了异常,那么可以通过 LogCat 来查看抛出异常的类型是什么。

(6) 右击 chapter3_2 工程,在弹出的快捷菜单中选择 Run As|Android Application 命令,单击界面上的 Hello World 按钮,观察日志信息的输出,如图 3-6 所示。

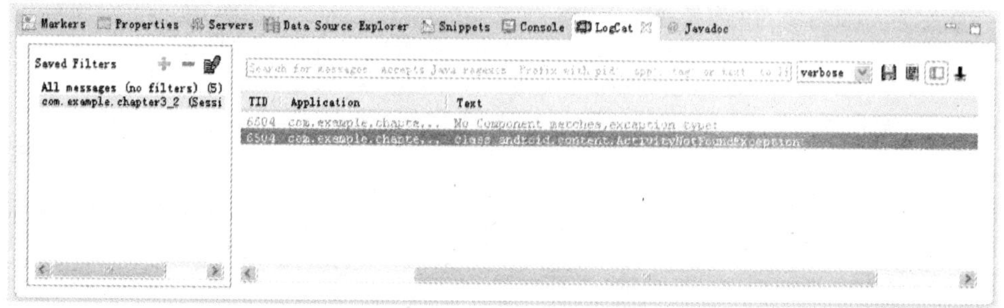

图 3-6　无匹配 Activity 组件 LogCat 输出

在隐式 Intent 对象当中笔者设置的 action 属性值与项目清单文件当中 ImplicitIntentTestActivity 活动配置的<intent-filter>节点的<action>子节点的 android:name 属性值是一致的，为什么程序当中会抛出 android.content.ActivityNotFoundException 的异常呢？因为 android 把所有传给 startActivity()的隐式 Intent 看做至少有一个 Category："android.intent.category.DEFAULT"。因此，想要接收隐式 Intent 的 Activity 必须在<intent-filter>中加入"android.intent.category.DEFAULT"。

(7) 编辑项目清单文件，源代码如下：

```
<manifest xmlns:android = "http://schemas.android.com/apk/res/android"
    package = "com.example.chapter3_2"
    android:versionCode = "1"
    android:versionName = "1.0" >
    <uses-sdk
        android:minSdkVersion = "8"
        android:targetSdkVersion = "15" />
    <application
        android:icon = "@drawable/ic_launcher"
        android:label = "@string/app_name"
        android:theme = "@style/AppTheme" >
        <activity
            android:name = ".MainActivity"
            android:label = "@string/title_activity_main" >
            <intent-filter>
                <action android:name = "android.intent.action.MAIN" />
                <category android:name = "android.intent.category.LAUNCHER" />
            </intent-filter>
        </activity>
        <!-- ImplicitIntentTestActivity 活动配置信息 -->
       <activity android:name = ".ImplicitIntentTestActivity"
               android:label = "@string/title_activity_main">
           <intent-filter>
               <action android:name = "com.example.chapter3_2.test"/>
               <category android:name = "android.intent.category.DEFAULT"/>
<!-- 由于要接受 startActivity(intent)的隐式 Intent -->
           </intent-filter>
       </activity>
    </application>
</manifest>
```

(8) 右击 chapter3_2 工程，在弹出的快捷菜单中选择 Run As|Android Applica-

tion 命令，运行效果如图 3 - 7 所示，单击界面上的 Hello World 按钮，运行效果如图 3 - 8 所示。

Hello world!

图 3 - 7　chapter3_2 运行效果图

隐式Intent测试

图 3 - 8　单击 Hello World 按钮运行效果图

在上述代码示例当中，由于 Intent 对象当中设置的 action 属性值以及通过调用 startActivity(intent)函数而默认含有的 category 属性值与 ImplicitIntentTestActivity 组件配置的<intent-filter>节点当中的<action>子节点的内容以及<category>子节点的内容都是一致的，所以可以找到匹配的 ImplicitIntentTestActivity 组件。现在如果修改<intent-filter>的配置内容，修改过后的源代码如下：

```
    <!-- ImplicitIntentTestActivity 活动配置信息 -->
        <activity android:name = ".ImplicitIntentTestActivity"
                android:label = "@string/title_activity_main">
            <intent-filter>
                <action android:name = "com.example.chapter3_2.test"/>
                <action android:name = "com.example.chapter3_2.test2"/>
                <category android:name = "android.intent.category.DEFAULT"/>
<!-- 由于要接受 startActivity(intent)的隐式 Intent -->
            </intent-filter>
        </activity>
```

通过源代码读者可以发现，笔者增加了一个<action>节点的配置，在这样的情况下，隐式 Intent 依旧可以找到匹配的 ImplicitIntentTestActivity 组件。因为只要 Intent 请求的 Action 和<intent-filter>中某一条<action>匹配，那么该 Intent 就通过了这条<intent-filter>的动作测试。

(9) 编辑 MainActivity.java 文件当中的 onClick(View v)函数，源代码如下：

```
@Override
public void onClick(View v)
{
if(R.id.implicit_intent_test == v.getId())
{//当 id 为 implicit_intent_test 的按钮被单击的时候
    Intent intent = new Intent();
    intent.setAction("com.example.chapter3_2.test");//隐式意图的 action 属性，与
                                                    //清单文件的<action>节点相对应
    intent.addCategory("com.freedie.category.test1");//为隐式意图添加 Category
    try
    {
      startActivity(intent);    //通过此方法发送的隐式视图的 Category 默认为
```

```
                                    //"android.intent.category.DEFAULT"
        }
        catch (Exception e)
        {//没有找到匹配的 Activity 的时候，会抛出异常
                Log.e(TAG, "No Component matches,exception type:\n" + e.getClass());
        }
    }
}
```

在上述代码段当中，通过调用 Intent 对象的 addCategory(String str)方法，为隐式意图对象添加了一个 Category，此时如果再次运行 chapter3_2 工程，单击主入口活动界面的 Hello World 按钮，将会同样抛出一个 android.content.ActivityNotFoundException 的异常。因为一个 Intent 对象要想通过 Category 测试，那么该 Intent 对象的每个 Category 都必须和<intent-filter>中的某一个<category>元素所匹配。由于在项目清单文件当中没有为 ImplicitIntentTestActivity 组件配置相应的<category>节点，所以导致匹配失败。

现在如果修改<intent-filter>的配置内容，修改过后的源代码如下：

```
    <!-- ImplicitIntentTestActivity 活动配置信息 -->
      <activity android:name = ".ImplicitIntentTestActivity"
              android:label = "@string/title_activity_main">
          <intent-filter>
              <action android:name = "com.example.chapter3_2.test"/>
              <action android:name = "com.example.chapter3_2.test2"/>
              <category android:name = "com.freedie.category.test1"/>
              <category android:name = "android.intent.category.DEFAULT"/>
<!-- 由于要接受 startActivity(intent)的隐式 Intent -->
          </intent-filter>
      </activity>
```

此时，如果再次运行 chapter3_2 工程，则会正确匹配。

上面只是对 action 以及 category 的匹配进行了相关的测试，下面笔者将对 data 的匹配进行测试。

(10) 编辑<intent-filter>的配置内容，修改过后的源代码如下：

```
    <activity android:name = ".ImplicitIntentTestActivity"
              android:label = "@string/title_activity_main">
          <intent-filter>
              <action android:name = "com.example.chapter3_2.test"/>
              <action android:name = "com.example.chapter3_2.test2"/>
              <category android:name = "com.freedie.category.test1"/>
              <category android:name = "android.intent.category.DEFAULT"/>
```

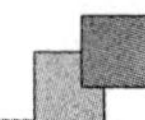

```
<!-- 由于要接受 startActivity(intent)的隐式 Intent -->
                    <data android:scheme = "http"  android:host = "www.baidu.com" an-
droid:port = "80" android:path = "/blog/abc.html"/>
                    <data android:mimeType = "audio/*"/>
                </intent-filter>
            </activity>
```

通过对<data>节点的各属性的设置，使得声明的 Activity 组件只接收 android:scheme 的值为 http，android:host 的值为 www.baidu.com，android:port 的值为 80，android:path 的值为/blog/abc.html 且 android:mimeType 的值为 audio/* 的隐式 Intent 对象。

编辑 MainActivity.java 文件当中的 onClick(View v)函数，源代码如下：

```
@Override
public void onClick(View v)
{
if(R.id.implicit_intent_test == v.getId())
{//当 id 为 implicit_intent_test 的按钮被单击的时候
    Intent intent = new Intent();
    intent.setAction("com.example.chapter3_2.test");//隐式意图的 action 属性，与
                                                    //清单文件的<action>节点相对应
    intent.addCategory("com.freedie.category.test1");//为隐式意图添加 Category
    intent.setDataAndType(Uri.parse("http://www.baidu.com:80/blog/abc.html"), "
audio/*");
    try
    {
            startActivity(intent);  //通过此方法发送的隐式视图的 Category 默认
                                    //为"android.intent.category.DEFAULT"
    }
    catch (Exception e)
    {//没有找到匹配的 Activity 的时候，会抛出异常
          Log.e(TAG, "No Component matches,exception type:\n" + e.getClass());
    }
  }
}
```

上述代码通过调用 setDataAndType(Uri data, String type)的方法设置隐式意图 Uri 以及 mimeType 的属性值。当 Intent 对象中的 Uri 和<intent-filter>相比较时，只有在 scheme、authority 以及 path 都匹配的时候，且 Intent 对象的 type 与<intent-filter>的 mimeType 也匹配的时候，才能叫做 Data 匹配成功。

运行工程 chapter3_2，由于各属性都安全匹配，所以正确启动了 ImplicitIntentTestActivity 组件。

3.2.4 意图打开内置应用程序组件

意图除了可以激活同一应用程序的一些组件，也可以激活一些其他应用程序的组件，特别是设备上内置的一些核心应用程序的组件。下面列举一些 Android 应用程序当中经常要调用的核心应用程序组件的代码：

1. 从 Google 搜索内容

```
Intent intent = new Intent();
intent.setAction(Intent.ACTION_WEB_SEARCH);
intent.putExtra(SearchManager.QUERY,"searchString")
startActivity(intent);
```

2. 浏览网页

```
Uri uri = Uri.parse("http://www.google.com");
Intent it   = new Intent(Intent.ACTION_VIEW,uri);
startActivity(it);
```

3. 显示地图

```
Uri uri = Uri.parse("geo:38.899533, -77.036476");
Intent it = new Intent(Intent.Action_VIEW,uri);
startActivity(it);
```

4. 拨打电话

```
Uri uri = Uri.parse("tel:xxxxxx");
Intent it = new Intent(Intent.ACTION_DIAL, uri);
startActivity(it);
```

5. 调用发短信的程序

```
Intent it = new Intent(Intent.ACTION_VIEW);
it.putExtra("sms_body", "The SMS text");
it.setType("vnd.android-dir/mms-sms");
startActivity(it);
```

6. 播放多媒体

```
Intent it = new Intent(Intent.ACTION_VIEW);
Uri uri = Uri.parse("file:///sdcard/song.mp3");
it.setDataAndType(uri, "audio/mp3");
```

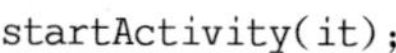

```
startActivity(it);
```

7. 安装 APK

```
Uri installUri = Uri.fromParts("package", "xxx", null);
Intent it = new Intent(Intent.ACTION_PACKAGE_ADDED, installUri);
startActivity(it);
```

8. 卸载 APK

```
Uri uri = Uri.fromParts("package", strPackageName, null);
Intent it = new Intent(Intent.ACTION_DELETE, uri);
startActivity(it);
```

3.3　广播接收者

在之前的章节当中，笔者已经为读者介绍了 Andorid 当中的最重要的组件之一：Activity，以及组件之间的桥梁：Intent，接下来将为读者介绍 Android 中另外一个非常重要的组件：Broadcast Receiver，也就是广播接收者。

所谓的广播接收者，事实上就是一种观察者模式的实现。在观察者模式当中，有两种角色，第一种角色叫做发布者，第二种角色叫做订阅者。如果订阅者对发布者进行了订阅，那么当发布者有相关更新的时候，就会发送消息通知订阅者：发布者已经进行了相关更新。广播接收者可以对其感兴趣的事件进行订阅，订阅一般有两种方式，一种是在 JAVA 文件当中，通过代码的方式进行订阅，一种是在项目清单文件当中，通过 XML 的方式进行订阅。订阅过后，当相关事件发生的时候，所有对此事件进行订阅的广播接收者，都会回调广播接收者的 onReceive 方法。在这里事件一般是以意图广播的形式发生的。

下面通过代码示例对这两种注册的方式，分别给予介绍。

3.3.1　XML 方式注册广播接收者

因为广播接收者需要对感兴趣的事件进行订阅，所以要进行相关的注册，下面笔者在项目清单文件当中通过 XML 的方式对广播接收者进行注册。

(1) 新建工程 chapter3_3，右击 com.example.chapter3_3 包名，在弹出的快捷菜单中选择 new|Class 命令，然后在弹出的 New Java Class 对话框当中的 Name 输入框中输入 MyBroadcastReceiver，并单击 Superclass 输入框右侧的 Browse 按钮，在弹出的 Superclass Selection 对话框当中，输入 MyBroadcastReceiver 要继承的父类，广播接收者都要继承 android.content.BroadcastReceiver 这个类，如图 3－9 所示。

(2) 编辑 AndroidManifest.xml 文件，为 MyBroadcastReceiver 进行相关注册，

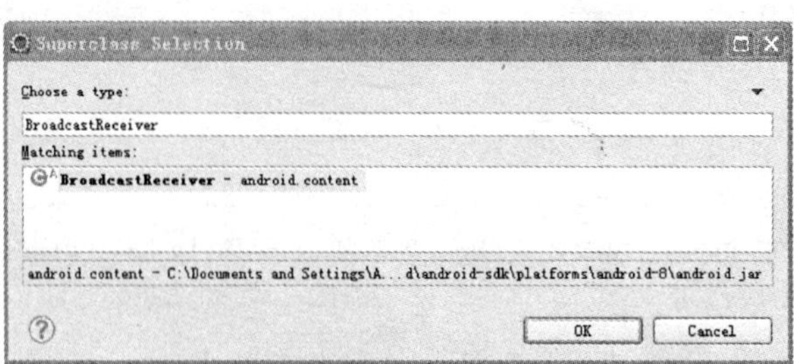

图 3-9　选择广播接收者继承的父类

源代码如下：

```
<manifest xmlns:android="http://schemas.android.com/apk/res/android"
    package="com.example.chapter3_3"
    android:versionCode="1"
    android:versionName="1.0" >
    <uses-sdk
        android:minSdkVersion="8"
        android:targetSdkVersion="15" />
    <application
        android:icon="@drawable/ic_launcher"
        android:label="@string/app_name"
        android:theme="@style/AppTheme" >
        <activity
            android:name=".MainActivity"
            android:label="@string/title_activity_main" >
            <intent-filter>
                <action android:name="android.intent.action.MAIN" />
                <category android:name="android.intent.category.LAUNCHER" />
            </intent-filter>
        </activity>
     <!-- 对 MyBroadcastReceiver 进行相关注册 -->
     <receiver android:name="com.example.chapter3_3.MyBroadcastReceiver">
        <intent-filter>
             <action android:name="com.freedie.broadcast"></action>
        </intent-filter>
  </receiver>
    </application>
</manifest>
```

由源代码当中读者可以看出，笔者要对 action 为 com.freedie.broadcast 的 Intent 对象进行监听。

(3) 修改主入口活动的界面布局文件 activity_main.xml，跟工程 chapter3_2 保持一

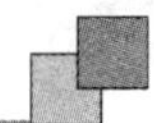

致，将 TextView 控件换成 Button 控件，当用户按下按钮的时候，发送相应的广播。

（4）编辑 MainActivity.java 文件，源代码如下：

```
public class MainActivity extends Activity implements OnClickListener
{
    Button testBtn;
    @Override
    public void onCreate(Bundle savedInstanceState)
    {
        super.onCreate(savedInstanceState);
        setContentView(R.layout.activity_main);
        testBtn = (Button)findViewById(R.id.test_button);
        testBtn.setOnClickListener(this);
    }
    @Override
    public void onClick(View v)
    {
        if(R.id.test_button == v.getId())
        {////当 id 为 test_button 的按钮被单击的时候
            Intent intent = new Intent();
            intent.setAction("com.freedie.broadcast");
            sendBroadcast(intent);//发送广播
        }
    }
    @Override
    public boolean onCreateOptionsMenu(Menu menu) {
        getMenuInflater().inflate(R.menu.activity_main, menu);
        return true;
    }
}
```

笔者为读者解释一下 onClick(View v)方法里面的代码，其中最核心的代码就是通过调用 sendBroadcast(intent)来发送广播意图。由于 MyBroadcastReceiver 对 action 为 com.freedie.broadcast 的广播意图进行了监听，所以当广播接收者接收到相关广播意图之后，onReceive(Context context, Intent intent)回调方法就会被执行。

（5）编辑 MyBroadcastReceiver.java 文件，源代码如下：

```
public class MyBroadcastReceiver extends BroadcastReceiver
{
    private static final String TAG = "MyBroadcastReceiver";
    @Override
    public void onReceive(Context context, Intent intent)
```

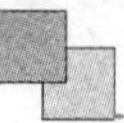


```
        {
            if("com.freedie.broadcast".equals(intent.getAction()))
            {//当意图 action 为 com.freedie.broadcast 时
                Log.d(TAG, "com.freedie.broadcast Intent was received");
            }
        }
    }
```

（6）右击工程 chapter3_3，在弹出的快捷菜单中选择 Run As|Android Application 命令，单击界面上的 Hello World 按钮，观察 LogCat 日志输出，如图 3-10 所示。

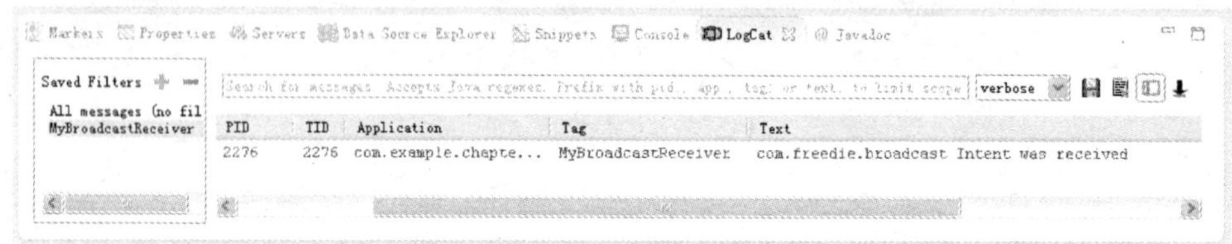

图 3-10　MyBroadcastReceiver 接收意图之后 LogCat 输出

3.3.2　代码方式注册广播接收者

除了可以使用项目清单文件对广播接收者进行注册外，还可以通过 JAVA 代码的方式对其进行注册。

（1）删除项目清单文件当中为 MyBroadcastReceiver 注册的<receiver>节点。

（2）编辑 MainActivity.java 文件，源代码如下：

```
public class MainActivity extends Activity implements OnClickListener
{
    Button testBtn;
    MyBroadcastReceiver myBroadcastReceiver;
    @Override
    public void onCreate(Bundle savedInstanceState)
    {
        super.onCreate(savedInstanceState);
        setContentView(R.layout.activity_main);
        testBtn = (Button)findViewById(R.id.test_button);
        testBtn.setOnClickListener(this);
    }
    @Override
    protected void onResume()
    {
        super.onResume();
        myBroadcastReceiver = new MyBroadcastReceiver();
```

```
        IntentFilter filter = new IntentFilter();
        filter.addAction("com.freedie.broadcast");
        registerReceiver(myBroadcastReceiver, filter);//对广播接收者进行注册
    }
    @Override
    public void onClick(View v)
    {
        if(R.id.test_button == v.getId())
        {//当 id 为 test_button 的按钮被单击的时候
            Intent intent = new Intent();
            intent.setAction("com.freedie.broadcast");
            sendBroadcast(intent);//发送广播
        }
    }
    @Override
    protected void onStop()
    {
        super.onStop();
        unregisterReceiver(myBroadcastReceiver);//一般在 onStop()方法中进行注销
    }
    @Override
    public boolean onCreateOptionsMenu(Menu menu)
    {
        getMenuInflater().inflate(R.menu.activity_main, menu);
        return true;
    }
}
```

运行工程 chapter3_3，单击主入口活动的 Hello World 按钮，LogCat 的输出结果与用 XML 方式注册的广播接收者一致。

除了对用户自定义的事件进行监听以外，广播接收者还广泛用于对系统的事件进行监听，比如收到短信之后，系统启动之后，电量变化之后，这些系统事件都会以广播 Intent 的方式发送出来。

综上所述，广播接收者（ BroadcastReceiver ）用于接收广播 Intent，广播 Intent 的发送通过调用 Context. sendBroadcast()来实现。通常一个广播 Intent 可以被订阅了此 Intent 的多个广播接收者所接收。广播是一种广泛运用的在应用程序之间传输信息的机制 。而 BroadcastReceiver 是对发送出来的广播进行过滤接收并响应的一类组件。BroadcastReceiver 自身并不实现图形用户界面，但是当它收到某个通知后，BroadcastReceiver 可以启动 Activity 作为响应，或者通过 NotificationMananger 提醒用户，或者启动 Service 等。

第 4 章

用户界面

通过前面章节的介绍，笔者相信读者已经清楚地知道，应用程序当中的 Activity 都必须拥有一个 XML 格式的用户界面布局文件。但是 Activity 本身是无法绘制出这些与用户进行交互的界面的，而是要通过 View 对象以及 ViewGroup 对象将 XML 格式所描述的用户界面绘制出来。在本章当中，笔者将会介绍在 Android 应用程序当中如何创建用户界面，以及用户是怎么与其进行交互的。除此之外，笔者还会介绍如何处理 Android 设备方向的变化。本章主要包含以下内容：

❑ 用户界面组件；

❑ 屏幕方向改变。

4.1 用户界面组件

在第 2 章当中笔者介绍过 Android 应用程序当中最基础的组件 Activity。每一个 Activity 实际上就是应用程序的一个界面，一个用户界面当中可以包含一些与用户进行交互的控件，例如 TextView 控件、Button 控件等。一般情况下，用户界面都是通过一个 XML 格式的文件来定义的。通过应用程序向导创建的工程，默认情况下都会为主入口 Activity 在 res/layout 目录下面生成一个叫做 activity_main.xml 的用户界面布局文件，源代码如下：

```
<RelativeLayout xmlns:android = "http://schemas.android.com/apk/res/android"
    xmlns:tools = "http://schemas.android.com/tools"
    android:layout_width = "match_parent"
    android:layout_height = "match_parent" >
    <TextView
        android:layout_width = "wrap_content"
        android:layout_height = "wrap_content"
        android:layout_centerHorizontal = "true"
```

```
        android:layout_centerVertical = "true"
        android:padding = "@dimen/padding_medium"
        android:text = "@string/hello_world"
        tools:context = ".MainActivity" />
</RelativeLayout>
```

定义好用户界面布局文件之后，在 Activity 的 onCreate()方法当中，通过调用 setContentView()来加载相应的 XML 格式的用户界面布局文件，源代码如下：

```
@Override
public void onCreate(Bundle savedInstanceState)
{
        super.onCreate(savedInstanceState);
        setContentView(R.layout.activity_main);//加载 res/layout 目录下的 activity_
                                               //main.xml 文件
}
```

通过 XML 的方式来定义用户界面是比较容易的，有的时候，用户界面是动态生成的，这个时候，不得不采用 Java 代码的方式来生成用户界面。

4.1.1 View 和 ViewGroup

一个 Activity 的用户界面是由 View 对象、ViewGroup 对象以及它们派生类的对象所组成的。例如：界面布局文件里面所定义的<TextView>节点，程序编译期间会将其编译成 TextView 类；再例如布局文件里面所定义的<Button>节点，程序编译期间会将其编译成 Button 类，这些类都是 android.view.View 的子类。View 代表了用户界面组件的一块可绘制的空间块。每一个 View 在屏幕上占据一个长方形的区域。在这个区域内，这个 View 对象负责图形绘制和事件处理。所有常用的继承于 android.view.View 的子类控件将会在后面章节中予以介绍。

继承于 android.view.View 的子类控件可以通过 ViewGroup 容器组织在一起，ViewGroup 类本身也是 android.view.View 的子类。Android 用户界面的一般结构可参考图 4-1。

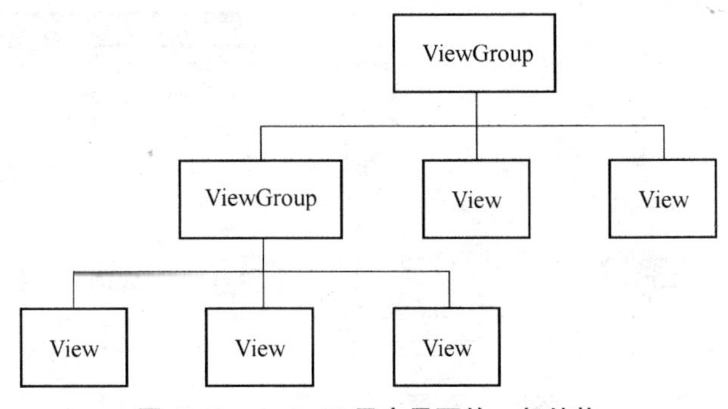

图 4-1 Android 用户界面的一般结构

可见，作为容器的 ViewGroup 可以包含作为叶子节点的 View，也可以包含作为更低层次的子 ViewGroup，而子 ViewGroup 又可以包含下一层的叶子节点的 View 和 ViewGroup。事实上，这种灵活的 View 层次结构可以形成非常复杂的 UI 布局，开发者可据此设计、开发非常精致的 UI 界面。一般来说，开发 Android 应用程序的 UI 界面都不会直接使用 View 和 ViewGroup，而是使用这两大基类的派生类。Android 应用程序当中使用的比较频繁的 ViewGroup 直接派生子类或者是间接派生子类有以下几个：LinearLayout、AbsoluteLayout、TableLayout、RelativeLayout、FrameLayout 及 ScrollView。下面就分别介绍这几个 UI 布局当中经常要使用到的 UI 元素。

4.1.2 LinearLayout(线性布局)

线性布局是最简单、Android 开发者使用得最多的布局类型之一，开发者用它来组织用户界面上的控件。线性布局的作用就像它的名字一样：它将控件组织成一个垂直或水平的形式。当布局方向设置为垂直时，它里面的所有子控件被组织在同一列；当布局方向设置为水平时，所有子控件被组织在一行。

为了说明 LinearLayout 是如何工作的，新建工程 chapter4_1，编辑 activity_main.xml 文件，源代码如下：

```
<LinearLayout xmlns:android = "http://schemas.android.com/apk/res/android"
    android:layout_width = "fill_parent"
    android:layout_height = "fill_parent"
    android:orientation = "vertical">
    <TextView
        android:layout_width = "fill_parent"
        android:layout_height = "wrap_content"
        android:text = "@string/hello_world"/>
    <Button android:layout_width = "fill_parent"
          android:layout_height = "wrap_content"
          android:text = "@string/hello_world"/>
</LinearLayout>
```

这个 XML 文件的根节点是<LinearLayout>，这个根节点当中包含两个子节点：<TextView>和<Button>。线性布局通过设置 android:orientation 的属性值为 vertical，表示线性布局的子控件被组织在同一列。

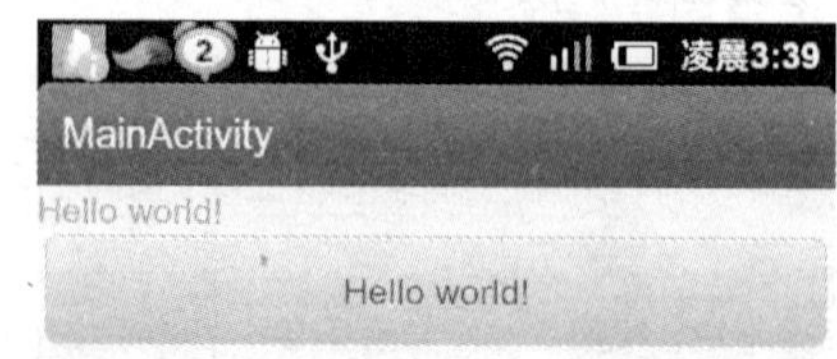

图 4-2 chapter4_1 线性布局运行效果

右击工程 chapter4_1，在弹出的快捷菜单中选择 Run As|Android Application，运行效果如图 4-2 所示。

在使用 XML 文件定义布局的时候，有如下一些比较常用的属性可以设置：

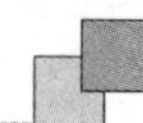

- layout_width：设置 View 或者 ViewGroup 的宽度。
- layout_height：设置 View 或者 ViewGroup 的高度。
- layout_marginTop：设置 View 或者 ViewGroup 上方应该空出的空间。
- layout_marginBottom：设置 View 或者 ViewGroup 下方应该空出的空间。
- layout_marginLeft：设置 View 或者 ViewGroup 左边应该空出的空间。
- layout_marginRight：设置 View 或者 ViewGroup 右边应该空出的空间。
- layout_gravity：用来设置当前 View 相对于父 view 的位置。例如一个 Button 在 LinearLayout 里，开发者如果想把该 Button 放在相对于父 View 靠左或靠右等位置就可以通过该属性进行设置。以 Button 为例，android：layout_gravity＝"right"则 Button 靠右。
- layout_x：设置 View 或者 ViewGroup 相对于 X 轴的位置。
- layout_y：设置 View 或者 ViewGroup 相对于 Y 轴的位置。

上面提到的这些常用的属性，有些属性只能当 View 处在特定的 ViewGroup 当中才可以使用，例如对于 layout_gravity 属性和 layout_weight 属性只能当 View 处于 LinearLayout 或者是 TableLayout 的时候才会起作用，而对于 layout_x 属性和 layout_y 属性只能当 View 处于 AbsoluteLayout 的时候才会起作用。

在前面的布局文件当中，<TextView>节点的 layout_width 属性被设置成 fill_parent，也就意味着这个 View 的宽度与它的父 View 的宽度是一致的，而它的 layout_height 属性被设置成 wrap_content，也就意味着这个 View 的高度是根据组件的内容来决定的。综上所述：fill_parent 的意思是水平方向或者是垂直方向填充父 View，而 wrap_content 的意思是根据组件的内容自动调整大小。除了使用 fill_parent 和 wrap_content 之外，还可以使用确切的大小来指定 View 的宽度和高度。

编辑 activity_main. xml 文件，源代码如下：

```
<LinearLayout xmlns:android = "http://schemas.android.com/apk/res/android"
    android:layout_width = "fill_parent"
    android:layout_height = "fill_parent"
    android:orientation = "vertical">
    <TextView
        android:layout_width = "100dp"
        android:layout_height = "wrap_content"
        android:text = "@string/hello_world"/>
    <Button android:layout_width = "160dp"
            android:layout_height = "wrap_content"
            android:text = "@string/hello_world"/>
</LinearLayout>
```

在指定 Android UI 元素大小的时候，有一些比较常用的单位需要了解：

- px：pixels（像素），屏幕的像素点，不同的设备显示的效果相同。

- ❑ dip：device independent pixels（设备独立像素），不同的设备有不同的显示效果，这个跟设备硬件有关，dip 是一个基于 densityDpi 的抽象单位，densityDpi 值表示的是每英寸有多少个显示点，dip 和 px 之间的换算关系是：px＝dip×（densityDpi/160），也就是当 densityDpi＝160 时，1dip＝1px，Android 的屏幕密度（densityDpi）是以 160 为基准的，屏幕密度为 160 时，是将一英寸分为 160 份，每一份是 1 像素。如果屏幕密度为 240 时，是将一英寸分为 240 份，每一份是 1 像素。
- ❑ in：英寸。
- ❑ mm：毫米。
- ❑ pt：磅，1/72 英寸。
- ❑ dp：等同于 dip。
- ❑ sp：同 dp 相似，但还会根据用户的字体大小偏好来缩放。

建议使用 sp 作为文本的单位，其他用 dip。在之前进行工程目录分析的时候，笔者曾经介绍过在 res 目录下面，会有 4 个与屏幕密度（densityDpi）有关的子目录，如图 4－3 所示。

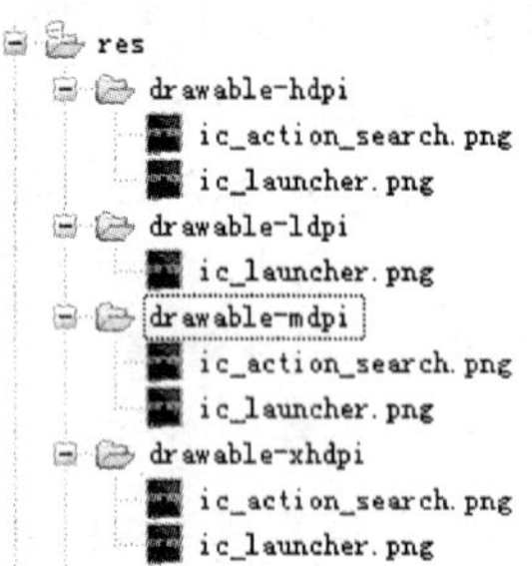

图 4－3　与屏幕密度相关的子目录

Android 会根据屏幕密度的不同去选择不同文件夹下面的图片资源。Android 定义和识别以下 4 种不同的屏幕密度：

- ❑ Low Density（ldpi）：120 dpi（低分辨率）。
- ❑ Medium Density（mdpi）：160 dpi（中等分辨率）。
- ❑ High Density（hdpi）：240 dpi（高分辨率）。
- ❑ Extra High Density（xhdpi）：320 dpi（极高分辨率）。

举例来说，对于 Google Nexus S 这样的设备，它的分辨率是 480（宽度）×800（高度）像素，它的屏幕宽度为 2.04 英寸，所以它的屏幕密度大概是 480/2.04，约等于 235 dpi。根据前面的 Android 可识别的屏幕密度列表，由于 235 dpi 最接近于高分辨率所定义的 240 dpi，所以对于运行于 Google Nexus S 的应用程序，Android 会使用 drawable-hdpi 文件夹下面的图片资源。而对于 HTC Hero 这样的设备，它的屏幕密度接近于 180 dpi，所以对于运行于 HTC Hero 的应用程序，Android 会使用 drawable-mdpi 文件夹下面的图片资源。

在之前的 activity_main.xml 文件当中，笔者设置了按钮的宽度为 160 dpi，如果 chapter4_1 运行在 Google Nexus S 设备之上，那么它真实的宽度应该是 160×（240/160）＝240 px。如果 chapter4_1 运行在 HTC Hero 设备之上，那么它真实的宽度应该是 160×（160/160）＝160 px。当屏幕密度为 160 dpi 的时候，1 dpi＝1 px。

如何通过代码方式获取屏幕的宽度、高度以及屏幕密度呢？

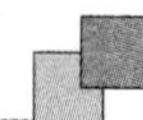

编辑 MainActivity 的 onCreate()方法，源代码如下：

```
@Override
 public void onCreate(Bundle savedInstanceState)
 {
     super.onCreate(savedInstanceState);
     setContentView(R.layout.activity_main);
     DisplayMetrics metric = new DisplayMetrics();
     getWindowManager().getDefaultDisplay().getMetrics(metric);
     int width = metric.widthPixels;//屏幕宽度(像素)
     int height = metric.heightPixels;//屏幕高度(像素)
     int densityDpi = metric.densityDpi;//屏幕密度 DPI(120,160,240,320)
     Log.d(TAG, "width:" + width + " height:" + height + " densityDPI:" + densityDpi);
  }
```

右击工程 chapter4-2，在弹出的快捷菜单中选择 Run As|Android Application，LogCat 日志输出信息如图 4-4 所示。

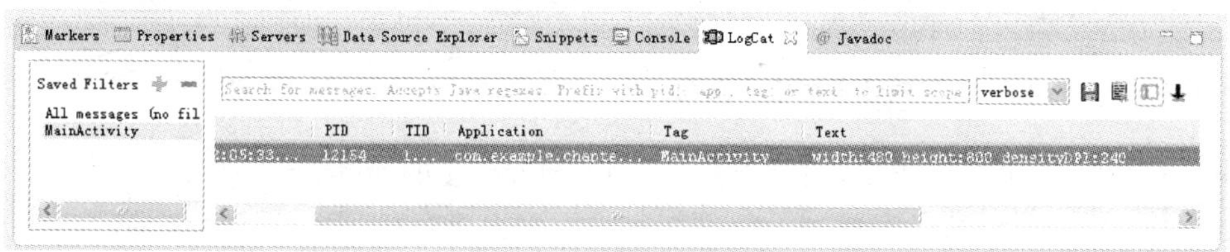

图 4-4　LogCat 输出屏幕宽度、高度以及 DensityDPI

由于 activity_main.xml 的<LinearLayout>节点的 android:orientation 的属性值为 vertical，所以 LinearLayout 的子视图 TextView 控件以及 Button 控件在同一列上进行排列。如果将 android:orientation 的属性值设置为 horizontal，运行工程 chapter4_2，效果如图 4-5 所示。

在进行 UI 布局的时候，android:layout_weight 的属性经常要使用到，下面介绍如何使用 android:layout_weight 属性进行 UI 界面的布局。编辑 activity_main.xml 文件，源代码如下：

```
<LinearLayout xmlns:android = "http://schemas.android.com/apk/res/android"
    android:layout_width = "fill_parent"
    android:layout_height = "fill_parent"
    android:orientation = "horizontal"
    >
  <Button android:id = "@ + id/imageViewLoginState"
        android:layout_width = "fill_parent"
        android:layout_height = "wrap_content"
        android:layout_weight = "1"
```

```
            android:text = "@string/hello_world"
            ></Button>
        <Button android:id = "@ + id/imageViewLoginState1"
            android:layout_width = "fill_parent"
            android:layout_height = "wrap_content"
            android:layout_weight = "2"
            android:text = "@string/hello_world"
            ></Button>
    </LinearLayout>
```

右击工程 chapter4_2,在弹出的快捷菜单中选择 Run As|Android Application,运行效果如图 4-6 所示。

图 4-5　chapter4-2 线性布局横行排列

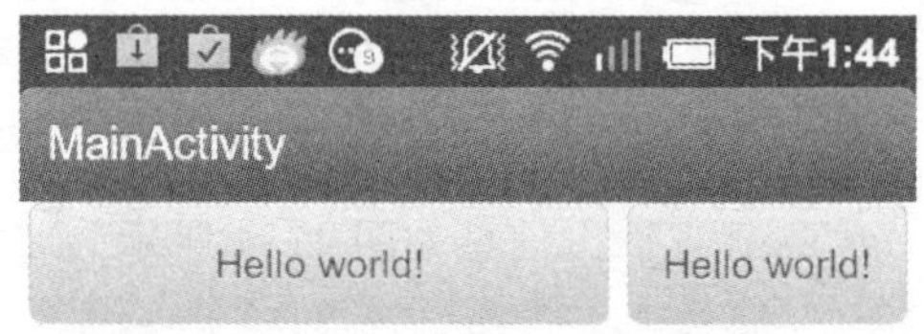

图 4-6　android layout_weight 属性使用

从图 4-6 上可以看出,由于设置了 LinearLayout 的 android:orientation 属性值为 horizontal,所以线性布局的子控件在水平方向上进行排列。此线性布局的两个子控件与之前的子控件相比较,多出了一个叫做 android:layout_weight 的属性的设置,这个属性表示的是控件的大小权重的设置,也就是占据页面的百分比,而且这两个按钮的 layout_width 属性值都是 fill_parent。

当线性布局的所有子控件都含有 android:layout_weight 属性,且这些子控件的 android:layout_width 的属性值都为 fill_parent 时,由于设置成了 fill_parent,所以这些子控件的宽度就要尽可能填充父控件的宽度,但是又设置了第一个按钮的 layout_weight 为 1,第二个按钮的 layout_weight 为 2,所以在这种情况下,第一个按钮填充父控件的优先级比较大,所以第一个按钮占比 2/3,第二个按钮占比 1/3。

当然读者也可以把第二个按钮设置为一个非常大的数字,比如 2 000,此时在 Graphical Layout 模式下可以看到第一个按钮基本填充满了整个宽度,而看不到第二个按钮的影子,事实上第二个按钮还是存在的,当把鼠标指向第一个按钮的后面就可以看到一个长长的竖条,那个就是第二个按钮,已经非常非常小了。因此在子控件的 layout_width 为 fill_parent 的时候,android:layout_weight 权重值越小表明这个控件所占的空间越大,但这个大也是有限度的,那就是 fill_parent。

再次编辑 activity_main.xml 文件,源代码如下:

```
<LinearLayout xmlns:android = "http://schemas.android.com/apk/res/android"
    android:layout_width = "fill_parent"
```

```
    android:layout_height = "fill_parent"
    android:orientation = "horizontal"
    >
  <Button android:id = "@ + id/imageViewLoginState"
        android:layout_width = "wrap_content"
        android:layout_height = "wrap_content"
        android:layout_weight = "1"
        android:text = "@string/hello_world"
        ></Button>
    <Button android:id = "@ + id/imageViewLoginState1"
        android:layout_width = "wrap_content"
        android:layout_height = "wrap_content"
        android:layout_weight = "2"
        android:text = "@string/hello_world"
        ></Button>
</LinearLayout>
```

笔者将两个子控件的宽度设置成 wrap_content，意思是适应内容的宽度，也就是这个控件要尽可能小，只要能把内容显示出来就可以了。也就是第一个按钮和第二个按钮都要尽可能小，只不过优先级不一样，第一个按钮的权重为 1，第二个按钮的权重为 2，由于都设置了 android:layout_weight 的这个属性，所以两者加起来的宽度也要填充整个父控件。

再根据两个按钮占比的大小，计算出第一个按钮占比 1/3，第二个按钮占比 2/3。因此在子控件的 layout_width 设置成 wrap_content 的时候，layout_weight 所代表的是控件要尽可能小，但这个小是有限度的，即 wrap_content。

再次右击工程 chapter4_1，在弹出的快捷菜单中选择 Run As|Android Application，运行效果如图 4-7 所示。

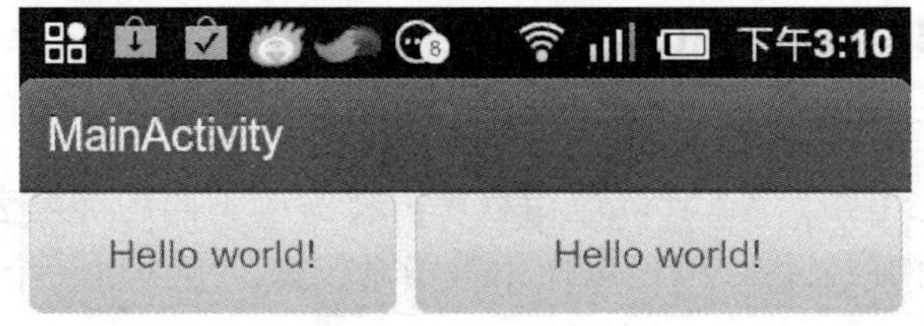

图 4-7 fill_parent 为 wrap_content 效果

android:layout_gravity 属性用来设置当前 View 相对于父 view 的位置。下面介绍如何综合使用 android:layout_gravity 属性和 android:layout_weight 属性进行 UI 界面的布局，编辑 activity_main. xml 文件，源代码如下：

```
<LinearLayout xmlns:android = "http://schemas.android.com/apk/res/android"
    android:layout_width = "fill_parent"
```

```
    android:layout_height = "fill_parent"
    android:orientation = "vertical">
  <Button
        android:layout_width = "wrap_content"
        android:layout_height = "wrap_content"
        android:layout_weight = "1"
        android:layout_gravity = "left"
        android:text = "@string/hello_world"
        ></Button>
    <Button
        android:layout_width = "wrap_content"
        android:layout_height = "wrap_content"
        android:layout_weight = "2"
        android:layout_gravity = "center"
        android:text = "@string/hello_world"
        ></Button>
    <Button
        android:layout_width = "wrap_content"
        android:layout_height = "wrap_content"
        android:layout_weight = "3"
        android:layout_gravity = "right"
        android:text = "@string/hello_world"
        ></Button>
</LinearLayout>
```

图 4-8　综合使用

右击工程 chapter4_1，在弹出的快捷菜单中选择 Run As | Android Application，运行效果如图 4-8 所示。

4.1.3　AbsoluteLayout（绝对布局）

绝对布局的子控件需要指定相对于此坐标布局的横纵坐标值，否则将会像框架布局那样被排在左上角。Android 应用程序需要适应不同的屏幕大小，而这种布局模型不能自适应屏幕尺寸大小，所以应用得相对较少。下面以一个例子简单说明绝对布局。

（1）新建工程 chapter4_2，编辑 activity_main.xml 文件，源代码如下：

```
<AbsoluteLayout xmlns:android = "http://schemas.android.com/apk/res/android"
    android:layout_width = "fill_parent"
    android:layout_height = "fill_parent"
    android:orientation = "vertical" >
    <!-- 定义一个文本框，使用绝对定位 -->
```

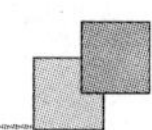

```
<TextView
    android:layout_width = "wrap_content"
    android:layout_height = "wrap_content"
    android:layout_x = "20dip"
    android:layout_y = "20dip"
    android:text = "用户名:" />
<!-- 定义一个文本编辑框,使用绝对定位 -->
<EditText
    android:layout_width = "wrap_content"
    android:layout_height =
"wrap_content"
    android:layout_x = "80dip"
    android:layout_y = "15dip"
    android:width = "200px" />
</AbsoluteLayout>
```

(2) 右击工程 chapter4_2,在弹出的快捷菜单中选择 Run As | Android Application 命令,运行效果如图 4-9 所示。

图 4-9 AbsoluteLayout 效果图

4.1.4 TableLayout(表格布局)

表格布局中每一行用一个 TableRow 对象来表示,当然也可以是一个 View 对象。TableRow 可以添加子控件,每添加一个为一列。表格布局当中有几个比较常用的属性:

- ❑ android:layout_column:控件在 TableRow 中所处的列。
- ❑ android:layout_span:该控件所跨越的列数。
- ❑ android:collapseColumns:将里面指定的列隐藏,若有多列需要隐藏,用逗号将列序号隔开。
- ❑ android:stretchColumns:设置指定的列为可伸展的列,该列会尽量伸展以填满所有可用的空间,若有多个列需要设置为可伸展,需要用逗号隔开。

(1) 新建工程 chapter4_3,编辑 activity_main.xml 文件,源代码如下:

```
<TableLayout xmlns:android = "http://schemas.android.com/apk/res/android"
    android:layout_width = "fill_parent"
    android:layout_height = "fill_parent"
    android:stretchColumns = "1"><!-- 列的标号从 0 开始 -->
    <TableRow>
        <TextView
            android:layout_column = "1"
            android:text = "open..."
```

```
            android:padding = "3dip"/>
        <TextView
            android:text = "ctrl + O"
            android:layout_marginRight = "10dip"
            android:gravity = "right"
            android:padding = "3dip"/>
    </TableRow>
    <TableRow>
        <TextView
            android:layout_column = "1"
            android:text = "save..."
            android:padding = "3dip"/>
        <TextView
            android:text = "ctrl + S"
            android:layout_marginRight = "10dip"
            android:gravity = "right"
            android:padding = "3dip"/>
    </TableRow>
    <TableRow>
        <TextView
            android:layout_column = "1"
            android:text = "save as..."
            android:padding = "3dip"/>
        <TextView
            android:text = "ctrl + shift + S"
            android:layout_marginRight = "10dip"
            android:gravity = "right"
            android:padding = "3dip"/>
    </TableRow>
    <View
        android:layout_height = "2dip"
        android:background = "#00cc33"/>
    <TableRow>
        <TextView
            android:text = " * "
            android:padding = "3dip"/>
        <TextView
            android:text = "import..."
            android:padding = "3dip"/>
    </TableRow>
    <TableRow>
        <TextView
            android:text = " * "
            android:padding = "3dip"/>
        <TextView
            android:text = "export..."
            android:padding = "3dip"/>
```

```
    <TextView
        android:text = "ctrl + E"
        android:layout_marginRight = "10dip"
        android:gravity = "right"
        android:padding = "3dip"/>
  </TableRow>
  <View
    android:layout_height = "2dip"
    android:background = "#00cc36"/>
  <TableRow>
    <TextView
        android:layout_column = "1"
        android:text = "exit..."
        android:padding = "3dip"
        />
  </TableRow>
</TableLayout>
```

在上述布局文件当中，android:gravity 用来设置 View 组件内容的对齐方式，举个例子，如果对 TextView 文本框控件设置 android:gravity＝"center"，则表明让 TextView 组件当中的文字在 TextView 组件当中居中显示，而之前笔者介绍过的 android:layout_gravity 属性是用来设置 View 组件相对于其父组件的对齐方式，例如设置 EditText 组件的 android:layout_gravity＝"center"，则表明让 EditText 组件在它的父组件当中居中显示。

android:padding 属性用来设置组件当中组件的内容与组件的边框之间的距离。其中 android:paddingLeft 用来设置组件当中的内容与组件的左边框之间的填充的距离，android:paddingRight 用来设置组件当中的内容与组件的右边框之间填充的距离，android:paddingTop 用来设置组件当中的内容与组件的上边框之间填充的距离，android:paddingBottom 用来设置组件当中的内容与组件的下边框之间填充的距离。

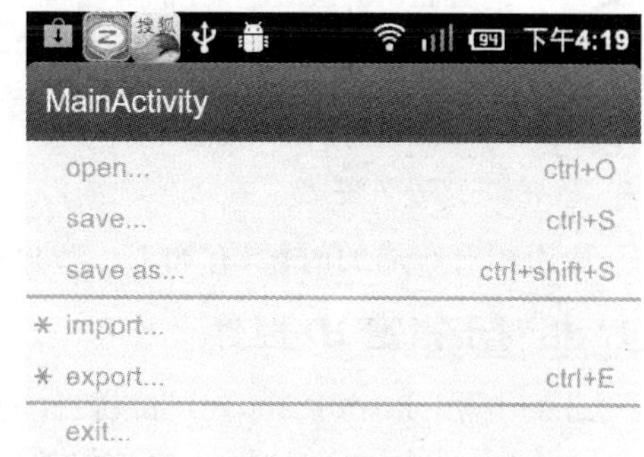

图 4－10　TableLayout 效果图

(2) 运行工程 chapter4_4，在弹出的快捷菜单中选择 Run As|Android Application 命令，运行效果如图 4－10 所示。

4.1.5　RelativeLayout(相对布局)

相对布局 RelativeLayout 允许子元素指定它们相对于其父元素或兄弟元素的位置，这是实际布局中最常用的布局方式之一。

新建工程 chapter4_4，编辑 activity_main.xml 文件，源代码如下：

```
<?xml version="1.0" encoding="utf-8"?>
<RelativeLayout xmlns:android="http://schemas.android.com/apk/res/android"
    android:layout_width="fill_parent"
    android:layout_height="fill_parent">
    <TextView
        android:id="@+id/label"
        android:layout_width="fill_parent"
        android:layout_height="wrap_content"
        android:text="Type here:"/>
    <EditText
        android:id="@+id/entry"
        android:layout_width="fill_parent"
        android:layout_height="wrap_content"
        android:layout_below="@id/label"
        android:background="@android:drawable/editbox_background"/>
    <Button
        android:id="@+id/ok"
        android:layout_width="wrap_content"
        android:layout_height="wrap_content"
        android:layout_alignParentRight="true"
        android:layout_below="@id/entry"
        android:layout_marginLeft="10dip"
        android:text="OK"/>
    <Button
        android:layout_width="wrap_content"
        android:layout_height="wrap_content"
        android:layout_alignTop="@id/ok"
        android:layout_toLeftOf="@id/ok"
        android:text="Cancel"/>
</RelativeLayout>
```

使用相对布局的时候，有如下一些比较常用的属性：

1. 相对于指定 ID 控件

- android:layout_above：将该控件的底部置于给定的 ID 控件之上。
- android:layout_below：将该控件的底部置于给定的 ID 控件之下。
- android:layout_toLeftOf：将该控件的右边缘与给定 ID 控件左边缘对齐。
- android:layout_toRightOf：将该控件的左边缘与给定 ID 控件右边缘对齐。
- android:layout_alignBaseLine：将该控件的 baseline 与给定 ID 控件的 baseline 对齐。
- android:layout_alignTop：将该控件的顶部边缘与给定 ID 控件顶部边缘对齐。

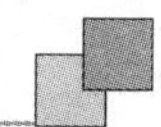

- ❑ android:layout_alignBottom:将该控件的底部边缘与给定 ID 控件底部边缘对齐。
- ❑ android:layout_alignLeft:将该控件的左边缘与给定 ID 控件左边缘对齐。
- ❑ android:layout_alignRight:将该控件的右边缘与给定 ID 控件右边缘对齐。

2. 相对于父控件

- ❑ android:layout_alignParentTop:若是 true,将该控件顶部与父控件顶部对齐。
- ❑ android:layout_alignParentBottom:若是 true,将该控件底部与父控件底部对齐。
- ❑ android:layout_alignParentLeft:若是 true,将该控件左部与父控件左部对齐。
- ❑ android:layout_alignParentRight:若是 true,将该控件右部与父控件右部对齐。

3. 对齐

- ❑ android:layout_centerHorizontal:如果是 true,将该控件置于水平居中。
- ❑ android:layout_centerVertical:如果是 true,将该控件置于垂直居中。
- ❑ android:layout_centerInParent:如果是 true,将该控件置于父控件的中央。

运行工程 chapter4_4,在弹出的快捷菜单中选择 Run As | Android Application 命令,运行效果如图 4-11 所示。

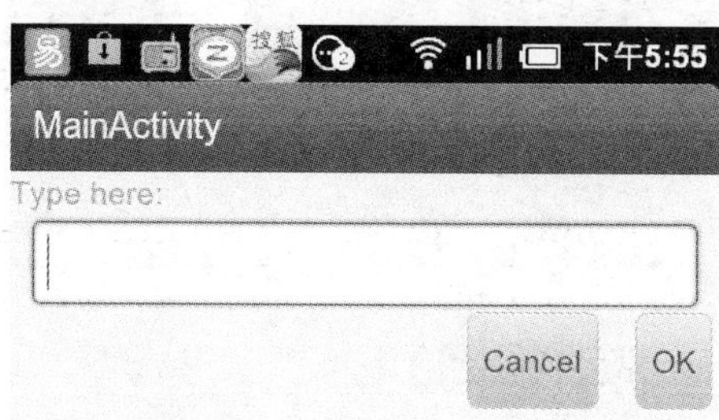

图 4-11 RelativeLayout 效果图

4.1.6 FrameLayout(单帧布局)

单帧布局尤为简单,这种布局下每个添加的子控件都被放在布局的左上角,并覆盖在前一子控件的上层。

(1) 新建工程 chapter4_5,编辑 activity_main.xml 文件,源代码如下:

```
<? xml version = "1.0" encoding = "utf-8"? >
<FrameLayout     xmlns:android = "http://schemas.android.com/apk/res/android"
```

```
    android:layout_width = "fill_parent"
    android:layout_height = "fill_parent">
    <TextView
        android:text = "big"
        android:layout_width = "wrap_content"
        android:layout_height = "wrap_content"
        android:textSize = "50sp"/>
    <TextView
        android:text = "middle"
        android:layout_width = "wrap_content"
        android:layout_height = "wrap_content"
        android:textSize = "20pt"/>
    <TextView
        android:text = "small"
        android:layout_width = "wrap_content"
        android:layout_height = "wrap_content"
        android:textSize = "10sp"/>
</FrameLayout>
```

(2) 右击工程 chapter4_5,在弹出的快捷菜单中选择 Run As|Android Application 命令,运行效果如图 4-12 所示。

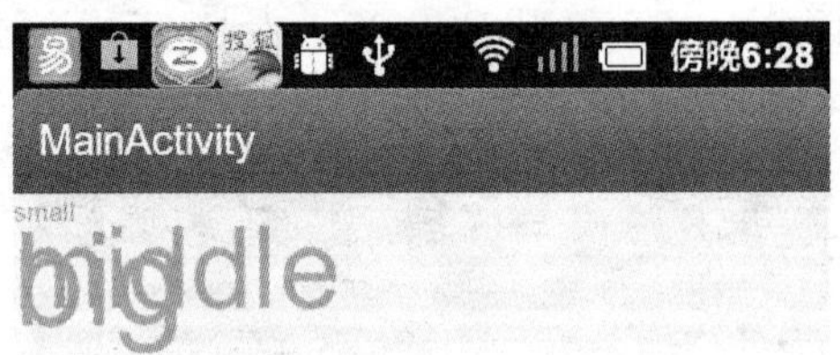

图 4-12 FrameLayout 效果图

4.1.7 ScrollView(滚动视图)

ScrollView 是一种特殊的 FrameLayout。当一屏显示不完时,用户可以通过垂直滚动来显示控件所包含的其余内容。ScrollView 控件只能包含一个子控件,这个子控件可以是 View 对象,也可以是 ViewGroup 对象。一般情况下 ScrollView 控件会包含一个 LinearLayout 作为它的子控件。

(1) 新建工程 chapter4_6,编辑 activity_main.xml 文件,源代码如下:

```
<? xml version = "1.0" encoding = "utf-8"? >
<ScrollView xmlns:android = "http://schemas.android.com/apk/res/android"
    android:layout_width = "fill_parent"
    android:layout_height = "fill_parent"
    android:scrollbars = "vertical">
```

```
<LinearLayout android:orientation = "vertical"
    android:layout_width = "fill_parent"
    android:layout_height = "fill_parent">
    <Button android:layout_width = "wrap_content"
        android:layout_height = "160dip"
        android:layout_gravity = "center_horizontal"
        android:text = "@string/hello_world"/>
  <Button android:layout_width = "wrap_content"
        android:layout_height = "160dip"
        android:layout_gravity = "center_horizontal"
        android:text = "@string/hello_world"/>
 <Button android: layout_width = "wrap_
content"
        android:layout_height = "160dip"
        android:layout_gravity = "center
_horizontal"
        android:text = "@string/hello_
world"/>
  <Button android:layout_width = "wrap_
content"
        android:layout_height = "160dip"
        android:layout_gravity = "center
_horizontal"
        android:text = "@string/hello_
world"/>
</LinearLayout>
</ScrollView>
```

(2) 右击工程 chapter4_6,在弹出的快捷菜单中选择 Run As|Android Application 命令,界面展示出来之后,由于一屏显示不下,垂直滚动效果如图 4－13 所示。

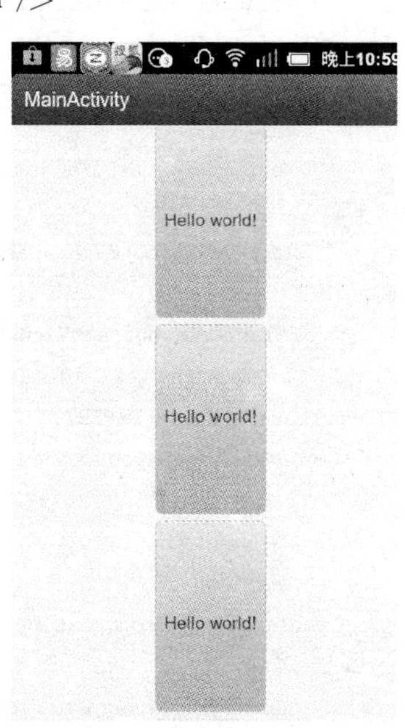

图 4－13　用户垂直滚动 ScrollView 控件效果

4.1.8　Java 代码方式布局

除了使用 XML 布局文件的方式来定义用户界面布局以外,有的时候,用户界面需要动态生成,这时候就需要通过 Java 代码的方式来定义用户界面的布局。

下面通过 Java 代码方式来定义布局,使得呈现出来的用户界面与使用默认的 activity_main. xml 布局文件所定义的用户界面保持一致。

(1) 新建工程 chapter4_7,编辑 MainActivity. java 文件,源代码如下:

```
public class MainActivity extends Activity
```

```
{
  @Override
  public void onCreate(Bundle savedInstanceState)
  {
      super.onCreate(savedInstanceState);
      RelativeLayout relativeLayout = new RelativeLayout(this);//对应 activity_
                                          //main.xml 文件根节点<RelativeLayout>
      TextView textView = new TextView(this); //对应 activity_main.xml 文件
                                          //<RelativeLayout>下<TextView>子节点
      textView.setText("Hello World");//设置 TextView 文本内容
      RelativeLayout.LayoutParams textViewLayoutParams = new RelativeLayout.Lay-
outParams(RelativeLayout.LayoutParams.WRAP_CONTENT,
      RelativeLayout.LayoutParams.WRAP_CONTENT);//对应于<TextView>的 layout_
                                          width 属性以及 layout_height 属性;
      textViewLayoutParams.addRule(RelativeLayout.CENTER_HORIZONTAL, RelativeLay-
out.TRUE);                     //相当于 android:layout_centerHorizontal = "true"
      textViewLayoutParams.addRule(RelativeLayout.CENTER_VERTICAL, RelativeLayout.
TRUE);                         //相当于 android:layout_centerVertical = "true"
      relativeLayout.addView(textView, textViewLayoutParams);
      setContentView(relativeLayout, new RelativeLayout.LayoutParams(RelativeLay-
out.LayoutParams.MATCH_PARENT,
        RelativeLayout.LayoutParams.MATCH_PARENT));//对应于 activity_main.xml 文件
                                 //根节点 layout_width 属性以及 layout_height 属性
  }
  @Override
  public boolean onCreateOptionsMenu(Menu menu)
  {
    getMenuInflater().inflate(R.menu.activity_main, menu);
    return true;
  }
}
```

在 onCreate()方法当中，首先实例化了一个 RelativeLayout 的对象，这个对象实际上就是用来代表 XML 布局文件当中的<RelativeLayout>节点，接着又实例化了一个 TextView 的对象，这个对象实际上是用来代表 XML 布局文件当中<RelativeLayout>根节点下面的<TextView>子节点，然后调用 TextView 对象的 setText()方法来设置文本控件上显示的文字。

在默认生成的 XML 布局文件当中，将 TextView 控件置于相对布局父控件的垂直和水平的中央位置，在 Java 代码当中通过 RelativeLayout. LayoutParams 这个内部类来代表控件的布局位置信息。RelativeLayout. LayoutParams 内部类当中定义的常量 RelativeLayout. LayoutParams. WRAP_CONTENT，实际上与布局文件当中

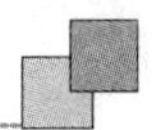

的 android:layout_width 的属性值 WRAP_CONTENT 是相对应的。通过调用 RelativeLayout. LayoutParams 对象的 addRule 方法来设置控件相对于父元素或者是兄弟元素的位置，布局位置设置好之后，调用前面声明的 RelativeLayout 对象将子控件 TextView 对象加入到相对布局当中。

最后调用 Activity 基类当中的 setContentView()方法将声明的 RelativeLayout 作为 Activity 的用户界面。

(2) 右击工程 chapter4_7，在弹出的快捷菜单中，选择 Run As|Android Application 命令，运行效果如图 4-14 所示。

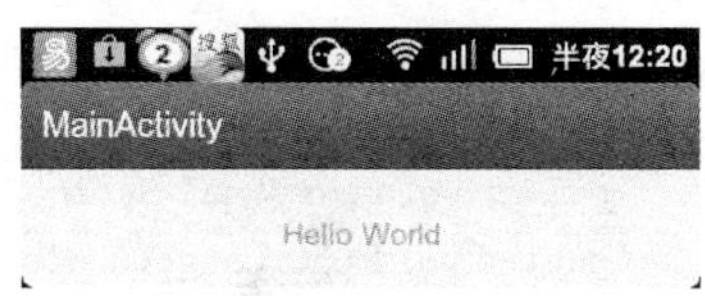

图 4-14　Java 代码方式布局效果

4.2　屏幕方向改变

现代手机设备的一个非常重要的特性就是可以旋转屏幕的方向，当然 Android 也不例外。Android 支持两种屏幕方向：portrait(竖屏)和 landscape(横屏)。默认情况下，当改变手持设备的屏幕方向时，当前的 anctivity 会自动地重绘它的内容。这是因为当屏幕方向发生变化的时候，onCreate()方法被触发了。

4.2.1　理解屏幕方向的改变

下面通过实例来观察当屏幕方向改变的时候，Activity 的声明周期的哪些回调方法被执行了，这些回调方法的执行顺序又是怎么样的。

(1) 新建工程 chapter4_8，编辑 res/layout/activity_main. xml 文件，源代码如下：

```
<LinearLayout xmlns:android = "http://schemas.android.com/apk/res/android"
    xmlns:tools = "http://schemas.android.com/tools"
    android:layout_width = "fill_parent"
    android:layout_height = "fill_parent"
    android:orientation = "vertical"
     >
    <EditText
        android:id = "@ + id/myEditText"
        android:layout_width = "fill_parent"
        android:layout_height = "wrap_content"/>
  <EditText
        android:layout_width = "fill_parent"
        android:layout_height = "wrap_content"/>
</LinearLayout>
```

此 XML 布局当中，采用的布局方式为线性布局。线性布局的两个 EditText 子控件排列在同一列上。其中第一个 EditText 控件声明了 android:id 这个属性的值，而第二个 EditText 控件并没有对属性 android:id 的值进行声明。

(2) 编辑 MainActivity.java 文件，源代码如下：

```
public class MainActivity extends Activity
{
    private static final String TAG = "MainActivity";
    public void onCreate(Bundle savedInstanceState)
    {//Activity 第一次创建时，onCreate()方法被回调
        super.onCreate(savedInstanceState);
        setContentView(R.layout.activity_main);
        Log.d(TAG, "onCreate() method was invoked");
    }
    @Override
    protected void onStart()
    {
    super.onStart();
    Log.d(TAG, "onStart() method was invoked");
    }
    @Override
    protected void onResume()
    {
    super.onResume();
    Log.d(TAG, "onResume() method was invoked");
    }
    @Override
    protected void onPause()
    {//Activity 部分不可见时，onPause()方法被回调
    super.onPause();
    Log.d(TAG, "onPause() method was invoked");
    }
    @Override
    protected void onStop()
    {//Activity 全部不可见时，onStop()方法被回调
    super.onStop();
    Log.d(TAG, "onStop() method was invoked");
    }
    @Override
    protected void onDestroy()
    {//Activity 销毁时，onDestroy()方法被回调
    super.onDestroy();
    Log.d(TAG, "onDestroy() method was invoked");
    }
```

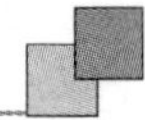

```
    @Override
    protected void onRestart()
    {
    super.onRestart();
    Log.d(TAG, "onRestart() method was invoked");
    }
    @Override
    public boolean onCreateOptionsMenu(Menu menu) {
        getMenuInflater().inflate(R.menu.activity_main, menu);
        return true;
    }
}
```

为了方便观察当屏幕方向改变的时候，LogCat 控制台的日志输出情况，笔者建立了一个名为 MainActivity 的过滤日志 TAG 为 MainActivity 的日志过滤器。如图 4－15 所示。

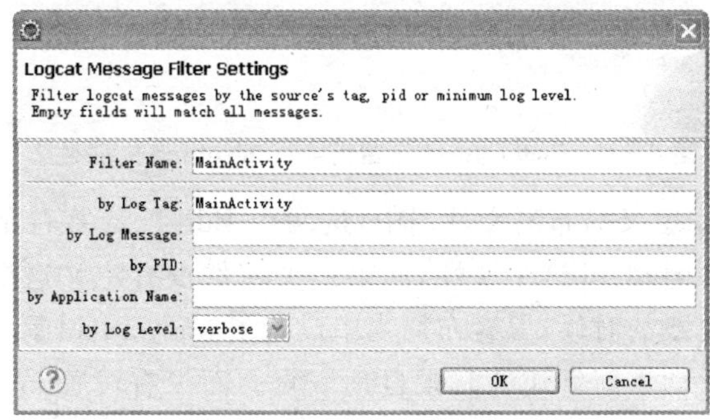

图 4－15　过滤 TAG 为 MainActivity 日志过滤器

(3) 右击工程 chapter4_8，在弹出的快捷菜单中选择 Run As|Android Application 命令，运行效果如图 4－16 所示，在两个 EditText 输入框当中，分别输入相应文字，效果如图 4－17 所示。

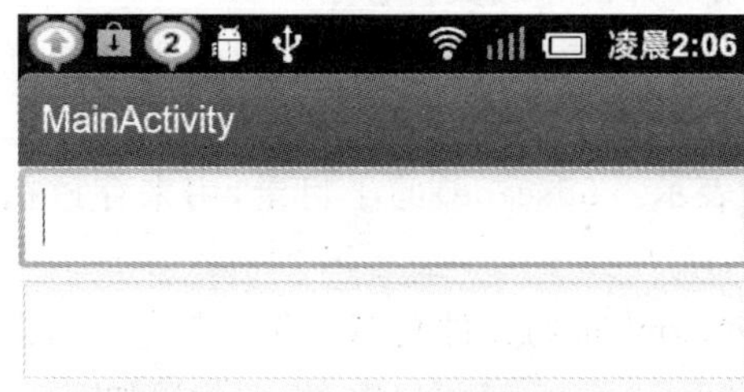

图 4－16　chapter4_8 运行效果

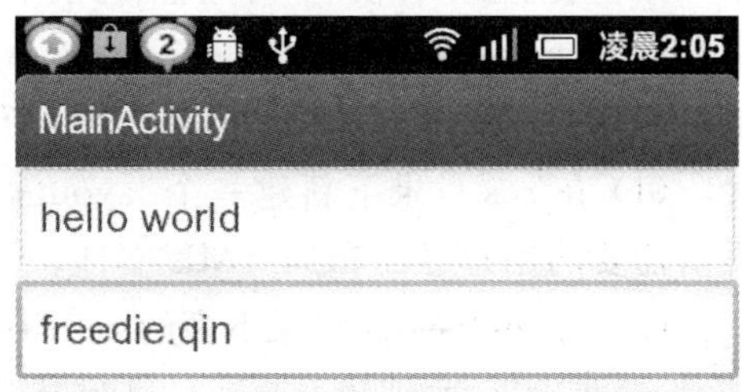

图 4－17　chapter4_8 输入文字运行效果

(4) 当屏幕由竖屏切向横屏的时候，LogCat 日志输出如图 4－18 所示。

从以上的 LogCat 输出信息来看，当屏幕方向发生变化的时候，Activity 会被销毁一

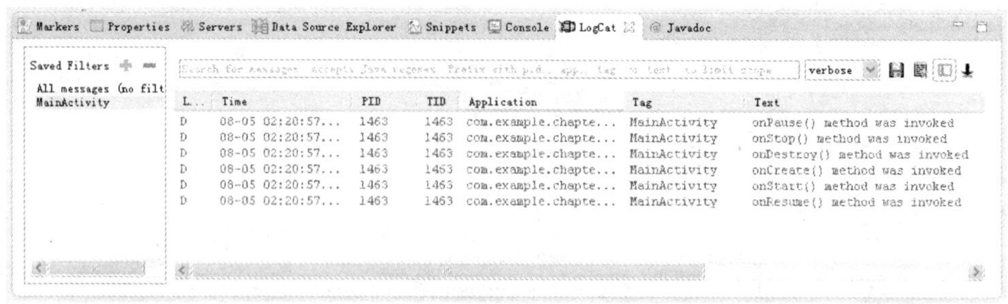

图 4-18　由竖屏切换到横屏时 LogCat 的输出

次,然后会被重新创建。通过上面的代码示例,除了认识到这一点以外,还有一点需要注意,那就是屏幕方向还没有发生变化的时候,笔者在两个文本框当中都输入了一些字符串,当屏幕方向从竖屏切换成横屏的时候,界面的运行效果如图 4-19 所示。

图 4-19　hello world 保留而 freedie. qin 消失的效果

通过 XML 方式定义的布局文件当中,这两个 EditText 控件的唯一差别就是第一个控件定义了 android:id 这个属性,而第二个控件没有定义这个属性。由此可以得出结论:当活动销毁的时候,只有布局当中设置了 android:id 属性的控件的值才会被保持下来,并在活动重新创建的时候自动恢复。而没有设置 android:id 这个属性的控件的值在活动被销毁的时候不会被保存下来。

4.2.2　适应方向改变

在有的应用场景下,需要应用程序竖屏时的用户界面与横屏时的用户界面完全不一致,也就意味着横、竖屏不能使用同一个用户界面布局文件。

接着使用之前的工程 chapter4_8,Android 当中提供了以下方法来处理横竖屏切换时如何使用不同的用户界面布局文件:

(1) 在 res 目录下新建一个 layout-land(表示 landscape)的子目录,用来存放横屏所需要的用户界面布局 XML 文件。

(2) 将 res/layout 目录下面的 activity_main. xml 布局文件复制一份至新建的 layout-land 目录下面,编辑 layout-land 目录下面的 activity_main. xml 文件,源代码如下:

```
<LinearLayout xmlns:android = "http://schemas.android.com/apk/res/android"
    xmlns:tools = "http://schemas.android.com/tools"
    android:layout_width = "fill_parent"
    android:layout_height = "fill_parent"
```

```
    android:orientation = "horizontal">
    <EditText
        android:id = "@ + id/myEditText"
        android:layout_width = "fill_parent"
        android:layout_height = "wrap_content"
        android:layout_weight = "1"/>
  <EditText
        android:layout_width = "fill_parent"
        android:layout_height = "wrap_content"
        android:layout_weight = "2"/>
</LinearLayout>
```

(3) 右击工程 chapter4_8,在弹出的快捷菜单中选择 Run As|Android Application 命令,在竖屏情况下运行效果如图 4-16 所示,在横屏情况下运行效果如图 4-20 所示。

图 4-20　chapter4_8 横屏显示效果

第 5 章 常用控件

第 4 章笔者已经介绍了组成 Android UI 界面当中几种常用的 ViewGroup 对象，包括 LinearLayout、RelativeLayout、TableLayout、AbsoluteLayout、FrameLayout 以及 ScrollView。除了这些 ViewGroup 对象之外，Android 应用程序的 UI 界面当中也少不了通过 ViewGroup 对象组合在一起的各种 View 对象，这些 View 对象也就是 Android 应用程序开发当中所指的常用控件。本章当中，笔者将会对 Android 应用程序当中用到的各种常用的控件进行一个完整而详细的介绍。这其中会包括一些基本的界面控件，比如：文本框（TextView）、编辑框（EditText）、按钮（Button）、图片按钮（ImageButton）、单选按钮（RadioButton）、单选按钮组（RadioGroup）、复选按钮（CheckBox）、状态开关按钮（ToggleButton）以及图像视图（ImageView），还会包括一些高级界面控件，比如：自动完成文本框（AutoCompleteTextView）、下拉列表（Spinner）、日期选择器（DatePicker）、时间选择器（TimerPicker）、进度条（ProgressBar）、拖动条（SeekBar）、星级评分条（RatingBar）、列表视图（ListView）以及网格视图（GridView）。所以，本章会包括以下两部分的内容：

❑ 基本界面控件；

❑ 高级界面控件。

5.1 基本界面控件

5.1.1 文本框（TextView）和编辑框（EditText）

TextView 的作用是在界面上显示文字，有以下一些常用的属性：

❑ android：textColor：用来设置文字的颜色。

❑ android：textSize：用来设置文字的大小，单位一般是 pt。

❑ android：text：用来设置显示的具体文字。

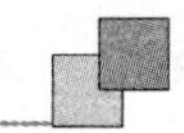

- ❑ android:ellipsize:用来设置当文本超出了 TextView 的长度时该如何处理。
- ❑ android:singleLine:用来设置 TextView 是否单行显示。
- ❑ android:maxLines:用来设置 TextView 最多显示的行数。
- ❑ android:autoLink:用来设置文本的链接属性。(none:不设置超链接;web:将文本中的 URL 地址转换为链接;phone:将文本中的电话转换为链接;email:将文本中的邮箱地址转换为链接)。
- ❑ android:hint:当文本框内容为空时,默认的提示文字。

EditText 的主要作用是用来接收用户的输入。由于 android. widget. EditText 继承于 android. widget. TextView,所以支持 TextView 的所有属性。EditText 有以下一些常用的属性:

- ❑ android:editable:用来设置编辑框是否可编辑。
- ❑ android:maxLength:用来设置最多可以输入多少个字符。
- ❑ android:singleLine:用来设置编辑框是否为单行模式。
- ❑ android:password:用来设置编辑框是否为密码框。
- ❑ android:hint:用来设置当编辑框为空时,默认的显示文字。
- ❑ android:digits:用来设置编辑框能接收的字符。
- ❑ android:phoneNumber:该属性用来设置编辑框是否只能接收电话号码。

(1) 新建工程 chapter5_1,编辑 res/layout/activity_main. xml 文件,源代码如下:

```
<LinearLayout xmlns:android = "http://schemas.android.com/apk/res/android"
    xmlns:tools = "http://schemas.android.com/tools"
    android:layout_width = "fill_parent"
    android:layout_height = "fill_parent"
    android:orientation = "vertical">
        <!-- 主要用来测试 TextView 的 android:singleLine 属性 -->
        <TextView
            android:id = "@ + id/tv"
            android:layout_width = "wrap_content"
            android:layout_height = "wrap_content"
            android:singleLine = "true"
            android:text = "这里是文本的内容这里是文本的内容这里是文本的内容这里
是文本的内容"
          android:textColor = "#FF0000"
          android:textSize = "15pt" >
        </TextView>
        <!-- 对此 EditText 监控焦点变化 -->
        <EditText   android:id = "@ + id/et"
            android:layout_width = "fill_parent"
```

```
            android:layout_height = "wrap_content"
            android:maxLength = "10"
            android:text = "请输入您的名字"
            android:textColor = "#DDDDDD" />
        <EditText
            android:id = "@ + id/et2"
            android:layout_width = "fill_parent"
            android:layout_height = "wrap_content"
            android:hint = "点击转换焦点" />
</LinearLayout>
```

(2) 编辑 MainActivity.java,源代码如下:

```
public class MainActivity extends Activity
{
    EditText et;  //声明 EditText 控件
    private Boolean isEmpty = true;//表明文本框是否为空
    @Override
    public void onCreate(Bundle savedInstanceState)
    {
        super.onCreate(savedInstanceState);
        setContentView(R.layout.activity_main);
        et = (EditText)findViewById(R.id.et);//找到 ID 为 tv 的 EditText
        //监听控件的焦点改变事件
        et.setOnFocusChangeListener(new OnFocusChangeListener()
    {
        @Override
        public void onFocusChange(View v, boolean hasFocus)
        {//编辑框焦点发生变化时回调
            //获取触发事件的 EditText
            EditText clickEditText = (EditText)v;
            if(hasFocus == false)
            {//EditText 失去焦点
                String text = clickEditText.getText().toString().trim();
                //编辑框文本内容不为空,且不等于"请输入您的名字"
                if(null != text && text.length() > 0 && !"请输入您的名字".equals
(text))
                {
                    isEmpty = false;//表明此时文本框不为空
                    clickEditText.setTextColor(Color.BLACK);
                    clickEditText.setText(text);
                }
                else
```

```
                {//编辑框文本内容为空,或者是内容为“请输入您的名字”
                    clickEditText.setText("请输入您的名字");
                    clickEditText.setTextColor(Color.GRAY);
                    isEmpty = true;
                }
            }
            else
            {//如果获得焦点
                clickEditText.setTextColor(Color.BLACK);
                if(isEmpty)
                {//如果控件处于未编辑状态,则清空“请输入您的名字”这几个字
                    clickEditText.setText("");
                }
            }
        }
    });
    //编辑框监控有新字符输入
    et.setOnKeyListener(new OnKeyListener()
    {
        @Override
        public boolean onKey(View v, int keyCode, KeyEvent event)
        {
          EditText clickEditText = (EditText)v;//获取触发事件的EditText
          String text = clickEditText.getText().toString().trim();//获取当前文本
          if(null != text && text.length() == 10)
{
              Toast.makeText(MainActivity.this, "最大长度为10个字符", Toast.
LENGTH_SHORT).show();//提示用户
            }
            return false;
        }
    });
}
    @Override
    public boolean onCreateOptionsMenu(Menu menu) {
        getMenuInflater().inflate(R.menu.activity_main, menu);
        return true;
    }
}
```

(3) 右击工程 chapter5_1,在弹出的快捷菜单中选择 Run As|Android Application 命令,运行效果如图 5-1 所示。

笔者在程序当中对编辑框的焦点变化事件进行了监听，当 ID 为 et 的编辑框获得焦点的时候，如果此时编辑框还处于未编辑状态，则清空编辑框的字符，当编辑框失去焦点的时候，会对编辑框的文本内容进行判断，如果编辑框的文本内容为空，则将编辑框的文本内容设置为“请输入您的名字”。

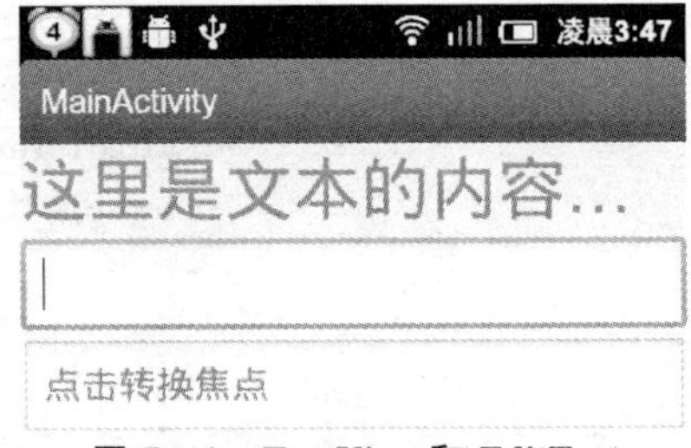

图 5-1　TextView 和 EditText

5.1.2　按钮（Button）和图片按钮（ImageButton）

Button 的主要作用是响应用户的单击事件，当用户单击按钮，按钮会触发一个 OnClick 事件。Button 比较常用的属性如下：

- ❑ android:text：该属性用来设置 Button 的显示文本。
- ❑ android:background：该属性用来设置 Button 的颜色或背景图片。

ImageButton 继承自 Button，两者的区别是 ImageButton 上不可以显示文字，而只能显示图片。ImageButton 比较常用的属性如下：

- ❑ android:src：该属性用来设置 ImageButton 上显示的图片。

（1）新建工程 chapter5_2，笔者将计算机上的一张 tomas_kinkade.jpg 图片复制到 res/drawable-hdpi 目录当中。

（2）编辑 res/layout/activity_main.xml 文件，源代码如下：

```
<LinearLayout xmlns:android = "http://schemas.android.com/apk/res/android"
    xmlns:tools = "http://schemas.android.com/tools"
    android:layout_width = "fill_parent"
    android:layout_height = "fill_parent"
    android:orientation = "vertical">
<!-- 普通按钮 -->
<Button
     android:id = "@ + id/my_button1"
     android:layout_width = "fill_parent"
     android:layout_height  = "wrap_content"
     android:text = "确定"
     ></Button>
<!-- 图片按钮 android:src 属性指定图片按钮展示的图片 -->
<ImageButton
     android:id = "@ + id/my_button2"
     android:layout_width = "fill_parent"
     android:layout_height  = "wrap_content"
     android:src = "@drawable/tomas_kinkade"
     ></ImageButton>
</LinearLayout>
```

(3) 编辑 MainActivity.java 文件,源代码如下:

```
public class MainActivity extends Activity implements OnClickListener
{
    Button myButton1;//声明一个普通按钮
    ImageButton myButton2;//声明一个图片按钮
    @Override
    public void onCreate(Bundle savedInstanceState)
    {
        super.onCreate(savedInstanceState);
        setContentView(R.layout.activity_main);
        myButton1 = (Button)findViewById(R.id.my_button1);//找到 Id 为 my_button1
                                                          //的按钮
        myButton2 = (ImageButton)findViewById(R.id.my_button2);//找到 Id 为 my_
                                                               //button2 的图片按钮
        myButton1.setOnClickListener(this);  //设置当前类为按钮的单击事件监听器
        myButton2.setOnClickListener(this);
    }
    @Override
    public void onClick(View v)
    {
        if(R.id.my_button1 == v.getId())
        {//如果是 Id 为 my_button1 的按钮
         //被单击
            Toast.makeText(MainActivity.this, "MyButton1 was clicked", Toast.LENGTH_SHORT).show();
        }
        else if(R.id.my_button2 == v.getId())
        {//如果是 Id 为 my_button2 的按钮 //被单击
            Toast.makeText(MainActivity.this, "MyButton2 was clicked", Toast.LENGTH_SHORT).show();
        }
    }
    public boolean onCreateOptionsMenu(Menu menu)
    {
        getMenuInflater().inflate(R.menu.activity_main, menu);
        return true;
```

图 5-2 Button 和 ImageButton

```
        }
    }
```

程序当中为 id 是 my_button1 的普通按钮以及 id 是 my_button2 的图片按钮的单击事件都进行了监听，当其中的一个按钮被单击的时候，都会进行相应的提示。

(3) 右击工程 chapter5_2，在弹出的快捷菜单中选择 Run As|Android Application 命令，运行效果如图 5-2 所示。

5.1.3 单选按钮(RadioButton)和单选按钮组(RadioGroup)

RadioButton 表示单个圆形单选框，而 RadioGroup 是可以容纳多个 RadioButton 的容器。它们之间的关系为：

- 每个 RadioGroup 中的 RadioButton 同时只能有一个被选中。
- 不同的 RadioGroup 中的 RadioButton 互不相干，即如果组 A 中有一个选中了，组 B 中依然可以有一个被选中。
- 大部分场合下，一个 RadioGroup 中至少有 2 个 RadioButton。
- 大部分场合下，一个 RadioGroup 中的 RadioButton 默认会有一个被选中，并建议读者将它放在 RadioGroup 中的起始位置。

(1) 新建工程 chapter5_3，编辑 res/layout/activity_main.xml 文件，源代码如下：

```
<?xml version="1.0" encoding="utf-8"?>
  <LinearLayout xmlns:android="http://schemas.android.com/apk/res/android"
     android:orientation="vertical"
     android:layout_width="fill_parent"
     android:layout_height="fill_parent"
     >
  <TextView
     android:layout_width="fill_parent"
     android:layout_height="wrap_content"
     android:text="请选择您的性别："
     android:textSize="9pt"
     />
  <RadioGroup android:id="@+id/radioGroup" android:contentDescription="性别"
android:layout_width="wrap_content" android:layout_height="wrap_content">
      <RadioButton android:layout_width="wrap_content" android:layout_height="
wrap_content" android:id="@+id/radioMale" android:text="男" android:checked="true"
></RadioButton>
      <RadioButton android:layout_width="wrap_content" android:layout_height="
wrap_content" android:id="@+id/radioFemale" android:text="女"></RadioButton>
```

```
  </RadioGroup>
<TextView
     android:id="@+id/tvSex"
     android:layout_width="fill_parent"
     android:layout_height="wrap_content"
     android:text="您的性别是:男"
     android:textSize="9pt"
     />
</LinearLayout>
```

(2) 编辑 MainActivity.java 文件,源代码如下:

```
public class MainActivity extends Activity
{
    TextView tv = null;//根据不同选项所要变更的文本控件
    @Override
    public void onCreate(Bundle savedInstanceState)
    {
        super.onCreate(savedInstanceState);
        setContentView(R.layout.activity_main);
        tv = (TextView)this.findViewById(R.id.tvSex);     //根据 ID 找到该文本控件
        RadioGroup group = (RadioGroup)this.findViewById(R.id.radioGroup);
                                                   //根据 ID 找到 RadioGroup 实例
        group.setOnCheckedChangeListener(new  RadioGroup.OnCheckedChangeListener()
        {//绑定一个匿名监听器
            public void onCheckedChanged(RadioGroup group, int checkedId)
            {
                int radioButtonId = group.getCheckedRadioButtonId();
                                                     //获取变更后的选中项的 ID
                 RadioButton rb = (RadioButton)MainActivity.this.findViewById(ra-
dioButtonId);                                 //根据 ID 获取 RadioButton 的实例
                tv.setText("您的性别是:" + rb.getText());      //更新文本内容,以
                                                             //符合选中项
            }
        });
    }
}
```

(3) 右击工程 chapter5_3,在弹出的快捷菜单中选择 Run As|Android Appcation 命令,运行效果如图 5-3 所示,选中"女"单选按钮过后,运行效果如图 5-4 所示。

图 5-3　RadioButton 选择男效果

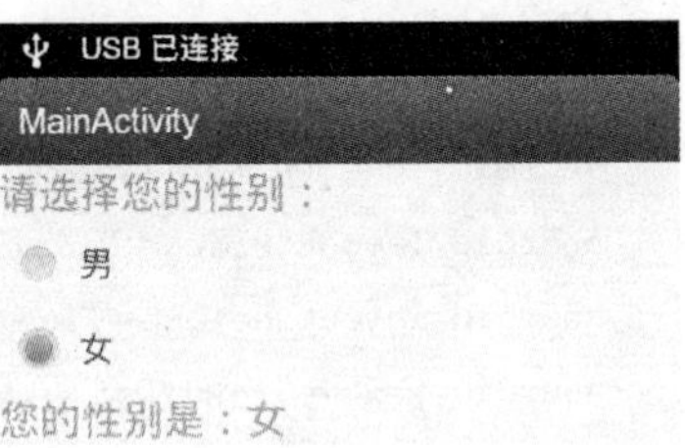

图 5-4　RadioButton 选择女效果

5.1.4　复选按钮(CheckBox)

CheckBox 和 Button 一样，也是一种古老的控件，它的优点在于，复选按钮当中可以有多个选项被选中，而不用用户去填写具体的信息，只需轻轻单击，我们往往利用它的这个特性，来获取用户的一些信息。

(1) 新建工程 chapter5_4，编辑 res/layout/activity_main.xml 文件，源代码如下：

```
<RelativeLayout xmlns:android = "http://schemas.android.com/apk/res/android"
    xmlns:tools = "http://schemas.android.com/tools"
    android:layout_width = "fill_parent"
    android:layout_height = "fill_parent" >
    <!-- android:checked = "false" 设置复选框没有选中 -->
    <CheckBox
        android:id = "@ + id/cb"
        android:layout_width = "wrap_content"
        android:layout_height = "wrap_content"
        android:layout_centerHorizontal = "true"
        android:layout_centerVertical = "true"
        android:checked = "false"
        android:text = "已婚" >
    </CheckBox>
</RelativeLayout>
```

(2) 编辑 MainActivity.java 文件，源代码如下：

```
public class MainActivity extends Activity
{
    CheckBox cb = null;//声明复选框
    @Override
    public void onCreate(Bundle savedInstanceState)
    {
        super.onCreate(savedInstanceState);
```

```
        setContentView(R.layout.activity_main);
    //获取 CheckBox 实例
        cb = (CheckBox)this.findViewById(R.id.cb);
        //绑定监听器
        cb.setOnCheckedChangeListener(new CompoundButton.OnCheckedChangeListener()
        {
          @Override//复选框选中状态发生变化时回调
          public void onCheckedChanged(CompoundButton buttonView, boolean isChecked)
          {
              Toast.makeText(MainActivity.this,
                  isChecked?"选中了":"取消了选中"  , Toast.LENGTH_LONG).show();
          }
    });
  }
}
```

(3) 右击工程 chapter5_4，在弹出的快捷菜单中选择 Run As|Android Application 命令，单击界面上的复选框效果如图 5-5 所示，取消复选框的选中效果如图 5-6 所示。

图 5-5　CheckBox 选中效果

图 5-6　CheckBox 取消选中效果

5.1.5　状态开关按钮(ToggleButton)

ToggleButton 有两种状态，开和关，通常用于切换程序中的某种状态。ToggleButton 的常用属性有如下几个：

❑ android:checked:该属性用来设置按钮是否被选中。

❑ android:textOff:该属性用来设置按钮没被选中时显示的文本。

❑ android:textOn:该属性用来设置按钮被选中时显示的文本。

(1) 新建工程 chapter5_5，编辑 res/layout/activity_main.xml 文件，源代码如下：

```
<?xml version = "1.0" encoding = "utf-8"?>
<LinearLayout xmlns:android = "http://schemas.android.com/apk/res/android"
    android:orientation = "vertical"
    android:layout_width = "fill_parent"
    android:layout_height = "fill_parent">
    <ToggleButton android:id = "@+id/toggleButton"
        android:layout_width = "140dip"
        android:layout_height = "wrap_content"
```

```
        android:textOn = "开灯"
        android:textOff = "关灯"
        android:layout_gravity = "center_horizontal" />
</LinearLayout>
```

（2）编辑 MainActivity. java 文件，源代码如下：

```
public class MainActivity extends Activity
{
    ToggleButton toggleButton;//声明开关按钮
    @Override
    public void onCreate(Bundle savedInstanceState)
    {
        super.onCreate(savedInstanceState);
        setContentView(R.layout.activity_main);
        toggleButton = (ToggleButton)findViewById(R.id.toggleButton);
                                                    //找到 Id 为 toggleButton 的开关按钮
        //绑定开关按钮的选中变化监听器
        toggleButton.setOnCheckedChangeListener(new OnCheckedChangeListener()
        {
            @Override//开关按钮选中状态发生变化时调用
            public void onCheckedChanged(CompoundButton buttonView, boolean isChecked)
            {
            //处理选中的变化
            if(isChecked)
            {//如果选中
                Toast.makeText(MainActivity.this, "选中了", Toast.LENGTH_LONG).
show();
            }
            else
            {   //如果没有选中
                Toast.makeText(MainActivity.this,"没有选中", Toast.LENGTH_LONG).
show();
            }
        }
    });
}
@Override
public boolean onCreateOptionsMenu(Menu menu) {
    getMenuInflater().inflate(R.menu.activity_main, menu);
    return true;
}
}
```

(3) 右击工程 chapter5_5,在弹出的快捷菜单中选择 Run As|Android Application 命令,界面的开关按钮被选中的时候效果如图 5-7 所示,被取消选中的时候效果如图 5-8 所示。

图 5-7 ToggleButton 选中效果

图 5-8 ToggleButton 没有选中效果

5.1.6 图像视图(ImageView)

ImageView 的主要功能就是显示图片,有如下几个常用的属性:

- ❑ android:src:该属性用来设置要显示的图片对象的 id。
- ❑ android:scaleType:该属性用来设置所显示的图片如何缩放或移动来适应 ImageView 控件所占的空间。属性值如下:
  - ➢ fitXY:对图片横向、纵向独立缩放。
  - ➢ fitStart:保持纵横比缩放,图片放在 ImageView 左上角。
  - ➢ fitCenter:保持纵横比缩放,图片放在 ImageView 中央。
  - ➢ fitEnd:保持纵横比缩放,图片放在 ImageView 右下角。
  - ➢ center:把图片放在 ImageView 中间,不进行任何缩放。
  - ➢ centerCrop:保持纵横比缩放,图片完全覆盖 ImageView。
  - ➢ centerInside:保持纵横比缩放,ImageView 能完全显示该图片。

(1) 新建工程 chapter5_6,笔者将计算机上的 google_logo.jpg,microsoft_logo.jpg 以及 pinterest_logo.jpg3 张图片复制到 res/drawable-hdpi 目录当中。

(2) 编辑 activity_main.xml 文件,源代码如下:

```
<LinearLayout xmlns:android = "http://schemas.android.com/apk/res/android"
    xmlns:tools = "http://schemas.android.com/tools"
    android:layout_width = "fill_parent"
    android:layout_height = "fill_parent"
    android:orientation = "vertical" >
    <ImageView
        android:id = "@ + id/imageview1"
        android:layout_width = "wrap_content"
        android:layout_height = "160dip"
  android:scaleType = "centerCrop"
        android:src = "@drawable/microsoft_logo" />
    <ImageView
```

```
        android:id = "@ + id/imageview2"
        android:layout_width = "wrap_content"
        android:layout_height = "160dip"
        android:layout_marginTop = "20dip"
        android:src = "@drawable/pinterest_logo" />
</LinearLayout>
```

(3) 编辑 MainActivity.java 文件,源代码如下:

```
public class MainActivity extends Activity
{
    ImageView image1;//声明第 1 个 ImageView
    ImageView image2;//声明第 2 个 ImageView
    @Override
    public void onCreate(Bundle savedInstanceState)
    {
        super.onCreate(savedInstanceState);
        setContentView(R.layout.activity_main);
        image1 = (ImageView) findViewById(R.id.imageview1);
                                        //找到 ID 为 imageview1 的 ImageView
        image2 = (ImageView) findViewById(R.id.imageview2);
                                        //找到 ID 为 imageview2 的 ImageView
        final int[] ids = new int[] { R.drawable.google_logo, R.drawable.microsoft_
logo, R.drawable.pinterest_logo };
        image1.setOnClickListener(new OnClickListener()
        {
            @Override
            public void onClick(View v)
            {
                int id = ids[0];
                image1.setImageResource(id);//更换 image1 显示的图片
                image2.setAlpha(100);//设置透明度,255 时为无透明度
            }
        });
    }
    @Override
    public boolean onCreateOptionsMenu(Menu menu)
    {
            getMenuInflater().inflate(R.menu.activity_main, menu);
            return true;
    }
}
```

(3) 右击工程 chapter5_6,在弹出的快捷菜单中选择 Run As|Android Applica-

tion 命令，运行效果如图 5 - 9 所示，单击第一个图片之后，效果如图 5 - 10 所示。

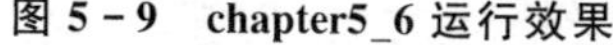

图 5 - 9　chapter5_6 运行效果

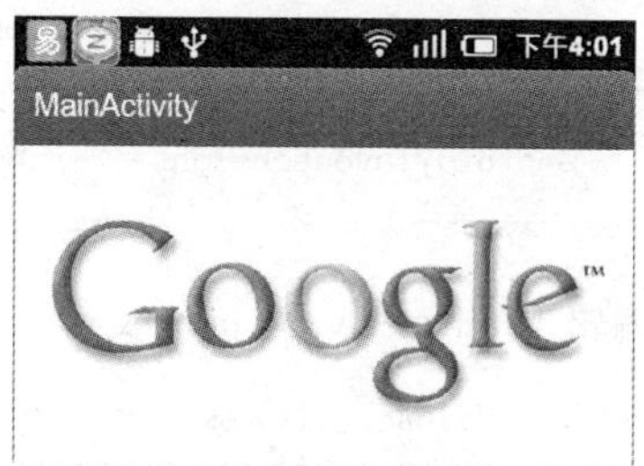

图 5 - 10　chapter5_6 单击第一个图片后运行效果

5.2　高级界面控件

5.2.1　自动完成文本框(AutoCompleteTextView)

AutoCompleteTextView 继承自 EditText，输入的时候带有自动文本补全提示功能，有以下几个常用的属性：

- android:completionHint：用来设置出现在下拉菜单中的提示标题。
- android:completionThreshold：用来设置至少输入几个字符才会显示提示。
- android:dropDownHeight：用来设置下拉菜单的高度。
- android:dropDownWidth：用来设置下拉菜单的宽度。
- android:popupBackgroud：用来设置下拉菜单的背景。
- android:dropDownHorizontalOffset：用来设置下拉菜单与文本框间的水平偏移。
- android:dropDownVerticalOffset：用来设置下拉菜单与文本框间的垂直偏移。

(1) 新建工程 chapter5_7，编辑 activity_main.xml 文件，源代码如下：

```
<? xml version = "1.0" encoding = "utf-8"? >
<LinearLayout
    xmlns:android = "http://schemas.android.com/apk/res/android"
    android:layout_width = "fill_parent"
    android:layout_height = "fill_parent">
```

```
    <AutoCompleteTextView
        android:id = "@ + id/myAutoCompleteTextView"
        android:layout_width = "fill_parent"
        android:layout_height = "wrap_content" />
</LinearLayout>
```

(2) 编辑 MainActivity.java 文件,源代码如下:

```
public class MainActivity extends Activity {
    private static final String[] myStr = {"aaa", "bbb", "aabb", "aad"};
    @Override
    public void onCreate(Bundle savedInstanceState)
    {
        super.onCreate(savedInstanceState);
        setContentView(R.layout.activity_main);
        //第二个参数使用 android 框架自带的一个布局文件
        ArrayAdapter<String> adapter = new ArrayAdapter<String>(this, android.
R.layout.simple_list_item_1, myStr);
        AutoCompleteTextView textView = (AutoCompleteTextView)findViewById(R.id.
myAutoCompleteTextView);
        textView.setAdapter(adapter);//设置自动补全列表的数据源
        textView.setThreshold(1);//输入 1 个字符之后自动补全提示
    }
    @Override
    public boolean onCreateOptionsMenu(Menu menu)
    {
        getMenuInflater().inflate(R.menu.activity_main, menu);
        return true;
    }
}
```

(3) 右击工程 chapter5_7,在弹出的快捷菜单中选择 Run As|Android Application 命令,在文本框当中输入字母 a 后,运行效果如图 5-11 所示。

图 5-11 chapter5_7 运行效果

5.2.2 下拉列表(Spinner)

用户单击 Spinner 之后,会弹出一个下拉列表式的选择框。Spinner 有以下几个常用的属性:

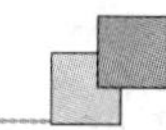

❑ android:prompt:该属性用来设置选择框的标题。

❑ android:entries:该属性用来设置下拉列表的列表项目所使用的数组资源。

(1) 新建工程 chapter5_8,编辑 res/layout/activity_main.xml 文件,源代码如下:

```
<?xml version="1.0" encoding="utf-8"?>
<LinearLayout
    xmlns:android="http://schemas.android.com/apk/res/android"
    android:layout_width="fill_parent"
    android:layout_height="fill_parent"
    android:orientation="vertical"
    >
    <TextView android:id="@+id/spinnerText"
              android:layout_width="fill_parent"
              android:layout_height="wrap_content"></TextView>
    <Spinner android:id="@+id/Spinner01"
           android:layout_width="fill_parent"
           android:layout_height="wrap_content"></Spinner>
</LinearLayout>
```

(2) 编辑 MainActivity.java 文件,源代码如下:

```
public class MainActivity extends Activity
{
    private static final String[] m = {"A型","B型","O型","AB型","其他"};
    private TextView view ;
    private Spinner spinner;
    private ArrayAdapter<String> adapter;
    @Override
    protected void onCreate(Bundle savedInstanceState)
  {
        //TODO Auto-generated method stub
        super.onCreate(savedInstanceState);
        setContentView(R.layout.activity_main);
        view = (TextView) findViewById(R.id.spinnerText);
        spinner = (Spinner) findViewById(R.id.Spinner01);
        //将可选内容与 ArrayAdapter 连接起来
        adapter = new ArrayAdapter<String>(this,android.R.layout.simple_spinner_
item,m);
        //设置下拉列表的风格
        adapter.setDropDownViewResource(android.R.layout.simple_spinner_dropdown_
item);
        //将 adapter 添加到 Spinner 中
```

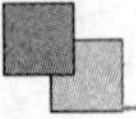


```
        spinner.setAdapter(adapter);
        //添加事件 Spinner 事件监听
        spinner.setOnItemSelectedListener(new SpinnerSelectedListener());
        //设置默认值
        spinner.setVisibility(View.VISIBLE);
    }
    //使用数组形式操作
    class SpinnerSelectedListener implements OnItemSelectedListener
    {
        //下拉列表的选项被选中的时候
        public void onItemSelected(AdapterView<?> arg0, View arg1, int arg2,
        long arg3)
        {
            view.setText("你的血型是:" + m[arg2]);
        }
          public void onNothingSelected(AdapterView<?> arg0)
        {       }
    }
}
```

(3) 右击工程 chapter5_8,在弹出的快捷菜单中选择 Run As|Android Application 命令,运行效果如图 5-12 所示,单击 Spinner 过后,效果如图 5-13 所示。

图 5-12　chapter5_8 运行效果

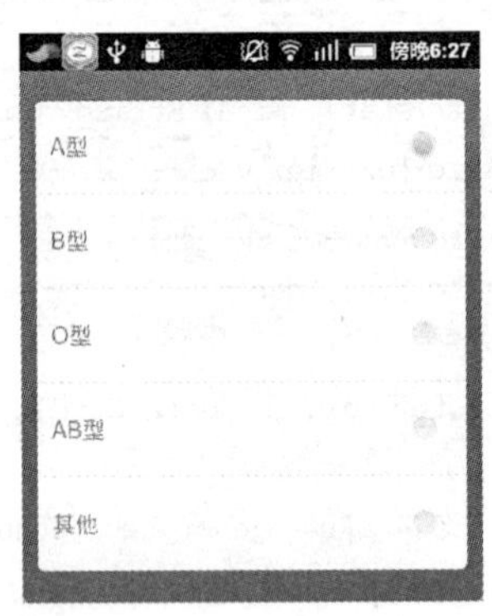

图 5-13　chapter5_8 单击 Spinner 运行效果

5.2.3 日期选择器(DatePicker)和时间选择器(TimePicker)

DatePicker 主要供用户选择日期,可以为 DatePicker 添加 onDateChangedListener 监听。TimePicker 主要供用户选择时间,如果程序需要获取用户选择的时间,可以为 TimePicker 添加 OnTimerChangedListenner 监听来实现。

(1) 新建工程 chapter5_9,编辑 res/layout/activity_main.xml 文件,源代码如下:

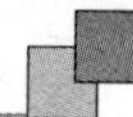

```
<?xml version="1.0" encoding="utf-8"?>
<LinearLayout xmlns:android="http://schemas.android.com/apk/res/android"
    android:orientation="vertical"
    android:layout_width="fill_parent"
    android:layout_height="fill_parent">
    <DatePicker android:id="@+id/datePicker"
        android:layout_width="wrap_content"
        android:layout_height="wrap_content"
        android:layout_gravity="center_horizontal"/>
    <EditText android:id="@+id/dateEt"
        android:layout_width="fill_parent"
        android:layout_height="wrap_content"
        android:cursorVisible="false"
        android:editable="false"/>
    <TimePicker android:id="@+id/timePicker"
        android:layout_width="wrap_content"
        android:layout_height="wrap_content"
        android:layout_gravity="center_horizontal"/>
    <EditText android:id="@+id/timeEt"
        android:layout_width="fill_parent"
        android:layout_height="wrap_content"
        android:cursorVisible="false"
        android:editable="false"/>
</LinearLayout>
```

(2) 编辑 MainActivity.java 文件，源代码如下：

```
public class MainActivity extends Activity
{
    private EditText dateEt = null;
    private EditText timeEt = null;
    @Override
    public void onCreate(Bundle savedInstanceState)
    {
        super.onCreate(savedInstanceState);
        setContentView(R.layout.activity_main);
        dateEt = (EditText)findViewById(R.id.dateEt);//找到 Id 为 dateEt 的 EditText
        timeEt = (EditText)findViewById(R.id.timeEt);//找到 Id 为 timeEt 的 EditText
        DatePicker datePicker = (DatePicker)findViewById(R.id.datePicker);
                                        //找到 Id 为 datePicker 的 DatePicker
        TimePicker timePicker = (TimePicker)findViewById(R.id.timePicker);
                                        //找到 Id 为 timePicker 的 timePicker
        Calendar calendar = Calendar.getInstance();
```

```
int year = calendar.get(Calendar.YEAR);//获取当前年
int monthOfYear = calendar.get(Calendar.MONTH);//获取当前月
int dayOfMonth = calendar.get(Calendar.DAY_OF_MONTH);//获取当前日
datePicker.init(year, monthOfYear, dayOfMonth, new OnDateChangedListener()
{
    public void onDateChanged(DatePicker view, int year,
            int monthOfYear, int dayOfMonth)
    {
        dateEt.setText("您选择的日期是:" + year + "年" + (monthOfYear + 1)
+ "月" + dayOfMonth + "日。");
    }
});
timePicker.setOnTimeChangedListener(new
OnTimeChangedListener()
{
    public void onTimeChanged(TimePicker
view, int hourOfDay, int minute)
    {
        timeEt.setText("您选择的时间是:"
+ hourOfDay + "时" + minute + "分。");
    }
});
    }
}
```

(3) 右击工程 chapter5_9，在弹出的快捷菜单中选择 Run As|Android Application 命令，选择 DatePicker 控件和选中 TimePicker 控件过后的运行效果如图 5-14 所示。

图 5-14 chapter5_9 运行效果

5.2.4 进度条(ProgressBar)和拖动条(SeekBar)

ProgressBar 的主要作用是向用户显示某个比较耗时的操作的进度。ProgressBar 有如下几个比较常用的属性：

- android:max：该属性用来设置进度条的最大值。
- android:progress：该属性用来设置进度条已经完成的进度。
- style：该属性用来设置进度条的样式，属性值有：
  - @android:style/Widget.ProgressBar.Horizontal：水平进度条。
  - @android:style/Widget.ProgressBar.Inverse：旋转的进度条。
  - @android:style/Widget.ProgressBar.Large：大进度条。
  - @android:style/Widget.ProgressBar.Small：小进度条。

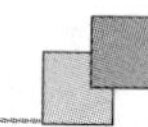

除了这些常用的属性之外，还有一些个比较常用的方法：

- setProgress(int)：该方法用来设置进度完成的百分比。
- incrementProgressBy(int)：该方法用来设置进度的增加或者减少。

SeekBar 允许用户拖动滑块来改变值，如调解音量等。SeeekBar 有如下几个比较常用的属性：

- android:max：该属性用来设置拖动条的最大值。
- android:progress：该属性用来设置拖动条的当前进度。
- android:thumb：该属性指定一个图片当做滑动条的滑块。

(1) 新建工程 chapter5_10，编辑 activity_main.xml 文件，源代码如下：

```
<?xml version = "1.0" encoding = "utf-8"?>
<LinearLayout xmlns:android = "http://schemas.android.com/apk/res/android"
    android:orientation = "vertical" android:layout_width = "fill_parent"
    android:layout_height = "fill_parent">
    <!--
        ProgressBar - 进度条控件
    -->
    <!--以下分别为大、中、小的进度条控件(圆圈状)-->
    <ProgressBar android:id = "@+android:id/progress_large"
        style = "?android:attr/progressBarStyleLarge" android:layout_width = "wrap_
content"
        android:layout_height = "wrap_content" />
    <ProgressBar android:id = "@+android:id/progress"
        android:layout_width = "wrap_content" android:layout_height = "wrap_content" />
    <ProgressBar android:id = "@+android:id/progress_small"
        style = "?android:attr/progressBarStyleSmall" android:layout_width = "wrap_
content"
        android:layout_height = "wrap_content" />
    <!--
        进度条控件(条状)的演示
            style - 进度条的样式,本例使用内置样式
            max - 进度的最大值
            progress - 第一进度位置
            secondaryProgress - 第二进度位置
    -->
    <ProgressBar android:id = "@+id/progress_horizontal"
        style = "?android:attr/progressBarStyleHorizontal" android:layout_width = "
200px"
        android:layout_height = "wrap_content" android:max = "100"
        android:progress = "50" android:secondaryProgress = "75" />
    <!--
        SeekBar - 可拖动的进度条控件
            max - 进度的最大值
            progress - 第一进度位置
            secondaryProgress - 第二进度位置
```

```
    -->
    <SeekBar android:id="@+id/seekBar" android:layout_width="fill_parent"
        android:layout_height="wrap_content" android:max="100"
        android:progress="50" android:secondaryProgress="75" />
    <TextView android:id="@+id/progress" android:layout_width="fill_parent"
        android:layout_height="wrap_content" />
    <TextView android:id="@+id/tracking" android:layout_width="fill_parent"
        android:layout_height="wrap_content" />
</LinearLayout>
```

(2) 编辑 MainActivity.java 文件，源代码如下：

```
public class MainActivity extends Activity implements OnSeekBarChangeListener
{
    SeekBar mSeekBar;//声明 SeekBar
    TextView mProgressText;//声明开始跟踪触摸 SeekBar 的文本
    TextView mTrackingText;//声明停止跟踪触摸 SeekBar 的文本
    @Override
    public void onCreate(Bundle savedInstanceState)
    {
        super.onCreate(savedInstanceState);
        setContentView(R.layout.activity_main);
        mSeekBar = (SeekBar)findViewById(R.id.seekBar);
                                                    //找到 id 为 seekBar 的 SeekBar 控件
        mSeekBar.setOnSeekBarChangeListener(this);//绑定拖动进度监听器
        mProgressText = (TextView)findViewById(R.id.progress);
                                                //找到 id 为 progress 的 TextView 控件
        mTrackingText = (TextView)findViewById(R.id.tracking);
                                                //找到 id 为 tracking 的 TextView 控件
    }
    //拖动 SeekBar 过后，进度发生改变时将要回调的方法
    @Override
    public void onProgressChanged(SeekBar seekBar, int progress, boolean fromUser)
    {
        mProgressText.setText(progress + "%");
    }
    //开始拖动回调方法
    @Override
    public void onStartTrackingTouch(SeekBar seekBar)
    {
        mTrackingText.setText("开始跟踪拖动");
    }
    //停止拖动回调方法
    @Override
    public void onStopTrackingTouch(SeekBar seekBar)
    {
        mTrackingText.setText("停止跟踪拖动");
```

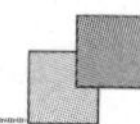

```
    }
    @Override
    public boolean onCreateOptionsMenu(Menu menu)
    {
        getMenuInflater().inflate(R.menu.activity_main, menu);
        return true;
    }
}
```

在上述代码当中，笔者没有对布局文件当中声明的 ProgressBar 控件进行处理，因为一般用 ProgressBar 的时候，会表示某一个正在进行的比较耗时的操作，而比较耗时的操作一般是在其他子线程当中运行的，笔者在后面介绍 Android 异步处理机制的时候，会通过实例来对 ProgressBar 控件进行操作。

(3) 右击工程 chapter5_10，在弹出的快捷菜单中选择 Run As|Android Application 命令，开始拖动 SeekBar 控件的时候效果如图 5-15 所示，结束拖动 SeekBar 控件的时候效果如图 5-16 所示。

图 5-15 chapter5_10 拖动 SeekBar 效果

图 5-16 chapter5_10 停止拖动 SeekBar 效果

5.2.5 星级评分条(RatingBar)

RatingBar 主要用来评分。RatingBar 有如下几个比较常用的属性：

- ❑ android:numStars：该属性用来设置评分条有多少个星级。
- ❑ android:rating：该属性用来设置评分默认的星级。
- ❑ android:stepSize：该属性用来设置每次需要改变多少星级。
- ❑ android:isIndicator：该属性用来设置是否允许用户改变。

(1) 新建工程 chapter5_11，编辑 res/layout/activity_main.xml 文件，源代码如下：

```
<?xml version="1.0" encoding="utf-8"?>
<LinearLayout xmlns:android="http://schemas.android.com/apk/res/android"
    android:orientation="vertical"
```

```
    android:layout_width = "fill_parent"
    android:layout_height = "fill_parent"
    >
<TextView  android:id = "@ + id/tv"
    android:layout_width = "fill_parent"
    android:layout_height = "wrap_content"
    android:text = ""
    />
<RatingBar android:id = "@ + id/rb"
  android:layout_width = "wrap_content"
  android:layout_height = "fill_parent"
  android:numStars = "3"
  android:rating = "1"
/>
<!--这里要注意了,layout_width 必须是 wrap_content,如果设成 fill_parent,
不管你在后面把满分设为几颗星,它都会把屏幕横向显示满为止-->
</LinearLayout>
```

(2) 编辑 MainActivity.java 文件,源代码如下:

```
public class MainActivity extends Activity
{
    RatingBar rb;//声明星级评分条
    TextView tv;//声明显示评分的文本
    @Override
    public void onCreate(Bundle savedInstanceState)
    {
        super.onCreate(savedInstanceState);
        setContentView(R.layout.activity_main);
        tv = (TextView)findViewById(R.id.tv);//找到 id 为 tv 的文本控件
        rb = (RatingBar)findViewById(R.id.rb);//找到 id 为 rb 的 RatingBar
        rb.setOnRatingBarChangeListener(rbLis);//绑定星级评分条评分变化的监听器
    }
    private OnRatingBarChangeListener rbLis = new OnRatingBarChangeListener()
    {
    //星级评分条评分变化回调的方法
    @Override
      public void onRatingChanged(RatingBar ratingBar, float rating, boolean fromUser)
      {
          tv.setText(String.valueOf(rb.getRating()));
      }
    };
    @Override
    public boolean onCreateOptionsMenu(Menu menu)
    {
        getMenuInflater().inflate(R.menu.activity_main, menu);
        return true;
```

```
    }
}
```

(3) 右击工程 chapter5_11，在弹出的快捷菜单中选择 Run As|Android Application 命令，运行效果如图 5－17 所示，当选择第 3 个星级的时候，运行效果如图 5－18 所示。

图 5－17　chapter5_11 运行效果

图 5－18　chapter5_11 选中星级运行效果

5.2.6　列表视图(ListView)

在 android 开发中 ListView 是比较常用的控件，它以列表的方式展示具体内容，并且能够根据数组的长度自适应显示。ListView 有如下几个比较常用的属性：

- ❑ android:divider:该属性用来设置 List 列表的分隔条(颜色或图片)。
- ❑ android:dividerHeight:该属性用来设置分隔条的高度。
- ❑ android:entries:指定要显示的数组资源。

ListView 里面的每个子项 Item 可以是一个简单的字符串，也可以是一个组合的控件。先介绍 ListView 的实现：

- ❑ 准备 ListView 要显示的数据。
- ❑ 使用一维或多维动态数组保存数据。
- ❑ 构建适配器(即 Item 数组)，动态数组有多少个元素就生成多少个 Item。
- ❑ 把适配器添加到 ListView，并显示出来。

(1) 新建工程 chapter5_12，编辑 res/layout/activity_main. xml 文件，源代码如下：

```
<?xml version = "1.0" encoding = "utf-8"?>
<LinearLayout
        android:id = "@ + id/LinearLayout01"
        android:layout_width = "fill_parent"
        android:layout_height = "fill_parent"
        xmlns:android = "http://schemas.android.com/apk/res/android">
        <ListView android:layout_width = "wrap_content"
                android:layout_height = "wrap_content"
                android:id = "@ + id/MyListView">
        </ListView>
```

```
</LinearLayout>
```

(2) 在 res/layout 目录下面，新建 XML 布局文件：my_listitem. xml，这个文件作为 ListView 每一项的布局文件。编辑 my_listitem. xml 文件，源代码如下：

```
<?xml version="1.0" encoding="utf-8"?>
<LinearLayout
        android:layout_width="fill_parent"
        xmlns:android="http://schemas.android.com/apk/res/android"
        android:orientation="vertical"
        android:layout_height="wrap_content"
        android:id="@+id/MyListItem"
        android:paddingBottom="3dip"
        android:paddingLeft="10dip">
        <TextView
                android:layout_height="wrap_content"
                android:layout_width="fill_parent"
                android:id="@+id/ItemTitle"
                android:textSize="30dip">
        </TextView>
        <TextView
                android:layout_height="wrap_content"
                android:layout_width="fill_parent"
                android:id="@+id/ItemText">
        </TextView>
</LinearLayout>
```

(3) 编辑 MainActivity. java 文件，源代码如下：

```
public class MainActivity extends Activity
{
    @Override
    public void onCreate(Bundle savedInstanceState)
    {
        super.onCreate(savedInstanceState);
        setContentView(R.layout.activity_main);
            //绑定 XML 中的 ListView,作为 Item 的容器
            ListView list = (ListView) findViewById(R.id.MyListView);
            //生成动态数组,并且转载数据
            ArrayList<HashMap<String, String>> mylist = new ArrayList<HashMap
<String, String>>();
            for(int i=0;i<30;i++)//30 次循环,将每一次循环产生的 map 对象加入
                                    //到 myList 对象当中
            {
              HashMap<String, String> map = new HashMap<String, String>();
              map.put("ItemTitle", "This is Title.....");
```

```
        map.put("ItemText", "This is text.....");
        mylist.add(map);
    }
    //生成适配器,数组===》ListItem
  SimpleAdapter mSchedule = new SimpleAdapter(this, //当前 MainActivity 对象
                                    mylist,//数据来源
                                R.layout.my_listitem,//ListItem 的
                                    XML 实现
                                //动态数组与 ListItem 对应的子项
                                  new String[] {"ItemTitle", "Item-
                                    Text"},
                                  //ListItem 的 XML 文件里面的两个
                                    TextView ID
                                  new int[] {R.id.ItemTitle,R.id.
                                    ItemText});
                                //添加并且显示
                                list.setAdapter(mSchedule);
                            }

                @Override
            public boolean onCreateOptionsMenu(Menu menu)
            {
                    getMenuInflater().inflate(R.menu.activ-
ity_main, menu);
                    return true;
                }
            }
```

图 5-19　chapter5_12 运行效果

（4）右击工程 chapter5_12，在弹出的快捷菜单中选择 Run As|Android Application 命令，运行效果如图 5-19 所示。

5.2.7　网格视图(GridView)

GridView 用于在界面上按行、列分布的方式显示多个组件。GridView 有如下几个常用的属性：

- ❑ android:columnWidth:该属性用来设置列的宽度。
- ❑ android:numColumns:该属性用来设置列数。
- ❑ android:stretchMode:该属性用来设置拉伸模式。
- ❑ android:horizontalSpacing:该属性用来设置各元素之间的水平间距。
- ❑ android:verticalSpacing:该属性用来设置各元素之间的垂直间距。

(1) 新建工程 chapter5_13，笔者将计算机上的 notepad.jpg 图片复制到 res/drawable-hdpi 目录当中。

编辑 activity_main.xml 文件，源代码如下：

```
<? xml version = "1.0" encoding = "utf-8"? >
<GridView xmlns:android = "http://schemas.android.com/apk/res/android"
    android:id = "@ + id/gridview"
    android:layout_width = "fill_parent"
    android:layout_height = "fill_parent"
    android:numColumns = "auto_fit"
    android:verticalSpacing = "10dp"
    android:horizontalSpacing = "10dp"
    android:columnWidth = "90dp"
    android:stretchMode = "columnWidth"
    android:gravity = "center"
/>
```

(2) 在 res/layout 目录下面，新建 XML 布局文件：night_item.xml，这个文件作为 GridView 每一个单元格显示的布局文件。编辑 night_item.xml 文件，源代码如下：

```
<? xml version = "1.0" encoding = "utf-8"? >
<RelativeLayout
        xmlns:android = "http://schemas.android.com/apk/res/android"
        android:layout_height = "wrap_content"
        android:paddingBottom = "4dip" android:layout_width = "fill_parent">
        <ImageView
              android:layout_height = "wrap_content"
              android:id = "@ + id/ItemImage"
              android:layout_width = "wrap_content"
              android:layout_centerHorizontal = "true">
        </ImageView>
        <TextView
              android:layout_width = "wrap_content"
              android:layout_below = "@ + id/ItemImage"
              android:layout_height = "wrap_content"
              android:text = "TextView01"
              android:layout_centerHorizontal = "true"
              android:id = "@ + id/ItemText">
        </TextView>
</RelativeLayout>
```

(3) 编辑 MainActivity.java 文件，源代码如下：

```
public class MainActivity extends Activity
{
```

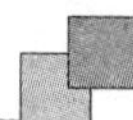

```
    @Override
    public void onCreate(Bundle savedInstanceState)
    {
            super.onCreate(savedInstanceState);
            setContentView(R.layout.activity_main);
            GridView gridview = (GridView) findViewById(R.id.gridview);
                                                    //找到 id 为 gridview 的 GridView 控件
            //生成动态数组,并且转入数据
            ArrayList<HashMap<String, Object>> lstImageItem = new ArrayList<
HashMap<String, Object>>();
            for(int i=0;i<10;i++)
            {
              HashMap<String, Object> map = new HashMap<String, Object>();
              map.put("ItemImage", R.drawable.notepad);//添加图像资源的 ID
       map.put("ItemText", "NO." + String.valueOf(i));//按序号做 ItemText
              lstImageItem.add(map);
            }
            //生成适配器的 ImageItem <====> 动态数组的元素,两者一一对应
            SimpleAdapter saImageItems = new SimpleAdapter(this, //没什么解释
                                                 lstImageItem,//数据来源
                                                 R.layout.night_item,//night_i-
                                                 tem 的 XML 实现

                                          //动态数组与 ImageItem 对应的子项
                                            new String[] {"ItemImage","Item-
                                            Text"},

                                   //ImageItem 的 XML 文件里面的一个 ImageView,两个
                                   TextView ID new int[] {R.id.ItemImage,R.id.Item-
                                   Text});
            //添加并且显示
            gridview.setAdapter(saImageItems);
            //添加消息处理
            gridview.setOnItemClickListener(new ItemClickListener());
        }

    //当 AdapterView 被单击(触摸屏或者键盘),则返回的 Item 单击事件
    class  ItemClickListener implements OnItemClickListener
    {
   public  void  onItemClick ( AdapterView <? > arg0,//The  AdapterView  where  the
click happened
                                  View arg1,//The view within the AdapterView that
```

```
                                        was clicked
                                        int arg2,//The position of the view in the adapter
                                        long arg3//The row id of the item that was clicked
                                        )
        {
            //在本例中 arg2 = arg3
            HashMap<String, Object> item = (HashMap<String, Object>) arg0.getItemAtPosition(arg2);
            //显示所选 Item 的 ItemText
            setTitle((String) item.get("ItemText"));
        }
    }
    @Override
    public boolean onCreateOptionsMenu(Menu menu)
    {
        getMenuInflater().inflate(R.menu.activity_main, menu);
        return true;
    }
}
```

(4) 右击工程 chapter5_13,在弹出的快捷菜单中选择 Run As|Android Application 命令,运行效果如图 5-20 所示。

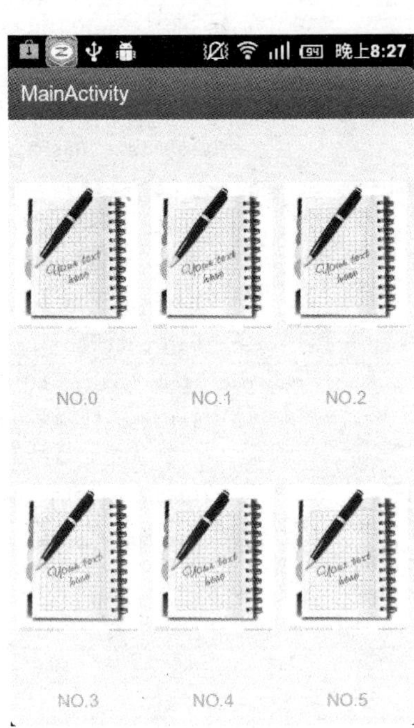

图 5-20　chapter5_13 运行效果

第 6 章

菜单、通知以及闹钟服务

菜单是许多应用程序不可或缺的一部分，在 Android 中更是如此，大多数搭载 Android 操作系统的手机都有一个 Menu 键，可见菜单在 Android 程序当中的特殊性。Android SDK 提供了 3 种类型的菜单，分别是选项菜单、上下文菜单以及子菜单。

在之前的章节当中，笔者已经介绍过如何通过对话框的方式来通知用户，程序发生了什么，但是这种方式有个特点就是：主动获取了用户的焦点。而在一些应用场景下，程序需要在不获取用户焦点的情况下去通知用户，比如用户正在玩游戏，这时候突然来了一条短信，如果用对话框来提醒，那用户会被活活气死。所以使用通知在状态栏对用户进行提醒，是一种用户体检比较好的方法。Android SDK 为此提供了一种叫做"Notification"的机制。Notification 不会干扰正常的操作，单击后还可以进入相关的界面查看详细的内容。闹钟服务是允许程序在指定的时间执行某项任务的一种服务。在本章当中，笔者将会对菜单通知以及闹钟服务进行具体介绍，因此，本章包括以下 3 部分内容：

❑ 菜单；

❑ 通知；

❑ 闹钟服务。

6.1 菜　单

菜单是 Android 应用程序中比较常见的元素之一，主要分为选项菜单（Options menu）、上下文菜单（Context menu）以及子菜单（Submenu）。

6.1.1 选项菜单

当 Activity 在前台运行时，如果用户按下手机上的 Menu 键，此时就会在屏幕低

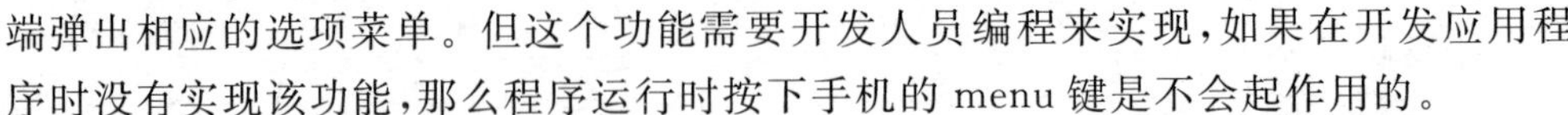

端弹出相应的选项菜单。但这个功能需要开发人员编程来实现，如果在开发应用程序时没有实现该功能，那么程序运行时按下手机的 menu 键是不会起作用的。

一般情况下，选项菜单最多显示 2 排每排 3 个菜单项，当菜单选项多于 6 个时，将只显示前 5 个和一个扩展菜单选项，单击扩展菜单选项将会弹出其余的菜单项。扩展菜单项中将不会显示图标，但是可以显示单选按钮和复选框。下面通过实例来演示，如何为 Activity 添加选项菜单。

(1) 新建工程 chapter6_1，编辑 MainActivity.java 文件，源代码如下：

```
public class MainActivity extends Activity
{
    @Override
    public void onCreate(Bundle savedInstanceState)
    {
        super.onCreate(savedInstanceState);
        setContentView(R.layout.activity_main);
    }
    //单击 Menu 时，系统调用当前 Activity 的 onCreateOptionsMenu 方法，并传一个实现了
    //一个 Menu 接口的 menu 对象供开发者使用
    @Override
    public boolean onCreateOptionsMenu(Menu menu)
    {/**
        * add()方法的 4 个参数，依次是：
        * 1)组别，如果不分组的话就写 Menu.NONE;
        * 2)Id，这个很重要，Android 根据这个 Id 来确定不同的菜单；
        * 3)顺序，菜单的排列顺序由这个参数决定；
        * 4)文本，菜单的显示文本。
        */
      menu.add(Menu.NONE, Menu.FIRST + 1, 5, "删除").setIcon(android.R.drawable.ic
_menu_delete);
        //setIcon()方法为菜单设置图标，这里笔者使用的是系统自带的图标
        //android.R 开头的资源是系统提供的
        menu.add(Menu.NONE, Menu.FIRST + 2, 2, "保存").setIcon(android.R.drawable.ic
_menu_edit);
        menu.add(Menu.NONE, Menu.FIRST + 3, 6, "帮助").setIcon(android.R.drawable.ic
_menu_help);
        menu.add(Menu.NONE, Menu.FIRST + 4, 1, "添加").setIcon(android.R.drawable.ic
_menu_add);
        menu.add(Menu.NONE, Menu.FIRST + 5, 4, "详细").setIcon(android.R.drawable.ic
_menu_info_details);
        menu.add(Menu.NONE, Menu.FIRST + 6, 3, "发送").setIcon(android.R.drawable.ic
_menu_send);
```

```
    return true;//return true 的话,菜单才会被显示
}
@Override
public boolean onOptionsItemSelected(MenuItem item) //菜单项被选择事件
{
    switch (item.getItemId())
    {
    case Menu.FIRST + 1:    //删除菜单
        Toast.makeText(this, "删除菜单被点击了", Toast.LENGTH_LONG).show();
        break;
    case Menu.FIRST + 2:    //保存菜单
        Toast.makeText(this, "保存菜单被点击了", Toast.LENGTH_LONG).show();
        break;
    case Menu.FIRST + 3:   //帮助菜单
        Toast.makeText(this, "帮助菜单被点击了", Toast.LENGTH_LONG).show();
        break;
    case Menu.FIRST + 4:   //添加菜单
        Toast.makeText(this, "添加菜单被点击了", Toast.LENGTH_LONG).show();
        break;
    case Menu.FIRST + 5:    //详细菜单
        Toast.makeText(this, "详细菜单被点击了", Toast.LENGTH_LONG).show();
        break;
    case Menu.FIRST + 6:    //发送菜单
        Toast.makeText(this, "发送菜单被点击了", Toast.LENGTH_LONG).show();
        break;
    }
    return false;
}
//菜单被显示之前调用的事件
@Override
public boolean onPrepareOptionsMenu(Menu menu)
{
    Toast.makeText(this, "选项菜单显示之前 onPrepareOptionsMenu 方法会被调用,
可以用此方法来根据打当时的情况调整菜单", Toast.LENGTH_LONG).show();
    return true;
}
//选项菜单被关闭事件,菜单被关闭有 3 种情形,menu 按钮被再次单击、back 按钮被单
//击或者用户选择了某一个菜单项
@Override
public void onOptionsMenuClosed(Menu menu)
{
    Toast.makeText(this, "选项菜单关闭了", Toast.LENGTH_LONG).show();
```

```
    }
}
```

上述代码当中，onCreateOptionsMenu()方法仅在选项菜单第一次显示的时候会被回调，紧接着 onPrepareOptionsMenu()方法会被回调。以后当用户按下 menu 键的时候，只有 onPrepareOptionsMenu()方法会被回调。所以每次选项菜单显示之前，可以在 onPrepareOptionsMenu()方法当中调整选项菜单。

(2) 右击工程 chapter6_1，在弹出的快捷菜单中选择 Run As|Android Application 命令，按住设备上的 menu 键过后，运行效果如图 6-1 所示，单击“保存”选项的效果如图 6-2 所示。

图 6-1 选项菜单运行效果

图 6-2 保存菜单项被单击运行效果

(3) 除了可以使用 Java 代码的方式来生成选项菜单以外，还可以采用定义 XML 文件的方式来定义 Activity 的选项菜单。在 res/menu 目录下面，新建一个 options_menu.xml 文件，编辑 options_menu.xml 文件，源代码如下：

```
<?xml version="1.0" encoding="utf-8"?>
<menu xmlns:android="http://schemas.android.com/apk/res/android">
    <item android:id="@+id/item01" android:icon="@android:drawable/ic_menu_add" android:title="添加"></item>
    <item android:id="@+id/item02" android:icon="@android:drawable/ic_menu_edit" android:title="编辑"></item>
    <item android:id="@+id/item04" android:icon="@android:drawable/ic_menu_info_details" android:title="详细"></item>
    <item android:id="@+id/item05" android:icon="@android:drawable/ic_menu_delete" android:title="删除"></item>
    <item android:id="@+id/item06" android:icon="@android:drawable/ic_menu_help" android:title="帮助"></item>
</menu>
```

(4) 编辑 MainActivity.java 的 onCreateOptionsMenu()方法，源代码如下：

```
//单击 Menu 时，系统调用当前 Activity 的 onCreateOptionsMenu 方法，并传一个实现了一
//个 Menu 接口的 menu 对象供开发者使用
```

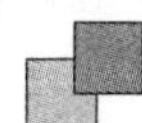

```
@Override
public boolean onCreateOptionsMenu(Menu menu)
{
        this.getMenuInflater().inflate(R.menu.options_menu, menu);
        return true;
}
```

(5) 采用 XML 文件的方式来定义选项菜单的运行效果与采用 Java 代码的方式生成选项菜单的效果一致。

6.1.2 上下文菜单

Android 的上下文菜单在概念上与 PC 软件的右键菜单类似。如果一个视图对象已经注册了相关的上下文菜单，执行一个在该视图对象上的"长按"(按住不动差不多两秒钟)动作，将出现一个提供相关功能的"浮动"菜单，这就是上下文菜单。

创建上下文菜单有以下几个步骤：

- ❑ 覆盖 Activity 的 onCreateContextMenu()方法，调用 Menu 对象的 add()方法添加菜单项。
- ❑ 覆盖 Activity 的 onContextItemSelected()方法，响应上下文菜单的单击事件。
- ❑ 调用 registerForContextMenu()方法为视图对象注册上下文菜单。

下面通过实例的方式来介绍如何为视图对象注册上下文菜单。

(1) 新建工程 chapter6_2，编辑 res/layout/activity_main.xml 文件，源代码如下：

```
<?xml version = "1.0" encoding = "utf-8"?>
<LinearLayout xmlns:android = "http://schemas.android.com/apk/res/android"
    android:layout_width = "fill_parent"
    android:layout_height = "fill_parent"
    android:orientation = "vertical" >
    <Button
        android:id = "@+id/myBtn"
        android:layout_width = "fill_parent"
        android:layout_height = "wrap_content"
        android:text = "@string/hello_world" />
</LinearLayout>
```

(2) 编辑 MainActivity.java 文件，源代码如下：

```
public class MainActivity extends Activity
{
    private static final int ITEM1 = Menu.FIRST;
    private static final int ITEM2 = ITEM1 + 1;
    private static final int ITEM3 = ITEM1 + 2;
```

```
    private Button myButton;//声明需要注册上下文菜单的按钮控件
    @Override
    public void onCreate(Bundle savedInstanceState)
    {
        super.onCreate(savedInstanceState);
        setContentView(R.layout.activity_main);
        myButton = (Button)findViewById(R.id.myBtn);//找到 Id 为 myBtn 的按钮控件
        registerForContextMenu(myButton);//向按钮控件注册上下文菜单
    }
    //创建上下文菜单
    @Override
    public void onCreateContextMenu(ContextMenu menu, View v, ContextMenuInfo menuInfo)
    {
    super.onCreateContextMenu(menu, v, menuInfo);
    //添加菜单项
    menu.add(Menu.NONE, ITEM1, 1, "红色背景");
    menu.add(Menu.NONE, ITEM2, 2, "绿色背景");
    menu.add(Menu.NONE, ITEM3, 3, "白色背景");
    }
    //处理上下文菜单的选项的单击
    @Override
    public boolean onContextItemSelected(MenuItem item)
    {
    switch(item.getItemId())
     {
    case ITEM1://红色背景
        myButton.setBackgroundColor(Color.RED);
        break;
    case ITEM2://绿色背景
        myButton.setBackgroundColor(Color.GREEN);
        break;
    case ITEM3://白色背景
        myButton.setBackgroundColor(Color.WHITE);
        break;
    }
    return true;
    }
    @Override
    public boolean onCreateOptionsMenu(Menu menu)
    {
        getMenuInflater().inflate(R.menu.activity_main, menu);
        return true;
    }
}
```

(3) 右击工程 chapter6_2，在弹出的快捷菜单中选择 Run As|Android Application 命令，长按界面中的按钮，运行效果如图 6-3 所示，选中红色背景选项后，运行

效果如图 6 - 4 所示。

图 6 - 3　长按按钮运行效果

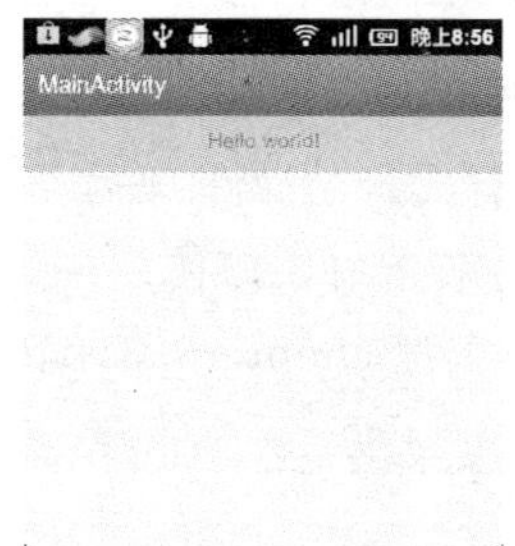

图 6 - 4　红色背景选项单击运行效果

（4）除了可以使用 Java 代码来生成上下文菜单以外，也可以采用 XML 文件的方式来定义上下文菜单，在 res/menu 目录下面，新建一个 context_menu.xml 文件，编辑 context_menu.xml 文件，源代码如下：

```
<menu xmlns:android = "http://schemas.android.com/apk/res/android">
  <item android:id = "@ + id/contextMenu1" android:title = "XML 创建的菜单子项 1">
</item>
  <item android:id = "@ + id/contextMenu2" android:title = "XML 创建的菜单子项 2">
</item>
  <item android:id = "@ + id/contextMenu3" android:title = "XML 创建的菜单子项 3">
</item>
</menu>
```

（5）编辑 MainActivity.java 文件的 onCreateContextMenu()方法，源代码如下：

```
//创建上下文菜单
@Override
public void onCreateContextMenu(ContextMenu menu, View v, ContextMenuInfo menuIn-
fo) {
  super.onCreateContextMenu(menu, v, menuInfo);
  //XML 方式创建的菜单项
      MenuInflater inflater = getMenuInflater();
      inflater.inflate(R.menu.context_menu, men-
u);
}
```

（6）右击工程 chapter6_2，在弹出的快捷菜单中选择 Run As|Android Application 命令，长按界面中的按钮，效果如图 6 - 5 所示。

图 6 - 5　XML 定义上下文菜单运行效果

6.1.3 子菜单

子菜单是可以被添加到其他菜单上的菜单,但是子菜单不能被添加到子菜单上。通常,当需要有大量的菜单项需要显示时,利用子菜单进行分类是一个很好的方法。下面通过实例来介绍如何使用子菜单。

(1) 新建工程 chapter6_3,编辑 MainActivity.java 文件,源代码如下:

```
public class MainActivity extends Activity
{
    public static final int MENU_BAIDU = 0;       //声明子菜单项"百度"的 ID
    public static final int MENU_GOOGLE = 1;      //声明子菜单项"谷歌"的 ID
    public static final int MENU_BING = 2;        //声明子菜单项"Bing"的 ID
    MenuItem baiduMenuItem = null;                //声明子菜单项"百度"
    MenuItem googleMenuItem = null;               //声明子菜单项"谷歌"
    MenuItem bingMenuItem = null;                 //声明子菜单项"Bing"
    @Override
    public void onCreate(Bundle savedInstanceState){
        super.onCreate(savedInstanceState);
        setContentView(R.layout.activity_main);
    }
    @Override
    public boolean onCreateOptionsMenu(Menu menu) {
    //一个 menu 可以包含多个子菜单
    SubMenu subMenu = menu.addSubMenu("搜索引擎");
    //使用 android 系统资源设置子菜单的图标
    subMenu.setIcon(android.R.drawable.ic_menu_search);
    //一个子菜单可以包含多个子菜单项
    baiduMenuItem = subMenu.add(Menu.NONE, MENU_BAIDU, 1, "百度");
                                                            //添加"百度"子菜单项
    googleMenuItem = subMenu.add(Menu.NONE, MENU_GOOGLE, 2 , "谷歌");
                                                            //添加"谷歌"子菜单项
    bingMenuItem = subMenu.add(Menu.NONE, MENU_BING, 3, "微软 Bing");
                                                            //添加"Bing"子菜单项
    googleMenuItem.setCheckable(true);
    googleMenuItem.setChecked(true);              //将 localMenuItem 菜单项设置为已选
    subMenu.setGroupCheckable(0, true, true);       //设置菜单项为单选菜单项,互斥的
    return true;
    }
}
```

上述代码演示了如何创建子菜单,一个 menu 可以包含多个子菜单,而每一个子菜单又可以包含多个菜单项。Android 当中可以为子菜单添加图标,但是为子菜单的菜单项设置的图标是不起作用的。

(3) 右击工程 chapter6_3,在弹出的快捷菜单中选择 Run As|Android Applica-

tion 命令，按住设备的 menu 键效果如图 6-6 所示，单击“搜索引擎”子菜单的效果如图 6-7 所示。

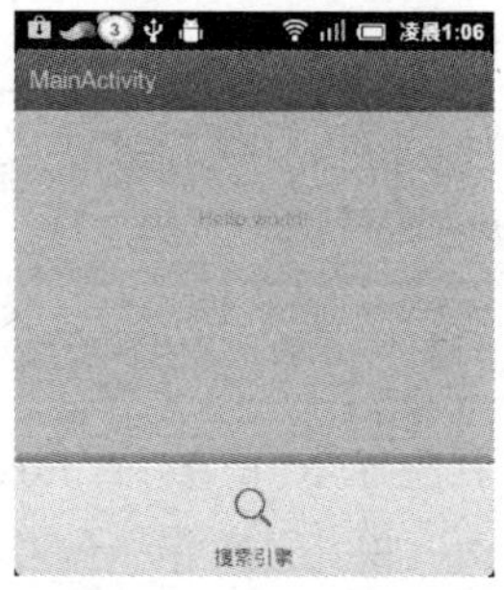

图 6-6　chapter6_3 按住 menu 键效果图

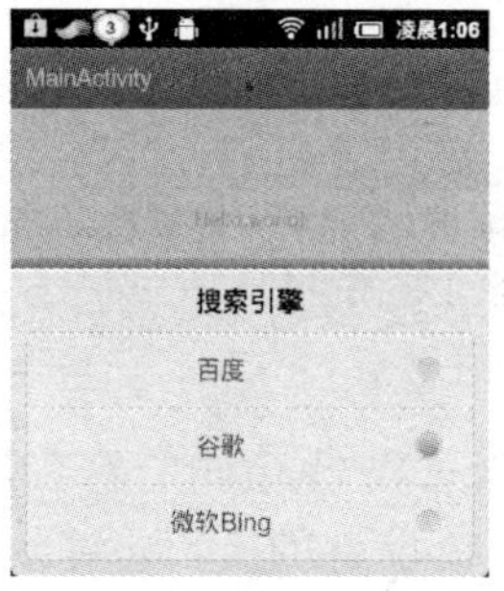

图 6-7　单击搜索引擎子菜单效果图

(4) 除了可以使用 Java 代码来生成子菜单之外，也可以采用 XML 文件的方式来定义子菜单。编辑 strings.xml 文件，源代码如下：

```
<resources>
    <string name = "app_name">chapter6_3</string>
    <string name = "hello_world">Hello world! </string>
    <string name = "menu_settings">Settings</string>
    <string name = "title_activity_main">MainActivity</string>
    <string name = "create_new">新建</string>
    <string name = "open">打开</string>
    <string name = "file">操作</string>
</resources>
```

(5) 在 res/menu 目录下面新建一个 sub_menu.xml 文件，编辑 sub_menu.xml 文件，源代码如下：

```
<? xml version = "1.0" encoding = "utf-8"? >
<menu xmlns:android = "http://schemas.android.com/apk/res/android">
    <item android:id = "@ + id/file"
          android:icon = "@android:drawable/ic_menu_view"
          android:title = "@string/file" >
        <!-- 两个子菜单项 -->
        <menu>
            <item android:id = "@ + id/create_new"
                  android:title = "@string/create_new" />
            <item android:id = "@ + id/open"
                  android:title = "@string/open" />
        </menu>
    </item>
</menu>
```

(6) 编辑 MainActivity.java 的 onCreateOptionsMenu()方法,源代码如下:

```
//创建选项菜单
    @Override
    public boolean onCreateOptionsMenu(Menu menu)
    {
    getMenuInflater().inflate(R.menu.sub_menu, menu);
    return true;
    }
```

(7) 右击工程 chapter6_3,在弹出的快捷菜单中选择 Run As|Android Application 命令,按住设备的 menu 键效果如图 6-8 所示,单击"操作"子菜单的效果如图 6-9 所示。

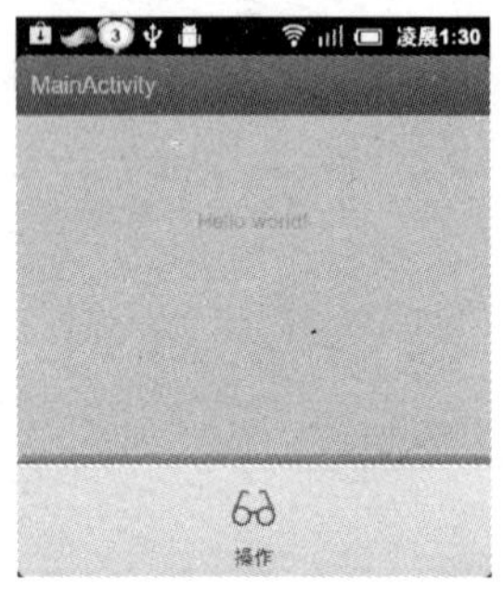

图 6-8 XML 定义子菜单按住 menu 键效果图

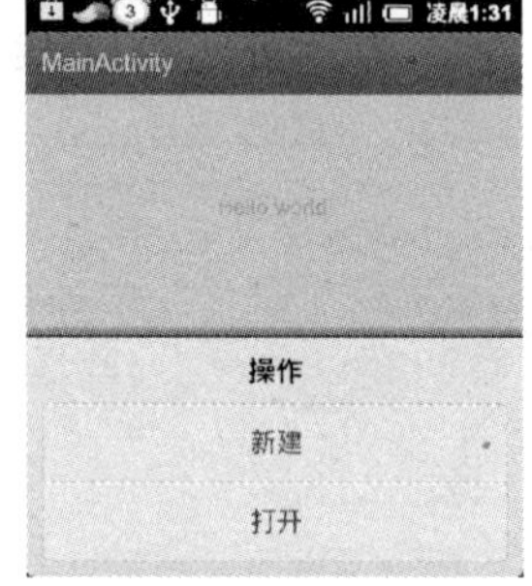

图 6-9 单击操作子菜单效果图

6.2 通 知

在消息通知时,笔者经常用会用到两个组件 Toast 和 Notification。特别是重要的和需要长时间显示的信息,用 Notification 就最合适不过了。当有消息通知时,状态栏会显示通知的图标和文字,通过下拉状态栏,就可以看到通知信息了,Android 这一创新性的组件赢得了用户的一致好评,就连苹果也开始模仿了。下面笔者就通过实例的方式来介绍一下在 Android 应用程序当中如何使用 Notification。

6.2.1 普通通知

下面这个实例,笔者将新建一个用户界面,界面上会放置两个按钮,当单击"发出通知"按钮的时候,会发出一个通知,而且这个通知会长时间地驻留在状态栏,单击"取消通知"按钮的时候,会将这个通知取消掉。

(1) 新建工程 chapter6_4,编辑 res/layout/activity_main.xml 文件,源代码如下:

```
<? xml version = "1.0" encoding = "utf-8"? >
```

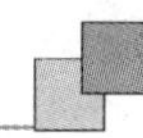

```
<LinearLayout xmlns:android = "http://schemas.android.com/apk/res/android"
    android:layout_width = "fill_parent"
    android:layout_height = "fill_parent"
    android:orientation = "vertical" >
    <!-- 发出通知按钮 -->
    <Button
        android:id = "@ + id/showButton"
        android:layout_width = "fill_parent"
        android:layout_height = "wrap_content"
        android:text = "showNotification" />
    <!-- 取消通知按钮 -->
    <Button
        android:id = "@ + id/cancelButton"
        android:layout_width = "fill_parent"
        android:layout_height = "wrap_content"
        android:text = "cancelNotification" />
</LinearLayout>
```

(2) 编辑 MainActivity.java 文件，源代码如下：

```
public class MainActivity extends Activity implements OnClickListener
{
    Button showButton;                                              //声明发出通知按钮
    Button cancelButton;                                            //声明取消通知按钮
    private NotificationManager mNotificationManager;              //声明通知管理器对象
    private Notification mNotification;                            //声明通知对象
    private final static int NOTIFICATION_ID = 0x0001;  //声明通知对象的标示 ID
    @Override
  public void onCreate(Bundle savedInstanceState)
    {
        super.onCreate(savedInstanceState);
        setContentView(R.layout.activity_main);
        showButton = (Button)findViewById(R.id.showButton);
                                                        //找到 Id 为 showButton 的按钮
        cancelButton = (Button)findViewById(R.id.cancelButton);
                                                      //找到 Id 为 cancelButton 的按钮
        showButton.setOnClickListener(this);  //绑定当前类为按钮的单击事件处理器
        cancelButton.setOnClickListener(this);
        //1.获取通知管理器对象
        mNotificationManager = (NotificationManager)getSystemService(NOTIFICATION_
SERVICE);
        //2.初始化通知对象
        mNotification = new Notification();
        //3.设置通知的图标，笔者使用系统资源图标
```

```
        mNotification.icon = android.R.drawable.ic_dialog_map;
        //4.设置通知在状态栏上的显示的文字
        mNotification.tickerText = "第一个通知";
        //5.把通知设置在"正在运行"栏目中,表示程序正在运行
        mNotification.flags = Notification.FLAG_ONGOING_EVENT;
        //6.设置通知显示的时间
        mNotification.when = System.currentTimeMillis();
        //7.设置通知时发出的默认的声音
        mNotification.defaults =  Notification.DEFAULT_SOUND;
    }
    @Override
    public void onClick(View v)
    {
        switch (v.getId())
        {
        case R.id.showButton: //发出通知按钮
            //设置通知的显示参数
            Uri uri = Uri.parse("tel:10000");
            Intent intent = new Intent(Intent.ACTION_DIAL, uri);
          PendingIntent pi = PendingIntent.getActivity(this, 0, intent, 0);
          //下拉状态栏时候,通知的显示参数
          mNotification.setLatestEventInfo(this, "电信宽带已经欠费", "请您尽快致
电 10000", pi);
          //发送通知
          mNotificationManager.notify(NOTIFICATION_ID, mNotification);
            break;
      case R.id.cancelButton: //取消通知按钮
          mNotificationManager.cancel(NOTIFICATION_ID);
          break;
      }
    }
    @Override
    public boolean onCreateOptionsMenu(Menu menu)
    {
        getMenuInflater().inflate(R.menu.activity_main, menu);
        return true;
    }
}
```

从上述代码当中,可以看出发出通知包括以下几个步骤:

① 获取 NotificationManager 对象,NotificationManager 对象有以下几个比较常用的方法:

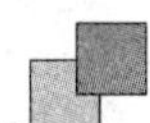

- cancel(int id)：取消以前显示的一个通知。
- cancelAll：取消以前所有的通知。
- notify(int id，Notification notification)：在状态栏上显示通知。

② 初始化要显示的 Notification 对象，其有以下几个比较常用的属性：

- icon：用来设置通知显示的图标。
- tickerText：用来设置通知显示在状态栏的文字信息。
- when：用来设置通知什么时刻在状态栏上显示出来。
- vibrate：用来设置通知在状态栏显示的时候，设备的震动模式。

③ 设置下拉状态栏的时候，通知的显示参数。使用 PendingIntent 来包装通知 Intent，使用 Notification 对象的 setLatestEventInfo()方法来设置通知的标题、通知内容等信息。

④ 发送通知。使用 NotificationManager 的 notify(int id，Notification notification)方法来发送通知。

(3) 右击工程 chapter6_4，在弹出的快捷菜单中选择 Run As|Android Application 命令，在显示的界面当中单击 showNotification 按钮运行效果如图 6－10 所示，下拉状态栏运行效果如图 6－11 所示。

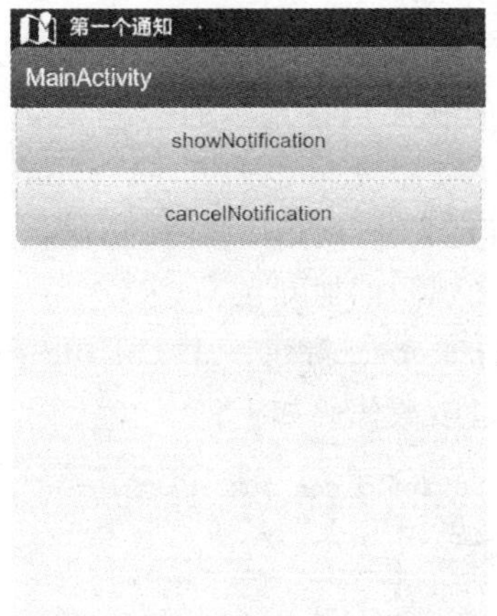

图 6－10　单击 showNotification 按钮运行效果图

图 6－11　下拉状态栏效果图

单击"电信宽带已经欠费"这条通知的效果如图 6－12 所示，单击 cancelNotification 按钮过后下拉状态栏的效果如图 6－13 所示。

图 6-12　单击通知条目的效果图

图 6-13　取消通知的效果图

6.2.2　自定义视图通知

默认情况下，下拉状态栏过后，通知窗口中的视图包括基本的标题和文本信息。这是由 Notification 对象的 setLastestInfo()方法设置的 contentTitle 和 contentText 参数指定的。不过可以使用 RemoteViews 来自定义一个展开视图的布局。下面笔者就通过实例的方式为读者介绍如何创建自定义视图的通知。

(1) 新建工程 chapter6_5，在 res/layout 目录下面新建一个 custom_notification.xml 文件，编辑 custom_notification.xml 文件，源代码如下：

```
<LinearLayout xmlns:android = "http://schemas.android.com/apk/res/android"
              android:orientation = "horizontal"
              android:layout_width = "fill_parent"
              android:layout_height = "fill_parent"
              android:padding = "3dp"
              >
   <!-- 自定义视图由 ImageView 和 TextView 水平排列组成 -->
  <ImageView android:id = "@ + id/image"
              android:layout_width = "wrap_content"
              android:layout_height = "fill_parent"
              android:layout_marginRight = "10dp"
              />
  <TextView android:id = "@ + id/text"
              android:layout_width = "wrap_content"
```

```
            android:layout_height = "fill_parent"
            android:textColor = "#000"
            />
</LinearLayout>
```

(2) 编辑 res/layout/main_activity.xml 文件，文件源代码与 chapter6_4 的 res/layout/activity_main.xml 文件源代码保持一致。

(3) 编辑 MainActivity.java 文件，源代码如下：

```
public class MainActivity extends Activity implements OnClickListener
{
    Button showButton;//声明发出通知按钮
    Button cancelButton;//声明取消通知按钮
    private NotificationManager mNotificationManager;      //声明通知管理器对象
    private Notification mNotification;                    //声明通知对象
    private final static int NOTIFICATION_ID = 0x0001;     //声明通知对象的标示 ID
    @Override
    public void onCreate(Bundle savedInstanceState)
    {
        super.onCreate(savedInstanceState);
        setContentView(R.layout.activity_main);
        showButton = (Button)findViewById(R.id.showButton);
                                                  //找到 Id 为 showButton 的按钮
        cancelButton = (Button)findViewById(R.id.cancelButton);
                                                  //找到 Id 为 cancelButton 的按钮
        showButton.setOnClickListener(this);  //绑定当前类为按钮的单击事件处理器
        cancelButton.setOnClickListener(this);
        //1.获取通知管理器对象
        mNotificationManager = (NotificationManager)getSystemService(NOTIFICATION_
SERVICE);
        //2.初始化通知对象
        mNotification = new Notification();
        //3.设置通知的图标,笔者使用系统资源图标
        mNotification.icon = android.R.drawable.ic_dialog_map;
        //4.设置通知在状态栏上的显示的文字
        mNotification.tickerText = "第一个通知";
        //5.把通知设置在"正在运行"栏目中,表示程序正在运行
        mNotification.flags = Notification.FLAG_ONGOING_EVENT;
        //6.设置通知显示的时间
        mNotification.when = System.currentTimeMillis();
        //7.设置通知时发出的默认的声音
        mNotification.defaults =  Notification.DEFAULT_SOUND;
```

```
        //8.先把程序的 package 名和布局资源 ID 传给 RemoteViews 的构造方法。
    //然后用 setImageViewResource()和 setTextViewText()定义 ImageView 和 TextView 的内容
    //分别把 View 对象的资源 ID、所赋的内容作为参数传入。最后,把 RemoteViews 对象
    //传给 Notification 的 contentView 属性
        RemoteViews contentView = new RemoteViews(getPackageName(), R.layout.custom
_notification);
        contentView.setImageViewResource(R.id.image, android.R.drawable.ic_dialog_
dialer);
        contentView.setTextViewText(R.id.text, "对不起,您的电话因余额不足而导致停机");
        mNotification.contentView = contentView;
        //下拉状态栏,单击通知条目的时候,发出的 Intent 设置
        Uri uri = Uri.parse("tel:10086");
        Intent contentIntent = new Intent(Intent.ACTION_DIAL, uri);
        PendingIntent pi = PendingIntent.getActivity(this, 0, contentIntent, 0);
        mNotification.contentIntent =  pi;
    }
    @Override
    public void onClick(View v)
    {
        switch (v.getId())
        {
        case R.id.showButton: //发出通知按钮
            //4.发送通知
            mNotificationManager.notify(NOTIFICATION_ID, mNotification);
              break;
          case R.id.cancelButton: //取消通知按钮
              mNotificationManager.cancel(NOTIFICATION_ID);
              break;
          }
    }
    @Override
    public boolean onCreateOptionsMenu(Menu menu)
    {
        getMenuInflater().inflate(R.menu.activity_main, menu);
        return true;
    }
}
```

注意:当建立一个自定义展开视图时,必须特别小心,保证自定义的布局能正常工作在不同的设备方向和分辨率下。这个建议对于所有在 Android 上创建的 View 布局都是适用的,但在这种情况下尤为重要,因为布局实际可用的屏幕区域非常有限。不要把自定义布局设计得过于复杂,并且一定要在各种环境配置下进行测试。

(4) 右击工程 chapter6_5，在弹出的快捷菜单中选择 Run As|Android Application 命令，单击界面上的 showNotification 按钮之后，会在状态栏上显示一个通知，下拉状态栏过后运行效果如图 6-14 所示，单击通知条目运行效果如图 6-15 所示。

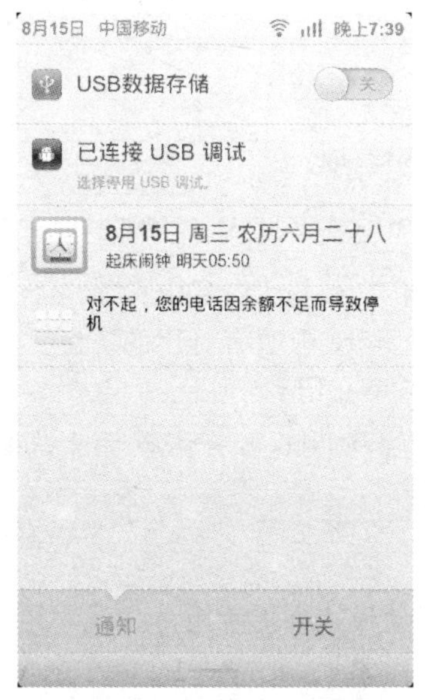

图 6-14　自定义视图通知

图 6-15　单击自定义视图通知效果

6.2.3　高级通知技术

通过前面两小节的介绍，相信读者对 Android 通知的使用应该有了一个比较清晰和具体的认识。下面笔者再为大家介绍一些比较具体的通知的高级使用技术。

1. 更新通知

应用程序可以在事件正在进行时更新状态栏的通知。比如，前一条短信还未读，可又来了一条新短信，短信程序为了正确显示未读短信的总数，可以更新已有的通知。此时，更新原有通知要比通过 NotificationManager 新增一条通知更合理些，因为这样可以有效地避免通知窗口的显示混乱。

因为 NotificationManager 对每个通知都用一个整数 ID 进行了唯一标识，新的通知内容可以用 setLatestEventInfo()方法方便地进行修改，然后再次调用 notify()显示出来。

2. 添加声音

可以使用默认提示音或者程序指定的声音来提醒用户。

要使用用户默认提示音，则需要给 defaults 属性添加 DEFAULT_SOUND，如以下代码：

```
mNotification.defaults |=  Notification.DEFAULT_SOUND;
```

要使用应用程序指定的声音来提醒用户的话，则不要传递一个 URI 对象给 Notification 对象的 sound 属性。如以下代码将保存在设备 SD 卡上面的音频文件作为提示音：

```
mNotification.sound = Uri.parse("file:///sdcard/ring.mp3");
```

下面代码从 MediaStore 类的 ContentProvider 中获取。

```
mNotification.sound = Uri.withAppendedPath(Audio.Media.INTERNAL_CONTENT_URI, "6");
```

这时，已知有资源 ID 为 6 的媒体文件，并且已添加到 Uri 内容中。如果不知道确切的 ID，则必须先用 ContentResolver 查询 MediaStore 中所有可用的资源。关于使用 ContentResolver 的详细信息笔者将在后面章节中介绍。

注意：如果 defaults 属性包含了 DEFAULT_SOUND，则默认提示音将覆盖 sound 属性里指定的声音。

3. 添加振动

可以用默认振动模式或者是程序当中指定的振动模式来提醒用户。

如果要用默认振动模式，则需要给 defaults 属性添加 DEFAULT_VIBRATE，如以下代码：

```
mNotification.defaults |=  Notification.DEFAULT_VIBRATE;
```

如果要使用自定义的振动模式，则需要给 vibrate 属性传递一个 long 类型的数组。如以下代码：

```
long[] vibrate = {0,100,200,300};
mNotification.vibrate = vibrate;
```

长整型数组定义了震动开和关交替的时间(毫秒)。第一个数是开始振动前的等待时间(震动关闭)，第二个数是第一次开启振动的持续时间，第三个数是下一次关闭时间，如此类推。振动模式的持续时间没有限制，但不能设置为重复振动。

注意：如果 defaults 属性包含了 DEFAULT_VIBRATE，则默认的震动模式将会覆盖 vibrate 属性里指定的模式。

4. 添加闪光

如果想要用 LED 闪光来提醒用户，可以执行默认闪光模式(如果可用的话)，也可以使用自定义的闪光的颜色和模式。

如果要使用默认的闪光设置，则需要给 defaults 属性添加 DEFAULT_LIGHTS，如以下代码：

```
mNotification.defaults |=  Notification.DEFAULT_LIGHTS;
```

如果要使用自定义的颜色和模式，则必须指定 ledARGB 属性（指颜色）、ledoffMS 属性（闪光关闭毫秒数）、ledOnMS 属性（闪光开启毫秒数），并在 flags 属性里加入 FLAG_SHOW_LIGHTS，如以下代码：

```
mNotification.ledARGB = 0xff00ff00;
mNotification.ledOnMS = 300;
mNotification.ledOffMS = 1000;
mNotification.flags |= Notification.FLAG_SHOW_LIGHTS;
```

上述代码实现了绿色光闪烁 300 ms 间歇 1 s 的闪光。每个设备的 LED 灯不可能支持所有颜色的发光，不同的设备所能支持的颜色也各不相同，因此硬件将按照最接近的颜色来发光。绿色是最常见的提醒色。

5. 其他特性

利用 Notification 的属性和标志位，可以给通知添加更多的特性。

下面列出了其中一些常用的特性：

- "FLAG_AUTO_CANCEL"标志：在 flags 属性中增加此标志，则在通知条目被单击后自动取消通知。
- "FLAG_INSISTENT"标志：在 flags 属性中增加此标志，则在用户响应前一直循环播放声音。
- "FLAG_ONGOING_EVENT"标志：在 flags 属性中增加此标志，则将通知放入通知窗口的"正在运行"（Ongoing）组中。表示应用程序正在运行，进程仍在后台运行，即使应用程序不可见（比如播放音乐或接听电话）。
- "FLAG_NO_CLEAR"标志：在 flags 属性中增加此标志，表示通知不允许被"清除通知"按钮清除。
- number 属性：表示通知所代表的事件数量。此数字显示在状态栏图标上。要利用此属性，必须在第一次创建通知时设为 1。（如果只是在更新通知时才把此值从 0 改成任意大于 0 的数，则数字不会显示出来）。

6.3 闹钟服务

Android 的闹钟服务是允许程序在指定的时间执行某项任务的一种服务。下面笔者通过实例的方式来介绍如何在 Android 应用程序当中使用 Android SDK 所提供的闹钟服务。

(1) 新建工程 chapter6_6，编辑 res/layout/activity_main.xml 文件，源代码如下：

```
<?xml version="1.0" encoding="utf-8"?>
<LinearLayout xmlns:android="http://schemas.android.com/apk/res/android"
```

```
    android:layout_width = "fill_parent"
    android:layout_height = "fill_parent"
    android:orientation = "vertical" >
    <!--  -->
    <CheckBox
        android:id = "@ + id/checkBox1_alarm"
        android:layout_width = "wrap_content"
        android:layout_height = "wrap_content"
        android:text = "是否重复响铃" >
    </CheckBox>
  <!-- 单击此按钮弹出时间设置选择框 -->
    <Button
        android:id = "@ + id/button_alarm1"
        android:layout_width = "wrap_content"
        android:layout_height = "wrap_content"
        android:text = "定时" >
    </Button>
</LinearLayout>
```

(2) 新建 MyAlarmBroadCast 广播接收者，编辑 MyAlarmBroadCast. java 文件，源代码如下：

```
public class MyAlarmBroadCast extends BroadcastReceiver
{
    private static final String TAG = "MyAlarmBroadCast";
    @Override
    public void onReceive(Context context, Intent intent)
    {
        Log.d(TAG, "收到广播");
    }
}
```

(3) 由于应用程序当中添加了广播接收者，所以需要在 AndroidManifest. xml 文件当中进行相应的注册，编辑 AndroidManifest. xml 文件，源代码如下：

```
<manifest xmlns:android = "http://schemas.android.com/apk/res/android"
    package = "com.example.chapter6_6"
    android:versionCode = "1"
    android:versionName = "1.0" >
    <uses-sdk
        android:minSdkVersion = "8"
        android:targetSdkVersion = "15" />
    <application
        android:icon = "@drawable/ic_launcher"
```

```
        android:label = "@string/app_name"
        android:theme = "@style/AppTheme" >
        <activity
            android:name = ".MainActivity"
            android:label = "@string/title_activity_main" >
            <intent-filter>
                <action android:name = "android.intent.action.MAIN" />
                <category android:name = "android.intent.category.LAUNCHER" />
            </intent-filter>
        </activity>
        <!-- 广播接受者注册 -->
      <receiver android:name = ".MyAlarmBroadCast"/>
    </application>
</manifest>
```

（4）编辑 MainActivity.java 文件，源代码如下：

```
public class MainActivity extends
Activity implements OnClickListener,OnTimeSetListener,OnCheckedChangeListener
{
    Calendar calendar  =  Calendar.getInstance();//代表当前时间的日历
    Button button1 ;
    CheckBox checkBox;
    boolean flag  =  true ;//是否只执行一次
    @Override
    protected void onCreate(Bundle savedInstanceState)
    {
        super.onCreate(savedInstanceState);
        setContentView(R.layout.activity_main);
        button1  =  (Button) findViewById(R.id.button_alarm1);
                                              //找到 id 为 button_alarm1 的按钮控件
        checkBox  =  (CheckBox) findViewById(R.id.checkBox1_alarm);
                                        //找到 id 为 checkBox1_alarm 的复选按钮控件
        checkBox.setOnCheckedChangeListener(this);
                                   //将当前 Activity 作为复选框选择改变事件的监听器
        button1.setOnClickListener(this);//将当前 Activity 作为按钮单击事件的监听器
    }
    @Override
    public void onClick(View v)
    {
        if(v == button1)
        {   //如果是 id 为 button_alarm1 的按钮被单击
            calendar.setTimeInMillis(System.currentTimeMillis());
            new TimePickerDialog(MainActivity.this, this,
```

```
                                    calendar.get(Calendar.HOUR_OF_DAY),
                                    calendar.get(Calendar.MINUTE), true).show();
                }
            }
            //当事件选择框置了新时间时触发
            @Override
            public void onTimeSet(TimePicker view, int hourOfDay, int minute)
            {
                //将时间设置为定时的时间
                calendar.set(Calendar.HOUR_OF_DAY, hourOfDay);
                calendar.set(Calendar.MINUTE, minute);
                Intent intent = new Intent(MainActivity.this,MyAlarmBroadCast.class);
                PendingIntent pendingIntent = PendingIntent.getBroadcast(getApplicationCon-
text(), 0, intent, 0);
                //获取全局定时器的服务管理器
                AlarmManager alarmManager = (AlarmManager) getSystemService(ALARM_SERVICE);
                if(flag)
                { //如果只执行一次的情况
                /**
                 * 指定的任务只会执行一次,如果该 pendingIntent 指定的任务已经被执行过了,
                   那么该方法直接会被 cancel 掉。
                 *  set(int type, long triggerAtTime, PendingIntent operation)。
                 *  type:指定定时模式。
                 *  triggerAtTime:触发任务的时间。该参数和定时模式息息相关。
                 *  operation:该参数指定一个广播 Intent,当时间到了时,系统会广播里面的
                    intent,触发相应的广播接收者执行某些操作,比如响铃……。
                 */
                alarmManager.set(AlarmManager.RTC_WAKEUP, calendar.getTimeInMillis(), pend-
ingIntent);
                }
                else
                {//不是执行一次的情况
                    /**
                     * 通过该方法指定的任务会一直间隔执行,第 3 个参数就指定了执行的时间
                       间隔。
                     * 如果我们想取消的话,请使用:alarmManager.cancel(pendingIntent);。
                     * 注意,这里的 pendingIntent 要和 setRepeating 方法中的一致。
                     */
                    alarmManager.setRepeating(AlarmManager.RTC_WAKEUP, calendar.getTimeIn-
Millis(), 5 * 1000, pendingIntent);
                }
            }
            @Override
            public void onCheckedChanged(CompoundButton buttonView, boolean isChecked)
```

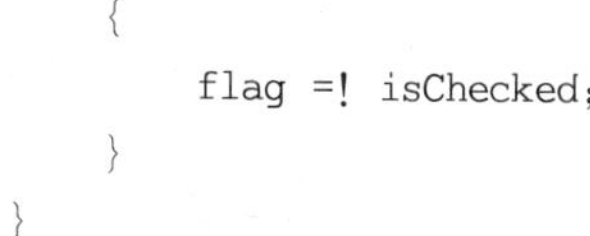

```
        {
            flag =! isChecked;
        }
    }
```

(5) 右击工程 chapter6_6，在弹出的快捷菜单中选择 Run As|Android Application 命令，在界面当中单击"定时"按钮设置任务执行的时间的效果如图 6-16 所示。

(6) 当时间到了 21:34 的时候，笔者定义的 MyAlarmBroadCast 广播接收者收到了相应的广播，LogCat 日志信息如图 6-17 所示。

图 6-16　设置定时任务执行时间的效果图

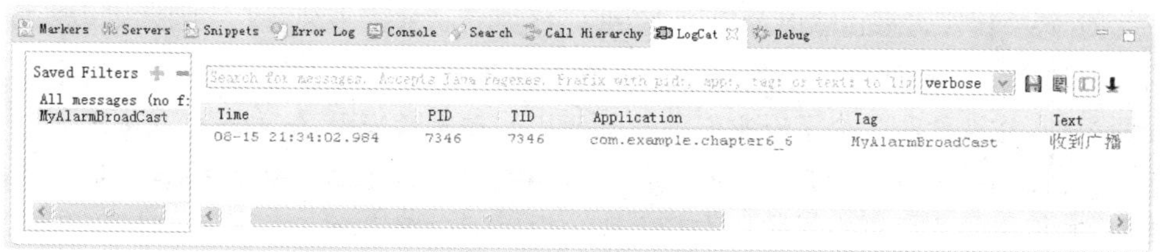

图 6-17　广播收到单次执行的任务运行效果图

(8) 选中"是否重复响铃"的复选框，且将任务执行时间设置为 21:45 后的 LogCat 日志效果如图 6-18 所示。

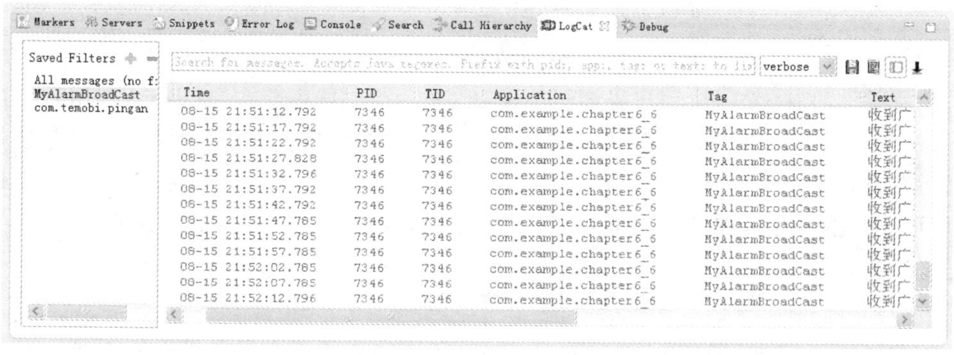

图 6-18　广播收到多次循环执行的任务的效果图

第 7 章

Android 事件处理

对于一个 Android 应用程序来说，事件处理是必不可少的，用户与应用程序之间的交互便是通过事件处理来完成的。在前面的章节当中，读者已经学习了很多用户界面组件，并且已经学会了对它们的熟练使用。在本章当中，笔者将要和读者一起去探索如何使用 Android 的事件处理，从而当相关用户界面事件发生之后，根据具体需要，完成相应的业务逻辑操作。本章主要包含以下内容：

- Android 事件处理概括介绍；
- 监听和处理用户单击事件；
- 监听和处理键盘事件。

7.1 Android 事件处理概述

无论是在 Java 图形用户界面程序的开发过程当中，还是在 Android 应用程序的开发过程当中，用户都是通过事件驱动的方式与设备进行交互的，这些事件都是通过各种各样的输入设备产生的，比如：触摸屏、键盘等。Android 平台提供了两种事件处理的机制，分别是基于监听器的事件处理机制以及基于回调的事件处理机制。

7.1.1 基于监听器的事件处理机制

对于基于监听器的事件处理而言，主要就是为 Android 用户界面组件绑定特定的事件监听器。

相比于基于回调的事件处理，这是更具"面向对象"性质的事件处理方式。在监听器模型中，主要涉及 3 类对象：

- 事件源(Event Source)：产生事件的来源，通常是指各种用户界面组件，比如按钮、文本框等。
- 事件(Event)：事件封装了用户界面组件上发生的特定事件的具体信息，如果

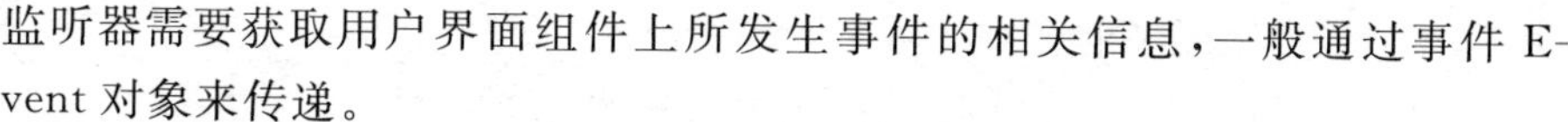

监听器需要获取用户界面组件上所发生事件的相关信息，一般通过事件 Event 对象来传递。

- 事件监听器(Event Listener)：负责监听事件源发生的事件，用来负责对不同的事件做出相应的处理。

基于监听器的事件处理机制是一种委托式的事件处理方式，事件源将整个事件委托给事件监听器，由事件监听器对事件进行响应处理。这种处理方式将事件源和事件监听器分离开来，有利于增强程序的可维护性，减少程序的耦合度。

下面举例说明：

View 类中的 OnLongClickListener 监听器定义如下(不需要传递事件)：

```
public interface OnLongClickListener
{
    boolean onLongClick(View v);
}
```

View 类中的 OnTouchListener 监听器定义如下(需要传递事件 MotionEvent)：

```
public interface OnTouchListener
{
    boolean onTouch(View v, MotionEvent event);
}
```

7.1.2　基于回调的事件处理机制

对于基于回调的事件处理而言，主要做法是重写 Android 用户界面组件特定的回调函数，Android 大部分用户界面组件都提供了事件响应的回调函数，开发者重写这些方法就可以了。

相比基于监听器的事件处理模型，基于回调的事件处理模型要简单些，该模型中，事件源和事件监听器是同一个对象，也就是说没有独立的事件监听器。当用户在 GUI 组件上触发某事件时，由该组件自身特定的函数负责处理该事件。通常通过重写 Override 用户界面组件的事件处理函数实现事件的处理。

下面举例说明：

View 类实现了 KeyEvent. Callback 接口中的一系列回调函数，因此，基于回调的事件处理机制通过自定义 View 来实现，自定义 View 时重写这些事件处理方法即可。

```
public interface Callback {
    //几乎所有基于回调的事件处理函数都会返回一个 boolean 类型值，该返回值用于
    //标识该处理函数是否能完全处理该事件
    //返回 true，表明该函数已完全处理该事件，该事件不会继续传播出去
    //返回 false，表明该函数未完全处理该事件，该事件会继续传播出去
```

```
        boolean onKeyDown(int keyCode, KeyEvent event);
        boolean onKeyLongPress(int keyCode, KeyEvent event);
        boolean onKeyUp(int keyCode, KeyEvent event);
        boolean onKeyMultiple(int keyCode, int count, KeyEvent event);
    }
```

综上所述，基于监听器的事件模型符合单一职责原则，事件源和事件监听器分开实现。Android 的事件处理机制保证基于监听器的事件处理会优先于基于回调的事件处理被触发。某些特定情况下，基于回调的事件处理机制会更好地提高程序的内聚性。

7.2 监听和处理用户单击事件

用户界面组件的单击事件是 Android 应用程序当中最常用到的事件类型。下面通过实例的方式为读者来介绍如何通过绑定监听器的方式来监听和处理用户界面组件的单击事件。可以采用匿名内部类、内部类或者是 Activity 本身来作为事件监听器类，下面一一介绍。

7.2.1 匿名内部类作为事件监听器类

下面的实例采用匿名内部类作为按钮控件的单击事件监听器类。

(1) 新建工程 chapter7_1，编辑 res/layout/activity_main.xml 文件，源代码参考本书配套资源。

(2) 编辑 MainActivity.java 文件，源代码如下：

```
public class MainActivity extends Activity
{
    TextView mTextView;//声明文本控件
    Button mMyButton;//声明按钮控件
    @Override
    public void onCreate(Bundle savedInstanceState)
    {
        super.onCreate(savedInstanceState);
        setContentView(R.layout.activity_main);
        mTextView = (TextView)findViewById(R.id.my_textview);//找到文本控件
        mMyButton = (Button)findViewById(R.id.my_button);//找到按钮控件
        //采用匿名内部类的方式作为按钮单击事件的监听器
        mMyButton.setOnClickListener(new OnClickListener()
        {
            @Override
            public void onClick(View v)
```

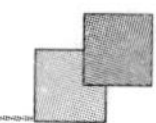

```
            {//当用户单击按钮时,触发 onClick 方法
                mTextView.setText("Button has been clicked");     //设置文本内容
            }
        });
    }
    @Override
    public boolean onCreateOptionsMenu(Menu menu)
    {
        getMenuInflater().inflate(R.menu.activity_main, menu);
        return true;
    }
}
```

在 Android 应用程序开发的过程当中,大多数的事件监听器都是以匿名内部类的形式存在的,这是由于事件监听器被复用的概率非常小,几乎可以忽略不计。一般情况下,声明的事件监听器只与某一个单独的用户界面组件事件相绑定,所以笔者在这里也建议读者使用匿名内部类作为事件监听器。

(3) 右击工程 chapter7_1,在弹出的快捷菜单中选择 Run As|Android Application 命令,运行效果如图 7－1 所示,单击 Hello World 按钮,运行效果如图 7－2 所示。

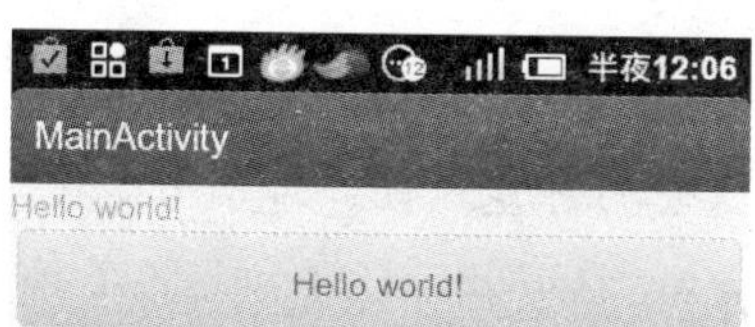

图 7－1　chapter7_1 运行效果图

图 7－2　单击按钮运行效果图

7.2.2　内部类作为事件监听器类

上面章节介绍了如何使用匿名内部类作为按钮控件的单击事件监听器类,下面通过代码来介绍如何使用内部类作为按钮控件的单击事件监听器类。

(1) 编辑 chapter7_1 工程的 MainActivity.java 文件,源代码如下:

```
public class MainActivity extends Activity
{
    TextView mTextView;//声明文本控件
    Button mMyButton;//声明按钮控件
    @Override
    public void onCreate(Bundle savedInstanceState)
    {
        super.onCreate(savedInstanceState);
```

```
        setContentView(R.layout.activity_main);
        mTextView = (TextView)findViewById(R.id.my_textview);//找到文本控件
        mMyButton = (Button)findViewById(R.id.my_button);//找到按钮控件
        mMyButton.setOnClickListener(new MyButtonClickListener());//为按钮绑定单击
                                                                  //事件监听器
    }
    //采用内部类方式作为按钮单击事件的监听器
    private final class MyButtonClickListener implements View.OnClickListener
    {
    @Override
    public void onClick(View v)
    {//当用户单击按钮时,触发 onClick 方法
        mTextView.setText("Button has been clicked");   //设置文本内容
    }
    }
    @Override
    public boolean onCreateOptionsMenu(Menu menu)
    {
        getMenuInflater().inflate(R.menu.activity_main, menu);
        return true;
    }
}
```

采用内部类作为事件监听器的一个好处就是它的可复用性。对于上述代码当中建立的 MyButtonClickListener 内部类实现了 View.OnClickListener 这个接口,这个内部类不仅可以与 id 为 my_button 的按钮控件单击事件相绑定,也可以与任何一个用户界面组件的单击事件相绑定。

(2) 运行效果与采用匿名内部类方式的运行效果一致。

7.2.3 Activity 本身作为事件监听器类

事件监听器不仅可以采用匿名内部类或者是内部类的形式,还可以使用 Activity 本身作为用户界面组件的事件监听器类。

(1) 编辑 chapter7_1 工程的 MainActivity.java 文件,源代码如下:

```
public class MainActivity extends Activity implements View.OnClickListener
{
    TextView mTextView;//声明文本控件
    Button mMyButton;//声明按钮控件
    @Override
    public void onCreate(Bundle savedInstanceState)
    {
```

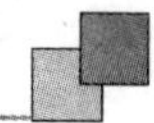

```
        super.onCreate(savedInstanceState);
        setContentView(R.layout.activity_main);
        mTextView = (TextView)findViewById(R.id.my_textview);//找到文本控件
        mMyButton = (Button)findViewById(R.id.my_button);//找到按钮控件
        mMyButton.setOnClickListener(this);//采用 Activity 本身作为按钮单击事件监
                                                   //听器
    }
    @Override
    public void onClick(View v)
    { //当用户单击按钮时,触发 onClick 方法
       mTextView.setText("Button has been clicked");  //设置文本内容
    }
    @Override
    public boolean onCreateOptionsMenu(Menu menu)
    {
        getMenuInflater().inflate(R.menu.activity_main, menu);
        return true;
    }
}
```

(2) 运行效果与采用匿名内部类以及内部类的方式一致。

7.3　监听和处理键盘事件

接下来读者将要熟悉 Android 应用程序当中键盘事件的监听和处理操作，主要是键盘按下(onKeyDown)事件和键盘释放(onKeyUp)事件，这些事件在游戏应用程序编写过程当中显得比较重要。

7.3.1　监听处理 onKeyDown 事件

下面通过实例为读者介绍如何处理键盘按下(onKeyDown)事件。

(1) 新建工程 chapter7_2，编辑 MainActivity.java 文件，源代码如下：

```
public class MainActivity extends Activity
{
    @Override
    public void onCreate(Bundle savedInstanceState)
    {
        super.onCreate(savedInstanceState);
        setContentView(R.layout.activity_main);
    }
    @Override
    public boolean onKeyDown(int keyCode, KeyEvent event) {//覆盖 Activity 基类的键盘
```

```
                                                        //按下事件
    if(KeyEvent.KEYCODE_BACK == keyCode)
    {//如果用户按下返回键
        new AlertDialog.Builder(this).setTitle("确认退出吗?").setPositiveButton("
确定", new DialogInterface.OnClickListener()
            {
                @Override
                public void onClick(DialogInterface dialog, int which)
                {
                    System.exit(0);//退出应用程序
                }
            }).setNegativeButton("取消", new DialogInterface.OnClickListener()
            {
                @Override
                public void onClick(DialogInterface dialog, int which)
                {
                    dialog.dismiss();//对话框消失
                }
            }).show();
        return true;//事件不再继续往下传递
    }
    else if(KeyEvent.KEYCODE_SEARCH == keyCode)
    {//如果用户按下搜索键
        Toast.makeText(this, "搜索键被按下", Toast.LENGTH_LONG).show();
        return true;
    }
    else if(KeyEvent.KEYCODE_MENU == keyCode)
    {//如果用户按下菜单键
        Toast.makeText(this, "菜单键被按下", Toast.LENGTH_LONG).show();
        return true;
    }
    return super.onKeyDown(keyCode, event);
    }
    @Override
    public boolean onCreateOptionsMenu(Menu menu)
    {
        getMenuInflater().inflate(R.menu.activity_main, menu);
        return true;
    }
}
```

在上述实例当中，笔者重写了 Activity 基类的 onKeyDown(int keyCode, KeyEvent event)方法，当然如果为相应用户界面组件添加 onKeyListener 的监听器，也可以对键盘按下的事件进行监听和处理，但是如果用户界面组件没有获得用户的

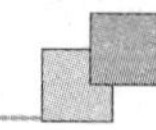

焦点，就不能对按下事件进行拦截和处理，所以还是重写 Activity 基类的 onKeyDown()方法会比较好一点。

onKeyDown(int keyCode, KeyEvent event)方法有两个参数。参数 keyCode，该参数表示被按下的键值的键盘码，手机键盘中每个按钮都会有其单独的键盘码来作为唯一的标识，应用程序都是通过键盘码才知道用户按下的是哪个键。参数 event，该参数表示按键事件的对象，其中包含了触发事件的详细信息，例如事件的状态、事件的类型、事件发生的时间等。当用户按下按键时，系统会自动将事件封装成 KeyEvent 对象供应用程序使用。

该方法的返回值为一个 boolean 类型的变量，当返回 true 时，表示已经完整地处理了这个事件，事件不再往下传递，而当返回 false 时，表示并没有完全处理完该事件，事件会被传递给其他的事件函数继续处理。

笔者在 onKeyDown()方法当中，对用户的按键进行了判断，如果用户按下的是搜索键或者是菜单键，界面上会给出一个相应的提示，如果用户按下的是返回键，用户界面上会弹出一个对话框，对话框当中有两个按钮，当用户单击“确定”按钮的时候，应用程序会退出，当用户单击“取消”按钮的时候，弹出的对话框会消失。

(2) 右击工程 chapter7_2，在弹出的快捷菜单中选择 Run As|Android Application 命令，当用户按下返回键的时候，运行效果如图 7-3 所示，当用户按下搜索键的时候，运行效果如图 7-4 所示。

图 7-3 onKeyDown 返回键按下时的运行效果图

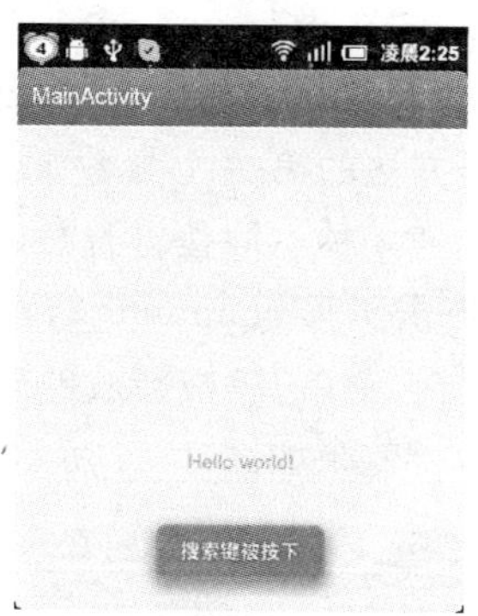

图 7-4 onKeyDown 搜索键按下时的运行效果图

7.3.2 监听处理 onKeyUp 事件

下面通过实例为读者介绍如何处理键盘释放(onKeyUp)事件。

(1) 编辑 chapter7_2 工程的 MainActivity.java 文件，重写 Activity 基类的 onKeyUp()方法，源代码如下：

```
@Override
public boolean onKeyUp(int keyCode, KeyEvent event)
```

```
{//覆盖 Activity 基类的 onKeyUp 键盘释放事件
if(KeyEvent.KEYCODE_BACK == keyCode)
{//如果返回键被释放
        Toast.makeText(this, "返回键被释放", Toast.LENGTH_LONG).show();
        return true;//事件被当前业务逻辑处理,不再继续往下传递
}
else if(KeyEvent.KEYCODE_MENU == keyCode)
{//如果菜单键被释放
        Toast.makeText(this, "菜单键被释放", Toast.LENGTH_LONG).show();
        return true;
}
else if(KeyEvent.KEYCODE_SEARCH == keyCode)
{//如果搜索键被释放
        Toast.makeText(this, "搜索键被释放", Toast.LENGTH_LONG).show();
        return true;
}
return super.onKeyUp(keyCode, event);
}
```

上述代码当中对 Activity 基类的 onKeyUp(int keyCode, KeyEvent event)方法进行了重写,用来对键盘释放的事件进行捕获和相应的处理,onKeyUp 方法的两个参数所表示的意思与 onKeyDown(int keyCode, KeyEvent event)方法的参数的意思是一致的,笔者在这里就不再赘述。笔者在 onKeyUp()方法中添加了相应的代码逻辑,根据方法的第一个参数 keyCode 对返回按键、菜单按键及搜索按键进行相应判断,且在这 3 种不同按键释放时对客户进行相应提示。

(2) 右击工程 chapter7_2,在弹出的快捷菜单中选择 Run As|Android Application 命令,当“菜单”键被释放的时候,运行效果如图 7－5 所示,当“搜索”按键被释放的时候,运行效果如图 7－6 所示。

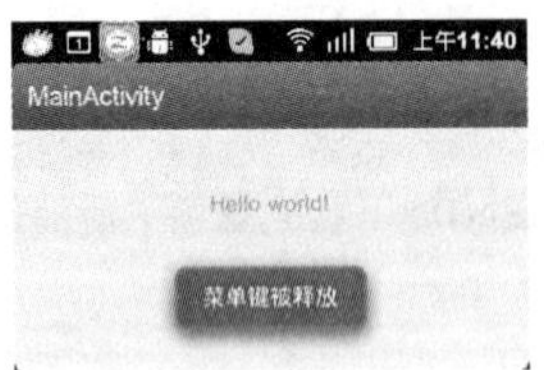

图 7－5 onKeyUp 菜单键释放时的运行效果图

图 7－6 onKeyUp 搜索键被释放时的运行效果图

7.4 自定义监听器

在实际项目开发中,经常需要自定义监听器来实现自定义业务流程的处理,而且

一般都不是基于 GUI 界面作为事件源的。这里以常见的 app 自动更新为例进行说明,在自动更新过程中,会存在两个状态:下载中和下载完成,而应用程序需要在这两个状态做不同的事情,"下载中"需要在 UI 界面上实时显示软件包下载的进度,"下载完成"后,取消进度条的显示。这里笔者通过实例来模拟这个过程,重点说明自定义监听器的事件处理流程。

(1) 新建工程 chapter7_3,编辑 res/layout/activity_main.xml 文件,源代码参考配套资料。

(2) 新建自定义事件监听器接口 DownloadListener.java 文件,源代码如下:

```
public interface DownloadListener
{
    public void onDownloading(int progress);//下载过程中的处理函数
    public void onDownloaded();//下载完成的处理函数
}
```

(3) 新建下载工具类 DownloadUtils.java 文件,源代码如下:

```
public final class DownloadUtils
{
    //单例模式
    private static DownloadUtils instance = new DownloadUtils();
    private boolean isDownloading = true;//是否正在下载
    private int progress = 0;//下载进度
    private DownloadUtils(){}
    public static DownloadUtils getInstance()
    {
        return instance;
    }
    public void download(DownloadListener listener)
    {
            while(isDownloading)//如果正在下载
              {
                listener.onDownloading(progress);   //下载进度发生变化,调用
                                                    //onDownloading()方法
                SystemClock.sleep(1000);//线程休眠 1 s
                progress + = 10;
                if(progress > 100)
                {//如果进度条的值大于 100
                    isDownloading = false;//下载完毕
                }
            }
            listener.onDownloaded();//下载完成,调用 onDownloaded()方法
```

```
    }
}
```

(4) 编辑 MainActivity.java 文件,源代码如下:

```
public class MainActivity extends Activity
{
    private static final String TAG = "MainActivity";
    Button downloadBtn;//声明下载按钮
    @Override
    public void onCreate(Bundle savedInstanceState)
    {
        super.onCreate(savedInstanceState);
        setContentView(R.layout.activity_main);
        downloadBtn = (Button)findViewById(R.id.download_btn);//找到 id 为 download
                                                               //_btn 的按钮
        downloadBtn.setOnClickListener(new OnClickListener()
        {//匿名内部类形式监听器
            @Override
            public void onClick(View v)
            {
                DownloadUtils.getInstance().download(new MyDownloadListener());
            }
        });
    }
    //内部类形式使用自定义监听器
    private final class MyDownloadListener implements DownloadListener
    {
    @Override
    public void onDownloading(int progress)
    {//下载中
        Log.d(TAG, "正在下载,进度:" + progress);
    }
    @Override
    public void onDownloaded()
    {//下载完成
        Log.d(TAG, "下载完成");
    }
}
@Override
public boolean onCreateOptionsMenu(Menu menu)
{
    getMenuInflater().inflate(R.menu.activity_main, menu);
```

```
            return true;
        }
    }
```

（5）右击工程 chapter7_3，在弹出的快捷菜单中选择 Run As|Android Application 命令，LogCat 的日志信息如图 7－7 所示。

Markers　Properties　Servers　Data Source Explorer　Snippets　Console　Search　LogCat

Saved Filters　All messages (no　MainActivity

Search for messages. Accepts Java regexes. Prefix with pid:, app:, tag: or text:　verbose

Time	PID	TID	Application	Tag	Text
08-20 15:00:13.863	561	561	com.example.chapter7_3	MainActivity	正在下载，进度：0
08-20 15:00:14.863	561	561	com.example.chapter7_3	MainActivity	正在下载，进度：10
08-20 15:00:15.863	561	561	com.example.chapter7_3	MainActivity	正在下载，进度：20
08-20 15:00:16.867	561	561	com.example.chapter7_3	MainActivity	正在下载，进度：30
08-20 15:00:17.867	561	561	com.example.chapter7_3	MainActivity	正在下载，进度：40
08-20 15:00:18.867	561	561	com.example.chapter7_3	MainActivity	正在下载，进度：50
08-20 15:00:19.867	561	561	com.example.chapter7_3	MainActivity	正在下载，进度：60
08-20 15:00:20.871	561	561	com.example.chapter7_3	MainActivity	正在下载，进度：70
08-20 15:00:21.871	561	561	com.example.chapter7_3	MainActivity	正在下载，进度：80
08-20 15:00:22.871	561	561	com.example.chapter7_3	MainActivity	正在下载，进度：90
08-20 15:00:23.875	561	561	com.example.chapter7_3	MainActivity	正在下载，进度：100
08-20 15:00:24.875	561	561	com.example.chapter7_3	MainActivity	下载完成

图 7－7　自定义事件监听器 Logcat 日志

7.5　基于回调的事件处理

前面的章节当中，笔者已经介绍了如何使用基于监听器的事件处理机制，并通过实例向读者详细介绍了如何使用监听器来监听和处理用户单击的事件以及键盘的按下和释放事件。本节将介绍如何使用基于回调的事件处理机制。

7.5.1　创建自定义视图

由于基于回调的事件处理机制，是当用户在 GUI 组件上触发某事件时，由该组件自身特定的函数负责处理该事件。因此，基于回调的事件处理机制通过自定义 View 来实现，自定义 View 时重写这些事件处理方法即可。

View 类是 Android 的一个超类，这个类几乎包含了所有的屏幕类型。每一个 View 都有一个用于绘图的画布，这个画布可以进行任意扩展。在游戏开发中往往需要自定义视图（View），这个画布的功能更能满足游戏开发中的需要。在 Android 中，任何一个 View 类都只需重写 onDraw 方法来实现界面显示，自定义的视图可以是复杂的 3D 实现，也可以是非常简单的文本形式等。下面通过实例的方法来向读者介绍如何创建自定义视图。

（1）新建工程 chapter7_4，新建一个继承自 android. view. View 的自定义视图类 MyView，编辑 MyView. java 文件，源代码如下：

```
public class MyView extends View
```

```
{
    public MyView(Context context)
    {
        super(context);
    }
    public MyView(Context context,AttributeSet attr)
    {
        super(context,attr);
    }
    private Paint myPaint;   //声明画笔对象
    private static final String myString1 = "中国移动互联网行业各年度投资情况";
    private static final String myString2 = "来源:清科研究中心 2011.08";
    @Override
    protected void onDraw(Canvas canvas)
    {
        //TODO Auto-generated method stub
        super.onDraw(canvas);
        myPaint = new Paint();
        //绘制标题
        myPaint.setColor(Color.BLACK); //设置画笔颜色
        myPaint.setTextSize(18);//设置文字大小
        canvas.drawText(myString1, 20, 20, myPaint);
        //绘制坐标轴
        canvas.drawLine(50, 100, 50, 500, myPaint);//纵坐标轴
        canvas.drawLine(50, 500, 400, 500, myPaint);//横坐标轴
        int[] array1 = new int[]{0, 50, 100, 150, 200, 250, 300, 350};
        //绘制纵坐标刻度
        myPaint.setTextSize(10);//设置文字大小
        canvas.drawText("单位:百万美元", 20, 90, myPaint);
        for (int i = 0; i < array1.length; i++)
        {
            canvas.drawLine(50, 500 - array1[i], 54, 500 - array1[i], myPaint);
            canvas.drawText(array1[i] + "", 20, 500 - array1[i], myPaint);
        }
        //绘制横坐标文字
        String[] array2 = new String[]{"2008年", "2009年", "2010年", "2011上半年"};
        for (int i = 0; i < array2.length; i++)
        {
            canvas.drawText(array2[i], array1[i] + 80, 520, myPaint);
        }
        //绘制条形图
        myPaint.setColor(Color.BLUE); //设置画笔颜色
```

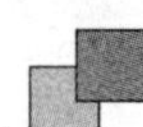

```
        myPaint.setStyle(Style.FILL); //设置填充
        canvas.drawRect(new Rect(90, 500 - 56, 110, 500), myPaint);//画 1 个矩形,前
                                          //两个参数是矩形左上角坐标,后两个参数是右下角坐标
        canvas.drawRect(new Rect(140, 500 - 98, 160, 500), myPaint);//第 2 个矩形
        canvas.drawRect(new Rect(190, 500 - 207, 210, 500), myPaint);//第 3 个矩形
        canvas.drawRect(new Rect(240, 500 - 318, 260, 500), myPaint);//第 4 个矩形
        //绘制出处
        myPaint.setColor(Color.BLACK); //设置画笔颜色
        myPaint.setTextSize(16);//设置文字大小
        canvas.drawText(myString2, 20, 560, myPaint);
    }
}
```

为了实现自定义 View,需要创建一个新的类,然后重写 onDraw 方法,在此需要注意,新创建的类 MyView 要继承 View 基类,同时还要加入有参数的两个构造方法 MyView(Context context)和 MyView(Contextcontext,AttributeSet attr),否则编译运行无法通过。

(2) 编辑 res/layout/activity_main.xml 文件,源代码如下:

```
<?xml version="1.0" encoding="utf-8"?>
<LinearLayout xmlns:android="http://schemas.android.com/apk/res/android"
    android:layout_width="fill_parent"
    android:layout_height="fill_parent"
    android:orientation="vertical" >
    <com.example.chapter7_4.MyView
        android:layout_width="fill_parent"
     android:layout_height="wrap_content"
     android:focusableInTouchMode="true"
        android:background="#FFFFFF"/>
</LinearLayout>
```

其中 android:focusableInTouchMode="true",可以使得自定义视图在触摸模式下获取用户焦点。

(3) 右击工程 chapter7_4,在弹出的快捷菜单中选择 Run As|Android Application 命令,运行效果如图 7-8 所示。

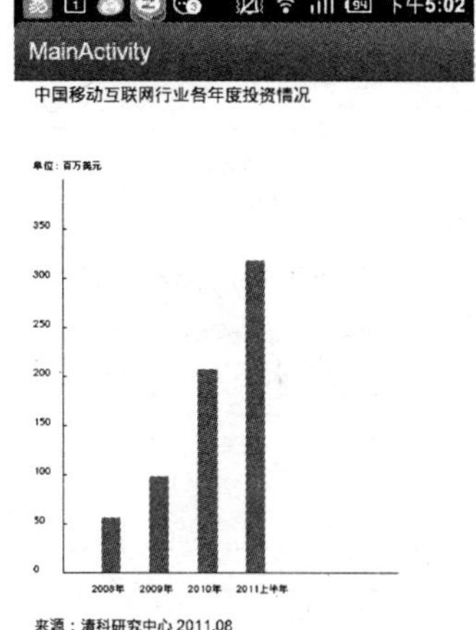

图 7-8 自定义视图

7.5.2 回调处理 onKeyDown 事件

对于基于回调方式的 onKeyDown 事件的处理,只需要重写 android.view.View 基类的 onKeyDown()方法,源代码如下:

```
@Override
    public boolean onKeyDown ( int keyCode,
KeyEvent event) {
    if(KeyEvent.KEYCODE_MENU == keyCode){//
如果菜单键被按下
        Toast.makeText(getContext(), "菜单
键被按下", Toast.LENGTH_LONG).show();
        return true;//事件不会继续往下传播
    }
    return super.onKeyDown(keyCode, event);
    }
```

右击工程 chapter7_4,在弹出的快捷菜单中选择 Run As|Android Application 命令,当触摸自定义视图控件过后,按下"菜单"键的运行效果如图 7-9 所示。

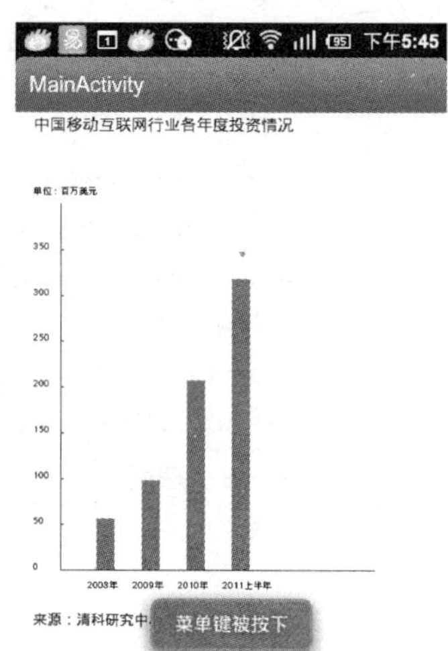

图 7-9　回调处理 onKeyDown 事件

7.5.3　回调处理 onKeyUp 事件

对于基于回调方式的 onKeyUp 事件的处理,也只需要重写 android.view.View 基类的 onKeyUp()方法,源代码如下:

```
@Override
    public boolean onKeyUp ( int keyCode,
KeyEvent event)
    {
        if (KeyEvent.KEYCODE_SEARCH == key-
Code)
    {//如果菜单键被释放
    Toast.makeText(getContext(), "搜索键被释
放", Toast.LENGTH_LONG).show();
    return true;//事件不会继续往下传播
    }
return super.onKeyUp(keyCode, event);
}
```

右击工程 chapter7_4,在弹出的快捷菜单中选择 Run As|Android Application 命令,当触摸自定义视图控件过后,释放"搜索"键时的运行效果如图 7-10 所示。

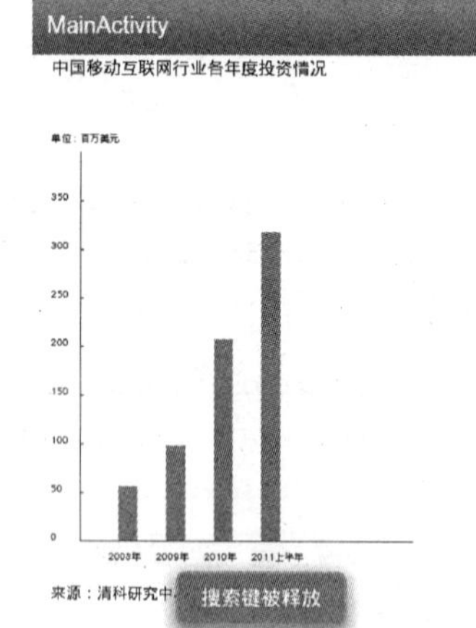

图 7-10　回调处理 onKeyUp 事件

7.5.4　回调处理触摸事件

智能机没有实体键盘,只有虚拟键盘,用户大部分的操作都是通过触摸屏幕来完

成的。因此对于触摸事件的捕获与处理就显得格外的重要，用户可以利用智能机的触摸屏的特性完成各种各样的手势操作。几乎任何一款线上的 Android 的应用程序里面，都有相关的触摸手势操作。利用触摸滑动来完成操作的最典型的例子就是在大多数应用程序安装之后第一次启动过程当中，用户一般都可以通过向左或者向右滑动来进行一些优美图片的切换，引入这些图片的主要目的就是介绍当前版本应用程序的一些功能性的特点，这就是当前比较流行的“引导型”设计。下面来看一下虾米音乐的客户端程序安装之后第一次的启动过程。如图 7－11 和图 7－12 所示，用户可以通过手指的滑动，切换不同的说明图片，当然虾米音乐的客户端程序的说明图片不止这两幅，笔者在这里只是简单举个例子，介绍一下。

客户端“引导型”设计图片简单扼要地说明了产品的主要功能和使用场景，使用户一目了然，非常清晰。不过对于这种“引导型”设计，有以下几点需要注意：

- ❑ 告诉用户现在处于指南的哪个位置。
- ❑ 每个页面只讲清楚一个功能。
- ❑ 最好可以选择性的跳过。
- ❑ 只能出现一次。

图 7－11 虾米音乐启动第一张引导图片

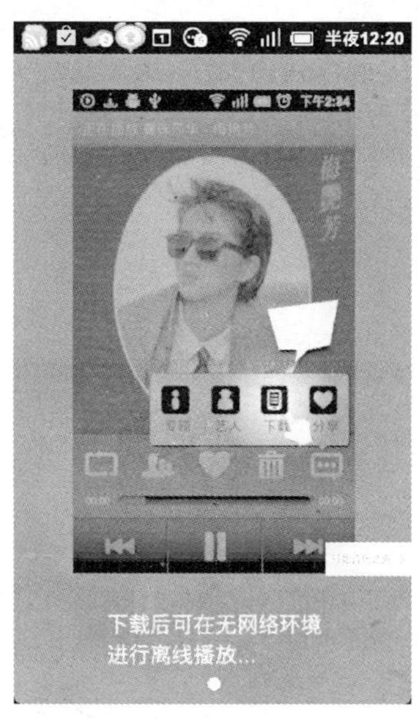

图 7－12 虾米音乐启动第二张引导图片

当用户界面组件被触摸到的时候，该组件的触摸(onTouch)事件会被回调。实际上真正触发的主要是 3 种不同的操作：

- ❑ 第一种是 MotionEvent.ACTION_DOWN，即手指刚刚碰到用户组件时。
- ❑ 第二种是 MotionEvent.ACTION_MOVE，也就是用户触摸用户组件的时候。

有的时候，可能还要滑动的操作，这个时候实际上真正的事件操作是 Motion-

Event. ACTION_MOVE。

- 第三种是 MotionEvent. ACTION_UP,即用户手指离开用户组件时。

由于大部分触摸事件都是针对于程序当中定义的自定义 View 控件的,所以在此之前笔者并没有介绍如何通过监听器的方式对触摸事件进行监听和处理。下面笔者通过实例的方式来为读者介绍如何通过重写自定义 View 控件的 onTouchEvent()方法来对用户的触摸事件进行监听和处理。

(1) 新建工程 chapter7_5,新建自定义视图类 CustomView,编辑 CustomView.java 文件,源代码如下:

```
public class CustomView extends View
{
    Paint mPaint;//声明画笔对象
    List<Dot> dots;//点集合
    public CustomView(Context context)
    {
        super(context);
        init();
    }
    public CustomView(Context context, AttributeSet attrs)
    {
        super(context, attrs);
        init();
    }
    public void init()
    {
          mPaint = new Paint();
          mPaint.setColor(Color.RED);   //设置绘制的颜色是红色
          mPaint.setAntiAlias(true);
          setBackgroundColor(Color.BLUE);  //视图对象背景色是蓝色
          dots = new ArrayList<CustomView.Dot>();  //初始化时里面没有显示任何
                                                   //的红点
    }
    @Override
    protected void onDraw(Canvas canvas)
    {
        for(Dot dot : dots)
        { //绘制所有视图对象里保存的 Dot 点信息
            canvas.drawCircle(dot.x, dot.y, dot.radius, mPaint);
        }
    }
  @Override
```

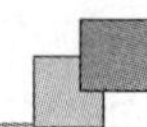

```
    public boolean onTouchEvent(MotionEvent event)
    {
        float x = event.getX();//获取随机单击位置的 X 坐标值
        float y = event.getY();//获取随机单击位置的 Y 坐标值
        float radius = (float) (1 + Math.random() * 9); //生成一个随机半径值范围
                                                        //是 1～10
        //将单击的坐标和随机生成的半径值构造成 Dot 对象存放到视图里
        dots.add(new Dot(x,y,radius));
        invalidate();//调用 onDraw()重绘 View 控件
        return super.onTouchEvent(event);
    }
  static class Dot
  {
    public float x;
    public float y;
    public float radius;
    public Dot(float x, float y, float radius)
    {
        this.x = x;
        this.y = y;
        this.radius = radius;
    }
  }
}
```

(2) 编辑 res/layout/activity_main.xml 文件，源代码如下：

```
<?xml version = "1.0" encoding = "utf-8"?>
<LinearLayout xmlns:android = "http://schemas.android.com/apk/res/android"
    android:layout_width = "fill_parent"
    android:layout_height = "fill_parent"       android:orientation = "vertical" >
        <com.example.chapter7_5.CustomView
          android:layout_width = "fill_parent"
          android:layout_height = "fill_parent"
          android:focusableInTouchMode = "true"
        android:background = "#FFFFFF"/>
</LinearLayout>
```

(3) 右击工程 chapter7_5，在弹出的快捷菜单中选择 Run As|Android Application 命令，在屏幕上触摸几次过后的效果如图 7-13 所示。

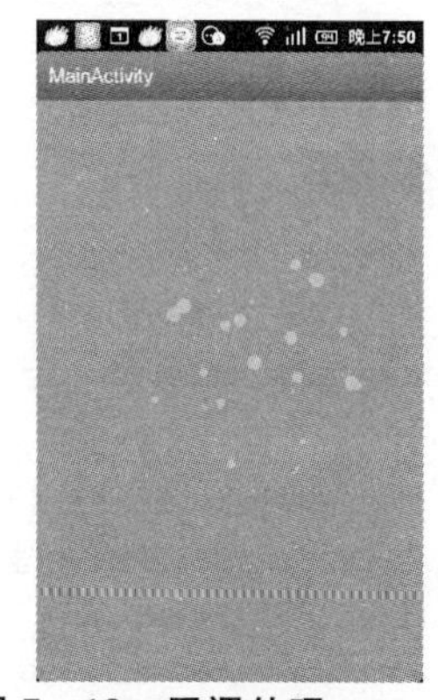

图 7-13 回调处理 onTouchEvent 事件

7.5.5 Android 的手势识别

除了能够对简单的触摸事件进行处理之外，为了能够提高应用程序的用户体验，有的时候应用程序还需要能够识别用户的手势，根据用户不同的手势去完成不同的操作。开发者只需要继承 SimpleOnGestureListener 类，然后重载感兴趣的手势即可。下面通过实例来为读者介绍如何在自定义 View 当中对用户手势进行识别。

(1) 新建工程 chapter7_6，新建一个继承自 SimpleOnGestureListener 的 MyGestureListener 类，编辑 MyGestureListener.java 文件，源代码如下：

```
public class MyGestureListener extends SimpleOnGestureListener
{
    private Context mContext; //声明上下文对象
    MyGestureListener(Context context)
    {
        mContext = context;
    }
    @Override
    public boolean onDown(MotionEvent e)
    {//单击，触摸屏按下时立刻触发
        Toast.makeText(mContext, "DOWN " + e.getAction(), Toast.LENGTH_SHORT).
show();
        return false;
    }
    @Override
    public void onShowPress(MotionEvent e)
    {//短按，触摸屏按下后片刻后抬起，会触发这个手势，如果迅速抬起则不会
        Toast.makeText(mContext, "SHOW " + e.getAction(), Toast.LENGTH_SHORT).
show();
    }
    @Override
    public boolean onSingleTapUp(MotionEvent e)
    {//抬起，手指离开触摸屏时触发(长按、滚动、滑动时，不会触发这个手势)
        Toast.makeText(mContext, "SINGLE UP " + e.getAction(), Toast.LENGTH_
SHORT).show();
        return false;
    }
    @Override
    public boolean onScroll(MotionEvent e1, MotionEvent e2,
            float distanceX, float distanceY)
    { //滚动，触摸屏按下后移动
        Toast.makeText(mContext, "SCROLL " + e2.getAction(), Toast.LENGTH_SHORT).
show();
        return false;
    }
    @Override
```

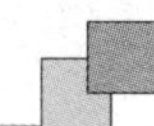

```
        public void onLongPress(MotionEvent e)
        { //长按,触摸屏按下后既不抬起也不移动,过一段时间后触发
            Toast.makeText(mContext, "LONG " + e.getAction(), Toast.LENGTH_SHORT).
show();
        }
        @Override
        public boolean onFling(MotionEvent e1, MotionEvent e2, float velocityX,
                float velocityY)
        { //滑动,触摸屏按下后快速移动并抬起,会先触发滚动手势,跟着触发一个滑动手势
            Toast.makeText(mContext, "FLING " + e2.getAction(), Toast.LENGTH_SHORT).
show();
            return false;
        }
        @Override
        public boolean onDoubleTap(MotionEvent e)
        { //双击,手指在触摸屏上迅速单击第二下时触发
            Toast.makeText(mContext, "DOUBLE " + e.getAction(), Toast.LENGTH_SHORT).
show();
            return false;
        }
        @Override
        public boolean onDoubleTapEvent(MotionEvent e)
        {//双击的按下跟抬起各触发一次
            Toast.makeText(mContext, "DOUBLE EVENT " + e.getAction(), Toast.LENGTH_
SHORT).show();
            return false;
        }
        @Override
        public boolean onSingleTapConfirmed(MotionEvent e)
        { //单击确认,即快速按下并抬起,但并不连续单击第二下
            Toast.makeText(mContext, "SINGLE CONF " + e.getAction(), Toast.LENGTH_
SHORT).show();
            return false;
        }
    }
```

(2) 创建一个自定义视图 MyView,编辑 MyView.java 文件,源代码如下:

```
    public class MyView extends View
    {
        private GestureDetector mGestureDetector; //声明手势监听者
        public MyView(Context context)
        {
            super(context);
            init(context);
        }
        public MyView(Context context, AttributeSet attrs)
```

```
    {
        super(context, attrs);
        init(context);
    }
    private void init(Context context)
    {
        mGestureDetector = new GestureDetector(context, new MyGestureListener(con-
text));//实例化手势监听者
        setLongClickable(true);//View 必须设置 longClickable 为 true,否则手势识别无
                                //法正确工作,只会返回 Down, Show, Long 三种手势
        this.setOnTouchListener(new OnTouchListener()
        {
            public boolean onTouch(View v, MotionEvent event)
            {
                return mGestureDetector.onTouchEvent(event);//采用手势监听者来
                                                             //识别用户手势
            }
        });}
}
```

对于自定义 View,如果要使用手势识别的话,以下两点必须注意:

① View 必须设置 longClickable 为 true,否则手势识别无法正常工作,只会返回 Down, Show, Long 三种手势 。

② 必须在 View 的 onTouchListener 中调用手势识别,而不能直接覆盖 onTouchEvent 方法,否则同样手势识别无法正常工作。

(3) 编辑 res/layout/activity_main.xml 文件,源代码如下:

```
<?xml version="1.0" encoding="utf-8"?>
<LinearLayout xmlns:android="http://schemas.android.com/apk/res/android"
    android:layout_width="fill_parent"
    android:layout_height="fill_parent"
    android:orientation="vertical" >
    <com.example.chapter7_6.MyView
        android:layout_width="fill_parent"
        android:layout_height="fill_parent"
        android:focusableInTouchMode="true"
        android:background="#FF0000"/>
</LinearLayout>
```

图 7-14 单击确认快速按下并抬起效果

(4) 右击工程 chapter7_6,在弹出的快捷菜单中选择 Run As|Android Application 命令,笔者单击确认快速按下并抬起后效果如图 7-14 所示。

第 8 章

数据存储

在本章当中，笔者将会为读者介绍在 Android 应用程序当中如何存储数据。作为 Android 应用程序开发当中非常重要的一环，数据存储从来都是开发过程中不可回避的关键技术。在 Android 当中，可供选择的存储方式有三种，分别是系统偏好设置(SharedPreferences)、文件存储以及 SQLite 数据库方式。Android 是基于 Linux 的系统，每个应用程序是以单独的进程方式而存在的，这些进程之间是不能互相访问的。本章介绍的数据存储技术允许应用程序创建和获取属于当前应用程序的私有数据，如果需要在不同的应用程序之间共享数据，需要使用内容提供者(Content Provider)来实现，有关内容提供者的知识，笔者将会在下一章当中介绍。本章主要包含以下内容：

- ❑ SharedPreferences(系统偏好设置)；
- ❑ 文件存储；
- ❑ SQLite 数据库。

8.1　SharedPreferences(系统偏好设置)

为了保存应用程序的系统配置信息，Android SDK 提供了 SharedPreferences 类，它是一个轻量级的存储类。SharedPreferences 使用 Map 数据结构来存储数据，以键值对的方式存储，采用了 XML 格式将数据存储到设备中。一般来讲，使用 SharedPreferences 主要是用来存储一些用户喜好的设置信息，比如是否愿意接收推送通知、是否愿意自动同步等。

8.1.1　SharedPreferences 数据存储

下面通过实例为读者演示如何使用 SharedPreferences 来完成数据的存储。

(1) 新建工程 chapter8_1，编辑 res/layout/activity_main. xml 文件，源代码参考配套资料。

（2）编辑 MainActivity.java 文件，源代码如下：

```
public class MainActivity extends Activity implements OnClickListener
{
    /**存储后的文件路径：/data/data/<package name>/shares_prefs + 文件名.xml */
    private  static final String FILE_NAME = "info";/**存储的文件名 */
    EditText usernameEdit;      //声明用户名输入框控件
    EditText ageEdit;           //声明年龄输入框控件
    Button saveBtn;             //声明保存个人信息按钮
    Button readBtn;             //声明读取个人信息按钮
    @Override
    public void onCreate(Bundle savedInstanceState)
    {
        super.onCreate(savedInstanceState);
        setContentView(R.layout.activity_main);
        usernameEdit = (EditText)findViewById(R.id.username);
                                        //找到 id 为 username 的输入框控件
        ageEdit = (EditText)findViewById(R.id.age);
                                            //找到 id 为 age 的输入框控件
        saveBtn = (Button)findViewById(R.id.saveBtn);
                                          //找到 id 为 saveBtn 的按钮控件
        readBtn = (Button)findViewById(R.id.readBtn);
                                          //找到 id 为 readBtn 的按钮控件
        saveBtn.setOnClickListener(this);          //绑定单击事件监听器
        readBtn.setOnClickListener(this);          //绑定单击事件监听器
    }
    @Override
    public void onClick(View v)
    {
        if(R.id.saveBtn == v.getId())
        {//保存个人信息
            saveInfo();
        }
        else if(R.id.readBtn == v.getId())
        {//读取个人信息
            readInfo();
        }
    }
    //保存个人信息
    private void saveInfo()
    {
        //获取 SharedPreferences 对象
        SharedPreferences infoPref = getSharedPreferences(FILE_NAME, Context.MODE_
PRIVATE);
        //获取 Editor 对象
        Editor editor = infoPref.edit();
        String username = usernameEdit.getText().toString();
```

```
                                                        //获取用户名
        int  age = Integer.parseInt(ageEdit.getText().toString());
                                                        //获取年龄
        editor.putString("username", username);
                                                        //存放字符串数据
        editor.putInt("age", age);      //存放整型数据
        editor.commit();                //保存
        Toast.makeText(this, "保存个人信息成功", Toast.LENGTH_LONG).show();
    }
    //读取个人信息
    private void readInfo()
    {
        //获取 SharedPreferences 对象
        SharedPreferences infoPref = getSharedPreferences(FILE_NAME, Context.MODE_
PRIVATE);
        String username = infoPref.getString("username", "");
                            //获取 key 为 username 所对应的值,第二个参数为默认值
        int age = infoPref.getInt("age", 0);
                                                //获取 key 为 age 所对应的值
        Toast.makeText(this, "用户名:" + username + " 年龄:" + age, Toast.LENGTH_
LONG).show();
    }
    @Override
    public boolean onCreateOptionsMenu(Menu menu)
    {
        getMenuInflater().inflate(R.menu.activity_main, menu);
        return true;
    }
}
```

上述代码当中,通过 Context.getSharedPreferences 方法获取 SharedPreferences 对象,getSharedPreferences(name,mode)方法的第一个参数用于指定该文件的名称,名称不用带后缀,后缀会由 Android 自动加上。方法的第二个参数指定文件的操作模式,共有 4 种操作模式:

- Context.MODE_PRIVATE:该模式为默认操作模式,代表该文件是私有数据,只能被该应用程序本身访问,在该模式下,写入的内容会覆盖原文件的内容,如果想把新写入的内容追加到原文件中,可以使用 Context.MODE_APPEND 模式。
- Context.MODE_APPEND:该模式会检查文件是否存在,存在就往指定的文件中追加内容,否则就创建新的文件。
- Context.MODE_WORLD_READABLE:表示当前文件可以被其他应用读取。
- Context.MODE_WORLD_WRITEABLE:表示当前文件可被其他应用写入。

说明：android 有一套自己的安全模型，当应用程序(.apk)在安装时系统就会分配给它一个 userid，当该应用要去访问其他资源比如文件的时候，就需要进行 userid 的匹配。默认情况下，任何应用创建的 Shared Preferences，文件以及 SQLite 数据库都应该是私有的(位于/data/data/<packagename>/)，其他程序则无法访问。如果要在应用程序之间共享数据，一般是通过内容提供者(Content Provider)来实现的。有关内容提供者的内容，笔者将在下一章当中介绍。

(3) 右击工程 chapter8_1，在弹出的快捷菜单中选择 Run As|Android Application 命令，在界面当中输入用户名和年龄之后，单击“保存个人信息”按钮的运行效果如图 8-1所示，清空输入框过后单击“读取个人信息”按钮的运行效果如图 8-2 所示。

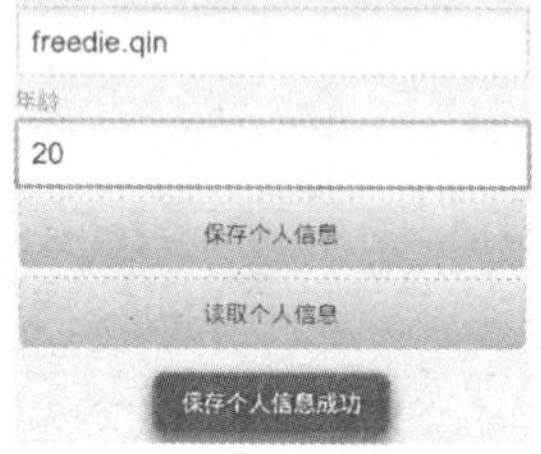

图 8-1　保存个人信息效果

图 8-2　读取个人信息效果

(4) 将 Android SDK 的 platform-tools 所在的路径添加到系统环境变量 Path 当中，添加过后，就可以使用 adb shell 命令，进入到设备或者模拟器的 Shell 环境中，在这个 Linux Shell 中，允许执行各种 Linux 命令。由于创建的 info.xml 文件位于/data/data/com.example.chapter8_1/shared_prefs 目录下，在控制台窗口输入 adb shell 命令过后，通过 Linux 的 cd 命令，进入到 info.xml 文件所在的目录后，再输入 Linux 查看文件内容的命令 cat info.xml，控制台窗口的输出如图 8-3 所示。

```
# cat info.xml
cat info.xml
<?xml version='1.0' encoding='utf-8' standalone='yes' ?>
<map>
<string name="username">freedie.qin</string>
<int name="age" value="20" />
</map>
#
```

图 8-3　cat info.xml 控制台输出

8.2　PreferenceActivity

在开发 Android 应用程序的时候，在选项菜单当中，一般都会有一个“设置”选项，用户可以根据自己的喜好来设置应用程序的一些运行参数，如图 8－4 所示。

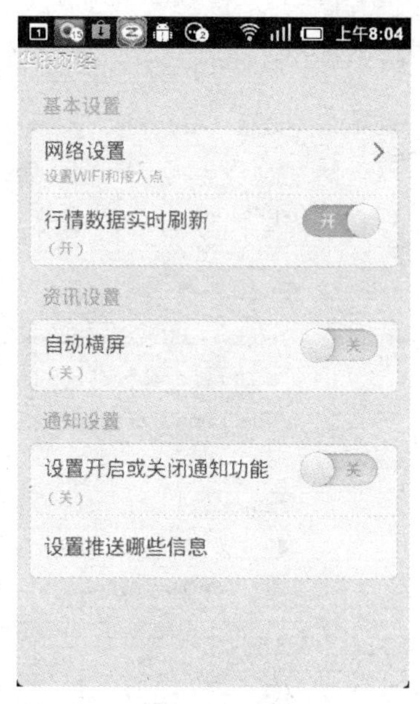

图 8－4　根据用户喜好设置界面

PreferenceActivity 的主要特点就是可以通过添加各种各样的 Preference UI 控件，在用户做完喜好设置过后，将这些 Preference UI 控件的状态持久化存储起来。举个例子，对于图 8－4当中的“设置开启或关闭通知功能”这一选项，用户在设置该选项为“开启”过后，退出应用然后下一次进入应用程序的时候，用户希望看到的是这一项是处于开启的状态，所以这就要求应用程序必须对用户每次的操作进行存储，然后在下一次启动应用的时候，根据存储的数据再来更新 UI 控件的状态。

有了 PreferenceActivity 以及各种各样的 Preference UI 控件过后，Android SDK 就帮助开发者自动实现了这些 Preference UI 控件状态的持久化存储。Android SDK 一共提供了 4 种 Preference UI 的控件，分别是 CheckBoxPreference 控件、EditTextPreference 控件、ListPreference 控件以及 RingtonePreference 控件。下面笔者就通过实例为读者一一介绍如何在项目当中使用这些 Preference UI 控件。

8.2.1　CheckBoxPreference

CheckBoxPreference 控件会以类似复选框控件的形式展现。

（1）新建工程 chapter8_2，在 res 目录下面新建 xml 子目录，右击 xml 子目录，在弹出的快捷方式中选择 new|other|Android XML File 命令，在弹出的对话框当中的 Resouce Type 下拉列表框当中选择 Preference 选项，在 File 输入框当中输入 checkboxpref.xml，单击 Finish 按钮。

（2）编辑 res/xml/checkboxpref.xml 文件，源代码如下：

```
<?xml version="1.0" encoding="utf-8"?>
<PreferenceScreen xmlns:android="http://schemas.android.com/apk/res/android" >
    <CheckBoxPreference  android:key="myCheckBox"
        android:summary="Check it on or off" android:title="CheckBox Prefer-
ence" />
</PreferenceScreen>
```

PreferenceScreen 节点与 Preference UI 控件的节点关系如同 ViewGroup 节点与 View 节点之间的关系一样。一个 PreferenceScreen 节点可以包含多个 Preference UI 控件的节点。

(3) 编辑 MainActivity.java 文件,源代码如下:

```
public class MainActivity extends PreferenceActivity
{
    @Override
    public void onCreate(Bundle savedInstanceState)
    {
        super.onCreate(savedInstanceState);
        addPreferencesFromResource(R.xml.checkboxpref);//加载 Preference 描述文件
        CheckBoxPreference checkBoxPrefs = (CheckBoxPreference) findPreference("my-
CheckBox");                                                    //根据 key 来寻找
        checkBoxPrefs.setOnPreferenceChangeListener(new OnPreferenceChangeListener()
        {//绑定 CheckboxPreference 喜好变更事件监听
            @Override
            public boolean onPreferenceChange(Preference preference, Object newValue)
            {
                Toast.makeText(MainActivity.this, "喜好设置发生变更,状态:" + ne-
wValue, Toast.LENGTH_LONG).show();
                return true;//返回 true 时更新相应的存储值
            }
        });
    }
}
```

加载包含 Preference UI 控件的 Activity 要求继承于 android.preference.PreferenceActivity 这个父类,然后在 onCreate()方法中调用 addPreferencesFromResource()方法就可以加载包含 Preference UI 控件的 XML 文件作为 Activity 的界面。

(3) 右击工程 chapter8_2,在弹出的快捷菜单中选择 Run As|Android Application 命令,运行效果如图 8-5 所示,改变选项设置过后效果如图 8-6 所示。

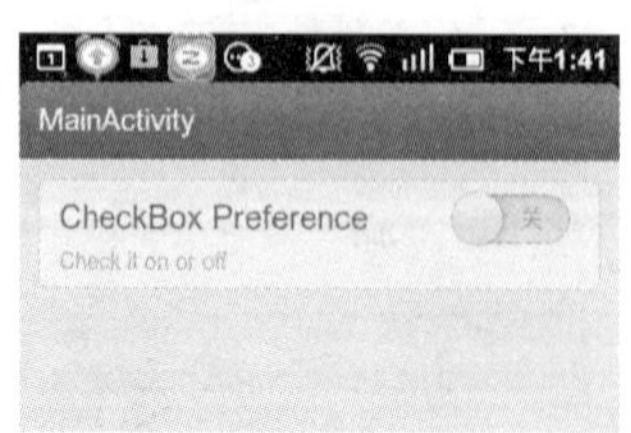

图 8-5 CheckBoxPreference 效果

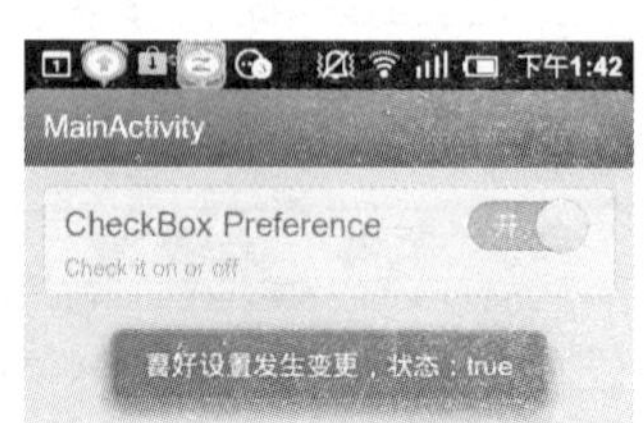

图 8-6 改变 CheckBoxPreference 选项设置效果

(4) 通过 PreferenceActivity 持久化的 Preference UI 控件的信息会保存在/data/data/包名/包名_ _preferences.xml 文件当中,通过进入到设备的 Shell 中,得到

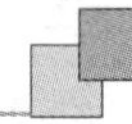

com. example. chapter8_2_preferences. xml 文件内容，如图 8－7 所示。

```
C:\Documents and Settings\Administrator>adb shell
# cd /data/data/com.example.chapter8_2
cd /data/data/com.example.chapter8_2
# ls
ls
lib             shared_prefs
# cd shared_prefs
cd shared_prefs
# ls
ls
com.example.chapter8_2_preferences.xml
# cat com.example.chapter8_2_preferences.xml
cat com.example.chapter8_2_preferences.xml
<?xml version='1.0' encoding='utf-8' standalone='yes' ?>
<map>
<boolean name="myCheckBox" value="true" />
</map>
#
```

图 8－7　com. example. chapter8_2_preferences. xml 内容

8.2.2　EditTextPreference

EditTextPreference 单击过后会弹出一个输入框，用户输入的内容会存储在 SharedPreferences 当中。

(1) 新建工程 chapter8_3，在 res 目录下面新建子目录 xml，在 xml 子目录当中新建 Preference 类型的 Android XML File：edittextpref. xml，编辑 edittextpref. xml 文件，源代码如下：

```
<? xml version = "1.0" encoding = "utf-8"? >
<PreferenceScreen  xmlns:android = "http://schemas.android.com/apk/res/android">
    <EditTextPreference android:key = "myEditText"
        android:title = "EditTextPreference"
        android:summary = "EditTextPreference Demo"
        android:defaultValue = "请输入用户名"
        android:dialogTitle = "用户名输入">
    </EditTextPreference>
</PreferenceScreen>
```

(2) 编辑 MainActivity. java 文件，源代码如下：

```
public class MainActivity extends PreferenceActivity
{
    private static final String TAG = "MainActivity";
    @Override
    public void onCreate(Bundle savedInstanceState)
    {
        super.onCreate(savedInstanceState);
        addPreferencesFromResource(R.xml.edittextpref);//加载 Preference UI 控件描
                                                       //述文件
    }
```

```
        @Override
        protected void onResume()
        {
        super.onResume();
        //Android 的偏好管理器来获取其所管理的 preferences,也就是/data/data/包名/包名_preferences.xml
        SharedPreferences sharedPrefs = PreferenceManager.getDefaultSharedPreferences(this);
        Log.d(TAG, sharedPrefs.getString("myEditText", "") );
        }
        @Override
        public boolean onCreateOptionsMenu(Menu menu)
        {
            getMenuInflater().inflate(R.menu.activity_main, menu);
            return true;
        }
    }
```

(3) 右击工程 chapter8_3,在弹出的快捷菜单中选择 Run As|Android Application 命令,运行效果如图 8-8 所示,单击 EditTextPreference 控件,输入相关信息效果如图 8-9 所示。

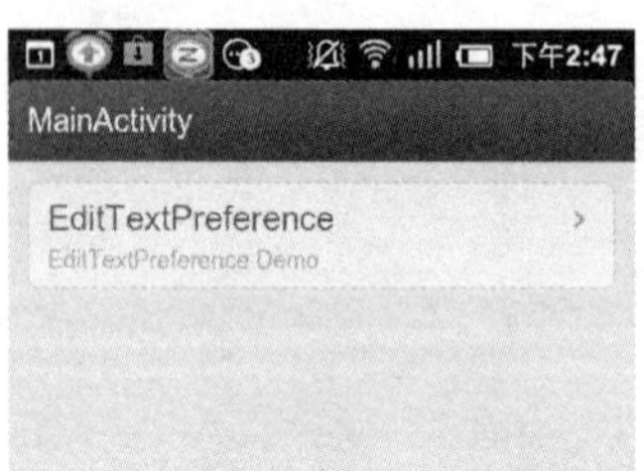

图 8-8　EditTextPreference 效果

图 8-9　单击 EditTextPreference 输入字符串效果

(4) 在 onResume()方法当中,通过调用 PreferenceManager 的静态方法 getDefaultSharedPreferences()来获取 Android 偏好管理器管理的 SharedPreferences 对象,然后调用 SharedPreferences 对象的 getString()方法获取指定键对应的值,而这个键就是在 edittextpref.xml 文件当中定义<EditTextPreference>节点的 android:key 属性所对应的值。运行效果如图 8-10 所示。

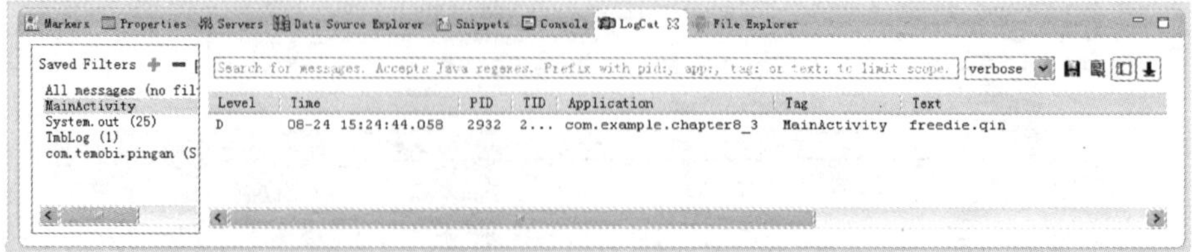

图 8-10　获取 Preference UI 控件存储的值

8.2.3 ListPreference

ListPreference 单击过后会弹出一个下拉列表框，用户选中的内容会存储在 SharedPreferences 当中。

(1) 新建工程 chapter8_4，编辑 res/value/strings.xml 文件，源代码参考光盘。

(2) 在 res 目录下面新建子目录 xml，在 xml 子目录当中新建 Preference 类型的 Android XML File：listpref.xml，编辑 edittextpref.xml 文件，源代码如下：

```
<?xml version = "1.0" encoding = "utf-8"?>
<PreferenceScreen xmlns:android = "http://schemas.android.com/apk/res/android" >
    <ListPreference
        android:dialogTitle = "选择金牌数最多国家"
        android:entries = "@array/golden_rank"
        android:entryValues = "@array/golden_rank_value"
        android:key = "golden_rank_key"
        android:title = "ListPreference Demo" >
    </ListPreference>
</PreferenceScreen>
```

(3) 编辑 MainActivity.java 文件，源代码如下：

```
public class MainActivity extends PreferenceActivity
{
    @Override
    public void onCreate(Bundle savedInstanceState)
    {
        super.onCreate(savedInstanceState);
        addPreferencesFromResource(R.xml.listpref);//加载 Preference UI 控件描述
                                                   //文件
    }
    @Override
    public boolean onCreateOptionsMenu(Menu menu)
    {
        getMenuInflater().inflate(R.menu.activity_main, menu);
        return true;
    }
}
```

(3) 右击工程 chapter8_3，在弹出的快捷菜单中选择 Run As|Android Application 命令，单击 ListPreference 控件过后的效果如图 8-11 所示，单击列表某一项过后效果如图 8-12 所示。

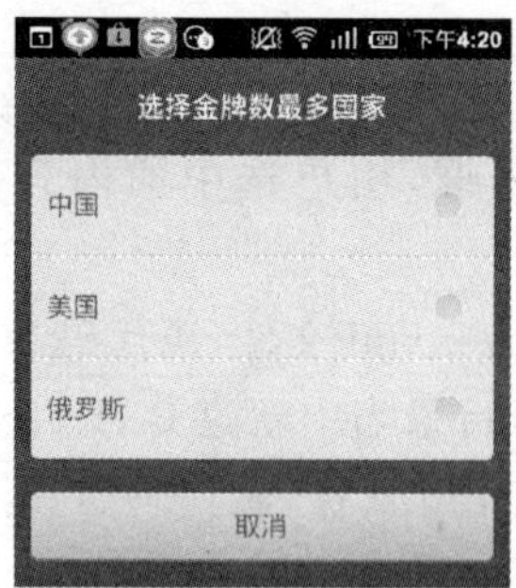

图 8-11　单击 ListPreference 效果

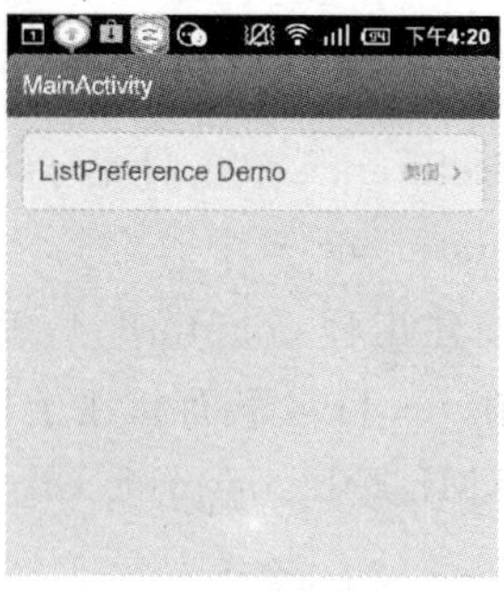

图 8-12　选中美国的效果

8.2.4　RingtonePreference

RingtonePreference 单击过后会弹出一个铃声的选择框，选中后会将选中的内容(uri 字符集)以字符串的形式存储在 SharedPreferences 中。

(1) 新建工程 chapter8_5，在 res 目录下面新建子目录 xml，在 xml 子目录当中新建 Preference 类型的 Android XML File：ringtone.xml，编辑 ringtone.xml 文件，源代码如下：

```
<? xml version = "1.0" encoding = "utf-8"? >
<PreferenceScreen
  xmlns:android = "http://schemas.android.com/apk/res/android">
    <RingtonePreference
        android:key = "myRingtone"
        android:summary = "设置系统铃声"
        android:title = "铃声设置"
        android:ringtoneType = "all"
        android:showSilent = "true" ></RingtonePreference>
</PreferenceScreen>
```

android:ringtoneType 属性有 4 个可选的值：铃声(ringtone)、通知(notification)、闹铃(alarm)及全部(all)。

(2) 编辑 MainActivity.java 文件，源代码如下：

```
public class MainActivity extends PreferenceActivity
{
    @Override
    public void onCreate(Bundle savedInstanceState)
    {
        super.onCreate(savedInstanceState);
        addPreferencesFromResource(R.xml.ringtone);//加载 Preference UI 控件描述
                                                    //文件
```

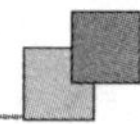

```
            RingtonePreference ringtonePreference = (RingtonePreference) findPreference
("myRingtone");                                    //根据 KEY 找到相应的 RingtonePreference
            ringtonePreference.setOnPreferenceChangeListener(new OnPreferenceChangeLis-
tener()
            {//绑定选项设置更改监听器
                @Override
                public boolean onPreferenceChange(Preference preference, Object newValue)
                {
                    Toast.makeText(MainActivity.this, "newValue:" + newValue, Toast.
LENGTH_LONG).show();
                    return true;//更新 SharedPreferences 值
                }
            });
        }
        @Override
        public boolean onCreateOptionsMenu(Menu menu)
        {
            getMenuInflater().inflate(R.menu.activity_main, menu);
            return true;
        }
    }
```

(3) 右击工程 chapter8_5,在弹出的快捷菜单中选择 Run As|Android Application 命令,单击 RingtonePreference 控件,效果如图 8-13 所示,更改选项设置过后,效果如图 8-14 所示(Android2.2 模拟器效果)。

图 8-13 单击 RingtonePreference 效果

图 8-14 RingtonePreference 选项更改后的效果

8.2.5 PreferenceCategory

PreferenceCategory 用来将 Preference UI 控件进行分组,可以将几个功能相关联的 Preference 控件组合在一起。

(1) 新建工程 chapter8_6，在 res 目录下面新建子目录 xml，在 xml 子目录当中新建 Preference 类型的 Android XML File：category. xml，编辑 category. xml 文件，源代码如下：

```
<? xml version = "1.0" encoding = "utf-8"? >
<PreferenceScreen xmlns:android = "http://schemas.android.com/apk/res/android">
    <!-- PreferenceCategory 为分组标签，把不同类型的参数设置进行分组 -->
    <PreferenceCategory android:title = "天气信息">
        <!-- ListPreference 为下拉列表参数设置标签，可以通过选择下拉列表中的项进
行参数设置 -->
        <ListPreference android:key = "country"
            android:title = "选择国家"
            android:dialogTitle = "国家" />
        <ListPreference android:key = "city"
            android:title = "选择城市"
            android:dialogTitle = "城市" />
  </PreferenceCategory>
  <PreferenceCategory android:title = "推送设置">
        <CheckBoxPreference android:key = "push"
            android:title = "推送开关"
            android:summaryOn = "接受推送"
            android:summaryOff = "不接受推送"
            android:defaultValue = "true" />
  </PreferenceCategory>
</PreferenceScreen>
```

(2) 编辑 MainActivity. java 文件，源代码如下：

```
public class MainActivity extends PreferenceActivity implements OnPreferenceChange-
Listener
{
    private static final String[] countryEntry = new String[]{"中国", "美国", "俄罗
斯"};
    private static final String[] countryEntryValue = new String[]{"1", "2", "3"};
    private static final String[] cityEntry = new String[]{"上海", "纽约", "莫斯
科"};
    private static final String[] cityEntryValue = new String[]{"101", "102", "
103"};
    ListPreference countryPref = null;//声明选择国家 ListPreference
    ListPreference cityPref = null;//声明选择城市 ListPreference
    @Override
    public void onCreate(Bundle savedInstanceState)
    {
```

```
        super.onCreate(savedInstanceState);
        addPreferencesFromResource(R.xml.category);//加载 Preference UI 控件描述
                                                   //文件
        countryPref = (ListPreference)findPreference("country");
        cityPref = (ListPreference)findPreference("city");
        countryPref.setEntries(countryEntry);//设置 ListPreference 显示列表
        countryPref.setEntryValues(countryEntryValue);//设置 ListPreference 显示列
                                                      //表对应的值
        cityPref.setEntries(cityEntry);
        cityPref.setEntryValues(cityEntryValue);
    }
    @Override
      public boolean onPreferenceChange(Preference preference, Object newValue)
    {
        if("country".equals(preference.getKey()))//如果 key 为"country"
        {
            Toast.makeText(this, "选择的国家:" + newValue, Toast.LENGTH_LONG).show();
        }
        else if("city".equals(preference.getKey()))//如果 key 为"city"
        {
          Toast.makeText(this, "选择的城市:" + newValue, Toast.LENGTH_LONG).show();
        }
        return true;
    }
    @Override
    public boolean onCreateOptionsMenu(Menu menu)
    {
        getMenuInflater().inflate(R.menu.activity_
main, menu);
        return true;
    }
}
```

(3) 右击工程 chapter8_6，在弹出的快捷菜单中选择 Run As|Android Application 命令，运行效果如图 8－15 所示。

图 8－15　chapter8_6 运行效果

8.3　文件存储

SharedPreferences 对象是以键值对的方式来保存简单的数据。在有的应用场景下，数据需要以文件的形式保存。Android 操作系统允许应用程序将需要保存的数据以文件形式保存在设备的内部存储里面，也允许应用程序将需要保存的数据以文件形式保存在设备的外部存储(SDCard)里面。下面笔者就通过实例介绍如何使用 Android 的内部存储以及外部存储。

8.3.1 内部存储

内部存储，就是将文件保存在设备的内部存储器中。默认情况下，这些文件是应用程序私有的，对其他应用程序是不透明的，对用户也是不透明的。当程序卸载过后，这些文件将会被删除。

(1) 新建工程 chapter8_7，编辑 res/layout/activity_main.xml 文件，源代码参考配套资料。

(2) 编辑 MainActivity.java 文件，源代码如下：

```
public class MainActivity extends Activity implements OnClickListener
{
    private static final String TAG = "MainActivity";
    EditText myEditText;//声明数据输入框
    Button saveBtn;//声明保存数据按钮
    Button readBtn;//声明读取数据按钮
    @Override
    public void onCreate(Bundle savedInstanceState)
    {
        super.onCreate(savedInstanceState);
        setContentView(R.layout.activity_main);
        myEditText = (EditText) findViewById(R.id.myEditText);
                                                    //找到 id 为 myEditText 的输入框
        saveBtn = (Button) findViewById(R.id.saveInternal);
                                                    //找到 id 为 saveInternal 的按钮
        readBtn = (Button) findViewById(R.id.loadInternal);
                                                    //找到 id 为 loadInternal 的按钮
        saveBtn.setOnClickListener(this);           //绑定单击事件监听器
        readBtn.setOnClickListener(this);
    }
    @Override
    public void onClick(View view)
    {
        int viewId = view.getId();
        if (R.id.saveInternal == viewId)
        {//保存数据按钮单击
            try
            {
                writeToInternalStorage("data.txt", myEditText.getText().toString
```

```
().getBytes());
                Toast.makeText(this,"写入数据成功",Toast.LENGTH_LONG).show();
            }
            catch (Exception e)
            {
                Toast.makeText(this,"写入数据失败",Toast.LENGTH_LONG).show();
            }
        }
        else if (R.id.loadInternal == viewId)
        {//读取数据按钮单击
            try
            {
                String readData = new String(readFormInternalStorage("data.txt"),
"UTF-8");
                 Toast.makeText(this,"读取的数据:" + readData,Toast.LENGTH_
LONG).show();
            }
            catch (Exception e)
            {
                Toast.makeText(this,"读取数据失败",Toast.LENGTH_LONG).show();
            }
        }
    }
    /**
     * 保存数据至内部存储
     * @param filename 文件名称
     * @param content 要写入的字节数据
     */
    private  void writeToInternalStorage(String filename, byte[] content)
  {
    if(null == content || 0 == content.length)
    {//对写入数据合法性进行校验
        return;
    }
    try {
        //根据传入的第一个参数,打开一个用来写数据的文件,如果文件不存在,则创建
        //这个文件
        FileOutputStream fos = openFileOutput(filename, Context.MODE_PRIVATE);
        fos.write(content);//向输出流写入数据
        fos.close();//关闭输出流
    }
    catch (Exception e)
```

```
    {//捕获写入异常
        Log.d(TAG, "写入失败:exception:" + e);
        throw new RuntimeException("write to internal storage exception");
    }
}
/**
 * 从内部存储读取数据
 * @param fileName  文件名称
 * @return
 */
private byte[] readFormInternalStorage(String fileName)
{
    int len = 1024;
    byte[] buffer = new byte[len];
    try
    {
        //打开一个用来读取数据的文件
        FileInputStream fis = openFileInput(fileName);
        ByteArrayOutputStream baos = new ByteArrayOutputStream();
        int nrb = fis.read(buffer, 0, len);//每次读取 1 024 个字节的数据存放在
                                           //buffer 当中
        while (nrb != -1)
    {     //还有数据可以读取
            baos.write(buffer, 0, nrb);
            nrb = fis.read(buffer, 0, len);//循环读取
        }
        buffer = baos.toByteArray();
        baos.close();
        fis.close();
    }
    catch (Exception e)
    {//捕获读取异常
        Log.d(TAG, "读取失败:exception:" + e);
        throw new RuntimeException("read from  internal storage exception");
    }
    return buffer;
}
@Override
public boolean onCreateOptionsMenu(Menu menu)
{
    getMenuInflater().inflate(R.menu.activity_main, menu);
    return true;
```

```
    }
}
```

上述代码演示了如何使用内部存储来存取数据和读取数据。其中使用内部存储来保存数据大致分为以下几个步骤：

- ❑ 调用 openFileOutput(String name, int mode)方法打开一个写入数据操作的文件，其中第一个参数用来指定文件的名称，如果这个文件不存在，则会创建这个文件。第二个参数指定文件的操作模式，在之前介绍 SharedPreferences 对象的时候已经介绍过，笔者在这里不做详细赘述。
- ❑ 调用 FileOutputStream 对象的 write()方法向文件写入数据。
- ❑ 调用 FileOutputStream 对象的 close()方法关闭输出流。

读取内部存储的数据的步骤与传统的 JAVA IO 读取文件的方式一致。

(3) 右击工程 chapter8_7，在弹出的快捷菜单中选择 Run As|Android Application 命令，在输入框中输入相关数据，单击“保存数据”按钮效果如图 8-16 所示，保存完成后，单击“读取数据”按钮效果如图 8-17 所示。

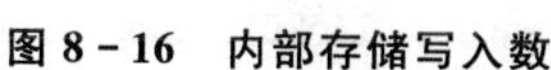

图 8-16　内部存储写入数

图 8-17　内部存储读取数据

(4) 内部存储的文件的路径为/data/data/包名/files/文件名，读者可以进入到设备的 Shell 环境当中，然后进入到文件存储的指定目录(/data/data/com. example. chapter8_7/files/data. txt)，最后通过 Linux 的 Cat 命令，查看指定文件的内容，这里笔者就不做相关演示了。

8.3.2　外部存储

外部存储，就是将文件保存在设备的外部存储器中，这里提到的外部存储器，一般就是通常所说的 SDCard 卡。内部存储里面的文件，一般来讲都是某一个应用程序所私有的，而外部存储的文件都是可读的，对任何一个应用程序来说都是透明的，这也是外部存储与内部存储比较大的一个差别。下面通过实例来演示如何使用 SDCard 卡进行数据的存储和读取。

(1) 新建工程 chapter8_8，由于需要访问 SDCard，需要申请访问 SDCard 的权限，编辑 AndroidManifest. xml 文件，源代码如下：

```
<manifest xmlns:android = "http://schemas.android.com/apk/res/android"
    package = "com.example.chapter8_8"   android:versionCode = "1"   android:version-
Name = "1.0" >
    <uses-sdk   android:minSdkVersion = "8" android:targetSdkVersion = "15" />

    <!-- 在 SDCard 卡创建和删除文件的权限 -->
    < uses-permission  android: name  =  " android. permission. MOUNT _ UNMOUNT _
FILESYSTEMS"/>
    <!-- 往 SDCard 卡当中写入数据的权限 -->
    <uses-permission android:name = "android.permission.WRITE_EXTERNAL_STORAGE"/>
    < application android: icon = " @ drawable/ic _ launcher"   android: label = " @
string/app_name"
        android:theme = "@style/AppTheme" >
        <activity android:name = ".MainActivity"
            android:label = "@string/title_activity_main" >
            <intent-filter>
                <action android:name = "android.intent.action.MAIN" />
                <category android:name = "android.intent.category.LAUNCHER" />
            </intent-filter>
        </activity>
    </application>
</manifest>
```

（2）编辑 res/layout/activity_main.xml 文件，源代码参考配套资料。

（3）编辑 MainActivity.java 文件，源代码如下：

```
public class MainActivity extends Activity implements OnClickListener
{
    private static final String TAG = "MainActivity";
    EditText dataEditText;//声明文本输入框
    Button saveExternalBtn; //声明保存数据按钮
    Button loadExternalBtn;//声明读取数据按钮
    @Override
    public void onCreate(Bundle savedInstanceState)
    {
        super.onCreate(savedInstanceState);
        setContentView(R.layout.activity_main);
        dataEditText = (EditText) findViewById(R.id.myEditText);
                                                //找到 id 为 myEditText 的按钮
        saveExternalBtn = (Button) findViewById(R.id.saveSDCard);
                                                //找到 id 为 saveSDCard 的按钮
        loadExternalBtn = (Button) findViewById(R.id.loadSDCard);
                                                //找到 id 为 loadSDCard 的按钮
```

```
        saveExternalBtn.setOnClickListener(this);        //绑定单击事件监听器
        loadExternalBtn.setOnClickListener(this);
    }
    @Override
    public void onClick(View v)
    {
        int viewId = v.getId();
        if (R.id.saveSDCard == viewId)
        {//保存数据按钮
            try
            {
                String content = dataEditText.getText().toString();
                saveToSDCard("data.txt", content.getBytes());
                Toast.makeText(this, "写入 SDCard 成功", Toast.LENGTH_LONG).show();
            }
            catch (Exception e)
            {
                Toast.makeText(this, "写入 SDCard 失败", Toast.LENGTH_LONG).show();
            }
        }
        else if (R.id.loadSDCard == viewId)
        {//读取数据按钮
            try
            {
                Toast.makeText(this, new String(loadFromSDCard("data.txt"), "UTF-
8"), Toast.LENGTH_LONG).show();
            }
            catch (Exception e)
            {
                Toast.makeText(this, "读取 SDCard 失败", Toast.LENGTH_LONG).show();
            }
        }
    }
    //检测 SDCard 是否可以写入
    private boolean isExternalStorageWriteable()
    {
        return Environment.MEDIA_MOUNTED.equals(Environment.getExternalStorageState
()) ? true : false;
    }
    //检测 SDCard 是否可以读取
    private boolean isExternalStorageReadable()
    {
```

```
        return Environment.MEDIA_MOUNTED.equals(Environment.getExternalStorageState
()) ||
         Environment.MEDIA_MOUNTED_READ_ONLY.equals(Environment.getExternalStor-
ageState()) ? true : false;
    }
    /**
     * 写入数据到 SDCard 卡
     *
     * @param fileName
     *            文件名称
     * @param content
     *            字节数组
     */
    private void saveToSDCard(String fileName, byte[] content)
    {
        if (isExternalStorageWriteable())
        {//检测 SDCard 是否可以写入
            File sdcardDir = Environment.getExternalStorageDirectory();
                                                        //获取 SDCard 的路径
            File baseDir = new File(sdcardDir, "chapter8_8");
            File file = new File(baseDir,  fileName);
            try
            {
                if(! baseDir.exists())
                {//如果 SDCard 不存在 chapter8_8 目录,则创建
                    baseDir.mkdir();
                }
                if(file.exists())
                {//如果 chapter8_8 目录不存在 data.txt,则创建
                    file.createNewFile();
                }
                FileOutputStream fos = new FileOutputStream(file);
                fos.write(content);
                fos.close();
            }
            catch (Exception e)
            {//捕获写入 SDCARD 异常
                Log.d(TAG, "写入 SDCard 出错,exception:" + e);
                throw new RuntimeException("write to sdcard exception");
            }
        }
    }
```

```
//从 SDCard 读取数据
private byte[] loadFromSDCard(String fileName)
{
  if (isExternalStorageReadable())
  {//检测 SDCard 是否可以读取
      int len = 1024;
      byte[] buffer = new byte[len];
      File dir = Environment.getExternalStorageDirectory();//获取 SDCard 的路径
      File file = new File(dir, "chapter8_8/" + fileName);
      try
      {
          FileInputStream fis = new FileInputStream(file);
          ByteArrayOutputStream baos = new ByteArrayOutputStream();
                                                          //写到内存的输出流
          int nrb = fis.read(buffer, 0, len);
          while (nrb != -1)
          {
              baos.write(buffer, 0, nrb);
              nrb = fis.read(buffer, 0, len);
          }
          buffer = baos.toByteArray();
          fis.close();
      }
    catch (Exception e)
      {//捕获读取 SDCard 异常
          Log.d(TAG, "读取 SDCard 出错:exception:" + e);
          throw new RuntimeException("read from sdcard exception");
      }
      return buffer;
    }
    throw new RuntimeException("sdcard is not readable");//抛出运行时异常
}
@Override
public boolean onCreateOptionsMenu(Menu menu)
{
      getMenuInflater().inflate(R.menu.activity_main, menu);
      return true;
  }
}
```

程序向 SDCard 卡中保存数据以及从 SDCard 中读取数据的时候，先对 SDCard 卡的状态进行相应的判断，笔者通过调用 Environment. getExternalStorageState()

来查看 SDCard 卡的状态，如果状态是 MEDIA_MOUNTED，则表示此时的 SDCard 卡既可以读取，也可以写入，如果状态是 MEDIA_MOUNTED_READ_ONLY，则表示此时的 SDCard 卡只可以读取，不可以写。除了提供 getExternalStorageState()方法来判断 SDCard 卡的状态之外，Environment 还提供了以下几个静态方法：

- getExternalStorageDirectory()：用来获取 SDCard 的路径，一般是"/mnt/sdcard/"。
- getDataDirectory()：用来获取 Android 中的 data 数据目录，一般是"/data/"。
- getDownloadCacheDirectory()：用来获取 Android 中下载的缓存目录，一般是"/cache/"。

(4) 右击工程 chapter8_8，在弹出的快捷菜单中选择 Run As|Android Application 命令，输入相关要保存的内容后，单击"保存数据"按钮，效果如图 8-18 所示，单击"读取数据"按钮，效果如图 8-19 所示。

图 8-18 写入 SDCard 效果

图 8-19 SDCard 读取效果

8.4 SQLite 数据库存储

SQLite 是一个非常流行的嵌入式数据库，它支持 SQL 语言，并且只利用很少的内存就会有很好的性能，由于它的开源性，任何项目都可以使用它，许多开源项目(Mozilla，PHP，Python)都使用了 SQLite。SQLite 基本符合 SQL-92 标准，和其他主要的 SQL 数据库没什么区别。它的优点是高效，Android 运行时环境包含了完整的 SQLite。

SQLite 和其他数据库最大的不同就是对数据库类型的支持，创建一个表的时候，可以在 CREATE TABLE 语句中指定某列的数据类型，但是可以把任何数据类型放入该列中。当某个值插入数据库时，SQLite 会检查它的类型，如果该类型与要插入的列的数据类型不匹配，则 SQLite 会尝试将插入的值转换为该列的类型。如果不能转换，则将该值作为其本身具有的类型存储。由于 JDBC 会消耗太多的系统资源，所以 JDBC 对于手机这种内存受限设备来说并不合适。Android 运行时集成了

SQLite,所以每个 Android 应用程序都可以使用 SQLite 数据库。对于熟悉 SQL 的开发人员来时,在 Android 开发中使用 SQLite 相当简单。

Android 不自动提供数据库,如果想要在 Android 应用程序当中使用 SQLite,则需要自己创建数据库,然后创建表、索引和填充数据,以及对数据进行操作。程序当中只需要新建一个继承于 SQLiteOpenHelper 类的子类,就可以轻松地创建数据库以及建表。SQLiteOpenHelper 类根据开发应用程序的需要,封装了创建和更新数据库使用的逻辑。SQLiteOpenHelper 的子类,至少需要实现以下 3 个方法:

- 构造函数,调用父类的构造函数。这个方法需要 4 个参数,上下文环境(例如,一个 Activity),数据库名字,一个可选的游标工厂(通常是 NULL),一个代表正在使用的数据库版本的整数。
- onCreate()方法,它需要一个 SQLiteDatabase 对象,根据需要对这个对象填充表和初始化数据。
- onUpgrade()方法,它需要 3 个参数,一个 SQLiteDatabase 对象,一个旧的版本号和一个新的版本号,这样就可以清楚地从旧的模型转变成新的模型。

(1) 新建工程 chapter8_9,新建一个继承于 android. database. sqlite. SQLiteOpenHelper 类的子类:MySQLiteOpenHelper,编辑 MySQLiteOpenHelper. java 文件,源代码如下:

```
public class MySQLiteOpenHelper extends SQLiteOpenHelper
{
    private static final String DATABASE_NAME = "test.db";
    private static final int DATABASE_VERSION = 1;
    public MySQLiteOpenHelper(Context context)
    {
        //CursorFactory 设置为 null,使用默认值
        super(context, DATABASE_NAME, null, DATABASE_VERSION);
    }
    //数据库第一次被创建时 onCreate 会被调用
    @Override
    public void onCreate(SQLiteDatabase db)
    {
        db.execSQL("CREATE TABLE IF NOT EXISTS person" +
                    "(_id INTEGER PRIMARY KEY AUTOINCREMENT, name VARCHAR, age INTE-
GER, info TEXT)");
    }
    //如果 DATABASE_VERSION 值被改为 2,系统发现现有数据库版本不同,即会调
    //用 onUpgrade
    @Override
    public void onUpgrade(SQLiteDatabase db, int oldVersion, int newVersion)
    {
```

```
        db.execSQL("ALTER TABLE person ADD COLUMN other STRING");
    }
}
```

上述代码当中，在 onCreate()方法当中调用 SQLiteDatabase 对象的 execSQL()方法来执行 DDL 语句，如果没有异常发生，这个方法没有返回值。数据库第一次创建时 onCreate()方法会被调用，可以执行创建表的语句，当系统发现版本变化之后，会调用 onUpgrade 方法，可以执行修改表结构等语句。

(2) 新建数据模型类：Person，编辑 Person.java 文件，源代码如下：

```
public class Person {
    public int _id; //id
    public String name;//姓名
    public int age; //年龄
    public String info;//信息
    public Person() {
    }
    public Person(String name, int age, String info)
    {
        this.name = name;
        this.age = age;
        this.info = info;
    }
}
```

(3) 新建业务处理类：DBManager，编辑 DBManager.java 文件，源代码如下：

```
public class DBManager
{
    private MySQLiteOpenHelper helper;
    private SQLiteDatabase db;
    public DBManager(Context context)
    {
        helper = new MySQLiteOpenHelper(context);
        //因为 getWritableDatabase 内部调用了 mContext.openOrCreateDatabase(mName,
        //0, mFactory);
        //所以要确保 context 已初始化,可以把实例化 DBManager 的步骤放在 Activity 的
        //onCreate 里
        db = helper.getWritableDatabase();
    }
    /**
     * 添加 Person 列表至数据库 person 表当中
     * @param persons
```

```
     */
    public void add(List<Person> persons)
    {
        db.beginTransaction();  //开始事务
        try
        {
        for (Person person : persons)
        {
            db.execSQL("INSERT INTO person VALUES(null, ?, ?, ?)", new Object[]{per-
son.name, person.age, person.info});
        }
        db.setTransactionSuccessful();  //设置事务成功完成
        }
        finally
        {
        db.endTransaction();  //结束事务
        }
    }
    /**
     * 根据名称删除数据
     * @param person
     */
    public void updateAge(Person person)
    {
        ContentValues cv = new ContentValues();
        cv.put("age", person.age);
        db.update("person", cv, "name = ?", new String[]{person.name});
    }
    /**
     * 删除 Person
     * @param person
     */
    public void deleteOldPerson(Person person)
    {
        db.delete("person", "age >= ?", new String[]{String.valueOf(person.
age)});
    }
    /**
     *以列表方式返回所有的 Person
     * @return List<Person>
     */
    public List<Person> query()
```

```
    {
        ArrayList<Person> persons = new ArrayList<Person>();
        Cursor c = queryTheCursor();
        while (c.moveToNext())
        {//如果游标可以指向下一个位置
        Person person = new Person();
        person._id = c.getInt(c.getColumnIndex("_id"));
        person.name = c.getString(c.getColumnIndex("name"));
        person.age = c.getInt(c.getColumnIndex("age"));
        person.info = c.getString(c.getColumnIndex("info"));
        persons.add(person);
      }
      c.close();
      return persons;
    }
    /**
     * 以游标的方式返回查询结果
    * @return     Cursor
    */
    public Cursor queryTheCursor()
    {
        Cursor c = db.rawQuery("SELECT * FROM person", null);
        return c;
    }
    /**
    * 关闭 SQLiteDatabase
    */
    public void closeDB()
    {
        db.close();//关闭 SQLiteDatabase 对象
    }
}
```

笔者在 DBManager 构造方法中实例化 DBHelper 并获取一个 SQLiteDatabase 对象，作为整个应用的数据库实例。在添加多个 Person 信息时，笔者采用了事务处理，确保数据完整性；最后笔者提供了一个 closeDB 方法，释放数据库资源，这个步骤在整个应用关闭时执行，这个环节容易被忘记，所以读者要特别注意。

获取数据库实例时使用了 getWritableDatabase()方法，也许读者会有疑问，在 getWritableDatabase()和 getReadableDatabase()中，为什么选择前者作为整个应用的数据库实例呢？首先来看一下 SQLiteOpenHelper 中的 getReadableDatabase()方法：

```
    public synchronized SQLiteDatabase getReadableDatabase() {
        if (mDatabase != null && mDatabase.isOpen())
        {
            //如果发现 mDatabase 不为空并且已经打开则直接返回
            return mDatabase;
        }
        if (mIsInitializing)
        {
            //如果正在初始化则抛出异常
            throw new IllegalStateException("getReadableDatabase called recursively");
        }
        //开始实例化数据库 mDatabase
        try
        {
            //注意这里是调用了 getWritableDatabase()方法
            return getWritableDatabase();
        }
        catch (SQLiteException e)
        {
            if (mName == null)
                throw e; //Can't open a temp database read-only!
            Log.e(TAG, "Couldn't open " + mName + " for writing (will try read-only):", e);
        }
        //如果无法以可读写模式打开数据库,则以只读方式打开
        SQLiteDatabase db = null;
        try
        {
            mIsInitializing = true;
            String path = mContext.getDatabasePath(mName).getPath();//获取数据库路径
            //以只读方式打开数据库
            db = SQLiteDatabase.openDatabase(path, mFactory, SQLiteDatabase.OPEN_REA-
DONLY);
            if (db.getVersion() != mNewVersion)
            {
                throw new SQLiteException("Can't upgrade read-only database from version "
 + db.getVersion() + " to " + mNewVersion + ": " + path);
            }
            onOpen(db);
            Log.w(TAG, "Opened " + mName + " in read-only mode");
            mDatabase = db;//为 mDatabase 指定新打开的数据库
            return mDatabase;//返回打开的数据库
            } finally
```

```
        {
            mIsInitializing = false;
            if (db != null && db != mDatabase)
                db.close();
        }
    }
```

在 getReadableDatabase()方法中，首先判断是否已存在数据库实例并且是打开状态，如果是，则直接返回该实例，否则通过调用 getWritableDatabase()方法，试图获取一个可读写模式的数据库实例，如果遇到磁盘空间已满等情况导致 getWritableDatabase()方法抛出异常的话，再以只读模式打开数据库，获取数据库实例并返回，然后为 mDatabase 赋值为最新打开的数据库实例。既然有可能调用到 getWritableDatabase()方法，再来看一下 getWritableDatabase()方法的具体实现：

```
public synchronized SQLiteDatabase getWritableDatabase() {
    if (mDatabase != null && mDatabase.isOpen() && ! mDatabase.isReadOnly())
{
        //如果 mDatabase 不为空且已打开并且不是只读模式，则返回该实例
        return mDatabase;
    }
    if (mIsInitializing)
    {
        throw new IllegalStateException("getWritableDatabase called recursively");
    }
    boolean success = false;
    SQLiteDatabase db = null;
    //如果 mDatabase 不为空则加锁，阻止其他的操作
    if (mDatabase != null)
        mDatabase.lock();
    try {
        mIsInitializing = true;
        if (mName == null) {
            db = SQLiteDatabase.create(null);
        } else {
            //打开或创建数据库
            db = mContext.openOrCreateDatabase(mName, 0, mFactory);
        }
        //获取数据库版本(如果是刚创建的数据库，则版本为 0)
        int version = db.getVersion();
        //比较版本(我们代码中的版本 mNewVersion 为 1)
        if (version != mNewVersion) {
            db.beginTransaction();//开始事务
```

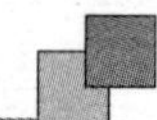

```
            try {
                if (version == 0) {
                    //执行我们的 onCreate 方法
                    onCreate(db);
                } else {
                    //如果我们应用升级了 mNewVersion 为 2,而原版本为 1,则执行
                    //onUpgrade 方法
                    onUpgrade(db, version, mNewVersion);
                }
                db.setVersion(mNewVersion);//设置最新版本
                db.setTransactionSuccessful();//设置事务成功
            } finally {
                db.endTransaction();//结束事务
            }
        }
        onOpen(db);
        success = true;
        return db;//返回可读写模式的数据库实例
    } finally {
        mIsInitializing = false;
        if (success) {
            //打开成功
            if (mDatabase != null) {
                //如果 mDatabase 有值则先关闭
                try {
                    mDatabase.close();
                } catch (Exception e) {
                }
                mDatabase.unlock();//解锁
            }
            mDatabase = db;//赋值给 mDatabase
        } else {
            //打开失败的情况:解锁、关闭
            if (mDatabase != null)
                mDatabase.unlock();
            if (db != null)
                db.close();
        }
    }
}
```

大家可以看到,几个关键的步骤是,首先判断 mDatabase 如果不为空且已打开

并不是只读模式则直接返回，否则如果 mDatabase 不为空则加锁，然后开始打开或创建数据库，比较版本，根据版本号来调用相应的方法，为数据库设置新版本号，最后释放旧的不为空的 mDatabase 并解锁，把新打开的数据库实例赋予 mDatabase，并返回最新实例。

看完上面的过程之后，大家或许就清楚了，如果不是在遇到磁盘空间已满等情况，getReadableDatabase()一般都会返回和 getWritableDatabase()一样的数据库实例，所以笔者在 DBManager 构造方法中使用 getWritableDatabase()获取整个应用所使用的数据库实例是可行的。当然如果真的担心这种情况会发生，那么可以先用 getWritableDatabase()获取数据实例，如果遇到异常，再试图用 getReadableDatabase()获取实例，当然这个时候获取的实例只能读不能写了。

(4) 编辑 res/layout/activity_main. xml 文件，源代码参考配套资料。

(5) 编辑 MainActivity. java 文件，源代码如下：

```
public class MainActivity extends Activity implements OnClickListener
{
    private DBManager mgr;//声明业务管理类
    private ListView listView;
    private Button addBtn, updateBtn, deleteBtn, queryBtn, queryTheCursorBtn;
    @Override
    public void onCreate(Bundle savedInstanceState)
    {
        super.onCreate(savedInstanceState);
        setContentView(R.layout.activity_main);
        //初始化 DBManager
        mgr = new DBManager(this);
        initUI();//初始化 UI 控件
    }
    private void initUI()
    {
    addBtn = (Button)findViewById(R.id.addBtn);
    updateBtn = (Button)findViewById(R.id.updateBtn);
    deleteBtn = (Button)findViewById(R.id.deleteBtn);
    queryBtn = (Button)findViewById(R.id.queryBtn);
    queryTheCursorBtn = (Button)findViewById(R.id.queryTheCursorBtn);
    addBtn.setOnClickListener(this);//绑定单击事件监听器
    updateBtn.setOnClickListener(this);
    deleteBtn.setOnClickListener(this);
    queryBtn.setOnClickListener(this);
    queryTheCursorBtn.setOnClickListener(this);
    istView = (ListView) findViewById(R.id.listView);
```

```
}
@Override
public void onClick(View v)
{
    int viewId = v.getId();
    switch(viewId)
  {
    case R.id.addBtn:               //增加数据
        add();
        break;
    case R.id.updateBtn:            //更新数据
        update();
        break;
    case R.id.deleteBtn:            //删除数据
        delete();
        break;
    case R.id.queryBtn:             //列表形式返回查询结果
        query();
        break;
    case R.id.queryTheCursorBtn:    //游标形式返回查询结果
        queryTheCursor();
        break;
  }
}
public void add()
{
ArrayList<Person> persons = new ArrayList<Person>();
Person person1 = new Person("Ella", 22, "lively girl");
Person person2 = new Person("Jenny", 22, "beautiful girl");
Person person3 = new Person("Jessica", 23, "sexy girl");
Person person4 = new Person("Kelly", 23, "hot baby");
Person person5 = new Person("Jane", 25, "a pretty woman");
persons.add(person1);
persons.add(person2);
persons.add(person3);
persons.add(person4);
persons.add(person5);
mgr.add(persons);
}
public void update()
{
Person person = new Person();
```

```
        person.name = "Jane";
        person.age = 30;
        mgr.updateAge(person);
        }
        public void delete()
        {
        Person person = new Person();
        person.age = 30;
        mgr.deleteOldPerson(person);
        }
        public void query()
        {
        List<Person> persons = mgr.query();
        ArrayList<Map<String, String>> list = new ArrayList<Map<String, String>
>();
        for (Person person : persons)
        {
            HashMap<String, String> map = new HashMap<String, String>();
            map.put("name", person.name);
            map.put("info", person.age + " years old, " + person.info);
            list.add(map);
        }
        SimpleAdapter adapter = new SimpleAdapter(this, list, android.R.layout.simple_
list_item_2,
                        new String[]{"name", "info"}, new int[]{android.R.id.text1, an-
droid.R.id.text2});
        listView.setAdapter(adapter);
        }
        public void queryTheCursor()
        {
        Cursor c = mgr.queryTheCursor();
        startManagingCursor(c);  //托付给 activity 根据自己的生命周期去管理 Cursor
                                 //的生命周期
        CursorWrapper cursorWrapper = new CursorWrapper(c)
        {
        @Override
        public String getString(int columnIndex)
        {
            //将简介前加上年龄
            if (getColumnName(columnIndex).equals("info"))
            {
                int age = getInt(getColumnIndex("age"));
```

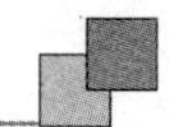

```
                return age + " years old, " + super.getString(columnIndex);
            }
            return super.getString(columnIndex);
          }
        };
        //确保查询结果中有"_id"列
            SimpleCursorAdapter adapter = new SimpleCursorAdapter(this, android.R.lay-
out.simple_list_item_2,
                      cursorWrapper, new String[]{"name", "info"}, new int[]{android.R.
id.text1, android.R.id.text2});
            ListView listView = (ListView) findViewById(R.id.listView);
            listView.setAdapter(adapter);
        }
        protected void onDestroy()
        {
        super.onDestroy();
        //应用的最后一个 Activity 关闭时应释放 DB
        mgr.closeDB();
        }
    }
```

这里需要注意的是 SimpleCursorAdapter 的应用，当程序当中使用这个适配器时，必须先得到一个 Cursor 对象，这里面有几个问题：如何管理 Cursor 的生命周期，如果包装 Cursor，Cursor 结果集都需要注意什么。如果手动去管理 Cursor 的话会非常麻烦，还有一定的风险，处理不当的话运行期间就会出现异常，幸好 Activity 提供了 startManagingCursor(Cursor cursor)方法，它会根据 Activity 的生命周期去管理当前的 Cursor 对象。就是说当 Activity 停止时它会自动调用 Cursor 的 deactivate 方法，禁用游标，当 Activity 重新回到屏幕时它会调用 Cursor 的 requery 方法再次查询，当 Activity 摧毁时，被管理的 Cursor 都会自动关闭释放。

如何包装 Cursor：可以使用 CursorWrapper 对象去包装 Cursor 对象，实现需要的数据转换工作，这个 CursorWrapper 实际上是实现了 Cursor 接口。查询获取到的 Cursor 其实是 Cursor 的引用，而系统实际返回的必然是 Cursor 接口的一个实现类的对象实例，可以用 CursorWrapper 包装这个实例，然后再使用 SimpleCursorAdapter 将结果显示到列表上。

Cursor 结果集需要注意些什么呢？一个最需要注意的是，在结果集中必须要包含一个“_id”的列，否则 SimpleCursorAdapter 就会翻脸不认人，为什么一定要这样呢？因为这源于 SQLite 的规范，主键以“_id”为标准。解决办法有 3 种：第一，建表时根据规范去做；第二，查询时用别名，例如：SELECT id AS _id FROM person；第三，在 CursorWrapper 里做文章。

```
    CursorWrapper cursorWrapper = new CursorWrapper(c)
    {
            @Override
            public int getColumnIndexOrThrow(String columnName) throws IllegalArgumen-
tException
            {
                if (columnName.equals("_id"))
                {
                    return super.getColumnIndex("
id");
                }
                    return super. getColumnIndex-
OrThrow(columnName);
            }
        };
```

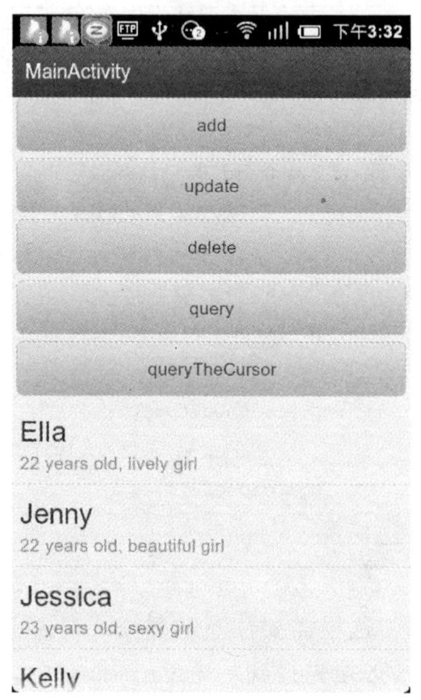

图 8-20 query 效果

(6) 右击工程 chapter8_9，在弹出的快捷菜单中选择 Run As|Android Application 命令，单击界面上的 add 按钮，然后单击 query 按钮，效果如图 8-20 所示。

一些其他操作的效果，笔者再这里就不做过多地演示了。

(7) 通过使用 SQLite 创建的数据库会存放在/data/data/包名/databases/目录下，笔者可以通过进入到设备的 Shell 环境中，查看相关数据库表结构以及表当中的数据。

第 9 章

内容提供者(Content Provider)

上一章介绍了如何在 Android 应用程序当中进行数据存储,提到了 Android 当中使用数据存储的 3 种方式,分别是 SharedPreferences 对象、文件存储以及 SQLite 数据库存储。其中使用 SQLite 数据库存储这种方式一般是用来存储结构比较复杂的数据,而所建立的数据库只能被创建它的应用程序所访问,对于其他应用程序而言是不能访问的。如何在不同应用程序之间共享数据,这就是内容提供者(Content Provider)起到的作用。Content Provider 是 Android 应用程序的 4 大组件之一,与 Activity,Broadcast Receiver 以及 Service 相同,使用前需要注册。作为应用程序之间唯一的共享数据的途径,Content Provider 主要的功能就是存储并检索数据以及向其他应用程序提供访问数据的接口。在本章当中,笔者将会和读者一起详细探讨如何使用 Android 提供的内置的内容提供者,以及如何建立自定义的内容提供者。本章主要包含以下内容:

- ❑ Android 内置内容提供者;
- ❑ 自定义内容提供者。

9.1 Android 内置内容提供者

在 Android 当中,内容提供者是作为在不同应用程序之间共享数据的最佳方式。当一个程序需要把自己的数据暴露给其他应用程序使用的时候,该程序就可以通过提供 Content Provider 来实现,其他应用程序就可以通过 Content Resolver 来操作 Content Provider 暴露的数据。

应用程序通过 Content Provider 开放了自己的数据,该应用程序不需要启动,其他应用程序就可以操作开放的数据,包括增删改查操作。

9.1.1 内置内容提供者

Android 提供了很多有用的内置 Content Provider，使得应用程序可以很方便地访问存储在这些内置 Content Provider 当中的数据，主要提供了以下几个常用的内置 Content Provider：

- Contacts：存储了联系人的详细信息。
- MediaStore：存储了音频、视频以及图片等媒体文件信息。
- Settings：存储了设备的设置信息。

要想查询内容提供者存储的数据，则必须以 URI 的形式指定查询字符串。URI 有两种形式，一种是查询某种数据类型的所有的值，一种是查询某种数据类型的一个特定的记录。举例如下：

- content://contacts/people/：这个 URI 查询字符串将会返回设备上所有联系人的信息。
- content://contacts/people/23：这个 URI 查询字符串只返回"_id"字段为 23 的联系人的信息。
- content://media/images：这个 URI 查询字符串将返回设备上所有图片的信息。

所以 URI 格式的查询字符串一般采用的格式是：<prefix>://<authority>/path/id：

- prefix：一般是 content。
- authority：实际上就是要查询的 Content Provider 的一个名称，用来唯一标识这个 Content Provider，外部应用需要根据这个标识来找到它。比如：对于内置的有关联系人信息相关的内容提供者，它的名称就是：contacts，有关 MediaStore 信息相关的内容提供者，它的名称就是：media。
- path：实际上指定了要查询的数据类型，比如要获取所有的联系人信息，则 URI 查询字符串指定的数据类型就是：people，要获取所有图片的信息，则 URI 查询字符串指定的数据类型就是：images。
- id：如果没有指定 id，则查询名称为<authority>指定，数据类型为<path>所指定的所有的数据信息，如果指定了相应的 id，则返回相应 id 所指定的数据信息。

9.1.2 使用内置内容提供者

下面笔者就通过实例介绍如何使用 Android 提供的有关联系人的内容提供者来进行数据的操作：

(1) 新建工程 chapter9_1，编辑 res/layout/activity_main.xml 文件，源代码参考配套资料。

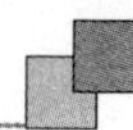

(2) 由于本实例需要用列表的方式来显示所有联系人的姓名以及电话号码,所以还需要为 ListView 的每一行定义一个布局文件。新建 Android 布局文件:contact_row. xml,编辑 contact_row. xml 文件,源代码如下:

```
<? xml version = "1.0" encoding = "utf-8"? >
<LinearLayout xmlns:android = "http://schemas.android.com/apk/res/android"
    android:layout_width = "fill_parent"
    android:layout_height = "wrap_content"
    android:orientation = "horizontal"
    android:weightSum = "3" >
    <!-- ID列 -->
    <TextView
        android:id = "@ + id/contact_id"
        android:layout_width = "fill_parent"
        android:layout_height = "wrap_content"
        android:layout_weight = "1"
        android:gravity = "center_horizontal"
        android:textSize = "18sp" />
    <!-- 姓名列 -->
    <TextView
        android:id = "@ + id/name"
        android:layout_width = "fill_parent"
        android:layout_height = "wrap_content"
        android:layout_weight = "1"
        android:gravity = "center_horizontal"
        android:textSize = "15sp" />
    <!-- 电话号码列 -->
    <TextView
        android:id = "@ + id/telphone"
        android:layout_width = "fill_parent"
        android:layout_height = "wrap_content"
        android:layout_weight = "1"
        android:gravity = "center_horizontal"
        android:text = "电话号码"
        android:textSize = "15sp" />
</LinearLayout>
```

(3) 由于需要读取联系人的数据,所以需要在项目清单文件当中添加相关权限:

```
<uses-permission   android:name = "android.permission.READ_CONTACTS"/>
```

(4) 新建数据模型类:

```
public class ContactShowData
```

```
{
    String id;                          //id
    String displayName;                 //名称
    String phoneNumber;                 //电话号码
    public ContactShowData(String id, String displayName)
    {
        this.id = id;
        this.displayName = displayName;
    }
    public ContactShowData(String id, String displayName, String phoneNumber)
    {
        this.id = id;
        this.displayName = displayName;
        this.phoneNumber = phoneNumber;
    }
    public String getId()
    {
        return id;
    }
    public void setId(String id)
    {
        this.id = id;
    }
    public String getDisplayName()
    {
        return displayName;
    }
    public void setDisplayName(String displayName)
    {
        this.displayName = displayName;
    }
    public String getPhoneNumber()
    {
        return phoneNumber;
    }
    public void setPhoneNumber(String phoneNumber)
    {
        this.phoneNumber = phoneNumber;
    }
}
```

(5) 由于需要使用自定义的布局文件作为 ListView 每一行的布局，笔者新建了

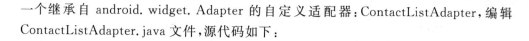

一个继承自 android. widget. Adapter 的自定义适配器：ContactListAdapter，编辑 ContactListAdapter. java 文件，源代码如下：

```
public class ContactListAdapter extends BaseAdapter
{
    private List<ContactShowData> contactList;
    private LayoutInflater layoutInflater;
    public ContactListAdapter(Context context, List<ContactShowData> contactList)
    {
        this.contactList = contactList;
        layoutInflater = LayoutInflater.from(context);
    }
    //返回数据项的总个数
    @Override
    public int getCount()
    {
        return contactList.size();
    }
    //返回指定位置的数据项
    @Override
    public Object getItem(int position)
    {
        return contactList.get(position);
    }
    @Override
    public long getItemId(int position)
    {
        return 0;
    }
    //绘制 ListView 控件每一行,返回 View 给 ListView 控件每一行
    @Override
    public View getView(int position, View convertView, ViewGroup parent)
    {
        ContactViewHolder viewHolder = null;
        if(null == convertView)
        {//如果重用的旧视图为 null
            viewHolder = new ContactViewHolder();
            convertView = layoutInflater.inflate(R.layout.contact_row, null);
            viewHolder.contactId = (TextView)convertView.findViewById(R.id.contact
_id);//ID
            viewHolder.name = (TextView)convertView.findViewById(R.id.name);
                                                                    //联系人姓名
             viewHolder.telphone = (TextView)convertView.findViewById(R.id.tel-
phone);
                                                                    //电话号码
            convertView.setTag(viewHolder);//加入 ViewHolderi 机制,优化 ListView
```

```
        }
        else
        {    //如果重用的旧视图不为 null
            viewHolder = (ContactViewHolder)convertView.getTag();
        }
        ContactShowData model = (ContactShowData) contactList.get(position);
        viewHolder.contactId.setText(model.getId());//设置 ID
        viewHolder.name.setText(model.getDisplayName());//设置联系人姓名
        viewHolder.telphone.setText(model.getPhoneNumber());//设置电话号码
        return convertView;
    }
    private final class ContactViewHolder
    {
        public TextView contactId;//ID
        public TextView name;//联系人姓名
        public TextView telphone;//电话号码
    }
}
```

Adapter 与 View 的连接主要依靠 getView 这个方法返回程序需要的自定义 View。ListView 是 Android 应用程序当中一个最常用的控件，所以如何让 ListView 流畅运行，获取良好的用户体验是非常重要的。由于 ListView 加载的数据会比较多，如果每次调用 getView 的时候，都要通过 LayoutInflater 对象去绘制布局文件，那这样的效率就比较低了。所以笔者采用 Google IO 大会上提供了优化 ListView 的方式，对 ListView 仅仅缓存了可视范围内的 View，随后的滚动都是对这些 View 进行数据更新。

(6) 编辑 MainActivity.java 文件，源代码如下：

```
public class MainActivity extends Activity
{
    ListView contactsListView;//联系人 ListView
    private List<ContactShowData> contactList;//声明存放数据模型的列表
    @Override
    public void onCreate(Bundle savedInstanceState)
      {
        super.onCreate(savedInstanceState);
        setContentView(R.layout.activity_main);
        contactList = new ArrayList<ContactShowData>();
        contactsListView = (ListView)findViewById(R.id.contact_list);
                                        //找到 ID 为 contact_list 的 ListView 控件
        fillListView();//填充 ListView 数据
    }
    private void fillListView()
    {
```

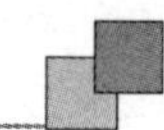

```
        ContentResolver contentResolver = getContentResolver();
        //获取所有联系人     URI:content://com.android.contacts/contacts
        Cursor cursor = contentResolver.query(ContactsContract.Contacts.CONTENT_URI,
null, null, null, null);
        while(cursor.moveToNext()){//循环遍历
            int idColumn = cursor.getColumnIndex(ContactsContract.Contacts._ID);
                                                                    //"_id"列的索引
             int displayNameColumn = cursor.getColumnIndex(ContactsContract.Contacts.
DISPLAY_NAME);                                               //"display_name"列索引
            String contactId = cursor.getString(idColumn);          //获得"_id"列值
            String displayName = cursor.getString(displayNameColumn);
                                                              //获得"display_name"列值
            int hasPhoneNumber = cursor.getInt(cursor.getColumnIndex(ContactsContract.
Contacts.HAS_PHONE_NUMBER));//获得"has_phone_number"列的值
              ContactShowData contactData =    new ContactShowData(contactId, dis-
playName);//显示界面的数据模型

            StringBuffer stringBuffer = new StringBuffer();//拼接电话号码
            if(hasPhoneNumber > 0)
            {   //"has_phone_number"列的值大于 0
                Cursor phoneCursor = contentResolver.query(ContactsContract.Common-
DataKinds.Phone.CONTENT_URI,
                        null, ContactsContract.Data.CONTACT_ID + " = " + contactId,
null, null);
            while(phoneCursor.moveToNext())
            {
                String phoneNumber = phoneCursor.getString(phoneCursor.getColumnIndex
(ContactsContract.CommonDataKinds.Phone.NUMBER));
               stringBuffer.append(phoneNumber).append(" ");
            }
        }
        contactData.setPhoneNumber(stringBuffer.toString());
        contactList.add(contactData);
        contactsListView.setAdapter(new ContactListAdapter(this, contactList));
        }
    }
    @Override
    public boolean onCreateOptionsMenu(Menu menu)
    {
          getMenuInflater().inflate(R.menu.activity_main, menu);
          return true;
        }
}
```

其中 ContentResolver 对象的 query 方法的声明如下：

```
public final Cursor query(Uri uri, String[] projection, String selection, String[]
selectionArgs, String sortOrder)
```

下面笔者对 query()方法的参数解释一下：

- uri：这个参数用来指定查询的 URI，比如：content://com. android. contacts/contacts。
- projection：这个参数用来指定结果中返回哪些列，如果传递的实参为 NULL 的话，则返回所有的列。
- selection：这个参数用来指定结果中返回哪些行，可以用 SQL 语句的 WHERE 子句来进行限制。
- selectionArgs：如果在 selection 参数当中指定了"?"占位符的话，则必须在 selectionArgs 当中指定相应的替代占位符的值，由于 selection 参数当中可能会指定多个"?"占位符，所以 selectionArgs 参数是一个字符串数组的类型。
- sortOrder：这个参数用来指定返回的结果如何排序，可以用 SQL 语句的 ORDER BY 子句来进行限制，如果传递的实参为 NULL 的话，则返回的数据不进行排序。

通过 query()方法获取的 Cursor 对象，可以用来访问结果集记录。如果在查询的 URI 当中指定了 id，那么这个集合最多只会包含一个值，否则，它可以包含多个值，当然如果没有匹配结果的话，则返回 NULL。程序可以直接读取返回的记录集当中的指定列的值，但是前提是知道这个列的数据类型，例如：getString()，getInt()或者是 getFloat()等方法。

(7) 右击工程，在弹出的快捷菜单中选择 Run As|Android Application 命令，运行效果如图 9-1 所示。

(8) 上述代码使用 ContentResolver 对象的 query()方法查询的结果集当中，返回了所有的列，下面笔者对返回的列的集合进行设置，编辑 MainActivity. java 文件的 fillListView()方法，源代码如下：

```
private void fillListView()
{
Uri allContacts = Uri.parse("content://com.android.contacts/contacts");
String[] projection = new String[]
        {ContactsContract.Contacts._ID,  //"_id"列
        ContactsContract.Contacts.DISPLAY_NAME,//"display_name"列
        ContactsContract.Contacts.HAS_PHONE_NUMBER}; //"has_phone_number"列
//managedQuery :由所在的 Activity 来管理 Cursor 的声明周期
Cursor contactsCursor = managedQuery(allContacts, projection, null, null, null);
ListAdapter listAdapter = new SimpleCursorAdapter(this, R.layout.contact_row,
            contactsCursor, new String[] { ContactsContract.Contacts._ID, Con-
tactsContract.Contacts.DISPLAY_NAME,  ContactsContract.Contacts.HAS_PHONE_NUMBER},
```

```
            new int[] { R.id.contact_id, R.id.name, R.id.telphone});
    contactsListView.setAdapter(listAdapter);
}
```

上述代码获得游标对象的时候,采用的是 Activity 当中的 managedQuery()这个方法,这个方法的声明如下:

```
public final Cursor managedQuery(Uri uri, String[] projection, String selection,
                                 String[] selectionArgs, String sortOrder)
```

这个方法的参数与 ContentResolver 对象的 query()方法参数所代表的意思是完全一致的。所不同的地方在于采用 ContentResolver 对象的 query()方法获取的 Cursor 对象是由使用者自行管理,而采用 managedQuery()方法获取的 Cursor 对象是由所在的 Activity 来管理。上述代码当中对结果集当中返回的列表进行了限制,只返回_id,display_name 以及 has_phone_number 这 3 列,然后通过 ListView 将这 3 列的值显示在界面上。

(9) 右击工程 chapter9_1,在弹出的快捷菜单中选择 Run As|Android Application 命令,运行效果如图 9-2 所示(第 3 列"1"表示至少含有一个电话号码,"0"表示不含有电话号码)。

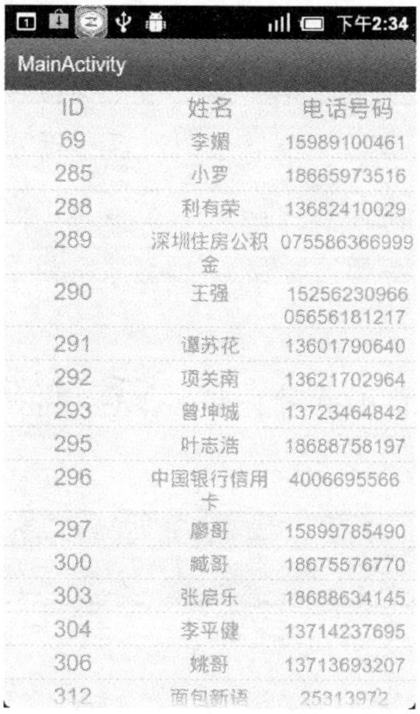

图 9-1　联系人 Content Provider 查询结果

图 9-2　Content Provider Projection 过滤

(10) 除了对返回的列进行过滤以外,还可以对返回的记录进行过滤,编辑 MainActivity.java 文件的 fillListView()方法,源代码如下:

```
private void fillListView(){
```

```
        Uri allContacts = Uri.parse("content://com.android.contacts/contacts");
        String selection = ContactsContract.Contacts.DISPLAY_NAME + " LIKE '李%'";
        //managedQuery :由所在的 Activity 来管理 Cursor 的声明周期
        Cursor contactsCursor = managedQuery(allContacts, null, selection, null, null);
         ListAdapter listAdapter = new SimpleCursorAdapter(this, R.layout.contact_row,
                              contactsCursor, new String[] { ContactsContract.Contacts._ID,
ContactsContract.Contacts.DISPLAY_NAME,  ContactsContract.Contacts.HAS_PHONE_NUMBER},
                             new int[] { R.id.contact_id, R.id.name, R.id.telphone});
         contactsListView.setAdapter(listAdapter);
    }
```

上述代码当中查询了显示名称以“李“字开头的所有联系人信息。

(11) 右击工程 chapter9_1,在弹出的快捷菜单中选择 Run As|Android Application 命令,运行效果如图 9-3 所示。

(12) 当然也可以对返回的记录进行排序处理,编辑 MainActivity.java 文件的 fillListView()方法,源代码如下:

```
    private void fillListView(){
            Uri allContacts = Uri.parse("content://com.android.contacts/contacts");
            String selection = ContactsContract.Contacts.DISPLAY_NAME + " LIKE '李%'";
            String sortType = ContactsContract.Contacts.DISPLAY_NAME + " ASC";
             Cursor contactsCursor = managedQuery(allContacts, null, selection, null,
sortType);
             ListAdapter listAdapter = new SimpleCursorAdapter(this, R.layout.contact_
row, contactsCursor, new String[] {ContactsContract.Contacts._ID, ContactsContract.Con-
tacts.DISPLAY_NAME,
                       ContactsContract.Contacts.HAS_PHONE_NUMBER }, new int[] { R.id.con-
tact_id, R.id.name, R.id.telphone });
            contactsListView.setAdapter(listAdapter);
        }
```

(13) 右击工程 chapter9_1,在弹出的快捷菜单中选择 Run As|Android Application 命令,运行效果如图 9-4 所示。

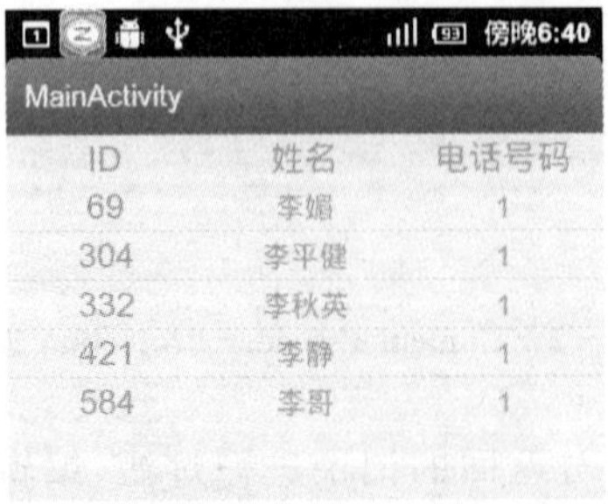

图 9-3 通过 Content Provider Selection 过滤后的效果

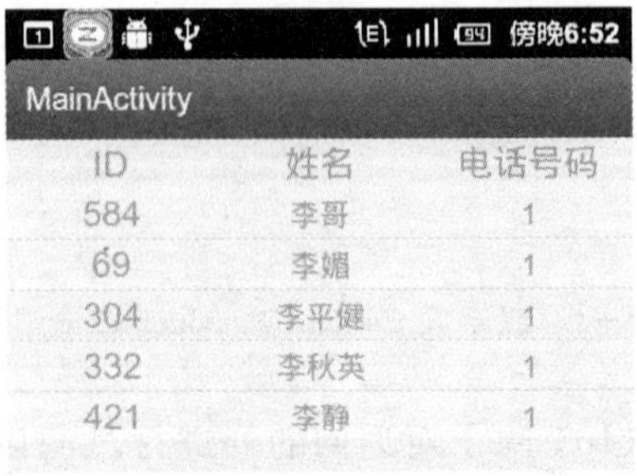

图 9-4 利用 Content Provider Sort 排序的效果

9.2　自定义内容提供者

使用 Android 提供的 SDK 来创建自定义的内容提供者还是比较简单的，只需要去继承 ContentProvider 这个抽象类，然后去实现这个抽象类当中的一些方法，这样就可以通过创建好的自定义内容提供者来完成在应用程序之间进行数据共享的任务。创建自定义的内容提供者的一般步骤如下：

(1) 建立应用程序内部的数据存储系统。程序内部数据存储系统可以由开发人员任意决定，一般来讲，大多数的内容提供者都通过 Android 的文件存储系统或 SQLite 数据库建立自己的数据存储系统，使用 Android 自带的嵌入式 SQLite 数据库更普遍一点。

(2) 继承 Android 的 android.content.ContentProvider 抽象类实现自己的 Content Provider 子类。开发一个继承自 ContentProvider 类的子类代码来扩展 ContentProvider 类，这个步骤的主要工作是将要共享的数据包装，并以 ContentResolver 和 Cursor 对象能够访问到的形式对外展示。具体来说需要实现 ContentProvider 类中的 6 个抽象方法。

- Cursor query(Uri uri, String[] projection, String selection, String[] selectionArgs, String sortOrder)：将查询的数据以 Cursor 对象的形式返回。
- Uri insert(Uri uri, ContentValues values)：向 Content Provider 中插入新数据记录，ContentValues 为数据记录的列名和列值映射。
- int update(Uri uri, ContentValues values, String selection, String[] selectionArgs)：更新 Content Provider 中已存在的数据记录。
- int delete(Uri uri, String selection, String[] selectionArgs)：从 Content Provider 中删除数据记录。
- String getType(Uri uri)：返回 Content Provider 中的数据(MIME)类型。
- boolean onCreate()：当 Content Provider 启动时被调用。

(3) 在 AndroidManifest.xml 文件中注册该 Content Provider 类，为 Content Provider 指定 Uri。

内容提供者跟其他 Android 组件一样，必须在应用程序的项目清单文件 AndroidManifest.xml 文件当中进行声明，否则该内容提供者对于 Android 系统将是不可见的。

下面笔者就通过实例演示如何创建自定义的内容提供者。

(1) 新建工程 chapter9_2，新建一个继承自 android.database.sqlite.SQLiteOpenHelper 的子类：DatabaseOpenHelper，编辑 DatabaseOpenHelper.java 文件，源代码如下：

```
public class DatabaseOpenHelper extends SQLiteOpenHelper
```

```
{
    public static final String DATABASE_NAME = "example";     //数据库名称
    public static final String TABLE_NAME = "book";           //表名称
    public static final int DATABASE_VERSION = 1;             //数据库版本号
    public static final String DATABASE_CREATE_DDL = "create table " + TABLE_NAME
            + "(_id integer primary key autoincrement, name text not null, author
text not null)";
    public DatabaseOpenHelper(Context context)\
    {
        super(context, DATABASE_NAME, null, DATABASE_VERSION);
    }
    @Override
    public void onCreate(SQLiteDatabase db)
    {
        db.execSQL(DATABASE_CREATE_DDL);
    }
    @Override
    public void onUpgrade(SQLiteDatabase db, int oldVersion, int newVersion)
    {
        db.execSQL("DROP TABLE IF EXISTS book");
        onCreate(db);
    }
}
```

(2) 新建一个继承自 android. content. ContentProvider 的子类：ExampleContentProvider，编辑 ExampleContentProvider. java 文件，源代码如下：

```
public class ExampleContentProvider extends ContentProvider
{
    private static final String PROVIDER_NAME = "com.example.chapter9_2.ExampleCon-
tentProvider";
    //定义 Content Provider URI
    public static final Uri CONTENT_URI = Uri.parse("content://" + PROVIDER_NAME +
"/books");
    public static final String _ID = "_id"; //图书 ID 列名
    public static final String NAME = "name"; //图书名称列名
    public static final String AUTHOR = "author"; //图书作者列名
    public static final int BOOKS = 1; //代表操作所有图书
    public static final int BOOK_ID = 2; //代表操作指定 ID 图书
    private static final UriMatcher uriMatcher; //Uri 匹配类
    static
    {
        //如果 URI 为 content://com.example.chapter9_2.ExampleContentProvider/
```

```
books,则匹配的 code 为 1
            uriMatcher = new UriMatcher(UriMatcher.NO_MATCH);
            uriMatcher.addURI(PROVIDER_NAME, "books", BOOKS);//
            //#号匹配任意数字,如果 URI 为 content://com.example.chapter9_2.Example-
ContentProvider/books/1,则匹配的 code 为 2
            uriMatcher.addURI(PROVIDER_NAME, "books/#", BOOK_ID);
        }
        //SQLite 数据库相关声明
        SQLiteDatabase exampleDB;
        @Override
        public boolean onCreate()
        {
            Context context = getContext(); //得到上下文对象
            DatabaseOpenHelper dbHelper = new DatabaseOpenHelper(context);
            exampleDB = dbHelper.getWritableDatabase();
            return (exampleDB == null) ? false : true;
        }
        @Override
        public int delete(Uri uri, String selection, String[] selectionArgs)
        {
            int count = 0;
            switch (uriMatcher.match(uri))
            {
            case BOOKS://删除所有书籍
                count = exampleDB.delete(DatabaseOpenHelper.TABLE_NAME, selection, se-
lectionArgs);
                break;
            case BOOK_ID://删除某一本书籍
                String id = uri.getPathSegments().get(1);
                count = exampleDB.delete(DatabaseOpenHelper.TABLE_NAME, _ID + " = " +
id + (! TextUtils.isEmpty(selection) ? " AND (" + selection + ')' : ""), selectionArgs);
                break;
            default:
                throw new IllegalArgumentException("Unknown URI " + uri);
            }
            getContext().getContentResolver().notifyChange(uri, null);
            return count;
        }
        @Override
        public String getType(Uri uri)
        {
            switch (uriMatcher.match(uri))
```

```
        {
        case BOOKS：//如果 uri 代表所有书籍
            return "vnd.android.cursor.dir/vnd.example.books";
        case BOOK_ID://如果 uri 代表某一本书籍
            return "vnd.android.cursor.item/vnd.example.books";
        default:
            throw new IllegalArgumentException("Unsupported uri:" + uri);
        }
    }
    @Override
    public Uri insert(Uri uri, ContentValues values)
    {
        //添加新书
        long rowID = exampleDB.insert(DatabaseOpenHelper.TABLE_NAME, "", values);
        //添加成功
        if (rowID > 0)
        {//如果 rowID 大于 0,代表添加成功
            Uri _uri = ContentUris.withAppendedId(CONTENT_URI, rowID);
            getContext().getContentResolver().notifyChange(_uri, null);
            return _uri;
        }
        throw new SQLException("Failed to insert row into " + uri);
    }
    @Override
    public Cursor query(Uri uri, String[] projection, String selection, String[] se-
lectionArgs, String sortOrder) {
        SQLiteQueryBuilder sqlBuilder = new SQLiteQueryBuilder();
        sqlBuilder.setTables(DatabaseOpenHelper.TABLE_NAME);
        if (uriMatcher.match(uri) == BOOK_ID)//查询某一本书籍
            //---if getting a particular book---
            sqlBuilder.appendWhere(_ID + " = " + uri.getPathSegments().get(1));
        if (sortOrder == null || sortOrder == "")
        {//如果 sortOrder 为空
            sortOrder = NAME;
        }
        Cursor c = sqlBuilder.query(exampleDB, projection, selection, selection-
Args, null, null, sortOrder);
        //---register to watch a content URI for changes---
        c.setNotificationUri(getContext().getContentResolver(), uri);
        return c;
    }
    @Override
```

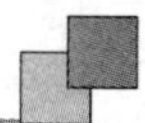

```
        public int update(Uri uri, ContentValues values, String selection, String[] selec-
tionArgs)
        {
            int count = 0;
            switch (uriMatcher.match(uri))
            {
            case BOOKS:     //更新所有书籍
                count = exampleDB.update(DatabaseOpenHelper.TABLE_NAME, values, selec-
tion, selectionArgs);
                break;
            case BOOK_ID://更新某一本书籍
                count = exampleDB.update(DatabaseOpenHelper.TABLE_NAME, values, _ID +
" = " + uri.getPathSegments().get(1) + (! TextUtils.isEmpty(selection) ? " AND (" + se-
lection + ')' : ""),
                        selectionArgs);
                break;
            default:
                throw new IllegalArgumentException("Unknown URI " + uri);
            }
            getContext().getContentResolver().notifyChange(uri, null);
            return count;
        }
    }
```

上述代码中 insert()方法以及 delete()方法在执行更新语句过后，都调用了 ContentResolver 对象的 notifyChange()方法，用来通知注册了的观察者有数据被更新了。默认情况下，CursorAdapter 的对象会收到这个通知。

(3) 由于需要在 AndroidManifest.xml 项目清单文件当中对自定义的 Content Provider 进行注册，编辑 AndroidManifest.xml 文件，源代码如下：

```
    <manifest xmlns:android = "http://schemas.android.com/apk/res/android"
        package = "com.example.chapter9_2"    android:versionCode = "1"     android:ver-
sionName = "1.0" >
        <uses-sdk    android:minSdkVersion = "8"   android:targetSdkVersion = "15" />
        <application   android:icon = "@drawable/ic_launcher"     android:label = "@
string/app_name"
                    android:theme = "@style/AppTheme" >
            <activity   android:name = ".MainActivity"   android:label = "@string/title
_activity_main" >
                <intent-filter>
                    <action android:name = "android.intent.action.MAIN" />
                    <category android:name = "android.intent.category.LAUNCHER" />
```

```
            </intent-filter>
        </activity>
        <!-- Content Provider 定义 -->
        <provider android:name = ".ExampleContentProvider"
                android:authorities = "com.example.chapter9_2.ExampleContentPro-
vider"></provider>
    </application>
</manifest>
```

定义好 Content Provider 过后，接下来笔者介绍如何使用自定义的内容提供者。

(4) 编辑 res/layout/activity_main.xml 文件，源代码如下：

```
<LinearLayout xmlns:android = "http://schemas.android.com/apk/res/android"
    xmlns:tools = "http://schemas.android.com/tools"
    android:layout_width = "fill_parent"
    android:layout_height = "fill_parent"
    android:orientation = "vertical" >
      <!-- 显示书名 -->
        <TextView
            android:layout_width = "fill_parent"
            android:layout_height = "wrap_content"
            android:text = "Title"/>
        <!-- 书名输入框 -->
         <EditText
            android:id = "@ + id/title"
            android:layout_width = "fill_parent"
            android:layout_height = "wrap_content"/>
          <!-- 显示作者 -->
        <TextView
            android:layout_width = "fill_parent"
            android:layout_height = "wrap_content"
            android:text = "Author"/>
        <!-- 作者输入框 -->
         <EditText
            android:id = "@ + id/author"
            android:layout_width = "fill_parent"
            android:layout_height = "wrap_content"/>
         <!-- 添加书籍按钮 -->
         <Button android:id = "@ + id/btnAdd"
                android:layout_width = "fill_parent"
                    android:layout_height = "wrap_content"
                    android:text = "添加书籍"/>
          <!-- 查询书籍按钮 -->
```

```
    <Button android:id = "@ + id/btnQuery"
            android:layout_width = "fill_parent"
                android:layout_height = "wrap_content"
                android:text = "书籍列表"/><LinearLayout
  android:layout_width = "fill_parent"
  android:layout_height = "wrap_content"
  android:orientation = "horizontal"
  android:weightSum = "3" >
  <!-- 显示 ID -->
  <TextView
      android:layout_width = "fill_parent"
      android:layout_height = "wrap_content"
      android:layout_weight = "1"
      android:gravity = "center_horizontal"
      android:text = "ID"
      android:textSize = "18sp" />
  <!-- 显示书名 -->
  <TextView
      android:layout_width = "fill_parent"
      android:layout_height = "wrap_content"
      android:layout_weight = "1"
      android:gravity = "center_horizontal"
      android:text = "姓名"
      android:textSize = "18sp" />
  <!-- 显示书的作者 -->
  <TextView
      android:layout_width = "fill_parent"
      android:layout_height = "wrap_content"
      android:layout_weight = "1"
      android:gravity = "center_horizontal"
      android:text = "作者"
      android:textSize = "18sp" />
</LinearLayout>
    <!-- 展示书籍列表 -->
    <ListView android:id = "@ + id/bookList"
          android:layout_width = "fill_parent"
          android:layout_height = "wrap_content"/>
</LinearLayout>
```

(5) 由于需要使用自定义的布局文件作为 ListView 每一行的布局，新建一个布局文件：book_row.xml，编辑 book_row.xml 文件，源代码如下：

```
<? xml version = "1.0" encoding = "utf-8"? >
```

```
<LinearLayout xmlns:android="http://schemas.android.com/apk/res/android"
    android:layout_width="fill_parent"
    android:layout_height="wrap_content"
    android:orientation="horizontal"
    android:weightSum="3" >
    <!-- ID 列 -->
    <TextView
        android:id="@+id/book_id"
        android:layout_width="fill_parent"
        android:layout_height="wrap_content"
        android:layout_weight="1"
        android:gravity="center_horizontal"
        android:textSize="18sp" />
    <!-- 书名列 -->
    <TextView
        android:id="@+id/name"
        android:layout_width="fill_parent"
        android:layout_height="wrap_content"
        android:layout_weight="1"
        android:gravity="center_horizontal"
        android:textSize="15sp" />
    <!-- 作者列 -->
    <TextView
        android:id="@+id/author"
        android:layout_width="fill_parent"
        android:layout_height="wrap_content"
        android:layout_weight="1"
        android:gravity="center_horizontal"
        android:textSize="15sp" />
</LinearLayout>
```

(6) 编辑 MainActivity.java 文件，源代码如下：

```
public class MainActivity extends Activity implements OnClickListener
{
    EditText titleEditText;//声明标题输入框
    EditText authorEditText;//声明作者输入框
    Button addBookBtn;//声明添加书籍按钮
    Button queryBookBtn;//声明查询书籍按钮
    ListView booksListView;//声明 ListView 控件
    @Override
    public void onCreate(Bundle savedInstanceState)
    {
```

```
        super.onCreate(savedInstanceState);
        setContentView(R.layout.activity_main);
        titleEditText = (EditText)findViewById(R.id.title);//找到标题输入框
        authorEditText = (EditText)findViewById(R.id.author);//找到作者输入框
        addBookBtn = (Button)findViewById(R.id.btnAdd);//找到添加书籍按钮
        queryBookBtn = (Button)findViewById(R.id.btnQuery);//找到查询书籍按钮
        booksListView = (ListView)findViewById(R.id.bookList);//找到 id 为 bookList
                                                               //的 ListView 控件
        booksListView.setOnItemClickListener(myListItemClickListener);
                                                  //为 ListView 绑定单击事件监听器
        addBookBtn.setOnClickListener(this);//为添加按钮绑定单击事件监听器
        queryBookBtn.setOnClickListener(this);//为查询书籍按钮绑定单击事件监听器
    }

    @Override
    public void onClick(View view)
    {
    int viewId = view.getId();
    if(R.id.btnAdd == viewId)
      {    //如果是添加书籍按钮
            String title = titleEditText.getText().toString();//书名
            String author = authorEditText.getText().toString();//书的作者
            //ContentValues 类和 Hashtable 比较类似,它也是负责存储一些名值对,但是
            //它存储的名值对当中的名是一个 String 类型,而值都是基本类型
            ContentValues values = new ContentValues();
            values.put(ExampleContentProvider.NAME, title);
            values.put(ExampleContentProvider.AUTHOR, author);
            Uri uri = getContentResolver().insert(ExampleContentProvider.CONTENT_
URI, values);
            Toast.makeText(getBaseContext(), uri.toString(), Toast.LENGTH_LONG).
show();
      }
    else if(R.id.btnQuery == viewId)
    {//如果是查询书籍按钮
    Uri allBooksUri = Uri.parse("content://com.example.chapter9_2.ExampleContent-
Provider/books");
        //managedQuery 返回的 Cursor 对象由 Activity 负责管理
        Cursor booksCursor = managedQuery(allBooksUri, null, null, null, null);
        ListAdapter booksAdapter = new SimpleCursorAdapter(this, R.layout.book_row,
booksCursor,
                new String[]{ExampleContentProvider._ID, ExampleContentProvider.
NAME, ExampleContentProvider.AUTHOR},
```

```
                    new int[]{R.id.book_id, R.id.name, R.id.author});
            booksListView.setAdapter(booksAdapter);
        }
    }
    /**
     * ListView 每一项单击事件监听器
     */
    private AdapterView.OnItemClickListener myListItemClickListener = new Adapter-
View.OnItemClickListener()
    {
    //处理 ListView 的 Item 的单击事件
    //parent 参数:当前的 AdapterView 对象,这里就是 ListView
    //view 参数:AdapterView 所绑定的 getView()方法返回的 View
    //position 参数:当前被点击的条目的索引号
    //id 参数:AdapterView 所绑定的 Adapter 的 getItemId()返回的值
        @Override
        public void onItemClick(AdapterView<? > parent , View view, int position,
long id)
        {
            TextView idTextView = (TextView)findViewById(R.id.book_id);
            String deleteId = idTextView.getText().toString();
            final Uri deleteBookUri = Uri.parse("content://com.example.chapter9_2.
ExampleContentProvider/books/" + deleteId);
             AlertDialog.Builder dialog = new AlertDialog.Builder(MainActivity.
this).setTitle("确认删除此条记录吗?");
            dialog.setPositiveButton("确定", new DialogInterface.OnClickListener()
              {
                @Override
                public void onClick(DialogInterface dialog, int which)
                {
                    getContentResolver().delete(deleteBookUri, null, null);
                    dialog.dismiss();
                }
            });
            dialog.setNegativeButton("取消", new DialogInterface.OnClickListener()
            {
                @Override
                public void onClick(DialogInterface dialog, int which)
                {
                    dialog.dismiss();
                }
            });
```

```
            dialog.show();
        }
    };
    @Override
    public boolean onCreateOptionsMenu(Menu menu)
    {
        getMenuInflater().inflate(R.menu.activity_main, menu);
        return true;
    }}
```

上述代码对自定义的内容提供者的添加数据、删除数据以及查询数据所暴露出的方法进行了相应操作，单击 ListView 的每一项，会首先弹出一个对话框，如果用户单击"确定"按钮，则会删除数据库里面相应的数据。

(7) 右击工程 chapter9_2，在弹出的快捷菜单中选择 Run As|Android Application 命令，在 Title 输入框和 Author 输入框当中输入相关信息，单击"添加书籍"按钮，效果如图 9-5 所示，单击"查询书籍"按钮，效果如图 9-6 所示。

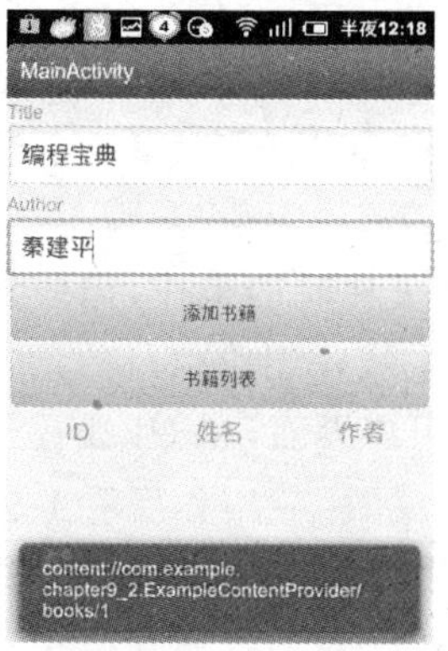

图 9-5　Content Provider 添加数据效果

图 9-6　Content Provider 查询数据效果

第 10 章

Android 异步处理机制

一个 Android 应用程序开始运行的时候，会单独启动一个进程。默认情况下，应用程序的 4 大组件：Activity、Broadcast Receiver、Content Provider 以及 Service 都在此进程当中运行。Android 每个应用程序的进程当中，都会有一个名称为 main 的主线程。这个主线程（也称为 UI 线程）很重要，Android 规定所有更新 UI 界面的操作必须在 UI 线程当中进行，否则，系统底层将会抛出相关异常。UI 主线程主要负责事件的分发以及屏幕的重绘。

Android 应用程序当中往往需要去执行一些比较耗时的操作，比如访问网络去下载图片，对于这样的耗时操作如果也是放在 UI 主线程当中执行的话，会导致 UI 线程不能及时地分发事件，也不能及时地重新绘制用户界面，在 UI 线程当中执行耗时操作的后果就是 UI 线程会被阻塞。从用户的角度看，应用程序看上去像是挂掉了。更糟糕的是，如果应用程序在 5 s 内没有响应用户输入的事件（例如，按下界面按钮或触摸屏幕），系统会向用户显示一个对话框，这个对话框称作应用程序无响应（ANR：Application Not Responding）对话框。用户可以选择让程序继续运行，但是，用户在使用应用程序的时候，并不希望每次都要处理这个对话框。因此，在程序里对响应性能的设计很重要，这样，系统不会显示 ANR 给用户。所以，需要将一些比较耗时的操作放到其他子线程当中去执行，执行完成过后，如果需要更新用户界面，再通知 UI 主线程去更新用户界面。因以，本章将会包含以下内容：

- ❑ 子线程；
- ❑ Handler 的使用；
- ❑ AsyncTask 的使用。

10.1 子线程

为了避免将 UI 主线程阻塞，所以需要将一些比较耗时的操作放到其余子线程

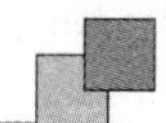

当中处理。在 Java 当中创建线程有两种方法。

10.1.1　实现 Runnable 接口

第一种方式:使用 Runnable 接口创建线程。

- Java 的线程是通过 java.lang.Thread 类来实现的。
- 每个线程都是通过某个特定的 Thread 对象所对应的 run()方法来完成其操作。方法 run()称为线程体。

例如:

```
public class Thread1
{
    public static void main(String[] args)
    {
        Runnable myRunnable = new MyRunnable();
        //Java 线程都是通过 Thread 类来实现
        Thread myThread = new Thread(myRunnable);
        myThread.start();//调用 myRunnable 对象的 start()方法
    }
}
//实现 Runnable 接口
class MyRunnable implements Runnable
{
    //调用 Thread 对象 start()方法执行的方法
    //真正的线程方法体
    @Override
    public void run()
    {
        for(int i = 1;i<20;i + + )
        {
            System.out.println(i);
        }
    }
}
```

程序的运行结果是:控制台输出了 1～19 这 19 个数字,当然也可以把 run()方法当中的函数体移到 main()函数当中执行。但是这两种方式的意义就会不一样,上述代码当中通过新建实现 Runnable 接口的一个内部类,从而程序当中会有两个线程同时执行,一个是 main 主线程,另外一个就是新建的 myThread 子线程。所以,如果要通过实现 Runnable 接口的方式来创建子线程一般会有以下几个步骤:

- 定义一个实现 Runnable 接口的类,实现接口当中定义的 run()方法,在 run()方法当中加入具体的处理逻辑。

❑ 创建 Runnable 接口实现类的对象。
❑ 创建一个 Thread 类的对象,需要封装前面 Runnable 接口实现类的对象。
❑ 调用 Thread 对象的 start()方法,启动线程。

10.1.2 继承 Thread 类

第二种方式:继承 Thread 类创建线程。

例如:

```
public class Thread2
{
    public static void main(String[] args)
    {
        Thread myThread = new MyThread();
        myThread.start();
    }
}
//继承 Thread 类
class MyThread extends Thread
{
        //调用 Thread 对象 start()方法执行的方法
        //真正的线程方法体
        @Override
        public void run()
        {
      for(int i = 1;i<20;i + + )
      {
            System.out.println(i);
      }
    }
}
```

所以,如果要通过继承 Thread 类的方式来创建子线程一般会有以下几个步骤:

❑ 首先定义一个类去继承 Thread 类,Thread 类实际上也实现了 Runnable 接口,这样新建的继承 Thread 类的子类实际上也实现了 Runnable 接口,重写 Thread 类的 run()方法,在 run()方法中加入具体的处理逻辑。
❑ 直接创建一个上述子类的对象,当然可以利用面向对象编程的多态性,将引用声明为父类的类型。
❑ 调用上述创建的子类的 start()方法,启动线程。

在 Java 中,类仅支持单继承,也就是说,当定义一个新的类的时候,它只能扩展一个外部类。这样,如果创建自定义线程类的时候是通过扩展 Thread 类的方法来

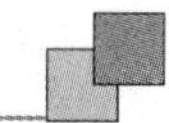

实现的，那么这个自定义类就不能再去扩展其他的类，也就无法实现更加复杂的功能。因此，如果自定义类必须扩展其他的类，那么就可以使用实现 Runnable 接口的方法来定义该类为线程类，这样就可以避免 Java 单继承所带来的局限性。

10.1.3 Android 创建子线程

下面笔者就通过实例介绍如何在 Android 应用程序当中通过创建子线程的方式完成一些比较耗时的操作。

(1) 新建工程 chapter10_1，编辑 res/layout/activity_main. xml 文件，源代码参考配套资料。

(2) 编辑 MainActivity. java 文件，源代码如下：

```
public class MainActivity extends Activity implements OnClickListener
{
    private static final String TAG = "MainActivity";
    private Button myButton;//声明按钮控件
    private TextView myTextView;//声明文本控件
    @Override
    public void onCreate(Bundle savedInstanceState)
    {
        Log.d(TAG, Thread.currentThread().getName() + "开始执行");
        super.onCreate(savedInstanceState);
        setContentView(R.layout.activity_main);
        myButton = (Button)findViewById(R.id.myButton);//找到按钮控件
        myTextView = (TextView)findViewById(R.id.myTextView);//找到文本控件
        myButton.setOnClickListener(this);//绑定单击事件监听器
        Log.d(TAG, Thread.currentThread().getName() + "结束执行");
    }
      @Override
    public void onClick(View view)
    {
        if(R.id.myButton == view.getId())
        {//如果 id 为 myButton 的按钮被单击
            Thread testThread = new Thread(new TestRunnable());//创建子线程
            testThread.start();//启动子线程
        }
    }
    //实现 Runnable 接口创建子类
    class TestRunnable implements Runnable
    {
        @Override
        public void run()
```

```
            {
                //模拟从网络下载图片需要耗时 10 s
                //SystemClock.sleep 的作用与 Thread.sleep()一致
                //Thread.currentThread().getName()获取当前运行线程名称
                Log.d(TAG, Thread.currentThread().getName() + "子线程开始执行");
                SystemClock.sleep(6 * 1000);//子线程休眠 10 s
                Log.d(TAG, Thread.currentThread().getName() + "子线程结束执行");
                //更新 UI 界面,子线程当中更新 UI 界面会抛异常
                myTextView.setText("子线程执行完毕");
            }
        }
        @Override
        public boolean onCreateOptionsMenu(Menu menu)
        {
            getMenuInflater().inflate(R.menu.activity_main, menu);
            return true;
        }
    }
```

上述代码当中,在按钮的单击事件处理当中,笔者新建了一个子线程,这个子线程是通过实现 Runnable 接口的方式来模拟从网络上下载数据等比较耗时的操作。SystemClock 类的 sleep()静态方法事实上与 Thread 类的 sleep()静态方法的作用是一致的,目的都是用来使当前线程休眠。休眠的时间是以毫秒为单位。Thread.currentThread().getName()是用来获取当前运行线程的名称。

在 TestRunnable 类的 run()方法体当中,笔者最后调用了 myTextView.setText()这个方法,用来更新文本控件的显示内容,不过正如笔者之前所介绍的那样,所有更新 UI 界面的操作必须放在 UI 主线程当中调用,所以如果在非 UI 线程当中对 UI 界面进行更新的话,系统会抛出相关异常。

(3) 右击工程 chapter10_1,在弹出的快捷菜单中选择 Run As|Android Application 命令,运行效果如图 10-1 所示,单击界面上的"子线程测试"按钮 LogCat 输出如图 10-2 所示。

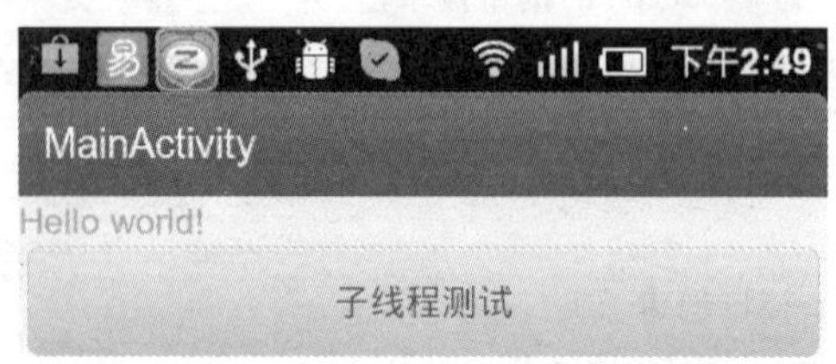

图 10-1 chapter10_1 运行效果

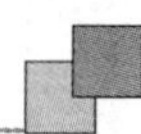

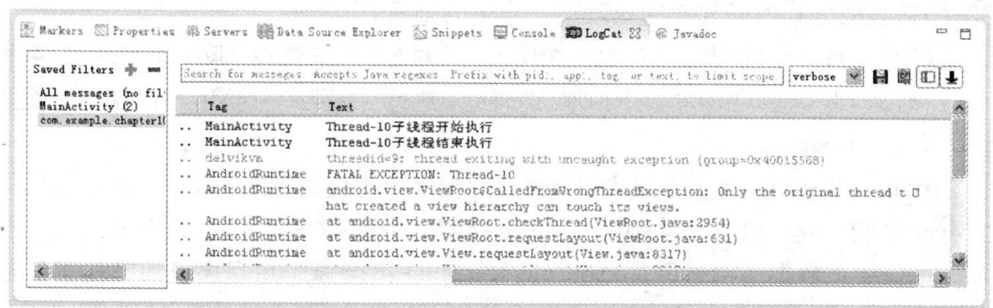

图 10-2　单击按钮 LogCat 输出

由于是在名称为 Thread-10 的子线程当中对 UI 界面进行更新，所以 Android 运行时会抛出一个 android. view. ViewRootMYMCalledFromWrongThreadException 异常。从错误消息中不难看出 Android 运行时禁止其他子线程来更新由 UI 线程所创建的 UI 界面。那么子线程的业务逻辑处理完成之后，如果通知 UI 主线程来更新相应的用户界面呢？下面笔者就会为读者介绍 Handler 的使用。

10.2　Handler 的使用

对于应用程序里面一些比较耗时的操作，需要放在子线程当中执行，在大部分情况下，子线程都会涉及 UI 界面的更新。而 Android 的 UI 主线程主要负责 UI 界面的显示、更新和控件交互，也就是说，更新 UI 界面只能在主线程中更新，子线程当中更新 UI 是被禁止的。Handler 对象就是用来解决这个问题的。

一般会在 UI 主线程当中定义 Handler 对象，它与子线程可以通过 Message 对象来传递数据，Handler 对象将 Message 加入到 MessageQueue 中，Looper 对象负责抽取出要处理的 Message 对象传递给 Handler 对象的 handleMessage()方法进行处理。

10.2.1　Android 消息机制

对于 Android 的消息处理，涉及到 Handler、Looper、Message、MessageQueue 等概念。笔者先来解释一下这几个概念：

- Message：消息，其中包含了消息的 ID、消息处理对象以及处理的数据等，由 MessageQueue 统一队列，终由 Handler 处理。
- Handler：处理者，负责 Message 的发送和处理。使用 Handler 时，需要实现 handlerMessage(Message msg)方法对特定的 Message 进行处理，例如更新 UI 界面等。
- MessageQueue：消息队列，用来存放 Handler 发送过来的消息，并按照先进先出的规则执行。当然存放 Message 并非实际意义的保存，而是将 Message 以链表的方式串联起来的,等待 Looper 的抽取。

- Looper:一个线程可以产生一个 Looper 对象,用来管理 MessageQueue,不断地从 MessageQueue 当中抽取 Message 执行。因此一个 MessageQueue 需要一个 Looper。

Android Runtime 在应用程序启动的时候,都会为应用程序的 UI 主线程创建一个 MessageQueue。也就是说每一个 UI 主线程当中都有一个 Looper 对象以及一个 MessageQueue 数据结构。但是 Android Runtime 不会为应用程序里面其他子线程创建与线程相绑定的 MessageQueue 数据结构,当然也不会创建抽取 MessageQueue 数据结构的 Looper 对象。

开发 Android 应用程序的时候,一般是在 UI 主线程当中创建 Handler 对象,这样创建的 Handler 对象就与 UI 主线程的 MessageQueue 数据结构以及负责抽取该 MessageQueue 数据结构的 Looper 对象绑定在一起。Handler 类允许发送消息和处理 MesageQueue 当中的消息以及 Runnable 对象。

Handler 对象负责将需要传递的信息封装成 Message 对象,将 Message 的 target 设定成自己(目的是为了在处理消息循环时,Message 对象能够找到正确的 Handler 对象),然后将 Message 对象放入到 MessageQueue 中。Looper 对象不断地从 MessageQueue 中获取下一个(next 方法)Message,然后通过 Message 中携带的 target 信息,交由正确的 Handler 处理(dispatchMessage 方法)。

10.2.2 Handler 更新 UI 界面

下面笔者就通过实例介绍子线程是如何使用 Handler 来与主线程进行交互,并且达到更新 UI 界面的效果的。

(1) 新建工程 chapter10_2,编辑 res/layout/activity_main.xml 文件,源代码参考配套资料。

(2) 编辑 MainActivity.java 文件,源代码如下:

```
public class MainActivity extends Activity implements OnClickListener
{
    ProgressBar myProgressBar;//声明进度条控件
    Button myStartButton;//声明控制进度条开始的按钮
    Button myEndButton;//声明控制进度条结束的按钮
    private boolean isRunning = true;
    @Override
    public void onCreate(Bundle savedInstanceState)
    {
        super.onCreate(savedInstanceState);
        setContentView(R.layout.activity_main);
        myProgressBar = (ProgressBar)findViewById(R.id.myProgressBar);
```

```
                                                    //找到进度条控件
    myProgressBar.setProgress(0);  //设置进度条控件当前进度
    myProgressBar.setMax(100);  //设置进度条控件最大进度
    myStartButton = (Button)findViewById(R.id.myStartButton);
                                                //找到控制进度条开始的按钮
    myEndButton = (Button)findViewById(R.id.myEndButton);
                                                //找到控制进度条结束的按钮
    myStartButton.setOnClickListener(this);         //绑定单击事件监听器
    myEndButton.setOnClickListener(this);
}
//UI 主线程当中定义的 Handler 对象会与 UI 主线程的 MessageQueue 数据结构以及负
//责抽取 MessageQueue
//数据结构的 Looper 对象绑定在一起
private Handler handler = new Handler()
{
//Looper 发现 MessageQueue 中有消息时,会按照 FIFO 的规则将消息传递给相应的
//Handler 对象去处理
public void handleMessage(android.os.Message msg)
{
      myProgressBar.incrementProgressBy(5);//每过一秒,进度条进度增加 5
};
};
//实现 Runnable 接口创建子类
class MyRunnable implements Runnable
{
@Override
public void run()
  {
    for(int i = 0; i < 20 && isRunning; i ++ )
    {//进行 20 次循环
        //线程休眠 1 s,也就是意味着,每过一 s,Handler 就发送一个 Message 对象
        //给 Looper
        SystemClock.sleep(1000);
        Message msg = handler.obtainMessage();//获取 Message 对象
        //发送消息,将消息放入到 MessageQueue 当中
        handler.sendMessage(msg);
    }
}
}
@Override
public void onClick(View view)
{
```

```
        int viewId = view.getId();
        if(R.id.myStartButton == viewId)
          {//如果是开始按钮
            myProgressBar.setProgress(0);
            isRunning = true;          //将 isRunning 标志设置为 true
            Thread myThread = new Thread(new MyRunnable());
            myThread.start();//启动子线程
        }
        else if(R.id.myEndButton == viewId)
        {     //如果是结束按钮
             isRunning = false;        //将 isRunning 标志设置为 false
        }
        }
        @Override
        public boolean onCreateOptionsMenu(Menu menu) {
            getMenuInflater().inflate(R.menu.activity_main, menu);
            return true;
        }
    }
```

Message 对象是一个描述消息的数据结构类，Message 类包含很多成员变量和方法，但对于简单的消息处理，一般仅需了解 3 项，分别是：

- what：这是用来区分消息类型的一个整型值。
- arg1：这是额外消息参数。
- arg2：同 arg1。

如果需要包含更多的数据信息，可以使用 Message 对象的 setData()方法将 Bundle 类的数据对象加入到 Message 当中，在 handleMessage()方法中则可以使用 Message 对象的 getData()方法取出该 Bundle 对象。

上述代码在子线程当中，每隔 1 s 就通过 Handler 对象将消息放入到绑定的 MessageQueue 当中，Looper 对象会根据先进先出的规则，不断地抽取 MessageQueue 里面的 Message 对象，并将抽取得到的 Message 对象传递给 Message 对象 target 属性指定的 Handler 对象。由于 Handler 对象是在 UI 主线程当中定义的，所以此 Handler 对象会与 UI 主线程当中的 MessageQueue 数据结构以及负责抽取 MessageQueue 数据结构的 Looper 对象绑定在一起。

当 Looper 对象将消息从消息队列当中抽取传递给 Handler 对象的时候，会回调 Handler 对象的 handleMessage()方法，在 handleMessage()方法当中，会为进度条控件的进度增加 5。这样一来，当用户单击界面的"重新开始"按钮过后，进度条每隔1 s 进度就会增加 5，20 s 过后，进度条的进度就达到 100。如果在此过程当中，用户单击了"停止"按钮，则进度条会暂停下来，如果再次单击"重新开始"按钮，则进度条的进

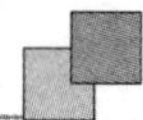

度又会从 0 开始，每隔 1 s 进度就会增加 5。

(3) 右击工程 chapter10_2，在弹出的快捷菜单中选择 Run As|Android Application 命令过后，单击用户界面的“重新开始”按钮，几秒钟过后，运行效果如图 10-3 所示。

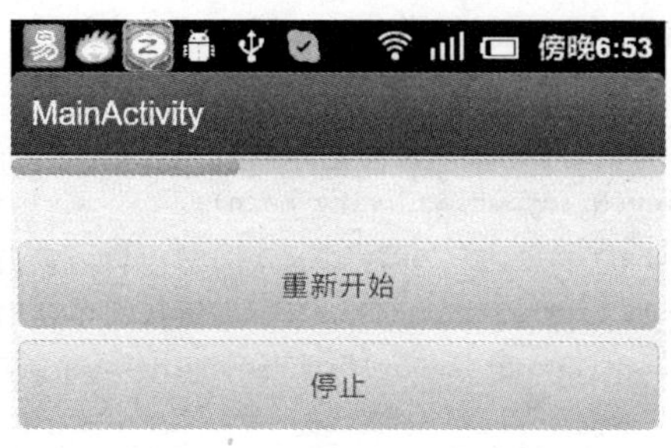

图 10-3　Handler 更新 ProgressBar

Handler 对象的 obtainMessage()方法用于从全局消息池当中获得一个已有的 Message 对象，系统为了加速线程间的消息传递，创建了一些全局的消息对象供各线程使用，使用 obtainMessage()方法比重新创建一个消息对象的效率要高。

使用 Handler 对象发送消息，还有以下几个方法：

- sendEmptyMessage(int what)：发送空消息，该消息仅包含 what 值。
- sendEmptyMessageAtTime(int what, long uptimeMillis)：在指定时间点发送空消息。
- sendEmptyMessageDelayed(int what, long delayMillis)：在指定的时间后发送空消息，指定的时间以 ms 为单位。
- sendMessageAtTime(Message msg, long uptimeMillis)：在指定的时间点发送该消息。
- sendMessageDelayed(Message msg, long delayMillis)：在指定的时间后发送该消息。

10.2.3　Handler 发送 Runnable 对象

Handler 对象还可以把一个 Runnable 对象发送到消息队列。从而当消息被处理时，能够执行 Runnable 对象的 run()方法体。

下面笔者就通过实例介绍如何将 Runnable 对象发送到消息队列上。

(1) 新建工程 chapter10_3，编辑 res/layout/activity_main.xml 文件，源代码参考配套资料。

(2) 编辑 MainActivity.java 文件，源代码如下：

```
public class MainActivity extends Activity implements  OnClickListener
{
```

```
private static final String TAG = "MainActivity";
Button myButton;//声明按钮控件
ProgressBar myProgressBar;//声明进度条控件
private Handler myHandler;
@Override
public void onCreate(Bundle savedInstanceState)
{
    super.onCreate(savedInstanceState);
    setContentView(R.layout.activity_main);
    myButton = (Button)findViewById(R.id.myButton);//找到按钮控件
    myButton.setOnClickListener(this);//绑定按钮单击事件监听器
    myProgressBar = (ProgressBar)findViewById(R.id.myProgressBar);
                                                    //找到进度条控件
    myHandler = new MyHandler();//主线程当中声明 Handler 对象
}
private class MyHandler extends Handler
{
@Override
public void handleMessage(Message msg)
{
    super.handleMessage(msg);
    Bundle bundle = msg.getData();
    int percent = bundle.getInt("percent");
    myProgressBar.setProgress(percent);
    if(percent >= 100)
{     //如果进度条的值大于等于 100,则返回
        return;
    }
    this.post(myRunnable);//继续将 myRunnable 对象发送到 MessageQueue 消息队列
}
}
@Override
public void onClick(View view)
{
if(R.id.myButton == view.getId())
{//如果 id 为 myButton 的按钮被单击
    myHandler.post(myRunnable);//将 myRunnable 对象发送到 MessageQueue 消息队列
}
}
//定义一个实现了 Runnable 接口的对象
private Runnable myRunnable = new Runnable()
{
```

```
        int percent = 0;//表明进度条进度的成员变量
            @Override
            public void run()
            {
                Log.d(TAG, "my Runnable run 方法被调用,线程名称:" + Thread.current-
Thread().getName());
                percent += 10;//每执行一次 run()方法,进度条进度增加 10
                Message msg = new Message();
                Bundle bundle = new Bundle();
                bundle.putInt("percent", percent);//通过 Bundle 对象设置 Message 数据
                msg.setData(bundle);
                SystemClock.sleep(1000);//线程休眠 1 s
                myHandler.sendMessage(msg);//myHandler 将 msg 消息对象放入到消息队列
            }
        };
        @Override
        public boolean onCreateOptionsMenu(Menu menu)
        {
            getMenuInflater().inflate(R.menu.activity_main, menu);
            return true;
        }
    }
```

上述代码当中,笔者在 onClick()回调函数当中,通过调用 Handler 对象的 post (Runnable r)方法将实现 Runnable 接口的 myRunnable 对象加入到 UI 线程的 MessageQueue 消息队列当中。等到与 UI 主线程绑定的 Looper 对象处理到此对象的时候,会执行 myRunnable 对象的 run()方法。

在 myRunnable 对象的 run()方法体当中,每次执行,都会将进度条的进度增加 10,然后将进度通过 Bundle 的形式设置给要传递的 Message 对象,并休眠 1 s。

(3) 右击工程 chpater10_3,在弹出的快捷菜单中选择 Run As|Android Application 命令,单击界面上的"发送 Runnable 对象"按钮,10 s 过后,运行效果如图 10－4 所示,LogCat 输出日志如图 10－5 所示。

图 10－4　Handler 发送 Runnable 对象效果

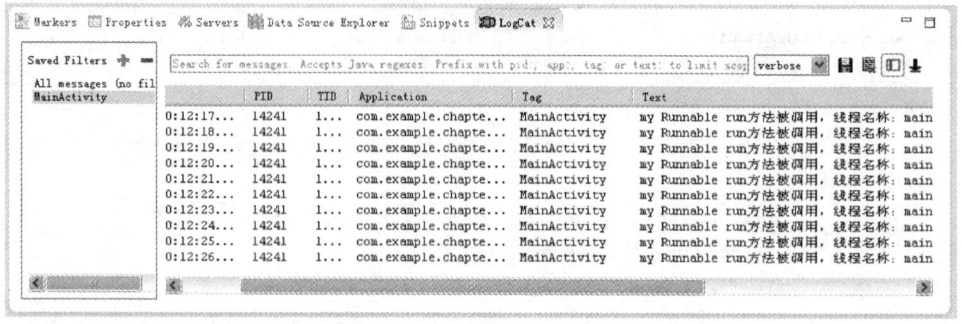

图 10 - 5　Handler 发送 Runnable 对象 LogCat 输出

从 Logcat 的输出结果来看，事实上 myRunnable 对象的 run()方法体是在 UI 主线程当中执行的。因为通过调用 Handler 对象的 post(Runnable r)方法将相应的实现 Runnable 接口的对象加入到 UI 主线程的 MessageQueue 消息队列当中，当与 UI 主线程绑定的 Looper 对象处理这些对象的时候，是直接调用这些对象的 run()方法的，而没有去调用启动子线程的 start()方法，所以这些对象的 run()方法体是在 UI 主线程当中执行的。如何让这些对象的 run()方法体在子线程当中执行呢?

(4) 重新编辑 MainActivity.java 文件，源代码如下：

```
public class MainActivity extends Activity implements  OnClickListener
{
    private static final String TAG = "MainActivity";
    Button myButton;//声明按钮控件
    ProgressBar myProgressBar;//声明进度条控件
    private Handler myHandler;
    @Override
    public void onCreate(Bundle savedInstanceState)
    {
        super.onCreate(savedInstanceState);
        setContentView(R.layout.activity_main);
        myButton = (Button)findViewById(R.id.myButton);//找到按钮控件
        myButton.setOnClickListener(this);//绑定按钮单击事件监听器
        myProgressBar = (ProgressBar)findViewById(R.id.myProgressBar);
                                                            //找到进度条控件
        HandlerThread handlerThread = new HandlerThread("myHandlerThread");
                                                        //声明 HandlerThread 对象
        handlerThread.start();
        myHandler = new MyHandler(handlerThread.getLooper());
                                                    //主线程当中声明 Handler 对象
    }
    private class MyHandler extends Handler
```

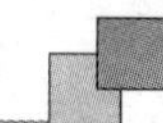

```
{
public MyHandler(Looper looper)
{
    super(looper);
}
    @Override
public void handleMessage(Message msg)
{
    super.handleMessage(msg);
    Bundle bundle = msg.getData();
    int percent = bundle.getInt("percent");
    myProgressBar.setProgress(percent);
    if(percent >= 100){//如果进度条的值大于等于 100,则返回
        return;
}
this.post(myRunnable);//继续将 myRunnable 对象发送到 MessageQueue 消息队列
}
}
@Override
public void onClick(View view)
{
if(R.id.myButton == view.getId())
{//如果 id 为 myButton 的按钮被单击
    myHandler.post(myRunnable);//将 myRunnable 对象发送到 MessageQueue 消息队列
}
}
//定义一个实现了 Runnable 接口的对象
private Runnable myRunnable = new Runnable()
{
int percent = 0;//表明进度条进度的成员变量
    @Override
    public void run()
    {
        Log.d(TAG, "my Runnable run 方法被调用,线程名称:" + Thread.current-
Thread().getName());
        percent += 10;//每执行一次 run()方法,进度条进度增加 10
        Message msg = new Message();
        Bundle bundle = new Bundle();
        bundle.putInt("percent", percent);//通过 Bundle 对象设置 Message 数据
        msg.setData(bundle);
        SystemClock.sleep(1000);//线程休眠 1 s
        myHandler.sendMessage(msg);
```

```
            }
        };
        @Override
        public boolean onCreateOptionsMenu(Menu menu)
        {
            getMenuInflater().inflate(R.menu.activity_main, menu);
            return true;
        }
    }
```

在上述代码当中的 onCreate()方法当中，声明了一个 HandlerThread 对象，并调用了此对象的 start()方法，由于 HandlerThread 事实上是继承于 Thread 类的一个子类，通过调用 HandlerThread 对象的 start()方法实际上是启动了这个线程。

然后为 MyHandler 类添加了一个参数为 Looper 类型的构造函数，目的是为了能够将 HandlerThread 对象的 getLooper()方法所得到的 Looper 对象传递给 MyHandler 的对象。

(6) 右击工程 chpater10_3，在弹出的快捷菜单中选择 Run As|Android Application 命令，单击界面上的"发送 Runnable 对象"按钮，10 s 过后，LogCat 输出日志如图 10-6 所示。

PID	TID	Application	Tag	Text
40:32...	1... 1...	com.example...	MainActivity	my Runnable run方法被调用，线程名称：myHandlerThread
40:33...	1... 1...	com.example...	MainActivity	my Runnable run方法被调用，线程名称：myHandlerThread
40:34...	1... 1...	com.example...	MainActivity	my Runnable run方法被调用，线程名称：myHandlerThread
40:35...	1... 1...	com.example...	MainActivity	my Runnable run方法被调用，线程名称：myHandlerThread
40:36...	1... 1...	com.example...	MainActivity	my Runnable run方法被调用，线程名称：myHandlerThread
40:37...	1... 1...	com.example...	MainActivity	my Runnable run方法被调用，线程名称：myHandlerThread
40:38...	1... 1...	com.example...	MainActivity	my Runnable run方法被调用，线程名称：myHandlerThread
40:39...	1... 1...	com.example...	MainActivity	my Runnable run方法被调用，线程名称：myHandlerThread
40:40...	1... 1...	com.example...	MainActivity	my Runnable run方法被调用，线程名称：myHandlerThread
40:41...	1... 1...	com.example...	MainActivity	my Runnable run方法被调用，线程名称：myHandlerThread

图 10-6 HandlerThread 使用

10.2.4 runOnUiThread 函数的使用

由于只有在 UI 主线程当中才能更新 UI 界面，所以在一些应用场景下需要强制将一些更新 UI 界面的逻辑放在 UI 主线程当中执行，除了可以使用 Handler 那一套模型处理之外，Activity 本身也提供了一个叫做 runOnUiThread 的函数。这个函数是 Activity 的成员函数，函数原型是：

```
public final void runOnUiThread(Runnable action)
```

在自定义的子线程当中执行对应的 Activity 的 runUiOnThread()方法，就可以

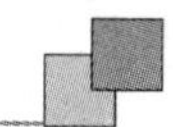

保证该方法的 Runnable 类型的参数对象的 run()方法会在 UI 主线程当中执行。下面笔者就通过实例介绍如何使用 runOnUiThread()这个函数。

(1) 新建工程 chapter10_4,编辑 res/layout/activity_main. xml 文件,源代码参考配套资料。

(2) 编辑 MainActivity. java 文件,源代码如下:

```
public class MainActivity extends Activity
{
    TextView timeTextView;//声明显示当前时间的文本控件
    @Override
    public void onCreate(Bundle savedInstanceState)
    {
        super.onCreate(savedInstanceState);
        setContentView(R.layout.activity_main);
        timeTextView = (TextView)findViewById(R.id.time);
                                                    //找到显示时间的文本控件
        //启动子线程
        new Thread(new Runnable() {
            @Override
            public void run()
              {
                    while(true)
                      {//死循环,不停地显示当前时间
                        updateTime();//更新当前时间
                        try
                        {
                            Thread.sleep(1000);//当前子线程休眠 1 s
                        }
                        catch(InterruptedException e){
                            e.printStackTrace();
                        }
                    }
              }
         }).start();
    }
    //更新时间的方法
    private void updateTime()
    {
    //UI 线程当中执行 run()方法体逻辑
    runOnUiThread(new Runnable()
        {
            @Override
```

```
                public void run()
                {
                    String str = new SimpleDateFormat("yyyy - MM - dd HH:mm:ss").format
(new Date());
                    timeTextView.setText(str);
                }
            });
        }
        @Override
        public boolean onCreateOptionsMenu(Menu menu)
        {
            getMenuInflater().inflate(R.menu.activity_main, menu);
            return true;
        }
    }
```

上述代码在 onCreate()方法中启动了一个子线程，在子线程中每隔 1 s 就去更新界面上的文本控件上面显示的时间，由于是在子线程当中去更新 UI 界面的元素，如果不做任何处理的话，Android Runtime 会抛出相关异常。上述代码在子线程当中更新 UI 界面的时候，调用了 Activity 类的 runOnUiThread()这个成员方法。

(3) 右击工程 chapter10_4，在弹出的快捷菜单中选择 Run As|Android Application 命令，运行效果如图 10 - 7 所示。

2012-09-02 03:39:20

图 10 - 7　runOnUiThread 函数使用

10.3　AsyncTask 的使用

无论是 Handler 的方式，还是 runUiOnThread()的方式，都会启动一些匿名的子线程，但是这种匿名线程的方式是存在缺陷的：第一，线程的开销较大，如果每个任务都要创建一个线程，那么应用程序的效率就会低很多；第二，线程无法管理，匿名线程创建过后就不受程序的控制了，如果要同时发送多个网路请求，系统将不堪重负。另外，如果使用 Handler 的话，还需要在代码当中引入 Handler，这会让代码看上去比较臃肿。

为了解决这个问题，Android SDK 引入了 AsyncTask，顾名思义就是执行异步任务，这个 AsyncTask 目的就是处理一些比较耗时的任务，给用户带来良好的用户体验，从编程的语法上也显得优雅了很多，不再需要子线程和 Handler 就可以完成异步任务并且刷新 UI 界面。

android. os. AsyncTask 类对线程间的通信进行了很好的封装，提供了简易的编程方式来使后台线程与 UI 线程进行通信，后台线程执行异步任务，并把操作结果通知给 UI 线程。下面笔者就通过实例介绍如何在应用程序当中使用 AsyncTask。

（1）由于本实例需要从网络上下载一张图片，并显示出来，所以需要在项目清单文件当中添加互联网访问的权限，编辑 AndroidManifest. xml 文件，源代码如下：

```
<manifest xmlns:android = "http://schemas.android.com/apk/res/android"
    package = "com.example.chapter10_5"
    android:versionCode = "1"
    android:versionName = "1.0" >
    <uses-sdk    android:minSdkVersion = "8"
                    android:targetSdkVersion = "15" />
    <!-- 互联网联网权限 -->
    <uses-permission android:name = "android.permission.INTERNET" />
    <application android:icon = "@drawable/ic_launcher"
        android:label = "@string/app_name"
        android:theme = "@style/AppTheme" >
        <activity
            android:name = ".MainActivity"
            android:label = "@string/title_activity_main" >
            <intent-filter>
                <action android:name = "android.intent.action.MAIN" />
                <category android:name = "android.intent.category.LAUNCHER" />
            </intent-filter>
        </activity>
    </application>
</manifest>
```

（2）编辑 res/layout/activity_main. xml 文件，源代码如下：

```
<?xml version = "1.0" encoding = "utf-8"?>
<LinearLayout xmlns:android = "http://schemas.android.com/apk/res/android"
    android:layout_width = "fill_parent"
    android:layout_height = "fill_parent"
    android:orientation = "vertical" >
    <ProgressBar
        android:id = "@+id/progressBar"
        style = "?android:attr/progressBarStyleHorizontal"
        android:layout_width = "fill_parent"
        android:layout_height = "wrap_content" >
```

```
    </ProgressBar>
    <Button
        android:id="@+id/button"
        android:layout_width="fill_parent"
        android:layout_height="wrap_content"
        android:text="下载 Baidu Logo" >
    </Button>
    <ImageView
        android:id="@+id/imageView"
        android:layout_width="fill_parent"
        android:layout_height="wrap_content" />
</LinearLayout>
```

(3) 编辑 MainActivity.java 文件，源代码如下：

```
public class MainActivity extends Activity
{
    private ImageView mImageView;                    //声明图片视图控件
    private Button mButton;                          //声明按钮控件
    private ProgressBar mProgressBar;                //声明进度条控件
    @Override
    public void onCreate(Bundle savedInstanceState)
    {
        super.onCreate(savedInstanceState);
        setContentView(R.layout.activity_main);
        mImageView = (ImageView) findViewById(R.id.imageView); //找到显示图片的图
                                                               //片视图控件
        mButton = (Button) findViewById(R.id.button); //找到按钮控件
        mProgressBar = (ProgressBar) findViewById(R.id.progressBar); //找到进度条
                                                                     //控件
        mButton.setOnClickListener(new OnClickListener()
        {//给按钮绑定单击事件监听器
            @Override
            public void onClick(View v)
            {//单击回调事件
                GetBaiduLogoTask task = new GetBaiduLogoTask();
                task.execute("http://www.baidu.com/img/baidu_sylogo1.gif");
            }
        });
    }
    //第一个泛型参数(String):doInBackground()回调方法的接受参数
    //第二个泛型参数(Integer):onProgressUpate()回调方法的接受参数
    //第三个泛型参数(Bitmap):onPostExecute()回调方法的接受参数,也是
    //doInBackground()回调方法的返回类型
    class GetBaiduLogoTask extends AsyncTask<String,Integer,Bitmap> {//继承 Async
                                                                     //Task
```

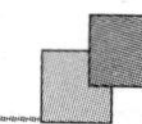

```
@Override
protected Bitmap doInBackground(String... params) {//处理后台执行的任务,在
                                                   //后台线程执行
    publishProgress(0);//将会调用 onProgressUpdate(Integer... progress)
                       //方法
    HttpClient hc = new DefaultHttpClient();
    publishProgress(30);
    HttpGet hg = new HttpGet(params[0]);//获取 Baidu 的 Logo
    final Bitmap bm;
    try
    {
        HttpResponse hr = hc.execute(hg);
        bm = BitmapFactory.decodeStream(hr.getEntity().getContent());
    }
    catch (Exception e)
    {
        return null;
    }
    publishProgress(100);  //更新进度,调用 onProgressUpdate()方法
    //mImageView.setImageBitmap(result); 不能在后台线程操作 ui
    return bm;
}
protected void onProgressUpdate(Integer... progress)
{//在调用 publishProgress 之后被调用,在 ui 线程执行
    mProgressBar.setProgress(progress[0]);//更新进度条的进度
}
protected void onPostExecute(Bitmap result)
{//后台任务执行完之后被调用,在 ui 线程执行
if(result != null)
  {  //如果 result 不为 null
        Toast.makeText(MainActivity.this, "成功获取图片", Toast.LENGTH_
LONG).show();
       mImageView.setImageBitmap(result);
}
else
{     //如果 result 为 null
     Toast.makeText(MainActivity.this, "获取图片失败", Toast.LENGTH_LONG).
show();
}
}

protected void onPreExecute ()
{//在 doInBackqround(Params...)之前被调用,在 ui 线程执行
mImageView.setImageBitmap(null);
mProgressBar.setProgress(0);//进度条复位
}
protected void onCancelled ()
```

```
        {//在 ui 线程执行
        mProgressBar.setProgress(0);//进度条复位
        }
    }
    @Override
    public boolean onCreateOptionsMenu(Menu menu)
    {
            getMenuInflater().inflate(R.menu.activity_main, menu);
            return true;
        }
    }
```

AsyncTask 抽象出后台线程运行的 5 个状态，分别是准备运行，正在后台运行，进度更新，完成后台任务和取消任务。对于这 5 个阶段，AsyncTask 提供了 5 个回调函数：

- 准备运行：onPreExecute()，该回调方法在任务被执行之后立即由 UI 线程调用。
- 正在后台运行：doInBackground(String… params)，该回调方法由后台线程在 onPreExecute()方法执行结束后立即调用。通常在此方法当中执行一些比较耗时的操作，该方法返回的结果将会传递到 onPostExecute()回调方法当中。在该回调方法内也可以使用 publishProgress(Integer… values)来回调 onProgressUpate(Integer… values)方法。
- 进度更新：onProgressUpdate(Integer… values)，该方法在 UI 线程当中执行，一般会在 doInBackground(String… params)回调函数当中显示调用 publishProgress(Integer… values)这个方法，从而完成对 onProgressUpdate(Integer… values)函数的调用。
- 完成后台任务：onPostExecute()，该方法在 UI 线程当中执行，会在 doInBackground()方法调用完成后执行。
- 取消任务：onCancelled()，该方法在调用 AsyncTask 对象的 cancel()方法时调用。

(4) 右击工程 chapter10_5，在弹出的快捷菜单中选择 Run As|Android Application 命令，单击界面上的“下载 Baidu Logo”按钮，运行效果如图 10－8 所示。

图 10－8　AsyncTask 获取图片

第 2 篇　Android 高级编程

第 11 章

服务(Service)

Service 是 Android 的 4 大组件之一,其余 3 大组件:Activity、Broadcast Receiver 以及 Content Provider 笔者在前面章节已经有过介绍。Service 不像 Activity 有可以操作的用户界面,Service 是在后台运行的。如果某个组件需要在运行时向用户呈现可操作的信息就应该选择 Activity,否则应该选择 Service。最常见的例子如:媒体播放器程序,它可以在转到后台运行的时候仍然保持播放歌曲。本章主要包含以下内容:

- ❑ Service 介绍;
- ❑ Local Service(本地服务);
- ❑ Remote Service(远程服务)。

11.1 Service 介绍

Service 是没有用户界面,在后台执行长时间操作的应用程序组件之一。其他的应用程序组件(如 Activity 和 Broadcast Receiver)可以启动服务,而且启动的服务即便是用户切换到其他应用程序也可以一直在后台保持运行状态。应用程序组件还可以绑定到服务上面,甚至可以通过 IPC(Interprocess Communication)机制来提供可供其他应用程序远程访问的服务。

11.1.1 Service 启动方式

1. startService()

应用程序组件(如 Activity),可以通过调用 Context. startService()来启动一个服务。服务一旦启动,便在后台执行,即便启动服务的应用程序组件被销毁了,也不影响服务的继续运行。通过 startService()方式启动的服务,通常来讲都会去执行一

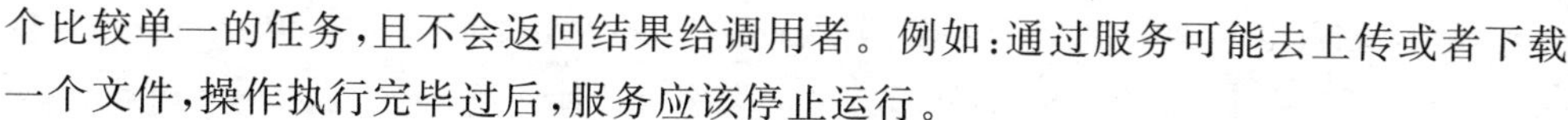

个比较单一的任务，且不会返回结果给调用者。例如：通过服务可能去上传或者下载一个文件，操作执行完毕过后，服务应该停止运行。

2. bindService()

应用程序组件可以通过调用 Context. bindService()绑定一个服务。通过 bindService()启动的服务可以返回结果给调用者，甚至可以通过 IPC 机制进行跨进程的通信。一旦应用程序组件与之绑定，则服务开始运行。多个组件可以同时绑定到同一个服务上面，直到所有的组件都与服务解绑之后，服务才会被销毁。

上述实际上介绍了两种不同种类的服务，但是可以让某一个服务同时具备两种功能。是否具备这两种功能仅仅取决于是否实现了 Service 的一些不同的回调方法。实现了 onStartCommand()方法就允许其他组件通过调用 startService()方法来启动这个服务；实现了 onBind()方法就允许其他组件绑定到这个服务上面，意味着可以接收到服务返回的结果。

默认情况下，Service 是运行在应用程序进程的 UI 主线程当中的。这也就意味着，如果需要在 Service 当中执行一些比较耗时的操作，应该在服务当中创建子线程去做这些比较耗时，耗 CPU 的操作。通过在 Service 中创建子线程的方式，可以有效地避免应用程序产生 ANR 的错误。

11.1.2　Service 基础

如果需要在应用程序当中创建服务，则必须创建一个继承于 android. app. Service 的子类。在创建的子类当中，需要去实现一些处理 Service 生命周期的重要的回调方法，例如：

- onStartCommand：当其他组件，例如 Activity，调用 startService()方法的时候，系统就会回调 onStartComand()这个方法。一旦这个方法被调用过后，服务就会启动并开始在后台运行。操作完成过后，可以通过调用 stopSelf()或者调用 Context. stopService()来停止服务的运行。
- onBind：当其他组件调用 bindService()方法的时候，系统就会回调 onBind()这个方法。在这个方法里面，需要给绑定到此服务的组件返回一个实现了 android. os. IBinder 接口的对象。如果服务不支持其他组件的绑定，则此方法返回 null 就可以了。
- onCreate：当服务第一次创建的时候，系统就会回调 onCreate()这个方法。系统会在 onStartCommand()或 onBind()方法之前调用这个方法，且仅仅调用一次。如果服务已经在运行了，则不会调用这个方法。
- onDestory：当服务不再被使用的时候，系统就会回调 onDestory()这个方法。在这个方法里面，一般需要去做一些释放资源的操作(例如：创建的子线程，注册的广播接收者)。对于 Service 来说，这是系统回调的最后一个方法。

如果 Service 是通过其他组件调用 startService()方法而运行的，那么 Service 会一直运行直到调用自己的 stopSelf()方法，或者是其他组件显示地调用 stopService()方法。

如果 Service 是通过其他组件调用 bindService()方法而运行的，那么只有当绑定到此服务的所有组件都与之解绑，这时候服务才会被销毁。

Android 系统在内存不足的情况下，为了保证跟用户正在交互的应用程序的正常使用，会强制地停止一些 Service 的运行。如果 Service 被绑定到正在与用户交互的 Activity 组件上，那么这种被强制停止运行的概率就会小很多。

11.2 本地服务

Android Service 事实上又分为两种类型，一种是本地服务，另外一种是远程服务。本地服务，指的是只用于当前应用程序内部的服务，而远程服务指的是通过 IPC 机制进行进程间通信的服务。

本地服务，大体上又可以分为两种，第一种是不需要跟组件进行交互的服务，也就是通过调用 Context. startService()方法启动的服务；另外一种是需要跟组件进行交互的服务，也就是通过调用 Context. bindService()方法启动的服务。

通过调用 Context. startService()方法启动的服务，与调用者之间没有任何关联，即使调用者退出了，服务也可以继续运行。在服务未被创建的时候，系统会首先调用服务的 onCreate()方法，接着调用服务的 onStartCommand()方法。如果在调用 Context. startService()方法前服务已经被创建，多次调用 Context. startService()方法并不会导致服务的多次创建，但会导致 onStartCommand()方法多次被调用。可以通过调用 Context. stopService()方法或调用服务本身的 stopSelf()方法来停止服务的运行，服务结束时会调用 onDestory()方法。

通过调用 Context. bindService()方法启动的服务，调用者与服务绑定在了一起。调用者一旦退出，服务也会相应地停止运行。在服务未被创建的时候，系统会首先调用服务的 onCreate()方法，接着调用服务的 onBind()方法，如果调用者与服务已经处于绑定的状态，多次调用 Context. bindService()方法并不会导致 onBind()方法被调用多次。采用 Context. bindService()方法启动的服务，只能通过调用 Context. unbindService()方法解除调用者与服务的绑定，服务结束时会调用 onDestory()方法。

这两种不同的方式启动的服务的生命周期如图 11－1 所示。

11.2.1 不需要与组件交互本地服务

下面笔者就通过实例介绍如何使用不需要与组件交互的本地服务。

(1) 新建工程 chapter11_1，编辑 res/layout/activity_main. xml 文件，源代码参

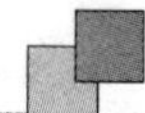

考配套资源。

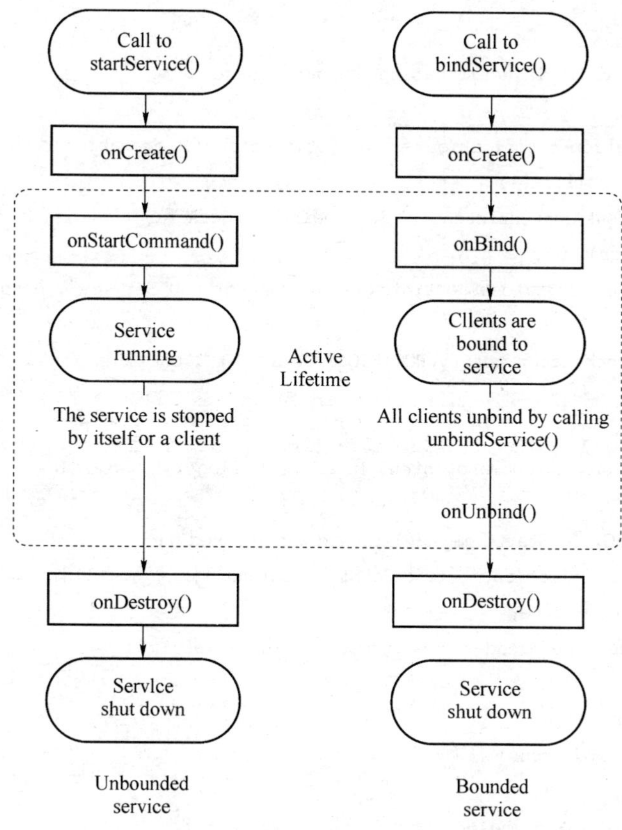

图 11-1　两种不同方式服务的生命周期

(2) 新建一个继承于 android. app. Service 的自定义 Service：MyLocalService，编辑 MyLocalService. java 文件，源代码如下：

```
public class MyLocalService extends Service
{
  private static final String TAG = "MyLocalService";
  private static  int NOTIFICATION_ID = 1;
  private boolean isRunning;//标识线程是否可以继续运行
  private NotificationManager notificationManager;
  //服务第一次创建时调用
  @Override
  public void onCreate()
  {
    super.onCreate();
    Log.d(TAG, "onCreate() method was invoked");
    //获取通知管理器
```

```
        notificationManager = (NotificationManager)getSystemService(Context.NOTIFICA-
TION_SERVICE);
    }
    private void displayNotification(String message)
    {
        Notification notification = new Notification(R.drawable.ic_launcher, message,
System.currentTimeMillis());
    PendingIntent contentIntent = PendingIntent.getActivity(this, 0, new Intent(this,
MainActivity.class), 0);
    notification.setLatestEventInfo(this, "Background Service", message, contentInt-
ent);
    notificationManager.notify(NOTIFICATION_ID ++ , notification);
    }
    @Override
    public int onStartCommand(Intent intent, int flags, int startId)
    {
      Log.d(TAG, "onStartCommand() method was invoked");
      displayNotification("Service onStartCommand");//提示通知消息
      isRunning = true;
      Thread serviceThread = new Thread(new Runnable()
      {
        @Override
        public void run()
          {
          while(isRunning)
          {//如果 isRunning 为 true,则始终循环
              Log.d(TAG, "service Thread executes again");
              SystemClock.sleep(10 * 1000);//休眠 10s
          }
        }
      });
      serviceThread.start();//启动子线程
      return super.onStartCommand(intent, flags, startId);
    }
    //不需要与组件绑定,所以 onBind()方法返回 null
    @Override
    public IBinder onBind(Intent arg0)
    {
      return null;
    }
    //服务停止运行时调用
    @Override
```

```
    public void onDestroy()
    {
        super.onDestroy();
        displayNotification("Service onDestory");
        isRunning = false;//停止子线程的运行
        Log.d(TAG, "onDestory() method was invoked");
    }
}
```

(3) 由于 Service 是应用程序 4 大组件之一，所以需要在项目清单文件当中加以声明，编辑 AndroidManifest.xml 文件，源代码如下：

```
<manifest xmlns:android = "http://schemas.android.com/apk/res/android"
    package = "com.example.chapter11_1"    android:versionCode = "1"
    android:versionName = "1.0" >
    <uses-sdk  android:minSdkVersion = "8"    android:targetSdkVersion = "15" />
    <application  android:icon = "@drawable/ic_launcher"
        android:label = "@string/app_name"    android:theme = "@style/AppTheme" >
        <activity
            android:name = ".MainActivity"
            android:label = "@string/title_activity_main" >
            <intent-filter>
                <action android:name = "android.intent.action.MAIN" />
                <category android:name = "android.intent.category.LAUNCHER" />
            </intent-filter>
        </activity>
        <!-- 声明 MyLocalService -->
        <service android:name = ".MyLocalService"></service>
    </application>
</manifest>
```

(4) 编辑 MainActivity.java 文件，源代码如下：

```
public class MainActivity extends Activity  implements OnClickListener
{
    Button startServiceBtn;//声明启动服务按钮
    Button endServiceBtn;//声明停止服务按钮
    @Override
    public void onCreate(Bundle savedInstanceState)
      {
        super.onCreate(savedInstanceState);
        setContentView(R.layout.activity_main);
        startServiceBtn = (Button)findViewById(R.id.startService);
                                                           //找到启动服务按钮
```

```
            endServiceBtn = (Button)findViewById(R.id.endService);
                                                                //找到停止服务按钮
            startServiceBtn.setOnClickListener(this);//绑定单击事件监听器
            endServiceBtn.setOnClickListener(this);
        }
        @Override
        public void onClick(View v)
        {
        int viewId = v.getId();
        if(R.id.startService == viewId)
        {//如果是启动服务按钮单击
            startService(new Intent(this, MyLocalService.class));//启动服务
        }
        else if(R.id.endService == viewId)
        {//如果是停止服务按钮单击
            stopService(new Intent(this, MyLocalService.class));
        }
        }
    }
```

(5) 打开 LogCat 视图，添加 MyLocalService 标签。右击工程 chapter11_1，在弹出的快捷菜单中选择 Run As|Android Application 命令，首先单击界面上的"启动本地服务"按钮，然后单击界面上的"停止本地服务"按钮，LogCat 输出如图 11－2 所示。

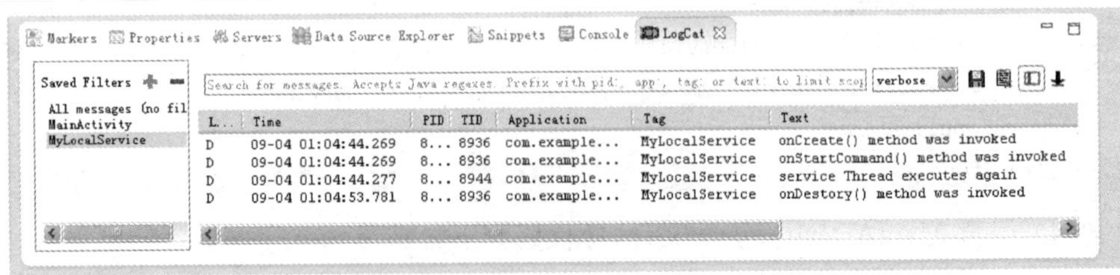

图 11－2　与组件不交互 Service Logcat 输出

(6) 上述实例演示了如何通过调用 Context.startService()的方式来启动一个服务，以及如何调用 Context.stopService()的方式来停止一个服务。当用户单击"启动本地服务"按钮的时候，MyLocalService 里面的 onCreate()及 onStartCommand()函数依次被调用。笔者在 onStartCommand()函数当中主要做了两件事情，第一件事情就是在通知栏显示一个通知，然后开启一个子线程，这个子线程的运行是通过 isRunning 这个 Boolean 类型的变量来控制的。当用户单击"停止本地服务"按钮的时候，MyLocalService 里面的 onDestory()函数会被调用。笔者在 onDestory()函数里

面也做了两件事情，第一件事情就是将 isRunning 的值设为 false，这样子线程当中的 run()方法体就不会再执行业务逻辑了，第二件事情也是在通知栏显示了一个通知。

(7) 对于已经启动了的 Service，如果再次调用 Context. startService()方法，也就是连续两次单击界面上的"启动本地服务"按钮，运行效果如图 11－3 所示。

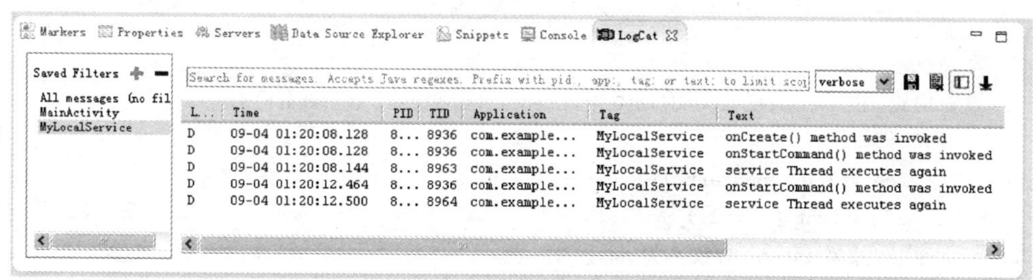

图 11－3　Context. startService()多次调用效果

也就意味着对于已经启动了的 Service，如果多次调用 Context. startService()方法，只会导致 Service 的 onStartCommand()方法会被多次调用。

11.2.2　本地服务结合广播接收者

除了 Activity 可以调用本地服务以外，广播接收者同样也可以调用本地服务。下面笔者就通过实例介绍如何通过广播接收者来调用本地服务。

(1) 新建工程 chapter11_2，编辑 res/layout/activity_main. xml 文件，源代码参考配套资料。

(2) 在 res 目录下，新建 raw 子目录，并将笔者硬盘上的 tmp. mp3 文件复制此 raw 子目录。

(3) 新建一个服务：MusicService，编辑 MusicService. java 文件，源代码如下：

```
public class MusicService extends Service
{
  private static final String TAG = "MusicService";
  private MediaPlayer mediaPlayer;
  @Override
  public IBinder onBind(Intent arg0)
  {
      return null;
  }
  @Override
  public void onCreate()
  {
      Log.v(TAG, "onCreate");
      Toast.makeText(this, "show media player", Toast.LENGTH_SHORT).show();
```

```
        if (mediaPlayer == null)
        {//如果 mediaPlayer 对象为 null
            mediaPlayer = MediaPlayer.create(this, R.raw.tmp);
            mediaPlayer.setLooping(false);
        }
    }
    @Override
    public int onStartCommand(Intent intent, int flags, int startId)
    {
        Log.v(TAG, "onStart");
        if (intent != null)
        {//如果 intent 对象不为 null
            Bundle bundle = intent.getExtras();
            if (bundle != null)
            {//如果 bundle 对象不为 null
                int op = bundle.getInt("op");
                switch (op)
                {
                case 1:
                    play();//播放音乐
                    break;
                case 2:
                    stop();//停止音乐播放
                    break;
                case 3:
                    pause();//暂停音乐播放
                    break;
                  }
              }
          }
          return super.onStartCommand(intent, flags, startId);
      }
      //播放音乐
      public void play()
      {
          if (! mediaPlayer.isPlaying())
          {//如果不在播放音乐
              mediaPlayer.start();
          }
      }
      //暂停音乐播放
      public void pause()
```

```
    {
        if (mediaPlayer != null && mediaPlayer.isPlaying())
        {//如果正在播放音乐
            mediaPlayer.pause();//暂停音乐播放
        }
    }
    //停止音乐播放
    public void stop()
    {
        if (mediaPlayer != null)
          {//如果 mediaPlayer 对象不为 null
            mediaPlayer.stop();
            try {
                mediaPlayer.prepare();  //在调用 stop 后如果需要再次通过 start 进
                                        //行播放,需要之前调用 prepare 函数
            }
            catch (IOException ex)
             {
                ex.printStackTrace();
            }
        }
    }
    @Override
    public void onDestroy()
    {
        Log.v(TAG, "onDestroy");
        if (mediaPlayer != null)
         {//如果 mediaPlayer 对象不为 null
            mediaPlayer.stop();
            mediaPlayer.release();
        }
    }
}
```

(4) 新建一个广播接收者:MusicBroadcastReceiver,编辑 MusicBroadcastReceiver.java 文件,源代码如下:

```
package com.example.chapter11_2;
import android.content.BroadcastReceiver;
import android.content.Context;
import android.content.Intent;
import android.os.Bundle;
//通过广播接收者来调用服务
```

```
public class MusicBroadcastReceiver extends BroadcastReceiver
{
    @Override
    public void onReceive(Context context, Intent intent)
    {
      if(intent != null)
      { //如果 intent 对象不为 null
          Bundle bundle = intent.getExtras();
          Intent it = new Intent(context, MusicService.class);
          it.putExtras(bundle);
          if(bundle != null)
            {//如果 bundle 对象不为 null
              int op = bundle.getInt("op");
              if(op == 4)
              {
                  context.stopService(it);          //停止服务的运行
              }
              else
              {
                  context.startService(it);         //开启服务
              }
            }
        }
    }
}
```

(5) 编辑 AndroidManifest.xml 文件,源代码如下:

```
<manifest xmlns:android="http://schemas.android.com/apk/res/android"
    package="com.example.chapter11_2"  android:versionCode="1"
    android:versionName="1.0" >
    <uses-sdk  android:minSdkVersion="8"  android:targetSdkVersion="15" />
    <application    android:icon="@drawable/ic_launcher"
        android:label="@string/app_name"  android:theme="@style/AppTheme" >
        <activity
            android:name=".MainActivity"
            android:label="@string/title_activity_main" >
            <intent-filter>
                <action android:name="android.intent.action.MAIN" />
                <category android:name="android.intent.category.LAUNCHER" />
            </intent-filter>
        </activity>
        <!-- 注册广播接收者 -->
```

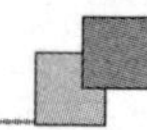

```
        <receiver android:name=".MusicBroadcastReceiver" >
            <intent-filter>
                <action android:name="com.example.chapter11_1.musicreceiver" />
            </intent-filter>
        </receiver>
        <!-- 注册服务 -->
        <service android:name=".MusicService"></service>
    </application>
</manifest>
```

(6) 编辑 MainActivity.java 文件,源代码如下:

```
public class MainActivity extends Activity implements OnClickListener
{
    private Button playBtn;        //播放音乐按钮
    private Button stopBtn;        //停止播放音乐按钮
    private Button pauseBtn;       //暂停播放音乐按钮
    private Button exitBtn;        //退出 Activity 按钮
    private Button closeBtn;       //停止服务按钮
    private Intent intent;
    @Override
    public void onCreate(Bundle savedInstanceState) {
        super.onCreate(savedInstanceState);
        setContentView(R.layout.activity_main);
        playBtn = (Button) findViewById(R.id.play);   //找到播放音乐按钮
        stopBtn = (Button) findViewById(R.id.stop);   //找到停止播放音乐按钮
        pauseBtn = (Button) findViewById(R.id.pause);//找到暂停播放音乐按钮
        exitBtn = (Button) findViewById(R.id.exit);   //找到退出 Activity 按钮
        closeBtn = (Button) findViewById(R.id.close);//找到停止服务按钮
        //为各按钮绑定单击事件监听器
        playBtn.setOnClickListener(this);
        stopBtn.setOnClickListener(this);
        pauseBtn.setOnClickListener(this);
        exitBtn.setOnClickListener(this);
        closeBtn.setOnClickListener(this);
    }
    @Override
    public void onClick(View v)
    {
        int op = -1;
        intent = new Intent("com.example.chapter11_1.musicreceiver");
        switch (v.getId()) {
        case R.id.play:                                //播放音乐
```

```
            op = 1;
            break;
        case R.id.stop:                                  //停止播放
            op = 2;
            break;
        case R.id.pause:                                 //暂停播放
            op = 3;
            break;
        case R.id.close:                                 //退出当前 Activity
            this.finish();
            break;
        case R.id.exit:                                  //process by MusicReceiver
            op = 4;
            this.finish();
            break;
        }
        Bundle bundle = new Bundle();
        bundle.putInt("op", op);
        intent.putExtras(bundle);
        sendBroadcast(intent);                           //发送广播
    }
    @Override
    public boolean onCreateOptionsMenu(Menu menu)
    {
        getMenuInflater().inflate(R.menu.activity_main, menu);
        return true;
    }
}
```

(7) 右击工程 chapter11_2，在弹出的快捷菜单中选择 Run As|Android Application 命令，单击界面上的 Play 按钮，运行效果如图 11-4 所示。

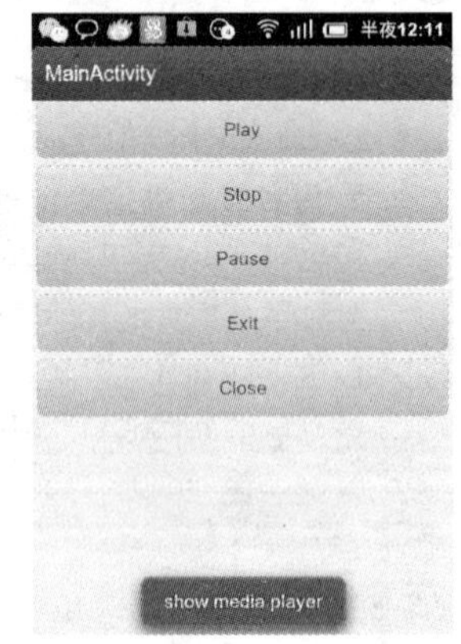

图 11-4　chapter11-2 play 单击按钮效果

当用户单击 Stop 按钮的时候，会停止播放音乐，当用户单击 Pause 按钮的时候，会暂停播放音乐，当用户单击 Exit 按钮的时候，会结束当前的 Activity，并且停止音乐播放服务的运行，当用户单击 Close 按钮的时候，只会结束当前的 Activity。读者可以试着多单击一下界面上的各个按钮，看看在单击不同的按钮的情况下，程序的运行效果。

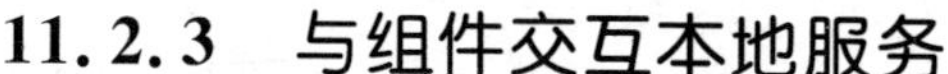

11.2.3 与组件交互本地服务

上面两个实例分别演示了 Activity 组件以及 Broadcast Receiver 组件是如何调用本地服务的，但是并没有与本地服务进行交互。下面笔者通过实例介绍通过 Activity 如何与本地服务进行交互。

(1) 新建工程 chapter11_3，编辑 res/layout/activity_main.xml 文件，源代码如下：

```
<LinearLayout xmlns:android = "http://schemas.android.com/apk/res/android"
    xmlns:tools = "http://schemas.android.com/tools"
    android:layout_width = "fill_parent"
    android:layout_height = "fill_parent"
    android:orientation = "vertical">
    <!-- 绑定按钮 -->
    <Button android:id = "@ + id/bind_button"
            android:layout_width = "fill_parent"
            android:layout_height = "wrap_content"
            android:text = "Bind Service"/>
    <!-- 接触绑定按钮 -->
    <Button android:id = "@ + id/unbind_button"
            android:layout_width = "fill_parent"
            android:layout_height = "wrap_content"
            android:text = "Unbind Service"/>
</LinearLayout>
```

(2) 新建一个服务：LocalService，编辑 LocalService.java 文件，源代码如下：

```
public class LocalService extends Service
{
    private static final String TAG  =  "LocalService";
    //返回给调用的客户端的 IBinder
    private final IBinder mBinder  =  new LocalBinder();
    //随机数生成器
    private final Random mGenerator  =  new Random();
    private static int NOTIFICATION_ID  =  1;
    //通知管理器
    private NotificationManager notificationManager;
    /**
     * 在这种情况下，调用者与 Service 由于都是在当前应用程序进程中执行，所以暂时不
需要 IPC */
    public class LocalBinder extends Binder
    {
```

```
        LocalService getService()
        {
            //返回当前 Service 的实例,使客户端可以调用 Service 里面声明的公共方
            //法,如 getRandomNumber()
            return LocalService.this;
        }
    }
    @Override
    public void onCreate()
    {
    super.onCreate();
    //获取通知管理器
    notificationManager = (NotificationManager) getSystemService(Context.NOTIFICA-
TION_SERVICE);
    displayNotification("Bind Service Created");
    Log.d(TAG, "onCreate() method was invoked");
    }
    @Override
    public IBinder onBind(Intent intent)
    {//服务与组件绑定时回调方法
    Log.d(TAG, "onBind() method was invoked");
        return mBinder;
    }
    /**客户端调用的方法*/
    public int getRandomNumber()
    {
    Log.d(TAG, "getRandomNumber() method was invoked");
      return mGenerator.nextInt(100);
    }
    @Override
    public void onDestroy()
    {//服务销毁时回调方法
    super.onDestroy();
    displayNotification("Bind Service Destoryed");
    Log.d(TAG, "onDestroy() method was invoked");
    }
    @Override
    public boolean onUnbind(Intent intent)
    {//服务与组件解除绑定时回调方法
    Log.d(TAG, "onUnbind() method was invoked");
    return super.onUnbind(intent);
    }
```

```
        //显示通知
        private void displayNotification(String message)
        {
        Notification notification = new Notification(R.drawable.ic_launcher, message,
System.currentTimeMillis());
            PendingIntent contentIntent = PendingIntent.getActivity(this, 0, new Intent
(this, MainActivity.class), 0);
            notification.setLatestEventInfo(this, "Background Service", message, conten-
tIntent);
            notificationManager.notify(NOTIFICATION_ID ++ , notification);
        }
    }
```

(3) 由于需要在项目清单文件当中声明 Service，编辑 AndroidManifest.xml 文件，源代码如下：

```
    <manifest xmlns:android = "http://schemas.android.com/apk/res/android"
        package = "com.example.chapter11_3"
        android:versionCode = "1"
        android:versionName = "1.0" >
        <uses-sdk  android:minSdkVersion = "8" android:targetSdkVersion = "15" />
        <application    android:icon = "@drawable/ic_launcher"
            android:label = "@string/app_name"
            android:theme = "@style/AppTheme" >
            <activity android:name = ".MainActivity"   android:label = "@string/title_
activity_main" >
                <intent-filter>
                    <action android:name = "android.intent.action.MAIN" />
                    <category android:name = "android.intent.category.LAUNCHER" />
                </intent-filter>
            </activity>
        <!-- Service 声明 -->
        <service android:name = ".LocalService"></service>
         </application>
    </manifest>
```

(4) 编辑 MainActivity.java 文件，源代码如下：

```
    public class MainActivity extends Activity implements OnClickListener
    {
        Button bindServiceBtn;//声明绑定服务按钮
        Button unbindServiceBtn;//声明解绑服务按钮
        boolean mBound = false;//是否绑定标识
        LocalService mService = null;
```

```
private ServiceConnection mConnection = new ServiceConnection()
{
@Override
public void onServiceDisconnected(ComponentName name)
{
    mBound = false;//解绑
}

@Override
public void onServiceConnected(ComponentName name, IBinder service)
{
    LocalBinder binder = (LocalBinder)service;
    mService = binder.getService();
    mBound = true;//绑定成功
    //获取随机数
    int num = mService.getRandomNumber();
    Toast.makeText(MainActivity.this, "随机数:" + num , Toast.LENGTH_LONG).
show();
    }
};
@Override
public void onCreate(Bundle savedInstanceState)
{
    super.onCreate(savedInstanceState);
    setContentView(R.layout.activity_main);
    bindServiceBtn = (Button)findViewById(R.id.bind_button);
                                                    //找到绑定服务按钮
    unbindServiceBtn = (Button)findViewById(R.id.unbind_button);
                                                    //未找到绑定服务按钮
    bindServiceBtn.setOnClickListener(this);//绑定单击事件监听器
    unbindServiceBtn.setOnClickListener(this);
}
@Override
public void onClick(View v)
{
int viewId = v.getId();
if(R.id.bind_button == viewId)
{  //如果是绑定按钮
   Intent intent = new Intent(this, LocalService.class);
   bindService(intent, mConnection, Context.BIND_AUTO_CREATE);
}
else if(R.id.unbind_button == viewId)
```

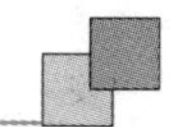

```
        {//如果是解绑按钮
            if(mBound)//如果服务与组件处于绑定状态
            {
                unbindService(mConnection);
                mBound = false;
            }
        }
    }
    @Override
    public boolean onCreateOptionsMenu(Menu menu)
    {
        getMenuInflater().inflate(R.menu.activity_main, menu);
        return true;
    }
}
```

上述代码当中，通过调用 Context. bindService()方法完成了 Activity 组件对于 Service 组件的绑定，绑定过后 Activity 组件就可以与 Service 组件进行交互。由于要通过调用 Context. bindService()方法对 Service 进行绑定，所以要在 LocalService 当中实现 onBind()这个方法，这个方法要求返回一个实现 IBinder 接口的对象，这个对象用来与客户端绑定的组件进行交互。

在 MainActivity 当中，笔者定义了一个实现了 ServiceConnection 的对象，用来对绑定的状态进行监听。当客户端 Activity 组件与 Service 组件绑定了的时候，onServiceConnected(ComponentName name, IBinder service)方法就会被调用，并且实现了 IBinder 接口的对象也会作为实参传入进来，当客户端 Activity 组件与 Service 组件解除绑定的时候，onServiceDisconnected()方法就会被调用。

(5) 右击工程 chapter11_3，在弹出的快捷菜单中选择 Run As|Android Application 命令，单击界面上的 Bind Service 按钮的运行效果如图 11-5 所示。

图 11-5　单击 Bind Service 按钮的运行效果

当 Bind Service 按钮被单击的时候，程序首先会去调用 Context. bindService()的方法与 LocalService 进行绑定，当绑定完成的时候，MainActivity 当中定义的实现 ServiceConnection 对象的 onServiceConnected()就会被调用，在 onServiceConnected()方法当中，笔者根据传进来的实现了 IBinder 接口的对象，此对象事实上是 LocalService 当中定义的 LocalBinder 对象，LocalBinder 类当中定义了一个叫做 getService()的方法，此方法用来返回 LocalService 本身。笔者据此拿到与 Activity 组件绑定的 Service 组件，并调用了 LocalSer-

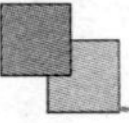

vice 当中定义的 getRandomNumber()方法来获取一个随机数，并通过“吐西”的方式显示出来。单击 Bind Service 按钮和 Unbind Service 按钮的时候，LogCat 控制台输出效果分别如图 11-6 和图 11-7 所示。

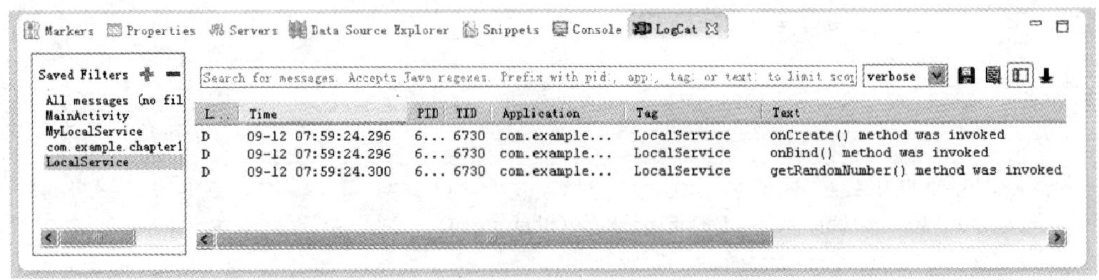

图 11-6　Context.bindService LogCat 输出

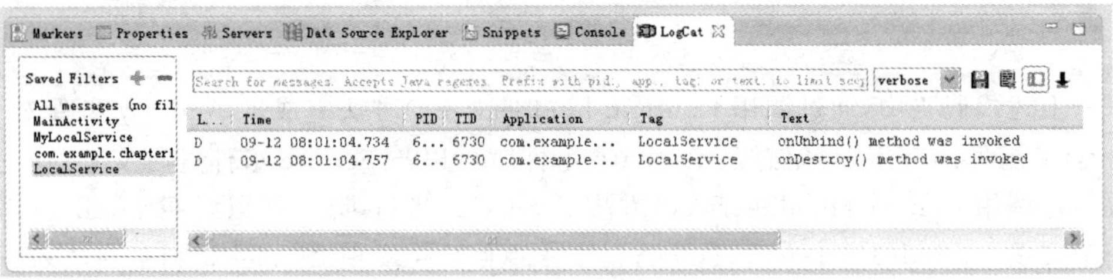

图 11-7　Context.unbindService LogCat 输出

11.2.4　Service 与 Thread 的区别

Thread 是程序执行的最小单元，它是分配 CPU 的基本单位，可以用 Thread 来执行一些异步的操作。子线程当中的 run()方法体，是在子线程当中运行的。Service 是 Android 的 4 大组件之一，主要用来在后台重复执行一些比较耗时的操作。Service 分为本地服务以及远程服务两种，对于本地服务来说，是在当前应用程序进程的 UI 主线程当中运行的，对于远程服务来说，是在被调用的远程服务所在的应用程序进程的 UI 主线程当中运行的，为了防止对 UI 线程的阻塞，一般在 Service 当中也要新建 Thread 来使得后台比较耗时的操作能够在非 UI 主线程当中执行。

如果在一个 Activity 组件当中，启动一个子线程的话，由于线程是程序执行的不同线索，所以子线程的执行是独立于 UI 主线程的，即便是启动子线程的 Activity 组件已经被销毁了，子线程仍可以继续执行下去。而且没有办法在不同的 Activity 当中对同一个子线程进行控制，因为 Android 当中的子线程一般都是通过匿名内部类的方式进行声明的。

对于一个 Service 来说，系统只会创建它的一个实例，也就是 Service 的 onCreate()方法只会在第一次创建的时候才会被调用。多次调用 Context.startService()方法不会启动多个 Service，只是相应 Service 对象的 onStartCommand()方法会被调用

多次，所以可以在不同组件中调用 Context. stopService()来销毁系统当中通过该方法启动的 Service。多次调用 Context. bindService()方法也不会启动多个 Service，只是相应 Service 对象的 onBind()方法会被调用多次，也可以在不同的组件当中通过调用 Context. unbindService()来销毁系统当中通过该方法启动的 Service。

总的来说，Thread 不在 UI 主线程当中运行，不易于管理，而 Service 默认是在主线程当中运行，易于管理，适合重复执行比较耗时的后台操作。

11.3　远程服务

每个 Android 应用程序都运行在自己独立的应用程序进程当中，默认情况下，一个进程是不能访问另外一个进程的内存空间的，如果想要在不同的进程之间进行通信，需要将对象分解成操作系统可以理解的基本单元，并且有序地通过进程边界。通过代码实现这个数据传输过程是冗长乏味的，Android 提供了 AIDL 工具来处理这项工作，使得不同的 Android 应用程序之间可以进行通信。

11.3.1　AIDL 介绍

AIDL(Android Interface Definition Language)是一种 IDL 语言，用于生成可以在 Android 设备上两个进程之间进行进程间通信的代码。如果一个应用程序进程的 Activity 组件去调用另外一个进程的 Service 组件，则可以使用 AIDL 生成可序列化的参数。AIDL IPC 机制是面向接口的，像 COM 或者是 Corba 一样，但是更加轻量级。它使用代理类在客户端和服务端传递数据。

11.3.2　远程服务实例

下面笔者就通过实例介绍如何在一个应用程序的 Activity 组件当中去调用另外一个应用程序提供的 Service 组件。

(1)创建工程 chapter11_4，由于需要借助于 Android 提供的 AIDL 工具来完成不同进程间的通信，所以首先需要定义 AIDL 接口，AIDL 接口和普通的接口没有什么特别，只不过它的扩展名为. aidl，保存在 Android 项目的主包下。如果其他应用程序需要通过 IPC 机制与此应用程序进行通信，则也需要在 Android 项目的主包中包含此. aidl 接口定义文件。所以，在工程的 src 根目录下新建 IStockService. aidl 接口文件，编辑 IStockService. aidl 文件，源代码如下：

```
package com.example.chapter11_4;
interface IStockService
{
    double getPrice(String ticket);
}
```

AIDL 使用简单的语法来声明接口，描述其方法以及方法的参数和返回值。这些参数和返回值可以是任何类型，甚至是其他 AIDL 生成的接口。其中对于 Java 编程语言的基本数据类型（int，long，char，boolean 等）、String 和 CharSequence，集合接口类型 List 和 Map，不需要使用 import 语句。

如果需要在 AIDL 中使用其他 AIDL 接口类型，即使是在相同包结构下，也需要使用 import 语句。AIDL 允许传递实现 Parcelable 接口的类，需要 import。需要特别注意的是，对于非基本数据类型，也不是 String 和 CharSequence 类型的，需要有方向指示，包括 in、out 和 inout，in 表示由客户端设置，out 表示由服务端设置，inout 是两者均可设置。AIDL 只支持接口方法，不能公开 static 变量。

（2）定义好 AIDL 接口文件过后，eclipse 工具会在工程的 gen 目录下生成一个叫做 IStockService.java 的文件，在 IStockService 接口当中定义了一个叫做 Stub 的静态抽象类，此类除了继承 android.os.Binder 这个类，还实现了 IStockService 这个接口。该代码为 eclipse 工具为工程自动生成，读者只需要了解如何去定义进程间通信的 AIDL 接口文件如何编写就足够了。

（3）定义好 AIDL 接口文件过后，需要定义相应的 Service。新建 Service：StockService，编辑 StockService.java 文件，源代码如下：

```
public class StockService extends Service
{
    private static String TAG = "StockService";
    //由于 IStockSerivice.Stub 继承于 android.os.Binder 类，也就是实现了 IBinder 这个
    //接口
    public class StockServiceImpl extends IStockService.Stub
    {
        @Override
        public double getPrice(String ticket) throws RemoteException
        {
            return 20.0;
        }
    }
    //服务第一次创建时调用
    @Override
    public void onCreate()
    {
        Log.d(TAG, "StockService OnCreate method was invoked");
        super.onCreate();
    }
    //其他客户端绑定到此服务时调用，可以是同一应用程序的组件
    //也可以是其他应用程序的组件
    @Override
```

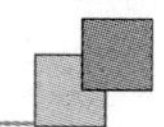

```
    public IBinder onBind(Intent intent)
    {
        //远程服务返回的 IBinder 对象与跟组件交互的本地服务类似,只不过返回的
        //IBinder 对象实现了一个 AIDL 接口
        return new StockServiceImpl();
    }
    @Override
    public int onStartCommand(Intent intent, int flags, int startId)
    {
        Log.d(TAG, "StockService onStartCommand method was invoked");
        return super.onStartCommand(intent, flags, startId);
    }
    //服务销毁时调用
    @Override
    public void onDestroy()
    {
        Log.d(TAG, "StockSerivce onDestory method was invoked");
        super.onDestroy();
    }
}
```

(4) 为了将此 Service 暴露给其他应用程序调用,需要在项目清单文件当中对 StockService 进行配置,编辑 AndroidManifest.xml 文件,源代码如下:

```
<manifest xmlns:android = "http://schemas.android.com/apk/res/android"
    package = "com.example.chapter11_4"
    android:versionCode = "1"
    android:versionName = "1.0" >
    <uses-sdk  android:minSdkVersion = "8"   android:targetSdkVersion = "15" />
    <application   android:icon = "@drawable/ic_launcher"   android:label = "@string/app_name"
                            android:theme = "@style/AppTheme" >
        <activity
            android:name = ".MainActivity"
            android:label = "@string/title_activity_main" >
            <intent-filter>
                <action android:name = "android.intent.action.MAIN" />
                <category android:name = "android.intent.category.LAUNCHER" />
            </intent-filter>
        </activity>
        <!-- 暴露给其他 Service 调用配置 -->
        <service android:name = ".StockService">
                <intent-filter>
```

```
                    <action android:name = "com.example.chapter11_4.IStockSer-
vice"/>
                </intent-filter>
            </service>
        </application>
    </manifest>
```

(5) 服务端准备好过后，就需要建立客户端，对服务端的 StockService 进行调用测试。新建工程 chapter11_5，编辑 res/layout/activity_main.xml 文件，源代码如下：

```
<LinearLayout xmlns:android = "http://schemas.android.com/apk/res/android"
    xmlns:tools = "http://schemas.android.com/tools"
    android:layout_width = "fill_parent"
    android:layout_height = "fill_parent"
    android:orientation = "vertical">
<!-- 绑定远程服务按钮 -->
    <Button
                android:id = "@ + id/bind_button"
                android:layout_width = "fill_parent"
                android:layout_height = "wrap_content"
                android:text = "Bind"/>
<!-- 调用远程服务按钮 -->
     <Button
                android:id = "@ + id/cal_button"
                android:layout_width = "fill_parent"
                android:layout_height = "wrap_content"
                android:text = "Cal"/>
   <!-- 解绑远程服务按钮 -->
      <Button
                android:id = "@ + id/unbind_button"
                android:layout_width = "fill_parent"
                android:layout_height = "wrap_content"
                android:text = "Unbind"/>
</LinearLayout>
```

(6) 需要将工程 chapter11_4 当中定义的 IStockService.aidl 接口文件复制到工程 chapter11_5 当中，当然在此之前，需要在工程 chapter11_4 当中建立好原先 AIDL 文件所在的包名，然后将 AIDL 接口文件放在此包名下所示。

(7) 编辑 MainActivity.java 文件，源代码如下：

```
public class MainActivity extends Activity implements OnClickListener
{
```

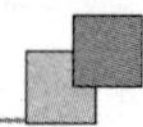

```
private static final String TAG = "MainActivity";
Button bindButton;                  //绑定按钮
Button calButton;                   //调用按钮
Button unBindButton;                //接触绑定按钮
private IStockService stockService = null;
//对与 Service 的绑定状态进行监听
private ServiceConnection serviceConnection = new ServiceConnection()
{
    //客户端 Activity 与远程 Service 绑定时调用
    @Override
    public void onServiceConnected(ComponentName name, IBinder service)
    {
        Log.d(TAG, "onServiceConnected called");
        stockService = IStockService.Stub.asInterface(service);
    }
    //客户端 Activity 与远程 Service 解除绑定时调用
    @Override
    public void onServiceDisconnected(ComponentName name)
    {
        Log.d(TAG, "onServiceDisconnected called");
        stockService = null;
    }
};
@Override
public void onCreate(Bundle savedInstanceState)
{
    super.onCreate(savedInstanceState);
    setContentView(R.layout.activity_main);
    bindButton = (Button)findViewById(R.id.bind_button);//找到绑定按钮
    calButton = (Button)findViewById(R.id.cal_button);//找到调用按钮
    unBindButton = (Button)findViewById(R.id.unbind_button);
                                                    //找到解除绑定按钮
    bindButton.setOnClickListener(this);//绑定按钮单击事件监听器
    calButton.setOnClickListener(this);
    unBindButton.setOnClickListener(this);
}
@Override
public void onClick(View view)
{
    int id = view.getId();
    switch(id)
    {
```

```
            case R.id.bind_button:          //绑定按钮
                bindService(new Intent(IStockService.class.getName()), serviceConnec-
tion, Context.BIND_AUTO_CREATE);
                break;
            case R.id.cal_button:           //调用按钮
                calRemoteService();
                break;
            case R.id.unbind_button:        //解除绑定按钮
                unbindService(serviceConnection);
                break;
            default:
                break;
            }
        }
        //与远程服务进行交互
        private void calRemoteService()
        {
            try
            {
                double val = stockService.getPrice("ALFA DeV");
                Toast.makeText(this, "Value From service is:" + val, Toast.LENGTH_
LONG).show();
            }
            catch(RemoteException e)
            {
                Log.e(TAG, e.getMessage());
            }
        }
        @Override
        public boolean onCreateOptionsMenu(Menu menu)
        {
            getMenuInflater().inflate(R.menu.activity_main, menu);
            return true;
        }
    }
```

上述代码当中也定义了一个 ServiceConnection 的对象对客户端 Activity 组件与远程 Service 之间绑定的状态进行监听。当客户端 Activity 组件与远程 Service 绑定的时候,通过调用以下代码就可以获取 IStockService 的一个实例。

```
stockService = IStockService.Stub.asInterface(service);
```

实际上由于远程 StockService 的 onBind()方法当中返回的是 StockServiceImpl

类的一个实例，所以得到这个实例过后，就可以调用 StockServiceImpl 这个类当中的 getPrice()这个方法。

(8) 由于工程 chapter11_5 的 MainActivity 组件需要对工程 chapter11_4 的远程服务：StockService 进行调用，所以需要先将工程 chapter11_4 在手机上运行起来。右击工程 chapter11_4，在弹出的快捷菜单中选择 Run As|Android Application 命令，将工程 chapter11_4 先在手机上运行起来。然后右击工程 chapter11_5，在弹出的快捷菜单中选择 Run As|Android Application 命令，在界面上单击 Bind 按钮过后，再单击界面上的 Cal 按钮，运行效果如图 11-8 所示。

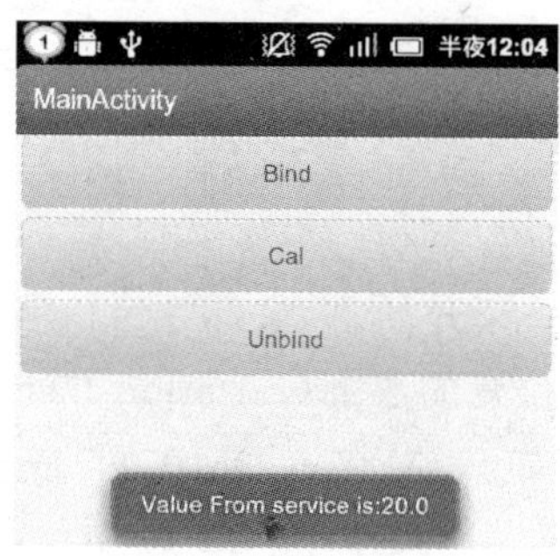

图 11-8　客户端调用远程 Service 结果

第 12 章

LBS

从移动互联网被引爆那一天开始，大家都在讨论所谓的 LBS，那么究竟 LBS 是什么呢？事实上 LBS 是基于位置的服务（Location Based Service），它是通过电信移动运营商的无线电通信网络（如 GSM 网、CDMA 网）或外部定位方式（如 GPS）获取移动终端用户的位置信息，在 GIS（Geographic Information System，地理信息系统）平台的支持下，为用户提供相应服务的一种增值业务。

LBS 包括两层含义：首先是确定移动设备或用户所在的地理位置；其次是提供与位置相关的各类信息服务。意指与定位相关的各类服务系统，简称“定位服务”，另外一种叫法为 MPS-Mobile Position Services，也称为“移动定位服务”系统。如找到手机用户的当前地理位置，然后在上海市 6 340 km^2 范围内寻找手机用户当前位置处 1 km 范围内的宾馆、影院、图书馆、加油站等的名称和地址。所以说 LBS 就是要借助互联网或无线网络，在固定用户或移动用户之间，完成定位和服务两大功能。在本章当中，笔者将会介绍如何获取用户的地理位置以及如何通过 Google Maps 将用户的地理位置信息在地图上显示出来。本章包括以下内容：

- 定位；
- Google Maps。

12.1 定 位

所谓“定位”，就是获取终端用户所处的地理位置信息。在本节当中，笔者将会介绍手机定位的几种方式，接着会通过实例来为读者一一介绍这几种定位方式的使用方法。

12.1.1 手机定位的方式

最简单的手机定位方式当然是通过智能手机的 GPS 模块。GPS 定位准确度是

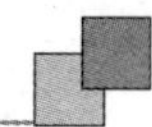

最高的，但是它的缺点也相当明显：①比较耗电；②绝大部分用户默认不开启 GPS 模块；③从 GPS 模块开启到获取第一次定位数据，可能需要比较长的时间；④室内无法使用。其中②和③两点都是比较致命的缺点。需要指出的是，GPS 是利用卫星通信的通道，在没有网络连接的情况下也能用。

另外一种常见的方式是基站定位。大致思路就是采集到手机上的基站 ID 号(cellid)和其他的一些信息(MNC、MCC、LAC 等)，然后通过网络访问一些定位服务，获取并返回相应的经纬度坐标。基站定位的准确度不如 GPS，但好处是能够在室内使用。

还有一种定位方式叫 WIFI 定位。和基站定位相似，这种方式是通过获取当前所用的 WIFI 的一些信息，然后访问网络上的定位服务以获得经纬度坐标。

12.1.2 GPS 定位

Android 系统让 Android 应用程序可以利用 android.location 包下面的类来使用移动设备提供的 GPS 定位服务，获取位置的相关信息。GPS 定位服务的核心类是 LocationManager，它提供相应的 API 来获取移动终端设备的经纬度信息。下面笔者就通过实例介绍如何使用 GPS 定位来获取移动终端设备的经纬度信息。

(1) 新建工程 chapter12_1，编辑 res/layout/activity_main.xml 文件，源代码参考配套资料。

(2) 由于是通过 GPS 定位的方式去获取用户所处的地理位置信息，所以还需要在项目清单文件当中添加相关的权限，编辑 AndroidManifest.xml 文件，源代码如下：

```
<manifest xmlns:android = "http://schemas.android.com/apk/res/android"
    package = "com.example.chapter12_1"
    android:versionCode = "1"
    android:versionName = "1.0" >
  <!-- 获取精确位置权限 -->
    <uses-permission android:name = "android.permission.ACCESS_FINE_LOCATION"/>
    <uses-sdk    android:minSdkVersion = "8"    android:targetSdkVersion = "15" />
    <application
        android:icon = "@drawable/ic_launcher"
        android:label = "@string/app_name"
        android:theme = "@style/AppTheme" >
        <activity
            android:name = ".MainActivity"
            android:label = "@string/title_activity_main" >
            <intent-filter>
                <action android:name = "android.intent.action.MAIN" />
                <category android:name = "android.intent.category.LAUNCHER" />
```

```
        </intent-filter>
    </activity>
  </application>
</manifest>
```

（3）编辑 MainActivity.java 文件，源代码如下：

```
public class MainActivity extends Activity implements OnClickListener
{
    Button  getLocationButton;//声明获取地理位置按钮
    Button stopGetLocation;  //声明停止获取地址位置按钮
    TextView showLocationTextView;//显示地理位置信息
    private LocationManager locationManager;//位置管理器
    private Location location ;//位置信息，包含经纬度
    //位置更新监听器
    private LocationListener myLocationListener = new LocationListener()
    {
        @Override
        public void onStatusChanged(String provider, int status, Bundle extras)
        {//状态变化时回调方法
        }
        @Override
        public void onProviderEnabled(String provider)
        {//可用时回调方法
        }
        @Override
        public void onProviderDisabled(String provider)
      {  //不可用时回调方法
         updateView(null);
      }
      @Override
      public void onLocationChanged(Location location)
       {//位置变化时回调方法
           updateView(location);
      }
      };
    @Override
    public void onCreate(Bundle savedInstanceState)
    {
        super.onCreate(savedInstanceState);
        setContentView(R.layout.activity_main);
```

```
        getLocationButton = (Button)findViewById(R.id.get_location);
                                                    //获取 id 为 get_location 的按钮
        stopGetLocation = (Button)findViewById(R.id.stop_location);
                                                    //获取 id 为 stop_location 的按钮
        showLocationTextView = (TextView)findViewById(R.id.show_location);
                                                    //获取 id 为 show_location 的按钮
        getLocationButton.setOnClickListener(this);//绑定单击事件监听器
        stopGetLocation.setOnClickListener(this);
        locationManager = (LocationManager)getSystemService(Context.LOCATION_SERV-
ICE);//获取位置管理器
        //指定由 GPS 定位来获取位置信息
        location = locationManager.getLastKnownLocation(LocationManager.GPS_PRO-
VIDER);
        updateView(location); //更新位置信息
    }
    @Override
    public void onClick(View v)
    {
    int id = v.getId();
    if(R.id.get_location == id)
    {//获取 id 为 get_location 的按钮
        getLocation();
    }
    else if(R.id.stop_location == id)
    {//获取 id 为 stop_location 的按钮
        locationManager.removeUpdates(myLocationListener);//移除位置监听器
    }
    }
    private void getLocation()
    {
        //设置监听器,当距离超过 minInstance,且时间超过 minTime 时更新
        //因为笔者希望位置信息实时更新,所以将第二个参数和第三个参数都设置成 0
        locationManager.requestLocationUpdates(LocationManager.GPS_PROVIDER, 0, 0,
myLocationListener);
    }
    /**
     * 更新位置信息
     * @param location
     */
    private void updateView(Location location)
```

```
        {
        if(location == null)
        { //如果 location 对象为 null
            showLocationTextView.setText("未定位到当前位置");
            return;
        }
        double latitude = location.getLatitude();//纬度
        double longitude = location.getLongitude();//经度
        showLocationTextView.setText("纬度:" + latitude + "\n" + "经度:" + longi-
tude);
        }
        @Override
        public boolean onCreateOptionsMenu(Menu menu)
        {
            getMenuInflater().inflate(R.menu.activity_main, menu);
            return true;
        }
    }
```

上述代码当中通过调用 Context.getSystemService(Context.LOCATION_SERVICE)方法来获取位置管理器(LocationManager)对象,有了 LocationManager 对象之后,就可以开始监听位置的变化了。可以通过调用 LocationManager 对象的 requestLocationUpdates(String provider, long minTime, float minDistance, LocationListener listener)来监听位置的变化。

对于第一个参数,有两个可选值,分别是 LocationManager.NETWORK_PROVIDER 以及 LocationManager.GPS_PROVIDER。前者用于移动网络获取位置,精度较低但速度很快;后者使用 GPS 定位,精度很高但一般需要 10～60 s 才能开始第 1 次定位,如果在室内则基本上无法定位。

在上述程序当中,笔者使用 GPS 进行定位,由于 GPS 在室内基本上无法定位,笔者开启了一个模拟器,并通过 Emulator Control 视图向模拟器发送经纬度信息的控制。

(4) 右击工程 chapter12_1,在弹出的快捷菜单中选择 Run As|Android Application 命令,将工程 chapter12_1 在开启的模拟器上运行。

(5) 单击 Window|Show View|other|Emulator Control,打开 Emulator Control 视图,单击视图 Manual 选项卡的 Send 按钮,如图 12-1 所示。

(6) 单击应用程序界面上的"GPS 定位"按钮,模拟器上运行的效果如图 12-2 所示。

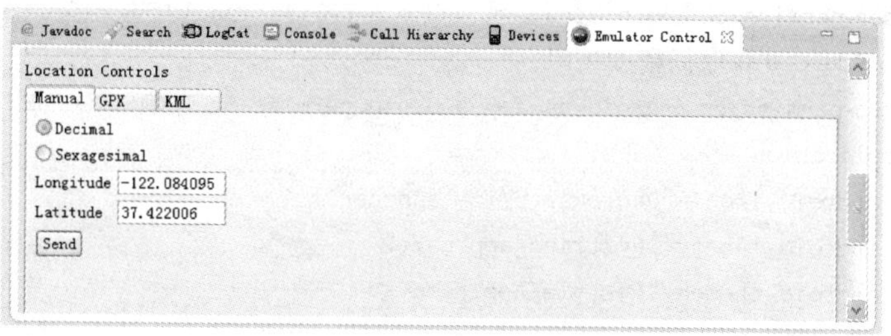

图 12-1 Emulator Control 视图

3:37 AM
MainActivity
GPS定位
停止
纬度：37.422005
经度：-122.084095

图 12-2 chapter12_1 在模拟器上运行的效果

12.1.3 基站定位

Android 操作系统下，基站定位其实很简单，先介绍其实现流程：

调用 Android SDK 的 Telephony API 获取 MCC，MNC，LAC，CID 等信息，然后通过 google 的 API 获得所在位置的经纬度，先来解释一下 MCC，MNC，LAC 以及 CID 的定义：

- MCC：Mobile Country Code，移动国家代码（中国的是 460）。
- MNC：Mobile Network Code，移动网络号码（中国移动为 00，中国联通为 01）。
- LAC：Location Area Code，位置区域码。
- CID：Cell Identity，基站编号，是一个 16 位的数据（范围是 0～65 535）。

下面就通过实例介绍如何使用基站定位获取设备的地理位置信息。

（1）新建工程 chapter12_2，编辑 res/layout/activity_main.xml 文件，源代码参考配套资料。

（2）编辑 AndroidManifest.xml 文件，需要添加两种权限，源代码如下：

```
<manifest xmlns:android = "http://schemas.android.com/apk/res/android"
    package = "com.example.chapter12_2"
    android:versionCode = "1"
    android:versionName = "1.0" >
    <uses-sdk    android:minSdkVersion = "8"    android:targetSdkVersion = "15" />
    <!-- Android 联网权限 -->
```

```
    <uses-permission android:name = "android.permission.INTERNET"/>
    <!-- Android 获取低精度权限 -->
    <uses-permission android:name = "android.permission.ACCESS_COARSE_LOCATION"/>
    <application
        android:icon = "@drawable/ic_launcher"
        android:label = "@string/app_name"
        android:theme = "@style/AppTheme" >
        <activity
            android:name = ".MainActivity"
            android:label = "@string/title_activity_main" >
            <intent-filter>
                <action android:name = "android.intent.action.MAIN" />
                <category android:name = "android.intent.category.LAUNCHER" />
            </intent-filter>
        </activity>
    </application>
</manifest>
```

(3) 新建 Java 类:SCell,用来代表基站信息,编辑 SCell.java 文件,源代码如下:

```
public class Scell
{
    private int MCC;//移动国家代码,中国是 460
    private int MNC;//移动网络号码,中国移动是 00,中国联通是 01
    private int LAC;//位置区域码
    private int CID;//基站编号
    public int getMCC()
    {
        return MCC;
    }
    public void setMCC(int mCC)
    {
        MCC = mCC;
    }
    public int getMNC()
    {
        return MNC;
    }
    public void setMNC(int mNC)
    {
        MNC = mNC;
    }
    public int getLAC()
```

```
    {
        return LAC;
    }
    public void setLAC(int lAC)
    {
        LAC = lAC;
    }
    public int getCID()
    {
        return CID;
    }
    public void setCID(int cID)
    {
        CID = cID;
    }
}
```

(4) 新建 Java 类:SItude,用来代表经纬度信息,编辑 SItude.java 文件,源代码如下:

```
public class Situde
{
    private String latitude;//纬度
    private String longitude;//经度
    public String getLatitude()
    {
        return latitude;
    }
    public void setLatitude(String latitude)
    {
        this.latitude = latitude;
    }
    public String getLongitude()
    {
        return longitude;
    }
    public void setLongitude(String longitude)
    {
        this.longitude = longitude;
    }
}
```

(5) 编辑 MainActivity.java 文件,源代码如下:

```
public class MainActivity extends Activity
{
    Button getLocationButton;//声明基站定位按钮
    TextView cellTextView;//声明显示基站信息文本
    TextView locationTextView;//声明显示位置信息文本
    @Override
    public void onCreate(Bundle savedInstanceState)
    {
        super.onCreate(savedInstanceState);
        setContentView(R.layout.activity_main);
        getLocationButton = (Button)findViewById(R.id.get_location);
                                              //找到 id 为 get_location 的按钮
        cellTextView = (TextView)findViewById(R.id.cell_msg);
                                                  //找到 id 为 cell_msg 的文本
        locationTextView = (TextView)findViewById(R.id.location_msg);
                                              //找到 id 为 location_msg 的文本
        //为按钮单击事件设定匿名内部类形式的监听器
        getLocationButton.setOnClickListener(new OnClickListener()
        {
            @Override
            public void onClick(View v)
            {//按钮单击时回调方法
                getLocationByCell();//基站定位获取位置信息
            }
        });
    }
    //基站方式获取位置信息
    private void getLocationByCell()
    {
    ProgressDialog mProgressDialog = new ProgressDialog(this);//声明进度条对话框
    mProgressDialog.setMessage("Now loading...");
    mProgressDialog.setProgressStyle(ProgressDialog.STYLE_SPINNER);
                                                              //设置对话框样式
    mProgressDialog.show();
    try
    {
        SCell cell = getCellInfo();//获取基站信息
        SItude itude = getITude(cell);//获取经纬度信息
        String location = getLocation(itude);//获取位置信息
        showResult(cell, location);
    }
    catch(Exception e)
```

```
        {
            cellTextView.setText(e.getMessage());//显示异常信息
        }
        mProgressDialog.dismiss();
        }
        /**
         * 获取基站信息
         * @return
         * @throws Exception
         */
        private SCell getCellInfo() throws Exception
        {
        SCell sCell = new SCell();
        TelephonyManager mTelephonyManager = (TelephonyManager) getSystemService(Con-
text.TELEPHONY_SERVICE);
        GsmCellLocation location = (GsmCellLocation)mTelephonyManager.getCellLocation();
        if(location == null)
        {//如果 location 对象为 null
            throw new Exception("Get Cell Failed");
        }
        //获取信息中的 getNetworkOperator 方法返回值赋值给字符串变量 operator
            String operator = mTelephonyManager.getNetworkOperator();
            int mcc = Integer.parseInt(operator.substring(0, 3));
            int mnc = Integer.parseInt(operator.substring(3));
            int cid = location.getCid();
            int lac = location.getLac();
            sCell.setMCC(mcc);
            sCell.setMNC(mnc);
            sCell.setCID(cid);
            sCell.setLAC(lac);
             return sCell;
        }
        /**
         * 根据基站信息,获取经纬度信息
         * @param cell
         * @return
         * @throws Exception
         */
        public SItude getITude(SCell cell) throws Exception
        {
        SItude itude = new SItude();
        HttpClient client = new DefaultHttpClient();
```

```
        //调用类 HttpPost 来准备向 google 发送 HttpPost 请求,参数为涉及的 url
            /* 无论是使用 HttpGet,还是使用 HttpPost,都必须通过如下 3 步来访问 HTTP
               资源。
                1.创建 HttpGet 或 HttpPost 对象,将要请求的 URL 通过构造方法传入 HttpGet
                  或 HttpPost 对象。
                2.使用 DefaultHttpClient 类的 execute 方法发送 HTTP GET 或 HTTP POST 请求,
                  并返回 HttpResponse 对象。
                3. 通过 HttpResponse 接口的 getEntity 方法返回响应信息,并进行相应的
                   处理。
                如果使用 HttpPost 方法提交 HTTP POST 请求,还需要使用 HttpPost 类的 setEn-
tity 方法设置请求参数。
            */
        HttpPost post = new HttpPost("http://www.google.com/loc/json");
        try
            {
                JSONObject holder = new JSONObject();
                //组装 holder 的内容
                holder.put("version", "1.1.0");
                holder.put("host", "maps.google.com");
                holder.put("address_language", "zh_CN");
                holder.put("request_address", true);
                holder.put("radio_type", "gsm");
                holder.put("carrier", "HTC");
                JSONObject tower = new JSONObject();
                tower.put("mobile_country_code", cell.getMCC());
                tower.put("mobile_network_code", cell.getMNC());
                tower.put("cell_id", cell.getCID());
                tower.put("location_area_code", cell.getLAC());
                JSONArray towerarray = new JSONArray();
                towerarray.put(tower);
                holder.put("cell_towers", towerarray);
                StringEntity query = new StringEntity(holder.toString());
                //发送查询请求
                post.setEntity(query);
                //获取传送 post 变量后 execute 动作的执行结果
                HttpResponse response = client.execute(post);
                //提取返回信息
                HttpEntity entity = response.getEntity();
                //申请缓存空间来显示返回信息的文本内容
                 BufferedReader buffReader = new BufferedReader(new InputStreamReader
(entity.getContent()));
                StringBuffer strBuff = new StringBuffer();
```

```
            //初始化提示信息字符串内容为空
            String result = null;
            //循环体按行添加返回信息到缓存空间,类似于解包输出
            while ((result = buffReader.readLine()) != null)
            {
                strBuff.append(result);
            }
            JSONObject json = new JSONObject(strBuff.toString());
            //提取缓冲空间中以 Location 作为文字标记的内容给变量 subjson
            JSONObject subjson = new JSONObject(json.getString("location"));
            itude.setLatitude(subjson.getString("latitude"));
            itude.setLongitude(subjson.getString("longitude"));
            //Log.i 表明输出 info 的日志,用于调试
            Log.i("Itude", itude.getLatitude() + itude.getLongitude());
        }
        catch (Exception e)
        {
            Log.e(e.getMessage(), e.toString());
            //如果捕获错误则进行抛错,throw 要与 try-catch-finally 合起来使用
            throw new Exception("Failed to get latitude/longitude:" + e.getMessage());
        }
        finally
        {
            //放弃 http 链接,清空 client 对象
            post.abort();
            client = null;
        }
        return itude;
    }
    /**
     * 获取位置信息
     * @param itude
     * @return
     * @throws Exception
     */
    private String getLocation(SItude itude) throws Exception
    {
    String resultString = "";
        String urlString = String.format("http://maps.google.cn/maps/geo? key =
abcdefg&q = %s, %s",
                itude.getLatitude(), itude.getLongitude());
        //记录拼装出来的 url 登录文本信息
```

```
        Log.i("URL", urlString);
        HttpClient client = new DefaultHttpClient();
        //定义 get 变量,用于提交 HttpGet 请求
        HttpGet get = new HttpGet(urlString);
        try
        {
            HttpResponse response = client.execute(get);
            HttpEntity entity = response.getEntity();
             BufferedReader buffReader = new BufferedReader(new InputStreamReader
(entity.getContent()));
            StringBuffer strBuff = new StringBuffer();
            String result = null;
            while ((result = buffReader.readLine()) != null)
                                              //如果 buffReader 读的一行内容不为 null
            {
                strBuff.append(result);
            }
            resultString = strBuff.toString();
            if (resultString != null && resultString.length() > 0)
                                                    //如果 resultString 对象不为空
            {
                JSONObject jsonObject = new JSONObject(resultString);
                 JSONArray jsonArray = new JSONArray(jsonObject.get("Placemark").
toString());
                resultString = "";
                for (int i = 0; i < jsonArray.length(); i++)
                                                        //循环迭代 jsonArray 对象
                {
                        resultString = jsonArray.getJSONObject(i).getString("ad-
dress");
                }
            }
        }
        catch (Exception e)
        {
            throw new Exception("Failed to get phy address" + e.getMessage());
        }
        finally
        {
            get.abort();
            client = null;
        }
```

```
        return resultString;
    }
    /**
     * 显示基站和位置信息
     * @param cell
     * @param location
     */
    private void showResult(SCell cell, String location)
    {
    cellTextView.setText(String.format("基站信息:mcc:%d,mnc:%d,lac:%d,cid:%d",  cell.getMCC(), cell.getMNC(), cell.getLAC(), cell.getCID()));
        locationTextView.setText("物理位置:" + location);
    }
    @Override
    public boolean onCreateOptionsMenu(Menu menu)
    {
        getMenuInflater().inflate(R.menu.activity_main, menu);
        return true;
    }
}
```

(6) 右击工程 chapter12_2,在弹出的快捷菜单中选择 Run As|Android Application 命令,单击界面上的"基站定位"按钮,运行效果如图 12-3 所示。

图 12-3 基站定位运行效果

12.1.4 WIFI 定位

现在在一些大中型城市里,WIFI 已经很普及,有私人或企业的 WIFI,亦有中国电信的 WIFI,通过 WIFI 进行定位,并不需要真正连接上指定的 WIFI 路由器,只需要探测到有 WIFI 存在即可,因此也可以使用 WIFI 进行定位,原理也和基站定位一样,必须要得到 WIFI 路由器的 SSID 和信号强度。

下面笔者就通过实例介绍如何使用 WiFi 定位。

(1) 新建工程 chapter12_3,编辑 res/layout/activity_main.xml 文件,源代码参考配套资料。

(2) 本实例当中,除了需要联网权限之外,还需要获取 WIFI 路由器状态的权限,所以需要添加相应的权限。编辑 AndroidManifest.xml 文件,源代码如下:

```
<manifest xmlns:android = "http://schemas.android.com/apk/res/android"
    package = "com.example.chapter12_3" android:versionCode = "1"
    android:versionName = "1.0" >
```

```
    <uses-sdk  android:minSdkVersion="8"  android:targetSdkVersion="15" />
  <!--Android 联网权限 -->
    <uses-permission  android:name="android.permission.INTERNET"/>
    <!-- 获取 WIFI 状态 -->
  <uses-permission  android:name="android.permission.ACCESS_WIFI_STATE"/>
    <application
        android:icon="@drawable/ic_launcher"
        android:label="@string/app_name"
        android:theme="@style/AppTheme" >
        <activity
            android:name=".MainActivity"
            android:label="@string/title_activity_main" >
            <intent-filter>
                <action android:name="android.intent.action.MAIN" />
                <category android:name="android.intent.category.LAUNCHER" />
            </intent-filter>
        </activity>
    </application>
</manifest>
```

(3) 编辑 MainActivity.java 文件，源代码如下：

```
public class MainActivity extends Activity
{
    Button wifiGetLocation;//声明 WIFI 定位按钮
    TextView showLocation;//声明显示地理信息位置文本
    @Override
    public void onCreate(Bundle savedInstanceState)
    {
        super.onCreate(savedInstanceState);
        setContentView(R.layout.activity_main);
        wifiGetLocation = (Button)findViewById(R.id.get_location);
                                                //找到 id 为 get_location 的按钮
        showLocation = (TextView)findViewById(R.id.show_location);
                                                //找到 id 为 show_location 的按钮
        wifiGetLocation.setOnClickListener(new OnClickListener()
          {//采用匿名内部类作为按钮单击事件监听器
              @Override
              public void onClick(View v)
              {  //按钮单击事件回调方法
                 try
                 {
                     showResult(execute(doWifi()));
```

```
                }
                catch (Exception e)
                {
                    showLocation.setText(e.getMessage());//显示异常信息
                }
            }
        });
    }
    /**
     * 组装发送给 http://www.google.com/loc/json 的请求参数
     * @return
     * @throws Exception
     */
    private JSONObject doWifi() throws Exception
    {
        JSONObject holder = new JSONObject();
        holder.put("version", "1.1.0");
        holder.put("host", "maps.google.com");
        holder.put("address_language", "zh_CN");
        holder.put("request_address", true);
        //获取 WIFI 管理器
        WifiManager wifiManager = (WifiManager)getSystemService(Context.WIFI_SERVICE);
        if (wifiManager.getConnectionInfo().getBSSID() == null)
        {//如果得到的 bssid 为 null,则抛出异常
            throw new RuntimeException("bssid is null");
        }
        JSONArray array = new JSONArray();
        JSONObject data = new JSONObject();
        data.put("mac_address", wifiManager.getConnectionInfo().getBSSID());
//WIFI 路由器的 BSSID
        data.put("signal_strength", 8);  //WIFI 路由器的信号强度
        data.put("age", 0);
        array.put(data);
        holder.put("wifi_towers", array);
        return holder;
    }

    /**
     * 发送 HTTP 请求,获取当前所在经纬度信息
     * @param params
     * @return
     * @throws Exception
```

```
 */
public String execute(JSONObject params) throws Exception
{
    //调用类 HttpPost 来准备向 google 发送 HttpPost 请求,参数为涉及的 url
    /* 无论是使用 HttpGet,还是使用 HttpPost,都必须通过如下 3 步来访问 HTTP 资源。
        1. 创建 HttpGet 或 HttpPost 对象,将要请求的 URL 通过构造方法传入 HttpGet
           或 HttpPost 对象。
        2. 使用 DefaultHttpClient 类的 execute 方法发送 HTTP GET 或 HTTP POST 请
           求,并返回 HttpResponse 对象。
        3. 通过 HttpResponse 接口的 getEntity 方法返回响应信息,并进行相应的
           处理。
        如果使用 HttpPost 方法提交 HTTP POST 请求,还需要使用 HttpPost 类的 setEn-
tity 方法设置请求参数。
     */
    HttpClient httpClient = new DefaultHttpClient();
    HttpPost post = new HttpPost("http://www.google.com/loc/json");
    StringEntity se = new StringEntity(params.toString());
    post.setEntity(se);
    //使用 HttpClient 执行 http 请求
    HttpResponse response = httpClient.execute(post);
    if (response.getStatusLine().getStatusCode() == 200)
                                                        //如果 http 请求返回码为 200
     {
        HttpEntity entity = response.getEntity();//提取返回信息
        BufferedReader br;
        br = new BufferedReader(new InputStreamReader(entity.getContent()));
        StringBuffer sb = new StringBuffer();
        String result = br.readLine();
        while (result != null) {   //如果 result 对象不为 null
            sb.append(result);
            result = br.readLine();
        }
        JSONObject json = new JSONObject(sb.toString());
        JSONObject lca = json.getJSONObject("location");
         return getLocation(lca.getString("latitude"), lca.getString("longi-
tude"));//获取位置信息
    }
    return null;
}
/**
 * 获取位置信息
 * @param itude
```

```
     * @return
     * @throws Exception
     */
    private String getLocation(String latitude, String longitude) throws Exception
    {
    String resultString = "";
        String urlString = String.format("http://maps.google.cn/maps/geo? key =
abcdefg&q = %s, %s",
                latitude, longitude);
        //记录拼装出来的 url 登录文本信息
        Log.i("URL", urlString);
        HttpClient client = new DefaultHttpClient();
        //定义 get 变量,用于提交 HttpGet 请求
        HttpGet get = new HttpGet(urlString);
        try
        {
            HttpResponse response = client.execute(get);
            HttpEntity entity = response.getEntity();
            BufferedReader buffReader = new BufferedReader(new InputStreamReader
(entity.getContent()));
            StringBuffer strBuff = new StringBuffer();
            String result = null;
            while ((result = buffReader.readLine()) != null)
            {
                strBuff.append(result);
            }
            resultString = strBuff.toString();
            if (resultString != null && resultString.length() > 0) //如果 result-
String 对象不为空
            {
                JSONObject jsonObject = new JSONObject(resultString);
                JSONArray jsonArray = new JSONArray(jsonObject.get("Placemark").
toString());
                resultString = "";
                for (int i = 0; i < jsonArray.length(); i++) //循环迭代 jsonArray 对象
                {
                    resultString = jsonArray.getJSONObject(i).getString("ad-
dress");
                }
            }
        }
        catch (Exception e)
```

```
        {
            throw new Exception("Failed to get phy address" + e.getMessage());
        }
        finally
        {
            get.abort();
            client = null;
        }
        return resultString;
    }
    /**
     * 显示基站和位置信息
     * @param cell
     * @param location
     */
    private void showResult(String location)
    {
      if(null == location || location.length() == 0)  //如果 location 对象为空
      {
        showLocation.setText("未定位到物理位置");
        return;
      }
      showLocation.setText("物理位置:" + location);
    }
    @Override
    public boolean onCreateOptionsMenu(Menu menu)
    {
        getMenuInflater().inflate(R.menu.activity_main, menu);
        return true;
    }
}
```

上述使用 WIFI 定位技术的代码与使用基站定位技术的代码基本一致，主要区别来自于向 http://www.google.com/loc/json 发送的 http 请求的参数不一致。如果采用基站定位的话，则需要获取最近的基站信息，包括 MCC，MNC，LAC 以及 CID，将这些数据发送到 google 的定位 WEB 服务，就能获取当前的位置信息；如果采用 WIFI 定位的话，则需要获取 WIFI 路由器的 BSSID 以及信号强度，同样也是将这些数据发送到 google 的定位 WEB 服务，从而获取用户设备当前的位置信息。

上述代码当中所有的网络请求耗时操作都是放在 UI 线程当中执行的，这在实际的 Android 应用程序开发过程当中也是不可取的，一般情况下，可以将这些耗时的网络操作放在独立的 AsyncTask 当中去完成，这样就不会造成 UI 主线程的阻塞。

读者也可以尝试将上述的基站定位的代码以及 WIFI 定位的代码放到 AsyncTask 当中去执行。

(4) 右击工程 chapter12_3，在弹出的快捷菜单中选择 Run As|Android Application 命令，单击界面上的 WiFi 定位按钮，运行效果如图 12-4 所示。

图 12-4 WIFI 定位运行效果

12.2 Google Maps

上面章节已经介绍了如何获取用户设备所在的地理信息位置，获取到地理位置的经纬度信息过后，就可以通过 Google 提供的地图外部库将用户设备的地理位置信息显示在地图上。通过获取设备的经纬度信息，然后与 Google Maps 相结合，可以非常方便地开发出定位、导航等应用。

Google 提供的地图外部库当中定义了一个名为 com. google. android. maps 的包，其中包含了一系列用于在 Google Map 上显示、控制和叠层信息的功能类。

com. google. android. maps 包当中有一个非常重要的类：com. google. android. maps. MapView。MapView 是一个可以展示 Google 地图的视图控件，它可以获取触摸事件来支持地图的移动(东南西北)以及缩放功能。它还支持多层 Overlay，可以在地图上进行一些标注来标识一些想要标识的地理位置。为了要在 MapView 控件上显示 Google 地图和位置的相关数据，必须先登录 Google 地图服务系统，申请一个 Google Maps API Key。

在默认的 Android SDK 中不包含 Google Map APIs。要使用 Google Map 必须下载支持 Google APIs 的 SDK。下面就介绍如何使用 Google Maps。

12.2.1 下载 Google APIs

由于 Android SDK 当中不包含 Google Map APIs，所以需要开发者手动下载。笔者下载的是针对 Android SDK 2.2 版本的 Google APIs，下载地址链接：

http://dl-ssl. google. com/android/repository/google_apis-8_r02. zip，下载完成后，将解压缩得到的文件夹的名称改成：google_apis_8，并且放到下载的 Android SDK 所在目录的 add-ons 子目录下在 windows 系统下面，一般默认是：C:\Documents and Settings\Administrator\Local Settings\Application Data\Android\android-sdk\add-ons 这个路径。

将下载得到的 Google APIs 放置到指定目录后，重启 Eclipse IDE。选择 Window|Preferences|Android 命令，如图 12-5 所示。

从图 12-5 当中可以看出，Target Name 那一栏多出了一个叫做 Google APIs 的 Target，这就意味着在新建应用程序的时候，可以将 Google APIs 这个 Target 指定为应用程序的 Build SDK 版本。

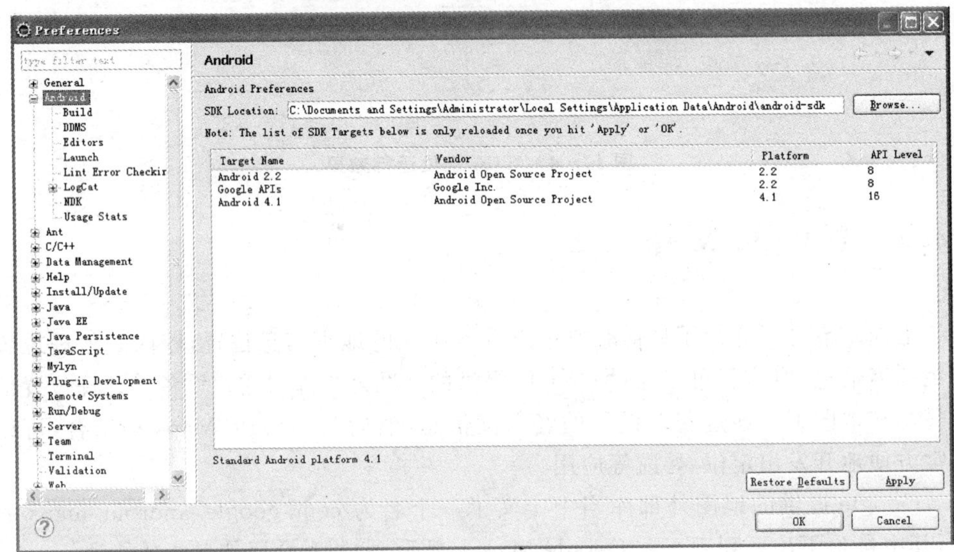

图 12-5 Google APIs target

12.2.2 获取 Google Maps API Key

要开发基于 Google Map 的应用，必须拥有 Google Maps API Key。申请步骤如下：

(1)打包 Android 应用程序的时候，都需要有一个签名文件对其进行签名。在调试的时候，一般用的是所谓的调试 keystore 文件，如果开发者的应用需要上架到 Google Play 应用商店的话，则需要使用正式的 keystore 文件对应用进行签名打包。选择 Window|Preferences|Android|Build 命令，如图 12-6 所示。

图 12-6 当中的 Default debug keystore 文本框当中的路径，就是默认签名文件所在的路径。

(2) 将 JDK 安装目录下 bin 子目录所在的路径添加到系统环境变量 path 中。

(3) 根据调试 keystore 文件所在的路径，使用 JDK 提供的 keytool 工具为 Android keystore 生成认证指纹。打开 CMD 窗口，然后输入：

keytool -list -keystore "C:\Documents and Settings\Administrator\. android\

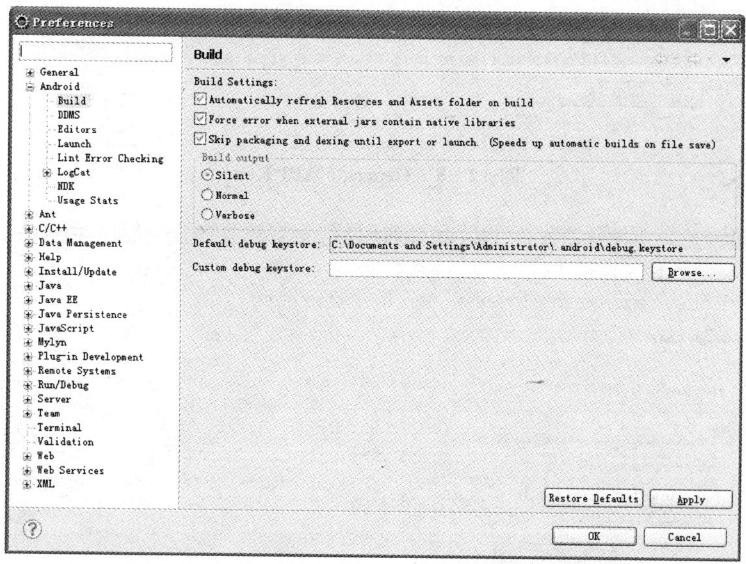

图 12－6　keystore 所在位置

debug.keystore" -storepass android -keypass android，按下 Enter 键，生成 MD5 签名，如图 12－7 所示。

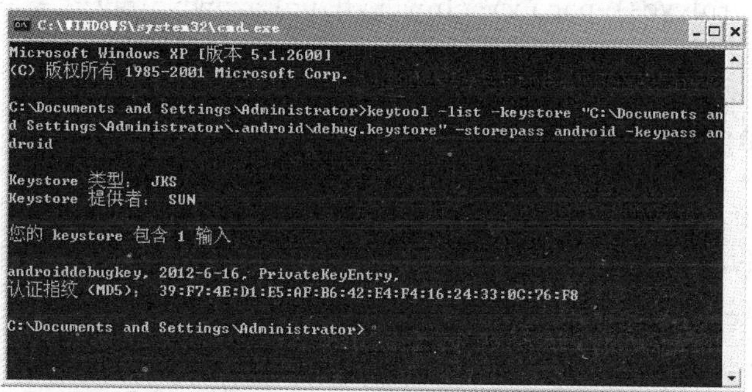

图 12－7　生成 MD5 签名

得到的 MD5 签名是：39：F7：4E：D1：E5：AF：B6：42：E4：F4：16：24：33：0C：76：F8，这个就是开发者申请 Google Maps API Key 的 MD5 签名。

（4）到 http://code.google.com/intl/zh-CN/android/maps-api-signup.html 页面提交上面得到的 MD5 签名即可获得 Google Maps API Key，如图 12－8 所示。

注意，登录这个网址前，请确保自己有 google 的账户，如果还没有 google 账户，到下面这个网址：https://www.google.com/accounts/Login? hl=zh-cn&continue=http://www.google.com.hk/ 注册一个。

（5）单击页面上的 Generate API Key 按钮，跳转到生成 Google Maps API Key 的页面，如图 12－9 所示。

☑ I have read and agree with the terms and conditions (printable version)

My certificate's MD5 fingerprint: 39:F7:4E:D1:E5:AF:B6:42:E4:F4:16:24:33:0C:76:F8

Generate API Key

图 12－8　Generate API Key

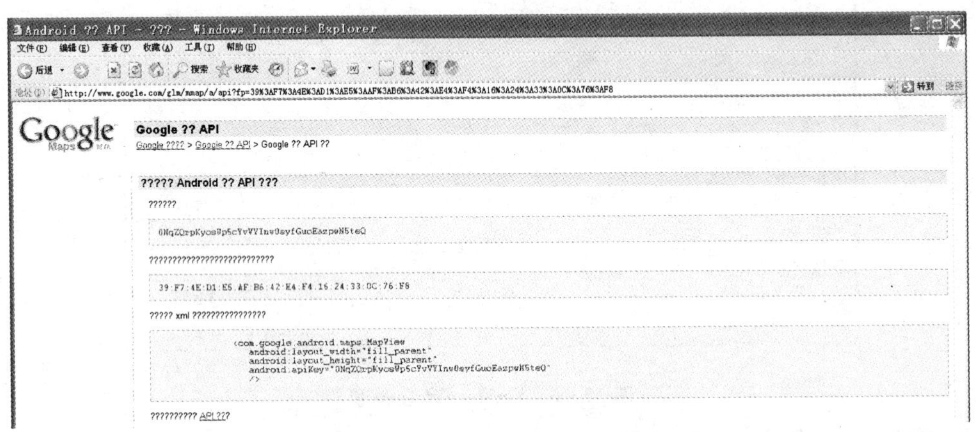

图 12－9　Google Maps API Key

其中 0NqZQrpKyosWp5cYvVYInw0syfGuoEazpwN5teQ 就是程序当中需要使用到的 Google Maps API Key。

12.2.3　MapView 的使用

Google Map 中提供了如下几个主要类来完成功能的开发：

- MapActivity：主要用来显示地图对应的 Activity，要显示地图的话，必须要继承 MapActivity，并重写相应的方法。
- MapView：用来显示地图，该控件必须要和 MapActivity 配合使用，来完成地图的各种功能。
- MapController：MapView 的控制器，用于控制地图的旋转、缩放等。
- Overlay：表示可以在地图上可绘制的图层。例如可以在地图上添加一些标记、符号、路线等。其本身完全透明。通过实现该类的 draw()方法来控制。
- GeoPoint：表示一个经纬度的位置对象。

1. MapView 常用方法

- void setBuildInZoomControls(boolean on)：设置是否启动内置放大和缩小地图的功能。
- List<Overlay> getOverlays()：获取当前 MapView 上的所有图层列表。
- void setSatellite(boolean on)：设置卫星模式显示地图。

- ❑ void setTraffic(boolean on)：设置交通模式显示地图。
- ❑ Projection getProjection()：得到一个投影，用来在屏幕的坐标和经纬度之间进行转换。

2. MapController 常用方法

- ❑ int setZoom(int zoomLevel)：设置地图放大缩小的倍数(1～21)。
- ❑ boolean zoomIn()：缩小地图。
- ❑ boolean zoomOut()：放大地图。
- ❑ void animateTo(GeoPoint point)：以动画的方式定位到指定的经纬度坐标。

3. GeoPoint 常用方法

- ❑ public GeoPoint(int latitudeE6, int longitudeE6)：构造方法。
- ❑ int getLatitudeE6()：获得维度。
- ❑ int getLongitudeE6()：获得经度。

下面就通过实例介绍如何使用 Google Maps。

(1) 新建工程 chapter12_4，特别要注意：在选择 Build SDK 版本的时候，要选择 Google APIs 这一选项。

(2) 编辑 res/layout/activity_main.xml 文件，源代码如下：

```
<LinearLayout xmlns:android = "http://schemas.android.com/apk/res/android"
    xmlns:tools = "http://schemas.android.com/tools"
    android:layout_width = "fill_parent"
    android:layout_height = "fill_parent"
    android:orientation = "vertical">
<!-- 显示 Google Maps 控件 -->
<com.google.android.maps.MapView
        android:id = "@ + id/mapview"
        android:layout_width = "fill_parent"
        android:layout_height = "fill_parent"
        android:apiKey = "0NqZQrpKyosWp5cYvVYInw0syfGuoEazpwN5teQ" />
</LinearLayout>
```

(3) 需要添加一些权限以及使用外部库的声明，编辑 AndroidManifest.xml 文件，源代码如下：

```
<manifest xmlns:android = "http://schemas.android.com/apk/res/android"
    package = "com.example.chapter12_4"
    android:versionCode = "1"
    android:versionName = "1.0" >
    <uses-sdk  android:minSdkVersion = "8"   android:targetSdkVersion = "15" />
    <!-- 联网权限 -->
```

```
    <uses-permission  android:name = "android.permission.INTERNET"/>
    <!-- 获取精确位置权限 -->
    <uses-permission  android:name = "android.permisssion.ACCESS_FINE_LOCATION"/>
    <application
        android:icon = "@drawable/ic_launcher"
        android:label = "@string/app_name"
        android:theme = "@style/AppTheme" >
        <activity
            android:name = ".MainActivity"
            android:label = "@string/title_activity_main" >
            <intent-filter>
                <action android:name = "android.intent.action.MAIN" />
                <category android:name = "android.intent.category.LAUNCHER" />
            </intent-filter>
        </activity>
        <!-- 使用 Google 地图外部库 -->
        <uses-library android:name = "com.google.android.maps"></uses-library>
    </application>
</manifest>
```

(4) 编辑 MainActivity.java 文件,源代码如下:

```
public class MainActivity extends MapActivity
{
    MapView mapView;  //声明地图控件
    @Override
    public void onCreate(Bundle savedInstanceState)
    {
        super.onCreate(savedInstanceState);
        setContentView(R.layout.activity_main);
        mapView = (MapView)findViewById(R.id.mapview);//找到 id 为 mapview 的控件
        //设置地图可以单击拖动
        mapView.setClickable(true);
        //打开放大缩小效果的功能
        mapView.setBuiltInZoomControls(true);
        //使用 MapController 定位地图的位置,首先可以去查询要显示的位置的经纬度
        //借助 MapController 和 GeoPoint 来实现定位效果
        MapController controller = mapView.getController();
        GeoPoint point = new GeoPoint((int)31.2303930 * 1000000, (int)121.4737040 *
1000000);
        //动画方式定位到指定经纬度
        controller.animateTo(point);
        //设置放大级别
```

```
        controller.setZoom(10);
        //将该经纬度设置于屏幕的中心
        controller.setCenter(point);
    }
    //使用选项菜单来切换地图的显示模式
    @Override
    public boolean onCreateOptionsMenu(Menu menu)
    {
      menu.add(0, Menu.FIRST, 1, "普通模式");
      menu.add(0, Menu.FIRST + 1,   2, "交通模式");
      menu.add(0, Menu.FIRST + 2, 3, "卫星模式");
     return true;
    }
    @Override
    public boolean onOptionsItemSelected(MenuItem item)
    {
    int itemId = item.getItemId();
    switch(itemId)
    {
    case Menu.FIRST:  //普通模式
        mapView.setStreetView(true);
        //一定要把其他的模式设置为 false,否则没有效果
        mapView.setTraffic(false);
        mapView.setSatellite(false);
        break;
    case Menu.FIRST + 1://交通模式
        mapView.setTraffic(true);
        mapView.setStreetView(false);
            mapView.setSatellite(false);
        break;
    case Menu.FIRST + 2://卫星模式
        mapView.setSatellite(true);
        mapView.setTraffic(false);
            mapView.setStreetView(false);
        break;
    }
    return true;
    }
    @Override
    protected boolean isRouteDisplayed()
    {
        return false;
```

```
    }
}
```

图 12-10　MapView 使用

上述实例当中，通过 MapView 控件来显示 Google Maps 地图，因为在使用 Google Maps 地图的过程当中不仅需要联网，还需要获取精确的地理位置，所以还需要在 AndroidManifest. xml 文件当中声明好相应的权限，以及对 Google 外部地图库的使用。通过选项菜单对 Google Maps 地图支持的 3 种模式进行切换，分别是普通模式、交通模式和卫星模式。通过 MapController，将上海市置于屏幕的中心显示。

（5）右击工程 chapter12_4，在弹出的快捷菜单中选择 Run As|Android Application 命令，运行效果如图 12-10 所示。

12.2.4　地图标记的使用

地图标记，可以理解为在 MapView 控件上创建自定义的图层，将要显示的标记放置在图层指定的位置中，就可以达到在地图上显示自定义标记的效果了。

一个 Overlay 对象就代表了显示在 MapView 控件之上的一个图层。

一个 Overlay 中可以包含多个地图标记。通常的使用步骤是：

- ❑ 在 MapView 上创建一个单独的图层。
- ❑ 创建标记对象。
- ❑ 将标记显示在指定图层的指定位置。
- ❑ 处理单击标记的事件。

可以使用 Overlay 的子类 ItemizedOverlay 来简化实现，通常需要编写一个继承于 ItemizedOverlay 的子类，并重写其中的方法。该类中有一个或者多个 OverlayItem，每一个 OverlayItem 代表一个标记。将需要添加的标记加入到该 ItemizedOverlay 对象中，再将该 ItemizedOverlay 对象加入到 MapView 对象的 Overlay 中，就可以在 MapView 上显示所有的标记了。下面就通过演示如何添加地图标记。

（1）新建工程 chapter12_5，编辑 res/layout/activity_main. xml 文件，源代码如下：

```
<LinearLayout xmlns:android = "http://schemas.android.com/apk/res/android"
    xmlns:tools = "http://schemas.android.com/tools"
    android:layout_width = "fill_parent"
    android:layout_height = "fill_parent"
    android:orientation = "vertical">
```

```
<!-- 显示 Google Maps 控件 -->
<com.google.android.maps.MapView
        android:id = "@ + id/my_map"
        android:layout_width = "fill_parent"
        android:layout_height = "fill_parent"
        android:apiKey = "0NqZQrpKyosWp5cYvVYInw0syfGuoEazpwN5teQ" />
</LinearLayout>
```

(2) 需要添加一些权限以及使用外部库的声明，编辑 AndroidManifest.xml 文件，源代码如下：

```
<manifest xmlns:android = "http://schemas.android.com/apk/res/android"
    package = "com.example.chapter12_5"
    android:versionCode = "1"
    android:versionName = "1.0" >
   <!-- 联网权限 -->
    <uses-permission  android:name = "android.permission.INTERNET"/>
    <!-- 获取精确位置权限 -->
    <uses-permission  android:name = "android.permisssion.ACCESS_FINE_LOCATION"/>
    <uses-sdk  android:minSdkVersion = "8"     android:targetSdkVersion = "15" />
    <application
        android:icon = "@drawable/ic_launcher"
        android:label = "@string/app_name"
        android:theme = "@style/AppTheme" >
        <activity
            android:name = ".MainActivity"
            android:label = "@string/title_activity_main" >
            <intent-filter>
                <action android:name = "android.intent.action.MAIN" />
                <category android:name = "android.intent.category.LAUNCHER" />
            </intent-filter>
        </activity>
         <!-- 使用 Google 地图外部库 -->
        <uses-library android:name = "com.google.android.maps"></uses-library>
    </application>
</manifest>
```

(3) 自定义一个继承自 ItemizedOverlay 的子类：MyOverlay，编辑 MyOverlay.java 文件，源代码如下：

```
public class MyOverlay extends ItemizedOverlay<OverlayItem>
{
    //定义要显示的所有标记集合
```

```
    private ArrayList<OverlayItem> overlayItems = new ArrayList<OverlayItem>();
    private Context context;//声明上下文对象的引用
    public MyOverlay(Drawable defaultMarker)
    {//必须重写该构造方法
        super(boundCenterBottom(defaultMarker));//设置标记图片放在某个经纬度的下
                                                //部中间
    }
    public MyOverlay(Drawable defaultMarker, Context context)
    {
        super(boundCenterBottom(defaultMarker));
        this.context = context;
    }
    //自定义方法用来向标记集合中添加标记
    public void addOverlayItems(OverlayItem item)
    {
        overlayItems.add(item);
        populate();//该方法很重要,将标记现实的功能
    }
    @Override
    protected OverlayItem createItem(int i)
    {//重载方法获取当前的第 i 个标记对象
        return overlayItems.get(i);
    }
    @Override
    public int size()
    {//重载方法获得标记的个数
        return overlayItems.size();
    }
    @Override
    protected boolean onTap(int index)
    {//单击标记的时候触发
        Toast.makeText(context,overlayItems.get(index).getTitle() +
                  overlayItems.get(index).getSnippet(),Toast.LENGTH_SHORT).
show();
        return super.onTap(index);
    }
}
```

(4) 将准备好的 china.jpg 图片文件复制到 res/drawable-hdpi 子目录下。

(5) 编辑 MainActivity.java 文件,源代码如下:

```
public class MainActivity extends MapActivity
{
```

```
    MapView mapView = null;//声明地图控件对象
    @Override
    public void onCreate(Bundle savedInstanceState) {
        super.onCreate(savedInstanceState);
        setContentView(R.layout.activity_main);
        mapView = (MapView)findViewById(R.id.my_map);
                                                    //找到 id 为 my_map 的地图控件对象
        GeoPoint beijingPoint = new GeoPoint((int)(39.9042140 * 1000000), (int)
(116.4074130 * 1000000));//北京市
        GeoPoint tianjinPoint = new GeoPoint((int)39.07 * 1000000, (int)117.11 *
1000000);//天津市
        MapController controller = mapView.getController();
        mapView.setClickable(true);
        mapView.setBuiltInZoomControls(true);//设置开启内置的放大缩小功能
        controller.setCenter(beijingPoint);//设置地图中心位置
        controller.setZoom(8);//设置缩放级别
        List<Overlay> overlays = mapView.getOverlays();//获取当前 MapView 的所有
                                                    //Overlay 集合
        //第一个 Overlay
        Drawable chinaDrawable = getResources().getDrawable(R.drawable.china);
        MyOverlay beijingOverlay = new MyOverlay(chinaDrawable, this);
                                                    //创建 Overlay 对象
        OverlayItem item = new OverlayItem(beijingPoint, "北京", "这是中国的首都");
        beijingOverlay.addOverlayItems(item);
        overlays.add(beijingOverlay);
        //第二个 Overlay
        Drawable tianjinDrawable = getResources().getDrawable(R.drawable.china);
        MyOverlay tianjianOverlay = new MyOverlay(tianjinDrawable, this);
                                                    //创建 Overlay 对象
        OverlayItem tianjianItem = new OverlayItem(tianjinPoint, "天津", "南开大学");
        tianjianOverlay.addOverlayItems(tianjianItem);
        overlays.add(tianjianOverlay);
    }
    @Override
    public boolean onCreateOptionsMenu(Menu menu) {
        getMenuInflater().inflate(R.menu.activity_main, menu);
        return true;
    }
    @Override
    protected boolean isRouteDisplayed()
    {
        return false;
```

```
    }
}
```

（6）右击工程 chapter12_5，在弹出的快捷菜单中选择 Run As|Android Application 命令，运行效果如图 12-11 所示，单击界面上的“北京”城市上方的地图标记，运行效果如图 12-12 所示。

图 12-11　地图标记显示

图 12-12　地图标记单击效果

第13章 网络编程

由于 Android 客户端应用程序经常需要与服务端后台程序进行通信，所以在 Android 应用程序开发过程当中，网络编程实际上起了很大的作用。在本章当中，笔者将会介绍 Android 客户端应用程序如何通过 Http 协议从网络上去下载文本以及二进制数据，如何去解析 XML 文档，以及 JSON 格式的数据。除此之外，还将为读者朋友们介绍如何使用 TCP Sockets 去链接服务端。

13.1 网络获取数据

Android 客户端应用程序经常需要从服务端下载一些文本或者是二进制格式的数据。本节将介绍如何使用 Http 协议从网络上获取数据。

13.1.1 从网络上下载图片

下面就通过实例介绍如何使用 Http 协议去网络上下载一张图片，并显示在 ImageView 控件当中。

(1) 新建工程 chapter13_1，编辑 res/layout/activity_main.xml 文件，源代码参考配套资料。

(2) 由于程序需要从网络上下载数据，所以需要添加联网权限。编辑 AndroidManifest.xml 文件，源代码如下：

```
<manifest xmlns:android = "http://schemas.android.com/apk/res/android"
    package = "com.example.chapter13_1"
    android:versionCode = "1"
    android:versionName = "1.0" >
    <!-- 联网权限 -->
    <uses-permission android:name = "android.permission.INTERNET"/>
```

```
    <uses-sdk  android:minSdkVersion="8" android:targetSdkVersion="15" />
    <application
        android:icon="@drawable/ic_launcher"
        android:label="@string/app_name"
        android:theme="@style/AppTheme" >
        <activity
            android:name=".MainActivity"
            android:label="@string/title_activity_main" >
            <intent-filter>
                <action android:name="android.intent.action.MAIN" />
                <category android:name="android.intent.category.LAUNCHER" />
            </intent-filter>
        </activity>
    </application>
</manifest>
```

(3) 编辑 MainActivity.java 文件,源代码如下:

```
public class MainActivity extends Activity
{
    private static final String TAG = "MainActivity";
    ImageView imageView;//声明图片视图控件
    @Override
    public void onCreate(Bundle savedInstanceState)
    {
        super.onCreate(savedInstanceState);
        setContentView(R.layout.activity_main);
        imageView = (ImageView)findViewById(R.id.my_img);//找到 id 为 my_img 的图片
                                                        //视图控件
        //开启异步任务,下载图片,并显示
        new DownloadImageTask().execute("http://www.baidu.com/img/baidu_sylogo1.
gif");
    }
    /**
     * 根据 URL 地址,返回相应的输入流对象
     * @param urlString
     * @return
     * @throws IOException
     */
    private InputStream openHttpConnection(String urlString) throws IOException
    {
    InputStream in = null;
    int response = -1 ;
```

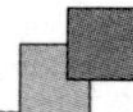

```
URL url = new URL(urlString);//根据 urlString 构造 URL 对象
URLConnection conn = url.openConnection();//打开 http 链接
if( ! (conn instanceof HttpURLConnection))
{
    throw new IOException("Not an HTTP Connection");
}
try
{
    HttpURLConnection httpConn = (HttpURLConnection)conn;
    httpConn.setInstanceFollowRedirects(true);//可以进行跳转
    httpConn.setRequestMethod("GET");//get 请求方式
    httpConn.connect();
    response = httpConn.getResponseCode();//获得服务端的返回码
    //如果返回码是 200,说明 http 请求正常返回
    if (response == HttpURLConnection.HTTP_OK) {
        in = httpConn.getInputStream();
    }
}
catch(Exception e)
{
    Log.e(TAG, "occurs errors:" + e.getLocalizedMessage());
    throw new IOException("Error connecting");
}
return in;
}
/**
 * 根据实参传递的图片 URL 访问地址,包装成相应的 Bitmap 对象返回
 * @param URL
 * @return
 */
private Bitmap downloadImage(String URL)
{
    Bitmap bitmap = null;
    InputStream in = null;
    try
    {
        in = openHttpConnection(URL);
        bitmap = BitmapFactory.decodeStream(in);
        in.close();
    }
    catch (IOException e)
    {
```

```
                Log.d(TAG, e.getLocalizedMessage());
            }
            return bitmap;
        }
        //声明异步任务对象,下载图片,防止阻塞 UI 线程
        private class DownloadImageTask extends AsyncTask<String, Void, Bitmap>
        {
            //doInBackground 方法会在子线程当中执行
            protected Bitmap doInBackground(String... urls)
            {
                return downloadImage(urls[0]);
            }
            //doInBackground 方法执行完成过后,onPostExecute 方法在 UI 主线程当中执行
            //任何更新 UI 界面的操作,只能在 UI 主线程当中执行
            protected void onPostExecute(Bitmap result)
            {
                imageView.setImageBitmap(result);
            }
        }
        @Override
        public boolean onCreateOptionsMenu(Menu menu)
        {
            getMenuInflater().inflate(R.menu.activity_main, menu);
            return true;
        }
    }
```

上述程序当中,通过 HttpURLConnection 对象来请求 WEB 资源。HttpURLConnection 对象不能直接构造,需要通过 URL.openConnection()方法来获得 HttpURLConnection 的实例。

由于本实例需要去网络上下载图片,属于比较耗时的操作,为了防止 UI 主线程的阻塞,笔者将此操作放在 AsyncTask 异步任务当中去执行。

(4) 右击工程 chapter13_1,在弹出的快捷菜单中选择 Run As | Android Application 命令,运行效果如图 13-1 所示。

图 13-1　从网络上下载图片

13.1.2　从网络上下载文本数据

下面就通过实例介绍如何使用 Http 协议去下载文本数据，并用 TextView 控件显示出来。

(1) 新建工程 chapter13_2，编辑 res/layout/activity_main.xml 文件，源代码参考配套资料。

(2) 由于程序需要从网络上下载数据，所以需要添加联网权限。编辑 AndroidManifest.xml 文件，源代码如下：

```
<manifest xmlns:android = "http://schemas.android.com/apk/res/android"
    package = "com.example.chapter13_2"
    android:versionCode = "1"
    android:versionName = "1.0" >
    <uses-sdk    android:minSdkVersion = "8"    android:targetSdkVersion = "15" />
    <!-- 联网权限 -->
    <uses-permission android:name = "android.permission.INTERNET"/>
    <application
        android:icon = "@drawable/ic_launcher"
        android:label = "@string/app_name"
        android:theme = "@style/AppTheme" >
        <activity
            android:name = ".MainActivity"
            android:label = "@string/title_activity_main" >
            <intent-filter>
                <action android:name = "android.intent.action.MAIN" />
                <category android:name = "android.intent.category.LAUNCHER" />
            </intent-filter>
        </activity>
    </application>
</manifest>
```

(3) 编辑 MainActivity.java 文件，源代码如下：

```
public class MainActivity extends Activity
{
    private static final String TAG = "MainActivity";
    TextView myTextView;//声明显示网络下载的文本数据的文本控件
    @Override
    public void onCreate(Bundle savedInstanceState)
    {
        super.onCreate(savedInstanceState);
        setContentView(R.layout.activity_main);
        myTextView = (TextView)findViewById(R.id.my_textview);
                                            //找到 id 为 my_textview 的文本控件
        //执行异步任务
```

```
        new DownloadTextAsyncTask().execute("http://www.baidu.com");
                                                    //下载 baidu 首页的网页内容
    }
private InputStream openHttpConnection(String urlString) throws IOException
{
    InputStream in = null;
    int response = -1 ;
    URL url = new URL(urlString);//根据 urlString 构造 URL 对象
    URLConnection conn = url.openConnection();//打开 http 链接
    if( ! (conn instanceof HttpURLConnection))
{
        throw new IOException("Not an HTTP Connection");
    }
    try{
        HttpURLConnection httpConn = (HttpURLConnection)conn;
        httpConn.setInstanceFollowRedirects(true);//可以进行跳转
        httpConn.setRequestMethod("GET");//get 请求方式
        httpConn.connect();
        response = httpConn.getResponseCode();//获得服务端的返回码
        //如果返回码是 200,说明 http 请求正常返回
        if (response == HttpURLConnection.HTTP_OK)
        {
            in = httpConn.getInputStream();
        }
    }
    catch(Exception e)
    {
        Log.e(TAG, "occurs errors:" + e.getLocalizedMessage());
        throw new IOException("Error connecting");
    }
    return in;
    }
    private String downloadText(String url){   //下载文本数据
    InputStream in = null;
    try
    {
        in = openHttpConnection(url);
    }
    catch(Exception e)
```

```
        {
            Log.e(TAG, "Open Http Connection Occurs Erros:" + e.getLocalizedMessage());
            return "";  //如果 openHttpConnection 抛出异常,则返回空字符串
        }
        //读取输入流
        BufferedReader bufferedReader = new BufferedReader(new InputStreamReader(in));
        String line = null;
        StringBuffer result = new StringBuffer();//返回结果
        try {
                while((line = bufferedReader.readLine()) != null)
                {
                        result.append(line);
                }
            }
            catch (IOException e)
            {
                Log.d(TAG, "Read InputStream occurs errors:" + e.getLocalizedMessage());
                return "";
            }
        return result.toString();
        }
        //使用异步任务处理耗时操作
        private class DownloadTextAsyncTask extends AsyncTask<String, Void, String>{
        @Override
            protected String doInBackground(String... params)
        {
            return downloadText(params[0]);
        }
        @Override
        protected void onPostExecute(String result)
        {
            myTextView.setText(result);//设置文本
        }}
    }
```

图 13-2 从网络上下载文本数据

(4) 右击工程 chapter13_2,在弹出的快捷菜单中选择 Run As|Android Application 命令,运行效果如图 13-2 所示。

13.2 XML 解析

XML 是 W3C 制定的一组规范，用来定义标记语言，其主要优点是它的可扩展性，从它的诞生到现在，已经得到了广泛的支持与应用。因为它架起了复杂的标准通用语言(SGML)与功能有限的超文本标记语言(HTML)之间的桥梁。Java 平台允许通过不同的方式使用 XML 可扩展标记语言(Extensible Markup Language)。大多数与 XML 相关的 Java API 在 Android 上得到了完全支持。XML 经常用作 Internet 上的一种通用的数据交换格式，它的平台无关性、语言无关性和系统无关性，给数据集成和交互带来了极大的方便。本节将会介绍 Android 上解析 XML 的 3 种技术 DOM、SAX 和 XML Pull。

13.2.1 DOM 解析技术

Android 完全支持 DOM 解析。DOM(Document Object Model)是文档对象模型。利用 DOM 中的对象可以对 XML 文档进行读取、搜索、修改、添加和删除操作。

1. DOM 的工作原理

利用 DOM 对 XML 文件进行操作时，首先要解析文件，将文件分为独立的元素、属性和注释等，然后以节点树的形式在内存中对 XML 文件进行表示，就可以通过节点树访问文档的内容，并根据需要修改文档。DOM 实现时首先为 XML 文档的解析定义一组接口，解析器读入整个文档，然后构造一个驻留内存的树结构，这样代码就可以使用 DOM 接口来操作整个树结构。

2. 常用的 DOM 接口和类

- Document：该接口定义分析并创建 DOM 文档的一系列方法，是文档树的根，是操作 DOM 的基础。
- Element：该接口继承 Node 接口，提供了获取、修改 XML 元素名字和属性的方法。
- Node：该接口提供处理并获取节点和子节点值的方法。
- NodeList：提供获得节点个数和当前节点的方法。这样就迭代地访问各个节点。
- DOMParser：该类是 Apache 的 Xerces 中的 DOM 解析类，可直接解析 XML 文件。

使用 DOM 操作 XML 的代码看起来比较直观、简单。但是，因为 DOM 需要将 XML 文件的所有内容读取到内存中，所以内存的消耗比较大。当然，如果 XML 文件的内容比较少，采用 DOM 是可行的。

下面就通过实例介绍如何采用 DOM 解析器去解析 XML 文件。

(1) 新建工程 chapter13_3，在 src 目录下新建一个 XML 文件：person.xml，编辑 person.xml 文件，源代码如下：

```
<?xml version="1.0" encoding="UTF-8"?>
<persons>
    <person>
        <name>刘翔</name>
        <age>28</age>
    </person>
    <person>
        <name>姚明</name>
        <age>30</age>
    </person>
</persons>
```

(2) 编辑 res/layout/activity_main.xml 文件，源代码如下：

```
<?xml version="1.0" encoding="utf-8"?>
<LinearLayout xmlns:android="http://schemas.android.com/apk/res/android"
    android:orientation="vertical"
    android:layout_width="fill_parent"
    android:layout_height="fill_parent">
<!-- 显示解析 person.xml 的结果 -->
<TextView
    android:id="@+id/textView"
    android:layout_width="fill_parent"
    android:layout_height="wrap_content"
    android:textSize="20sp"/>
</LinearLayout>
```

(3) 新建数据模型类：Person，编辑 Person.java 文件，源代码如下：

```
public class Person
{
    private String name;//姓名
    private int age;//年龄
    public String getName()
    {
        return name;
    }
    public void setName(String name)
    {
        this.name = name;
    }
```

```
        public int getAge()
        {
            return age;
        }
        public void setAge(int age)
        {
            this.age = age;
        }
        @Override
        public String toString()
        {
            return "[姓名 = " + name + ",年龄 = " + age + "]";
        }
    }
```

(4) 编辑 MainActivity.java 文件,源代码如下:

```
    public class MainActivity extends Activity
    {
        private String result = "";
        @Override
        public void onCreate(Bundle savedInstanceState)
        {
            super.onCreate(savedInstanceState);
            setContentView(R.layout.activity_main);
            InputStream inputStream = this.getClassLoader().getResourceAsStream("per-
son.xml");
            try
            {
              DocumentBuilderFactory factory = DocumentBuilderFactory.newInstance();
              DocumentBuilder builder = factory.newDocumentBuilder();
              Document document = builder.parse(inputStream);
              Element root = document.getDocumentElement();   //获取根节点
              List<Person>  personList =  parse(root);
              for (Person person : personList)
              {//迭代 personList 列表
                  result += person.toString() + "\n";
              }
              TextView textView = (TextView) findViewById(R.id.textView);
              textView.setText(result);//设置文本
          }
          catch (Exception e)
          {
```

```
            e.printStackTrace();
        }
    }
    /**
     *
     * @param element 将要进行遍历的节点
     */
    private List<Person> parse(Element root)
    {
        List<Person> persons = new ArrayList<Person>();
        /*得到 person 的节点列表*/
        NodeList nodes = root.getElementsByTagName("person");
        /* getNodeName:获取 tagName,例如<book>thinking in android</book>这个
Element 的 getNodeName 返回 book
            * getNodeType:返回当前节点的确切类型,如 Element、Attr、Text 等
            * getNodeValue:返回节点内容,如果当前为 Text 节点,则返回文本内容;否则会
返回 null
            * getTextContent:返回当前节点以及其子代节点的文本字符串,这些字符串会拼
成一个字符串给用户返回。例如
            * 对<book><name>thinking in android</name><price>12.23</price>
</book>调用此方法,则会返回"thinking in android12.23"
            */
        for(int i = 0; i < nodes.getLength(); i++)
        {//迭代 person 节点列表
            Element personElement = (Element)nodes.item(i);//element /text == node
            Person person = new Person();
            /*得到子节点*/
            NodeList childnodes = personElement.getChildNodes();
            for(int j = 0; j<childnodes.getLength(); j++ )
            {//迭代每一个 person 节点子节点列表
                Node childNode = (Node)childnodes.item(j);
                if(childNode.getNodeType() == Node.ELEMENT_NODE)
                {//如果是节点类型
                    Element childElement = (Element)childNode;
                    if("name".equals(childElement.getNodeName()))
                    {//如果节点名称是 name
                    person.setName(childElement.getFirstChild().getNodeValue());
                }
                else if("age".equals(childElement.getNodeName()))
                    {//如果节点名称是 age
                        person.setAge(new Short(childElement.getFirstChild().get-
NodeValue()));
```

```
                    }
                }
            }
            persons.add(person);
        }
        return persons;
    }
}
```

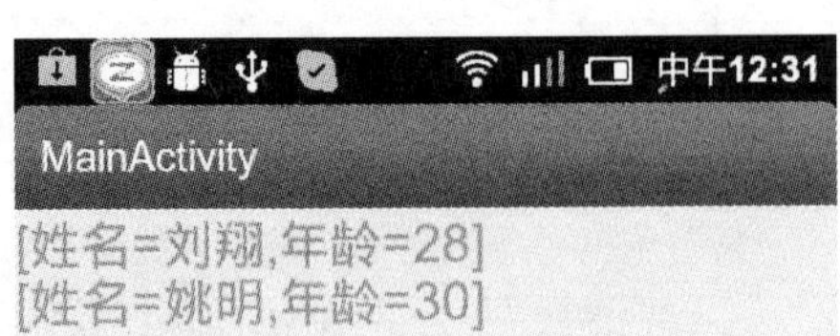

图 13-3　Dom 解析运行效果

(5) 右击工程 chapter13_3，在弹出的快捷菜单中选择 Run As|Android Application 命令，运行效果如图 13-3 所示。

13.2.2　SAX 解析技术

SAX(Simple API for XML)，是一个公共的基于事件的 XML 文档解析标准。它以事件作为解析 XML 文件的模式，将 XML 文件转换为一系列的事件，由不同的事件处理器来决定如何处理。

1. SAX 的工作原理

SAX 的工作原理简单地说就是对文档进行顺序扫描，当扫描到文档(document)开始与结束、元素(element)开始与结束等地方时通知事件处理函数，由事件处理函数做出相应的处理，然后继续同样的扫描，直至文档结束。

2. 常用的 SAX 接口和类

- ❑ Attributes：用于得到属性的个数、名字和值。
- ❑ ContentHandler：定义与文档本身关联的事件(例如，开始和结束标记)。大多数应用程序都注册这些事件。
- ❑ DTDHandler：定义与 DTD 关联的事件。它没有定义足够的事件来完整地报告 DTD。如果需要对 DTD 进行语法分析，请使用可选的 DeclHandler。DeclHandler 是 SAX 的扩展。不是所有的语法分析器都支持它。
- ❑ EntityResolver：定义和装入实体关联的事件。只有少数几个应用程序注册这些事件。
- ❑ ErrorHandler：定义错误事件。许多应用程序注册这些事件以便用它们自己的方式报错。
- ❑ DefaultHandler：它提供了这些接口的默认实现。在大多数情况下，为应用程序扩展 DefaultHandler 并覆盖相关的方法要比直接实现一个接口更容易。

下面就通过实例介绍如何使用 SAX 解析器来解析 XML 文件。

(1) 新建工程 chapter13_4，编辑 res/layout/activity_main.xml 文件，源代码参考配套资料。

（2）在 src 目录下新建 XML 文件：products. xml，编辑 products. xml 文件，源代码如下：

```
<? xml version = "1.0" encoding = "UTF-8"? >
<products>
    <product id = "12">
        <name>海尔冰箱</name>
        <price>1000</price>
    </product>
    <product id = "13">
        <name>格力空调</name>
        <price>2000</price>
    </product>
    <product id = "14">
        <name>TCL 彩电</name>
        <price>4000</price>
    </product>
    <product id = "15">
        <name>三星手机</name>
        <price>5000</price>
    </product>
</products>
```

（3）新建数据模型类：Product，编辑 Product. java 文件，源代码如下：

```
public class Product {
    private String id;                          //产品 ID
    private String productName;//产品名称
    private double productPrice;//产品价格
    public String getId()
    {
        return id;
    }
    public void setId(String id)
    {
        this.id = id;
    }
    public String getProductName()
    {
        return productName;
    }
    public void setProductName(String productName)
    {
```

```
        this.productName = productName;
    }
    public double getProductPrice()
    {
        return productPrice;
    }
    public void setProductPrice(double productPrice)
    {
        this.productPrice = productPrice;
    }
    @Override
    public String toString() {
        return "[id = " + id + ",name = " + productName + ",price" + productPrice
+ "]";
    }
}
```

(4) 新建 Java 类:SAXParseService,编辑 SAXParseService.java 文件,源代码如下:

```
public class SAXParseService
{
    public List<Product> getProductList(InputStream is)throws Exception
    {
        SAXParserFactory factory = SAXParserFactory.newInstance();
        SAXParser parser = factory.newSAXParser();
        ProductHandler handler = new ProductHandler();
        parser.parse(is, handler);
        return handler.getList();
    }
    private final class ProductHandler extends DefaultHandler
    {
        private List<Product> list = null;     //保存 XML 解析的结果
        private Product product;
        private String tag;
        public List<Product> getList()
        {
                return list;
        }
        //XML 文件开始解析时调用此方法
        @Override
        public void startDocument() throws SAXException
        {
                list = new ArrayList<Product>();//开始解析时,实例化 list
```

```
        }
        //解析到 Element 的开头时调用此方法
        @Override
        public void startElement(String uri, String localName, String qName,Attrib-
utes attributes) throws SAXException
        {
                if("product".equals(localName))
                {//如果开始节点名称是"product"
                        product = new Product();
                        product.setId(attributes.getValue(0));//提取 Product 节
                                                        //点 ID 属性值
                }
                tag = localName;
        }
        //取得 Element 的开头结尾中间夹的字符串
        @Override
        public void characters(char[] ch, int start, int length)throws SAXException
        {
                if(tag != null)
                {
                    String textdata = new String(ch,start,length);
                    if("name".equals(tag))
                    {       //提取产品名称信息
                            product.setProductName(textdata);
                    }
                    else if("price".equals(tag))
                    {//提取产品价格信息
                            product.setProductPrice(Double.parseDouble(textdata));
                    }
                }
        }
        @Override
        public void endElement(String uri, String localName, String qName)throws SAXEx-
ception
        {
            tag = null;
            if("product".equals(localName))
            {//如果结束节点名称是 product
                    list.add(product);
                    product = null;
            }
        }
```

```
        }
    }
```

(5) 编辑 MainActivity.java 文件,源代码如下:

```
public class MainActivity extends Activity
{
    ListView productListView;//声明显示产品列表的ListView
    @Override
    public void onCreate(Bundle savedInstanceState)
    {
        super.onCreate(savedInstanceState);
        setContentView(R.layout.activity_main);
        productListView = (ListView)findViewById(R.id.my_list);//找到id为my_list
//的列表视图
        InputStream inputStream = this.getClassLoader().getResourceAsStream("prod-
ucts.xml");
        try
        {
            List<Product> productList = new SAXParseService().getProductList(in-
putStream);
            productListView.setAdapter(new ArrayAdapter(this, android.R.layout.sim-
ple_list_item_1, productList));
        }
        catch (Exception e)
        {
            e.printStackTrace();
        }
    }
}
```

(6) 右击工程 chapter13_4,在弹出的快捷菜单中选择 Run As|Android Application 命令,运行效果如图 13-4 所示。

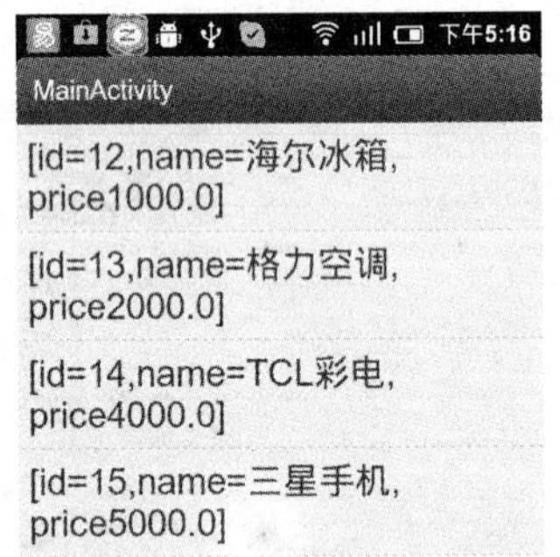

图 13-4 SAX 解析运行效果

13.2.3 Pull 解析技术

Pull 解析器是一个开源的 Java 项目,既可以用于 Android,也可以用于 JavaEE。如果用在 JavaEE 需要把其 jar 文件放入类路径中,因为 Android 已经集成进了 Pull 解析器,所以无需添加任何 jar 文件。Android 系统本身使用到的各种 xml 文件,其内部也是采用 Pull 解析器进行解析的。Pull 解析器的运行方式与 SAX 解析器相似。它提供了类似的事件,如:开始元素和结束元素事件,使用 parser.next()可以进入下一个元素并触发相应事件。跟 SAX 不同的是,Pull 解析器产生的事件是一个

数字，而非方法，因此可以使用一个 switch 对感兴趣的事件进行处理。当元素开始解析时，调用 parser. nextText()方法可以获取下一个 Text 类型节点的值。

下面就通过实例介绍如何使用 Pull 解析技术来解析 XML 文件。

(1) 新建工程 chapter13_5，在 src 目录下建立 XML 文件：content. xml，编辑 content. xml 文件，源代码参考配套资料。

(2) 新建一个数据模型类：Content，编辑 Content. java 文件，源代码如下：

```
public class Content {
    private String contentId;//栏目 ID
    private String name;          //栏目名称
    public String getContentId() {
        return contentId;
    }
    public void setContentId(String contentId) {
        this.contentId = contentId;
    }
    public String getName() {
        return name;
    }
    public void setName(String name) {
        this.name = name;
    }
    @Override
    public String toString() {
        return "[contentID = " + contentId + ",name = " + name + "]";
    }
}
```

(3) 新建一个 Java 类：ContentPullService，编辑 ContentPullService. java 文件，源代码如下：

```
public class ContentPullService
{
    /**
     * pull 解析 XML 文件
     * @param inputStream
     * @return
     * @throws Throwable
     */
     public static List<Content> getContentList(InputStream inputStream)
throws Throwable
     {
       List<Content> contentList = null;
```

```
Content content = null;
XmlPullParser parser = Xml.newPullParser();
parser.setInput(inputStream, "utf-8");
int event = parser.getEventType();
//采用循环方式,直到读取到 XML 文件的末尾
while(event != XmlPullParser.END_DOCUMENT)
{
    switch (event)
    {
        case XmlPullParser.START_DOCUMENT:  //开始文档
          contentList = new ArrayList<Content>();
          break;
        case XmlPullParser.START_TAG:       //开始节点
          if("Content".equals(parser.getName()))
          {
              content = new Content();
          }
          if(content != null)
          {
              //parser.getName():得到当前节点名称
              if("ContentID".equals(parser.getName()))
                                          //如果当前节点名称是 ContentID
              {
                  String contentId  = parser.nextText().toString();
                  content.setContentId(contentId);
              }
              else if("Name".equals(parser.getName()))
                                          //如果当前节点名称是 Name
              {
                  String name  =  parser.nextText().toString();
                  content.setName(name);
                  }
              }
              break;
          case XmlPullParser.END_TAG:   //结束节点
              if("Content".equals(parser.getName()))
                                          //如果当前节点名称是 Content
              {
                contentList.add(content);
                content = null;
              }
              break;
          default:
```

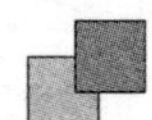

```
                    break;
                }
                event = parser.next();
            }
            return contentList;
        }
    }
```

(4) 编辑 MainActivity.java 文件,源代码如下:

```
public class MainActivity extends Activity
    {
        private static final String TAG = "MainActivity";
        @Override
        public void onCreate(Bundle savedInstanceState)
         {
            super.onCreate(savedInstanceState);
            setContentView(R.layout.activity_main);
            InputStream inputStream = this.getClassLoader().getResourceAsStream("
content.xml");
            try
            {
                List<Content> contentList = ContentPullService.getContentList
(inputStream);
                for(Content content : contentList)
                {//迭代 contentList 列表对象
                    Log.d(TAG, content.toString());
                }
            }
            catch (Throwable e)
            {
                e.printStackTrace();
            }
        }
        @Override
        public boolean onCreateOptionsMenu(Menu menu)
        {
            getMenuInflater().inflate(R.menu.activity_main, menu);
            return true;
        }
    }
```

(5) 右击工程 chapter13_5,在弹出的快捷菜单中选择 Run As|Andorid Appli-

cation 命令,运行效果如图 13-5 所示。

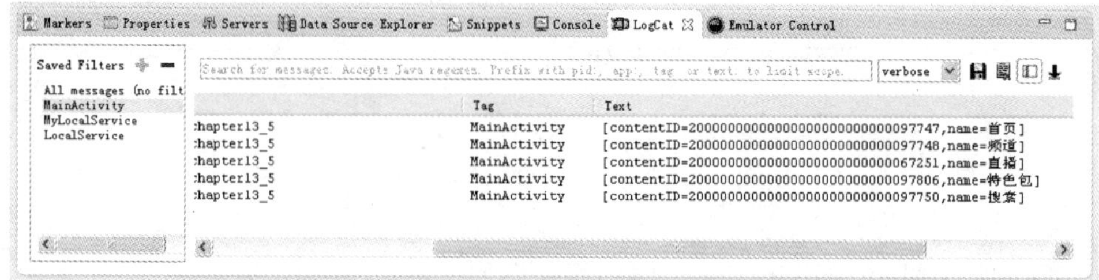

图 13-5　Pull 解析运行效果

13.3　JSON 数据解析

JSON 即 Javascript Object Notation,是一种轻量级的数据交换格式,与 XML 一样,是广泛被采用的客户端与服务端交互的解决方案。

JSON 对象:JSON 中的对象以"{"开始,以"}"结束。对象中的每一个 Item 都是一个 key-value 对,表现为 key:value 的形式,key-value 对之间使用逗号分隔。如:{"name":"freedie","age":24:"male":true,"address":{"city":"beijing", "country":"china"}}。JSON 对象的 key 只能是字符串类型的,而 value 可以是 string,number,false,true,null,Object 对象甚至是数组类型,也就是说可以存在嵌套的情况。

JSON 数组:JSON 数组以"["开始,以"]"结束,数组中的每一个元素可以是 string,number,false,true,null,Object 对象甚至是数组,数组间的元素使用逗号分隔。比如:

```
[
  {
    "precision": "zip",
    "Latitude":  37.7668,
    "Longitude": -122.3959,
    "Address":   "",
    "City":      "SAN FRANCISCO",
    "State":     "CA",
    "Zip":       "94107",
    "Country":   "US"
  },

  {
    "precision": "zip",
    "Latitude":  37.371991,
    "Longitude": -122.026020,
    "Address":   "",
    "City":      "SUNNYVALE",
    "State":     "CA",
    "Zip":       "94085",
```

```
    "Country":  "US"
  }
]
```

下面就通过实例介绍在 Android 应用程序开发过程中如何对 JSON 格式的数据进行解析。

(1) 新建工程 chapter13_6，编辑 res/layout/activity_main.xml 文件，源代码参考配套资料。

(2) 由于需要访问网络，所以需要添加相关权限。编辑 AndroidManifest.xml 文件，源代码如下：

```
<manifest xmlns:android = "http://schemas.android.com/apk/res/android"
    package = "com.example.chapter13_6"
    android:versionCode = "1"
    android:versionName = "1.0" >
    <!-- 联网权限 -->
    <uses-permission  android:name = "android.permission.INTERNET"/>
    <uses-sdk  android:minSdkVersion = "8"   android:targetSdkVersion = "15" />
    <application
        android:icon = "@drawable/ic_launcher"
        android:label = "@string/app_name"
        android:theme = "@style/AppTheme" >
        <activity
            android:name = ".MainActivity"
            android:label = "@string/title_activity_main" >
            <intent-filter>
                <action android:name = "android.intent.action.MAIN" />
                <category android:name = "android.intent.category.LAUNCHER" />
            </intent-filter>
        </activity>
    </application>
</manifest>
```

(3) 新建数据模型类：CityBean，编辑 CityBean.java 文件，源代码如下：

```
public class CityBean implements Serializable
{
    public CityBean(){}
    public CityBean(String cityName, Long lat, Long lon)
    {
        super();
        this.cityName = cityName;
        this.lat = lat;
        this.lon = lon;
    }
```

```
    private String cityName;  //城市名称
    private Long lat; //经度
    private Long lon; //维度
    public String getCityName()
    {
        return cityName;
    }
    public void setCityName(String cityName) {
        this.cityName = cityName;
    }
    public Long getLat()
    {
        return lat;
    }
    public void setLat(Long lat)
    {
        this.lat = lat;
    }
    public Long getLon()
    {
        return lon;
    }
    public void setLon(Long lon)
    {
        this.lon = lon;
    }
    @Override
    public String toString()
    {
        return "名称=" + cityName + ";维度=" + lat + ";经度=" + lon;
    }
}
```

(4) 编辑 MainActivity.java 文件,源代码如下:

```
public class MainActivity extends Activity
{
    private static final String TAG = "MainActivity";
    private String url = "http://www.google.com/ig/cities? country=CN";
                                                        //获取中国城市信息的 url 地址
    ListView citiesListView;//声明显示中国城市信息的列表控件
    @Override
    public void onCreate(Bundle savedInstanceState)
```

```
{
  super.onCreate(savedInstanceState);
  setContentView(R.layout.activity_main);
  citiesListView = (ListView)findViewById(R.id.cities_listview);
                                  //找到 id 为 cities_listview 的 ListView 控件
  setListAdapter(); //填充数据
}
private void setListAdapter()
 {
  try {
        InputStream is = openHttpConnection(url);//获取指定 url 地址的输入流
                                                //对象
        List<CityBean> cityList = new ArrayList<CityBean>();
        StringBuilder sBuilder = new StringBuilder();//存放 JSON 字符串
        BufferedReader bReader = new BufferedReader(new InputStreamReader(is));
        for (String s = bReader.readLine(); s != null; s = bReader.readLine())
        {
            sBuilder.append(s);
        }
        //对 JSON 字符串进行解析
        JSONObject jsonObject = new JSONObject(sBuilder.toString());
        JSONArray jsonArray = jsonObject.getJSONArray("cities");
        for (int i = 0; i < jsonArray.length(); i++)
          {//迭代 jsonArray 对象
            CityBean cityBean = new CityBean();
            JSONObject jsonObj = (JSONObject) jsonArray.opt(i);
            cityBean.setCityName(jsonObj.getString("name"));    //城市名称
            cityBean.setLat(jsonObj.getLong("lat"));             //维度信息
            cityBean.setLon(jsonObj.getLong("lon"));             //经度信息
            cityList.add(cityBean);
        }
        citiesListView.setAdapter(new ArrayAdapter(this, android.R.layout.sim-
ple_list_item_1, cityList));
    }
    catch (Exception e)
    {
        e.printStackTrace();
    }
}
/**
 * 根据 URL 地址,返回相应的输入流对象
 * @param urlString
 * @return
```

```
     * @throws IOException
     */
    private InputStream openHttpConnection(String urlString) throws IOException
    {
    InputStream in = null;
    int response = -1 ;
    URL url = new URL(urlString);//根据 urlString 构造 URL 对象
    URLConnection conn = url.openConnection();//打开 http 链接
    if( ! (conn instanceof HttpURLConnection))
    {
        throw new IOException("Not an HTTP Connection");
    }
    try{
        HttpURLConnection httpConn = (HttpURLConnection)conn;
        httpConn.setInstanceFollowRedirects(true);//可以进行跳转
        httpConn.setRequestMethod("GET");//get 请求方式
        httpConn.connect();
        response = httpConn.getResponseCode();//获得服务端的返回码
        //如果返回码是 200,说明 http 请求正常返回
        if (response == HttpURLConnection.HTTP_OK)
        {
            in = httpConn.getInputStream();
        }
    }
    catch(Exception e)
    {
      Log.e(TAG, "occurs errors:" + e.getLocalizedMessage());
      throw new IOException("Error connecting");
    }
    return in;
    }
    @Override
    public boolean onCreateOptionsMenu(Menu menu)
    {
        getMenuInflater().inflate(R.menu.activity_main, menu);
        return true;
    }
}
```

其中 JSONObject 代表的就是一个 JSON 对象，JSONArray 代表的就是一个 JSON 数组对象。

(5) 右击工程 chapter13_6，在弹出的快捷菜单中选择 Run As|Android Application 命令，运行效果如图 13-6 所示。

晚上10:36
MainActivity
名称=Baoding;维度=38849998;经度=115569999
名称=Beijing;维度=39930000;经度=116279998
名称=Changchun;维度=43900001;经度=125220001
名称=Changsha;维度=28229999;经度=112870002
名称=Chengdu;维度=30670000;经度=104019996
名称=Chongqing;维度=29520000;经度=106480003
名称=Dalian;维度=38900001;经度=121629997

图 13-6　JSON 数据解析效果

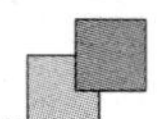

13.4　HttpClient

Apache HttpClient 是 Apache 的一个开源项目，弥补了 Java. net. *（如 HttpURLConnection）包的灵活性不足的特点，为客户端的基于 Http 协议的编程提供了高效、功能丰富的工具包的支持。Android SDK 引入了 Apache HttpClient 的同时还提供了对它的一些封装和扩展，例如设置默认的 Http 超时时间以及缓存大小等。

HttpClient 的一般使用步骤如下：

- 使用 DefaultHttpClient 类实例化 HttpClient 对象。
- 创建 HttpGet 或者是 HttpPost 对象，将要请求的 URL 通过构造方法传入 HttpGet 或 HttpPost 对象。
- 调用 execute 方法发送 Http Get 或 Http Post 请求，并返回 HttpResponse 对象。
- 调用 HttpResponse 接口的 getEntity 方法返回响应信息，并进行相应的处理。

13.4.1　HttpClient 发送 HttpGet 请求

下面就通过实例介绍如何使用 HttpClient 发送 Http Get 请求。

(1) 新建工程 chapter13_7，由于需要访问网络，所以需要添加相关权限。编辑 AndroidManifest. xml 文件，源代码如下：

```
<manifest xmlns:android = "http://schemas.android.com/apk/res/android"
    package = "com.example.chapter13_7"
    android:versionCode = "1"
    android:versionName = "1.0" >
    <!-- 访问网络权限 -->
    <uses-permission android:name = "android.permission.INTERNET"/>
    <uses-sdk android:minSdkVersion = "8"     android:targetSdkVersion = "15" />
    <application
        android:icon = "@drawable/ic_launcher"
        android:label = "@string/app_name"
        android:theme = "@style/AppTheme" >
        <activity
            android:name = ".MainActivity"
            android:label = "@string/title_activity_main" >
            <intent-filter>
                <action android:name = "android.intent.action.MAIN" />
                <category android:name = "android.intent.category.LAUNCHER" />
            </intent-filter>
        </activity>
```

```
    </application>
</manifest>
```

(2) 编辑 MainActivity.java 文件,源代码如下:

```
public class MainActivity extends Activity
{
    private static final String TAG = "MainActivity";
    private static final String URL = "http://developer.android.com";
    @Override
    public void onCreate(Bundle savedInstanceState)
    {
        super.onCreate(savedInstanceState);
        setContentView(R.layout.activity_main);
        try
        {
            //得到 HttpClient 对象
            HttpClient getClient = new DefaultHttpClient();
            //得到 HttpGet 对象
            HttpGet request = new HttpGet(URL);
            //客户端使用 GET 方式执行 Http 请求,获得服务器端的响应
            HttpResponse response = getClient.execute(request);
            //判断请求是否成功
            if(response.getStatusLine().getStatusCode() == HttpStatus.SC_OK)
            {
              Log.i(TAG, "请求服务器端成功");
              //获得输入流
              InputStream  inStream = response.getEntity().getContent();
              //读取输入流
               BufferedReader bufferedReader = new BufferedReader(new InputStream-
Reader(inStream));
              String line = null;
              StringBuffer result = new StringBuffer();//返回结果
              try
              {
                  while((line = bufferedReader.readLine()) != null)
                  {
                      result.append(line);
                  }
                  Toast.makeText(this, result, Toast.LENGTH_LONG).show();
              }
              catch (IOException e)
              {
```

```
                    Log.d(TAG, "Read InputStream occurs errors:" + e.getLocalizedMes-
sage());
                }
                //关闭输入流
                inStream.close();
            }
            else
            {
                Log.i(TAG, "请求服务器端失败");
            }
        }
        catch (Exception e)
        {
            e.printStackTrace();
        }
    }
    @Override
    public boolean onCreateOptionsMenu(Menu menu)
    {
        getMenuInflater().inflate(R.menu.activity_main, menu);
        return true;
    }
}
```

HTTP GET 的核心代码如下：

- DefaultHttpClient client = new DefaultHttpClient();
- HttpGet get = new HttpGet(String url);//此处的 URL 为 http://..../path? arg1=value&....argn=value
- HttpResponse response = client.execute(get); //模拟请求
- int code = response.getStatusLine().getStatusCode();//返回响应码
- InputStream in = response.getEntity().getContent();//服务器返回的数据

(3) 右击工程 chapter13_7，在弹出的快捷菜单中选择 Run As | Android Application 命令，运行效果如图 13-7 所示。

图 13-7　HttpClient Http Get 请求效果

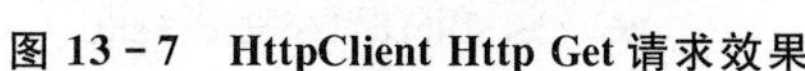

13.4.2 HttpClient 发送 HttpPost 请求

下面通过实例介绍如何使用 HttpClient 发送 Http Post 请求。

(1) 新建工程 chapter13_8，由于需要访问网络，所以需要添加相关网络权限。编辑 AndroidManifest.xml 文件，源代码如下：

```
<manifest xmlns:android = "http://schemas.android.com/apk/res/android"
    package = "com.example.chapter13_8"
    android:versionCode = "1"
    android:versionName = "1.0" >
    <!-- 访问网络权限 -->
    <uses-permission android:name = "android.permission.INTERNET"/>
    <uses-sdk  android:minSdkVersion = "8"   android:targetSdkVersion = "15" />
    <application
        android:icon = "@drawable/ic_launcher"
        android:label = "@string/app_name"
        android:theme = "@style/AppTheme" >
        <activity
            android:name = ".MainActivity"
            android:label = "@string/title_activity_main" >
            <intent-filter>
                <action android:name = "android.intent.action.MAIN" />
                <category android:name = "android.intent.category.LAUNCHER" />
            </intent-filter>
        </activity>
    </application>
</manifest>
```

(2) 编辑 MainActivity.java 文件，源代码如下：

```
public class MainActivity extends Activity
{
        private static final String TAG  =  "MainActivity";
        @Override
        public void onCreate(Bundle savedInstanceState)
        {
          super.onCreate(savedInstanceState);
          setContentView(R.layout.activity_main);
          BufferedReader in  =  null;
          try
          {
            HttpClient client  =  new DefaultHttpClient();
            HttpPost request  =  new HttpPost("http://code.google.com/android/");
```

```
            //使用 NameValuePair 来保存要传递的 Post 参数
            List<NameValuePair> postParameters = new ArrayList<NameValuePair>();
            //添加要传递的参数
            postParameters.add(new BasicNameValuePair("id", "123456"));
            postParameters.add(new BasicNameValuePair("username", "freedie.qin"));
            //实例化 UrlEncodedFormEntity 对象
              UrlEncodedFormEntity formEntity = new UrlEncodedFormEntity(postParam-
eters);
            //使用 HttpPost 对象来设置 UrlEncodedFormEntity 的 Entity
            request.setEntity(formEntity);
            HttpResponse response = client.execute(request);
              in = new BufferedReader(new InputStreamReader(response.getEntity().
getContent()));
            StringBuffer string = new StringBuffer("");
            String lineStr = "";
            while ((lineStr = in.readLine()) != null)
            {
                string.append(lineStr + "\n");
            }
            in.close();
              Toast.makeText(this, string.toS-
tring(), Toast.LENGTH_LONG).show();
        }
        catch(Exception e)
        {
            Log.e(TAG, "Http ConnectLoion occurs
erros:" + e.getLocalizedMessage());
        }
    }
}
```

(3) 右击工程 chapter13_8，在弹出的快捷菜单中选择 Run As|Android Application 命令，运行效果如图 13-8 所示。

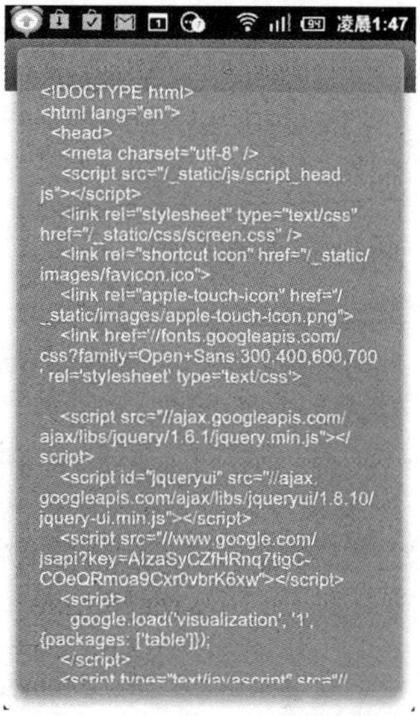

图 13-8　HttpClient Http Post 请求效果

13.5　Android 调用 WebService 查询号码归属地

WebService 是由企业发布的完成其特定商务需求的在线应用服务，其他公司或应用软件能够通过 Internet 来访问并使用这项在线服务。在程序当中，一般要使用 SOAP 协议向 WebService 发送 Http 请求，然后处理 WebService 返回的信息。下面就通过查询号码归属地这个实例来向读者介绍在 Android 应用程序当中如何调用

WebService。

本实例当中，笔者使用 http://www.webxml.com.cn/zh_cn/index.aspx 的最新 Web Services 中的一种查询国内手机号码归属地的 Web Service。

这项查询手机号码归属地的 Web Service 的 SOAP 协议的请求格式如下：

```
POST /WebServices/MobileCodeWS.asmx HTTP/1.1
Host: webservice.webxml.com.cn
Content-Type: application/soap+xml; charset=utf-8
Content-Length: length
<?xml version="1.0" encoding="utf-8"?>
<soap12:Envelope xmlns:xsi="http://www.w3.org/2001/XMLSchema-instance" xmlns:xsd
="http://www.w3.org/2001/XMLSchema" xmlns:soap12="http://www.w3.org/2003/05/soap-en-
velope">
  <soap12:Body>
    <getMobileCodeInfo xmlns="http://WebXml.com.cn/">
      <mobileCode>string</mobileCode>
      <userID>string</userID>
    </getMobileCodeInfo>
  </soap12:Body>
</soap12:Envelope>
```

响应格式如下：

```
HTTP/1.1 200 OK
Content-Type: application/soap+xml; charset=utf-8
Content-Length: length
<?xml version="1.0" encoding="utf-8"?>
<soap12:Envelope xmlns:xsi="http://www.w3.org/2001/XMLSchema-instance" xmlns:xsd
="http://www.w3.org/2001/XMLSchema" xmlns:soap12="http://www.w3.org/2003/05/soap-en-
velope">
  <soap12:Body>
    <getMobileCodeInfoResponse xmlns="http://WebXml.com.cn/">
      <getMobileCodeInfoResult>string</getMobileCodeInfoResult>
    </getMobileCodeInfoResponse>
  </soap12:Body>
</soap12:Envelope>
```

(1) 新建工程 chapter13_9，编辑 res/layout/activity_main.xml 文件，源代码参考配套资料。

(2) 由于本实例需要访问网络，所以需要添加相关网络权限。编辑 AndroidManifest.xml 文件，源代码如下：

```
<manifest xmlns:android="http://schemas.android.com/apk/res/android"
```

```
package = "com.example.chapter13_9"
android:versionCode = "1"
android:versionName = "1.0" >
<!-- 联网权限 -->
<uses-permission  android:name = "android.permission.INTERNET"/>
<uses-sdk   android:minSdkVersion = "8"   android:targetSdkVersion = "15" />
<application
    android:icon = "@drawable/ic_launcher"
    android:label = "@string/app_name"
    android:theme = "@style/AppTheme" >
    <activity
        android:name = ".MainActivity"
        android:label = "@string/title_activity_main" >
        <intent-filter>
            <action android:name = "android.intent.action.MAIN" />
            <category android:name = "android.intent.category.LAUNCHER" />
        </intent-filter>
    </activity>
</application>
</manifest>
```

(3) 在 src 目录下新建一个 XML 文件:send.xml,编辑 send.xml 文件,源代码如下:

```
<? xml version = "1.0" encoding = "utf-8"? >
<soap12:Envelope xmlns:xsi = "http://www.w3.org/2001/XMLSchema-instance" xmlns:xsd
 = "http://www.w3.org/2001/XMLSchema" xmlns:soap12 = "http://www.w3.org/2003/05/soap-en-
velope">
  <soap12:Body>
    <getMobileCodeInfo xmlns = "http://WebXml.com.cn/">
      <mobileCode>MYMnumber</mobileCode>
      <userID></userID>
    </getMobileCodeInfo>
  </soap12:Body>
</soap12:Envelope>
```

(4) 编辑 MainActivity.java 文件,源代码如下:

```
public class MainActivity extends Activity
{
    EditText phoneNumber;//声明电话号码输入框
    TextView showLocation;//声明显示归属地文本框
    Button sendHttpRequest;//声明查询号码归属地按钮
    @Override
    public void onCreate(Bundle savedInstanceState)
```

```
{
    super.onCreate(savedInstanceState);
    setContentView(R.layout.activity_main);
    phoneNumber = (EditText) findViewById(R.id.phone_number);
                                //找到 id 为 phone_number 的 EditText 控件
    showLocation = (TextView) findViewById(R.id.locationTV);
                                //找到 id 为 locationTV 的 TextView 控件
    sendHttpRequest = (Button) findViewById(R.id.my_button);
                                //找到 id 为 my_button 的 Button 控件
    sendHttpRequest.setOnClickListener(new OnClickListener()
            {//绑定单击事件监听器
              @Override
              public void onClick(View v)
              {//按钮单击事件回调方法
                  String number = phoneNumber.getText().toString();//号码
                  try
                  {
                      String location = getLocation(number);
                      showLocation.setText(location);//设置文本
                  }
                  catch (Exception e)
                  {
                      e.printStackTrace();
                  }
              }
            });
}
//获取号码归属地信息
private String getLocation(String number) throws Exception
{
    //读取本地准备好的文件，用输入的号码替换原来的占位符
    InputStream in = this.getClassLoader().getResourceAsStream("send.xml");
    byte[] data = load(in);
    String content = new String(data);
    content = content.replace("$number", number);
    //创建连接对象，设置请求头，按照 Webservice 服务端提供的要求来设置
     URL url = new URL("http://webservice.webxml.com.cn/WebServices/Mobile-
CodeWS.asmx");
    HttpURLConnection conn = (HttpURLConnection) url.openConnection();
    conn.setConnectTimeout(5000);
    conn.setRequestProperty("Host", "webservice.webxml.com.cn");
    conn.setRequestProperty("Content-Type", "application/soap+xml; charset=utf-8");
```

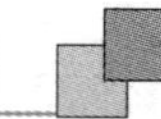

```
            conn.setRequestProperty("Content-Length", content.getBytes().length + "");
            conn.setRequestMethod("POST");
            //输出数据
            conn.setDoOutput(true);
            conn.getOutputStream().write(content.getBytes());
            //获取服务端传回的数据,解析 XML,得到结果
            XmlPullParser parser = Xml.newPullParser();
            parser.setInput(conn.getInputStream(), "UTF-8");
            for (int type = parser.getEventType(); type != XmlPullParser.END_DOCUMENT;
type = parser.next())
            //如果开始节点名称是 getMobileCodeInfoResult
            if (type == XmlPullParser.START_TAG && parser.getName().equals("getMobile-
CodeInfoResult")) {
                    return parser.nextText();
                }
            return "没有找到此号码";
        }

        //将输入流转换成字节数组
        private byte[] load(InputStream in) throws IOException
        {
            ByteArrayOutputStream baos = new ByteArrayOutputStream();
            byte b[] = new byte[1024];
            int len = -1;
            while ((len = in.read(b)) != -1)
            {//循环读取输入流的字节信息
                baos.write(b, 0, len);
            }
            baos.close();
            return baos.toByteArray();
        }
        @Override
        public boolean onCreateOptionsMenu(Menu menu)
        {
            getMenuInflater().inflate(R.menu.activity
_main, menu);
            return true;
        }
    }
```

(5) 右击工程 chapter13_9,在弹出的快捷菜单中选择 Run As|Android Application 命令,运行效果如图 13-9 所示。

图 13-9　归属地查询效果

13.6 Android Tcp Socket

在 Android 的网络通信中，通常会使用 Socket 进行设备之间的数据通信，使用 Http 来对网络数据进行请求。

1. Socket(套接字)

Socket 是用于描述网络上的一个设备中的一个进程或者应用程序的，Socket 由 IP 地址和端口号两部分组成。IP 地址用来定位设备，端口号用来定位应用程序或者进程，例如我们常见的运行在 80 端口上的 HTTP 协议。Socket 的常见格式为：192.168.1.1:1234。

那么应用程序是如何通过 Socket 来与网络中的其他设备进行通信的呢？通常情况下，Socket 通信有两部分，一部分为监听的 Server 端，一部分为主动请求连接的 Client 端。Server 端会一直监听 Socket 中的端口直到有请求为止，当 Client 端对该端口进行连接请求时，Server 端就给予应答并返回一个 Socket 对象，以后在 Server 端与 Client 端的数据交换就可以使用这个 Socket 来进行操作了。

2. Android 中使用 Socket 进行数据通信

- ServerSocket：建立服务端(Server)时，需要使用 ServerSocket 对象，这个对象会自动对其构造函数中传入的端口号进行监听，并在收到连接请求后，使用 ServerSocket.accept()方法返回一个连接的 Socket 对象。
- Socket：不管建立客户端(Client)还是在进行其他数据交换方面的操作时，都需要使用 Socket 类。Socket 类在进行初始化时需要出入 Server 端的 IP 地址和端口号，并返回连接到 Server 端的一个 Socket 对象，如果连接失败，那么将返回异常。同 ServerSocket 一样，也是自动进行连接请求的。通过上面两个步骤后，Server 端和 Client 端就可以连接起来了，但是仅仅连接起来是没有任何作用的，数据交换才是我们的目的，这时候就需要用到 IO 流中的 OutputStream 类和 InputStream 类。
- OutputStream：当应用程序需要对流进行数据写操作时，可以使用 Socket.getOutputStream()方法对返回的数据流进行操作。
- InputStream：当应用程序要从流中取出数据时，可以使用 Socket.getInputStream()方法对返回的数据流进行操作。

下面就通过实例介绍如何在 Android 应用程序当中使用 Tcp Socket。

(1) 新建工程 chapter13_10，编辑 res/layout/activity_main.xml 文件，源代码如下：

```
<?xml version = "1.0" encoding = "utf-8"?>
<LinearLayout xmlns:android = "http://schemas.android.com/apk/res/android"
```

```
    android:layout_width = "fill_parent"
    android:layout_height = "fill_parent"
    android:orientation = "vertical" >
    <!-- 内容输入框 -->
        <EditText
            android:id = "@ + id/my_editText"
            android:layout_width = "fill_parent"
            android:layout_height = "wrap_content"
            android:hint = "input the message and click the send button" >
        </EditText>
        <Button
            android:id = "@ + id/my_button"
            android:layout_width = "fill_parent"
            android:layout_height = "wrap_content"
            android:text = "send" >
        </Button>
</LinearLayout>
```

(2) 由于客户端需要联网才能访问到服务端，所以需要添加相关网络权限，编辑 AndroidManifest.xml 文件，源代码如下：

```
<manifest xmlns:android = "http://schemas.android.com/apk/res/android"
    package = "com.example.chapter13_10"
    android:versionCode = "1"
    android:versionName = "1.0" >
    <!-- 联网权限 -->
    <uses-permission  android:name = "android.permission.INTERNET"/>
    <uses-sdk  android:minSdkVersion = "8"   android:targetSdkVersion = "15" />
    <application
        android:icon = "@drawable/ic_launcher"
        android:label = "@string/app_name"
        android:theme = "@style/AppTheme" >
        <activity
            android:name = ".MainActivity"
            android:label = "@string/title_activity_main" >
            <intent-filter>
                <action android:name = "android.intent.action.MAIN" />
                <category android:name = "android.intent.category.LAUNCHER" />
            </intent-filter>
        </activity>
    </application>
</manifest>
```

(3) 编辑 MainActivity.java 文件，源代码如下：

```
public class MainActivity extends Activity
{
    private EditText editText;
```

```
        private Button button;
        /** Called when the activity is first created. */
        @Override
        public void onCreate(Bundle savedInstanceState)
        {
          super.onCreate(savedInstanceState);
          setContentView(R.layout.activity_main);
          editText = (EditText)findViewById(R.id.my_ editText);//找到 id 为 my_ editText
                                                              //的编辑框
          button = (Button)findViewById(R.id.my_button);      //找到 id 为 my_button 的
                                                              //按钮
          button.setOnClickListener(new OnClickListener()
          {
            @Override
            public void onClick(View v) //按钮的单击事件回调方法
            {
              Socket socket = null;
              String message = editText.getText().toString() + "\r\n" ;
              try {
                  //创建客户端 socket,由于笔者所用的手机设备连的
                  //WIFI 与计算机所在的网络属于同一局域网,所以可以访问
                  socket = new Socket("192.168.0.102",18888);
                  PrintWriter out = new PrintWriter(new BufferedWriter(new OutputStre-
amWriter
                          (socket.getOutputStream())),true);
                  //发送数据
                  out.println(message);
                  //接收数据
                  BufferedReader in = new BufferedReader(new InputStreamReader(socket.
getInputStream()));
                  String msg = in.readLine();
                  if (null != msg)
                  {//如果 msg 对象不为 null
                      editText.setText(msg);
                      System.out.println(msg);
                  }
                  else
                  {//如果 msg 对象为 null
                      editText.setText("data error");
                  }
                  out.close();
                  in.close();
              } catch (UnknownHostException e) {
                  e.printStackTrace();
              } catch (IOException e) {
                  e.printStackTrace();
              }
```

```
            finally
            {
              try
               {
                 if (null != socket)
               {  //如果 socket 对象不为 null
                  socket.close();
               }
            }catch (IOException e)
            {
              e.printStackTrace();
            }
          }
        }
      });
    }
  }
```

(4) 新建 Java Project:Server,在 src 子目录下新建 Server 类,编辑 Server.java 文件,源代码如下:

```
public class Server implements Runnable
{
    @Override
    public void run() {
        Socket socket = null;
        try
          {
            ServerSocket server = new ServerSocket(18888);
            //循环监听客户端链接请求
            while(true)
            {
              System.out.println("start...");
              //接收请求
              socket = server.accept();
              //接收客户端消息
              BufferedReader in = new BufferedReader(new InputStreamReader(socket.
getInputStream()));
              String message = in.readLine();
              System.out.println("accepted  message is: " + message);
              //发送消息,向客户端
              PrintWriter out = new PrintWriter(new BufferedWriter(new OutputStre-
amWriter(socket.getOutputStream())),true);
              out.println("Server:" + message);
              //关闭流
              in.close();
              out.close();
```

```
            }
        }
        catch (IOException e)
        {
            e.printStackTrace();
        }finally
        {
            if (null != socket)
            {
                try
                {
                    socket.close();
                }
                catch (IOException e)
                {
                    e.printStackTrace();
                }
            }
        }
    }
    //启动服务器
    public static void main(String[] args)
    {
        Thread server = new Thread(new Server());
        server.start();
    }
}
```

(5) 右击工程 Server,在弹出的快捷菜单中选择 Run As|Java Application 命令。

(6) 右击工程 chapter13_10,在弹出的快捷菜单中选择 Run As|Android Application 命令,在文本输入框当中输入:Hello World,Server 工程控制台输出如图 13-10 所示。

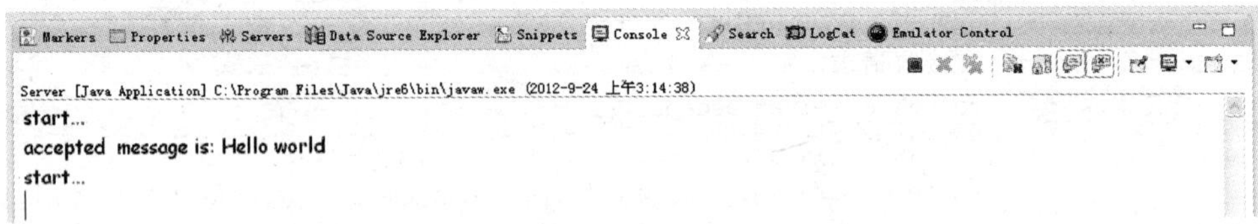

图 13-10 Tcp Socket Server 控制台输出

工程 chapter13_10 界面效果如图 13-11 所示。

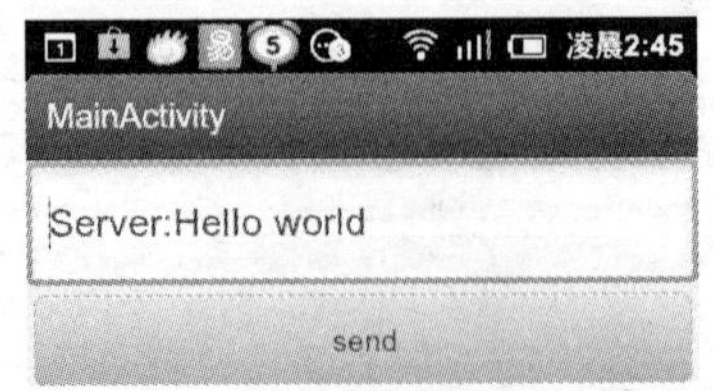

图 13-11 Tcp Socket 客户端效果

第 14 章 多媒体

Android 的多媒体框架支持多种通用媒体的播放以及录制，因此能够很容易地在程序中集成音频、视频和图片信息。在本章当中笔者主要给读者介绍如何在Android应用程序当中进行音视频的播放、音视频的录制以及 TTS 的使用。

14.1 音频播放

14.1.1 MediaPlayer 的介绍

OpenCore 为 Android 用户提供了强大的多媒体开发运用功能，为了简化音频、视频系统的开发和播放，Android 提供了一个综合的 MediaPlayer 类以简化对多媒体的操作，通过 MediaPlayer 类可以对音视频文件进行播放。

使用 MediaPlayer 进行音频的播放，主要有以下几个步骤：

1. 如何获得 MediaPlayer 类的实例

① 可以直接使用 new 的方式，如：

```
MediaPlayer mediaPlayer = new MediaPlayer();
```

② 也可以调用 MediaPlayer 类的 create 静态方法，如：

```
MediaPlayer mediaPlayer = MediaPlayer.create(this, R.raw.test);//这时就不用调用
setDataSource 方法了
```

2. 如何设置要播放的文件

MediaPlayer 要播放的文件主要包括 3 个来源：

❑ 应用中自带的 resource 资源。

例如：MediaPlayer.create(this, R.raw.test);

❑ 存储在 SDCard 卡或其他文件路径下的媒体文件。

例如：mediaPlayer. setDataSource("/sdcard/test. mp3")；

❑ 网络上的媒体文件。

例如：mediaPlayer. setDataSource("http://www. citynorth. cn/music/confucius. mp3")；

3. 对播放器的控制

Android 通过控制播放器状态的方式来控制媒体文件的播放，其中：

① prepare()和 prepareAsync()提供了同步和异步两种方式设置播放器进入 prepare 状态，需要注意的是，如果 MediaPlayer 实例是由 create 静态方法创建的，那么第一次启动播放前就不需要调用 prepare()方法了，因为在 create 静态方法中已经调用过 prepare()方法了。

② start()方法是真正启动文件播放的方法。

③ pause()方法和 stop()方法比较简单，起到暂停播放和停止播放的作用。

④ seekTo()方法是定位方法，可以让播放器从指定的位置开始播放，需要注意的是该方法是个异步方法，也就是说该方法返回时并不意味着定位完成。

⑤ release()方法可以释放播放器占用的资源，一旦确定不再使用播放器时应当尽早调用它释放资源。

⑥ reset()方法可以使播放器从 Error 状态中恢复过来，重新回到 Idle 状态。

MediaPlayer 对音视频的播放流程是用一个状态机来处理的，如图 14－1 所示。

4. 设置播放器的监听器

MediaPlayer 提供了一些设置不同监听器的方法来更好地对播放器的工作状态进行监听，如：

setOnComPletionListener(MediaPlayer. OnCompletionListener listener)

设置播放器时需要考虑到播放器可能出现的情况，设置好监听和处理逻辑，以保持播放器的健壮性。

14.1.2 MediaPlayer 播放音频

下面就通过实例介绍如何在 Android 应用程序当中使用 MediaPlayer 对象来播放音频文件。

(1) 新建工程 chapter14_1，编辑 res/layout/activity_main. xml 文件，源代码参考配套资料。

(2) 将用于测试的 test. mp3 文件复制到工程的 assets 目录。

(3) 编辑 MainActivity. java 文件，源代码如下：

```
public class MainActivity extends Activity implements OnClickListener
{
```

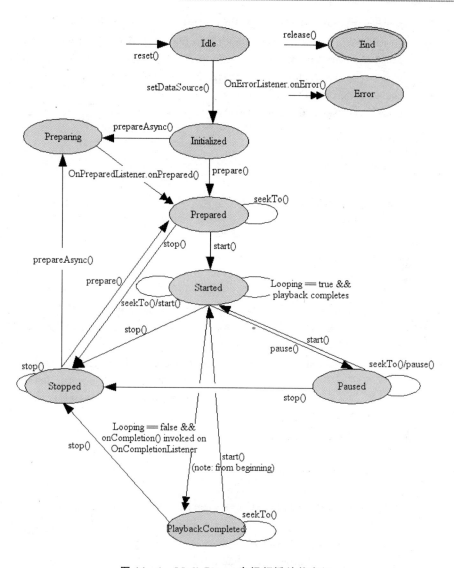

图 14－1　MediaPlayer 音视频播放状态机

```
private static final String TAG = "MainActivity";
private Button playBtn;//声明播放按钮
private Button pauseBtn;//声明暂停按钮
private Button stopBtn;//声明停止按钮
private MediaPlayer mediaPlayer;
@Override
public void onCreate(Bundle savedInstanceState)
{
    super.onCreate(savedInstanceState);
    setContentView(R.layout.activity_main);
```

```
        playBtn = (Button) findViewById(R.id.play);//找到 id 为 play 的播放按钮
        pauseBtn = (Button) findViewById(R.id.pause);//找到 id 为 pause 的暂停按钮
        stopBtn = (Button) findViewById(R.id.stop);//找到 id 为 stop 的停止按钮
        playBtn.setOnClickListener(this);//绑定单击事件监听器
        pauseBtn.setOnClickListener(this);
        stopBtn.setOnClickListener(this);
    }
    @Override
    public void onClick(View v)
    {
        int id = v.getId();
        switch (id)
        {
        case R.id.play://播放按钮
            try
            {
                AssetManager assetManager = this.getAssets();
                AssetFileDescriptor fileDescriptor = assetManager.openFd("test.mp3");
                mediaPlayer = new MediaPlayer();
                  mediaPlayer.setDataSource(fileDescriptor.getFileDescriptor(),
                                            fileDescriptor.getStartOffset(),
                                            fileDescriptor.getLength());
                mediaPlayer.prepare();
                mediaPlayer.start();
            }
            catch (Exception e)
            {
                Log.e(TAG, "play music occurs errors:" + e.getMessage());
            }
            //设置媒体播放器监听器
            mediaPlayer.setOnCompletionListener(new OnCompletionListener()
            {
                @Override
                public void onCompletion(MediaPlayer mp)
                {
                    mp.release();
                }
            });
            break;
        case R.id.pause://暂停按钮
            if (mediaPlayer != null)
            {
```

```
                mediaPlayer.pause();
            }
            break;
        case R.id.stop://停止按钮
            if (mediaPlayer != null)
            {
                mediaPlayer.stop();
            }
            break;
        }
    }
    @Override
    protected void onDestroy()
    {
        super.onDestroy();
        if(mediaPlayer != null)
        {
             mediaPlayer.release();
        }
    }
    @Override
    public boolean onCreateOptionsMenu(Menu menu) {
      getMenuInflater().inflate(R.menu.activity_main, menu);
      return true;
    }
}
```

(4) 右击工程 chapter14_1，在弹出的快捷菜单中选择 Run As | Android Application 命令，运行效果如图 14－2 所示。

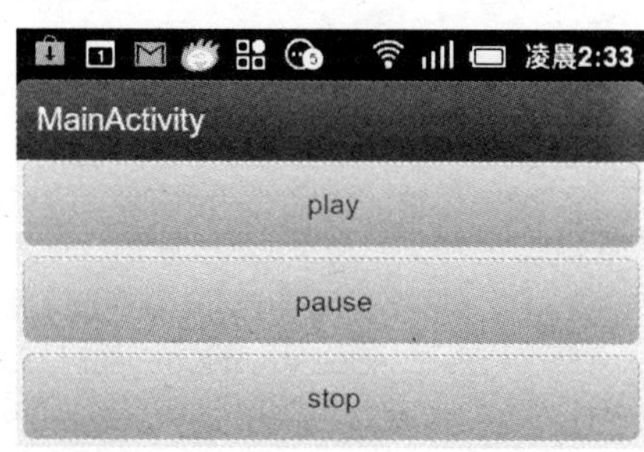

图 14－2　MediaPlayer 播放音频

14.2　视频播放

Android 提供了 3 种方式来实现视频的播放。

- ❑ 使用其自带的播放器。指定 Action 为 ACTION_VIEW，Data 为 Uri，Type 为其 MIME 类型。
- ❑ 使用 VideoView 来播放。在布局文件中使用 VideoView 结合 MediaController 来实现对其控制。
- ❑ 使用 MediaPlayer 类和 SurfaceView 来实现。这种方式最灵活，也最复杂。

14.2.1 自带播放器播放视频

下面就通过实例演示如何使用自带的播放器来播放视频文件。

(1) 新建工程 chapter14_2,将用于测试的 test.mp4 视频文件复制到 SDCard 卡的根目录下。

(2) 编辑 res/layout/activity_main.xml 文件,源代码如下:

```
<LinearLayout xmlns:android = "http://schemas.android.com/apk/res/android"
    xmlns:tools = "http://schemas.android.com/tools"
    android:layout_width = "fill_parent"
    android:layout_height = "fill_parent" >
<!-- 调用自带播放器播放视频 -->
    <Button
        android:id = "@ + id/play_video"
        android:layout_width = "fill_parent"
        android:layout_height = "wrap_content"
        android:text = "播放视频"/>
</LinearLayout>
```

(3) 编辑 MainActivity.java 文件,源代码如下:

```
public class MainActivity extends Activity
{
  Button playVideoButton;//声明发送播放视频意图的按钮
  @Override
  public void onCreate(Bundle savedInstanceState)
  {
      super.onCreate(savedInstanceState);
      setContentView(R.layout.activity_main);
      playVideoButton  = (Button)findViewById(R.id.play_video);
                                                          //找到 id 为 play_video 的按钮
      playVideoButton.setOnClickListener(new OnClickListener()
      {
          @Override
          public void onClick(View v)
          {
            Uri uri  =  Uri.parse(Environment.getExternalStorageDirectory().getPath
() + "/test.mp4");
            //调用系统自带的播放器
            Intent intent  =  new Intent(Intent.ACTION_VIEW);
            intent.setDataAndType(uri, "video/mp4");
            startActivity(intent);
```

```
            }
        });
    }
    @Override
    public boolean onCreateOptionsMenu(Menu menu) {
        getMenuInflater().inflate(R.menu.activity_main, menu);
        return true;
    }
}
```

（4）右击工程 chapter14_2，在弹出的快捷菜单中选择 Run As|Android Application 命令，单击界面上的“播放视频”按钮，如果有多个应用程序可以处理这样的意图，那么就会弹出一个可供选择的应用程序列表，用户可以根据自己的喜好，来确定到底需要哪一个应用程序来处理程序发出的意图。如果只有单个应用程序满足 MainActivity 发出的意图，那么就会跳转到相应的应用程序播放视频文件界面。具体的播放效果，笔者在这里就不贴图展示了。

14.2.2 VideoView 播放视频

下面通过实例介绍如何使用 Android 提供的 VideoView 控件来播放视频。

（1）新建工程 chapter14_3，编辑 res/layout/activity_main.xml 文件，源代码如下：

```
<?xml version="1.0" encoding="utf-8"?>
<LinearLayout xmlns:android="http://schemas.android.com/apk/res/android"
    android:orientation="vertical"
    android:layout_width="fill_parent"
    android:layout_height="fill_parent"
    >
<VideoView android:id="@+id/video_view"
            android:layout_width="fill_parent"
            android:layout_height="wrap_content"
            android:layout_gravity="center"/>
</LinearLayout>
```

（2）将用于测试的 test.mp4 文件复制到 SDCard 根目录下。

（3）编辑 MainActivity.java 文件，源代码如下：

```
public class MainActivity extends Activity implements OnErrorListener, OnCompletion-
Listener
{
    private static final String TAG = "MainActivity";
    private VideoView videoView;//声明视频播放的控件
```

```
    private MediaController mediaController;
    private Uri videoUri;                    //播放的视频路径 Uri
    private int positionWhenPaused = -1;//暂停播放时的位置
    @Override
    public void onCreate(Bundle savedInstanceState)
    {
            super.onCreate(savedInstanceState);
            setContentView(R.layout.activity_main);
            videoView = (VideoView)findViewById(R.id.video_view);
                                        //找到 id 为 video_view 的视频播放控件
            videoUri  = Uri.parse(Environment.getExternalStorageDirectory() + "/
test.mp4");
            mediaController = new MediaController(this);
            videoView.setMediaController(mediaController);
        }
        @Override
    protected void onStart()
    {
    super.onStart();
    videoView.setVideoURI(videoUri);
    videoView.start();//开始播放
    }
    @Override
    protected void onPause()
    {
    super.onPause();
    positionWhenPaused = videoView.getCurrentPosition();
    videoView.stopPlayback();//为了节省系统资源,当 Activity 处于 Paused 状态时,停止
                            //VideoView 的播放
    Log.d(TAG, "positionWhenPaused=" + positionWhenPaused);
    Log.d(TAG, "duration=" + videoView.getDuration());
    }
    @Override
    protected void onResume()
    {
    super.onResume();
    if(positionWhenPaused >= 0)
    {//如果暂停播放的位置大于等于 0
        videoView.seekTo(positionWhenPaused);//恢复视频的播放
        positionWhenPaused = -1;
    }
    }
```

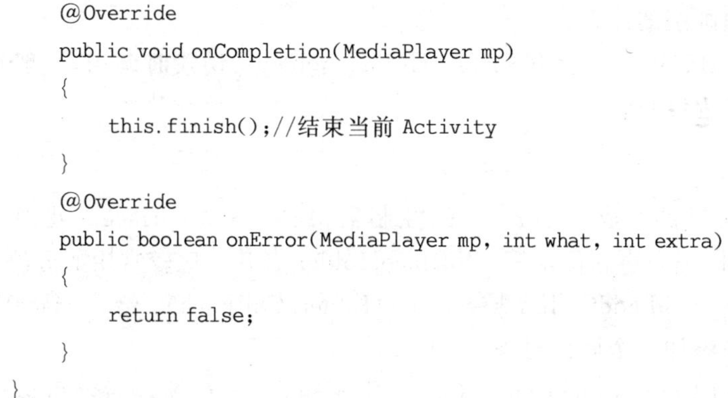

```
    @Override
    public void onCompletion(MediaPlayer mp)
    {
        this.finish();//结束当前 Activity
    }
    @Override
    public boolean onError(MediaPlayer mp, int what, int extra)
    {
        return false;
    }
}
```

在上述代码当中，只调用了 VideoView 类的 setVideoURI()方法设置要播放的文件，然后再调用 start()方法就可以播放视频文件了。为了节省系统资源，一般需要在 Activity 的 onPause()方法当中，暂停视频的播放。在 Activity 的 onResume()方法恢复视频的播放。

可以通过实现 MediaPlayer. OnErrorListener 接口来监听 MediaPlayer 上的报错信息；可以通过实现 MediaPlayer. OnCompletionListener 接口来得到视频播放完成的通知回调。

(4) 右击工程 chapter14_3，在弹出的快捷菜单中选择 Run As | Android Application 命令，运行效果笔者在这里就不贴图展示了。

14.2.3　MediaPlayer 结合 SurfaceView 播放视频

SurfaceView 是一个继承了 View，但是与一般的 View 有很大区别的类。这是由于 SurfaceView 的绘制方法与 View 的绘制方法不一样造成的。

1. 定义

SurfaceView 可以直接从内存或者 DMA 等硬件接口取得图像数据，是个非常重要的绘图容器。

它的特性是：可以在主线程之外的线程中向屏幕上绘图。这样可以避免画图任务繁重的时候造成主线程的阻塞，从而提高了程序的反应速度。在游戏开发中经常要用到 SurfaceView，游戏中的背景、任务、动画等尽量在画布 Canvas 中画出。

2. 实现

由于需要对 SurfaceView 进行监听，所以需要实现 SurfaceHolder. Callback 这个接口。这个接口需要实现 3 个方法：

- ❑ public void surfaceChanged(SurfaceHolder holder,int format,int width,int height) 在 surface 的大小发生变化时调用。
- ❑ public void surfaceCreated(SurfaceHolder holder)：在 surface 创建时调用，一

般在这里调用画图的线程。

- public void surfaceDestoryed(SurfaceHolder holder)：销毁时调用，一般在这里将画图的线程停止。

3. SurfaceHolder

SurfaceHolder 可以被看成是 surface 的控制器，用来操纵 surface。处理它在 Canvas 上画的效果、控制大小和像素等。SurfaceHolder 有几个比较常用的方法：

- public abstract void addCallBack(SurfaceHolder.CallBack)：给 SurfaceView 当前的持有者添加一个回调对象。
- public abstract Canvas lockCanvas()：锁定画布，一般在锁定后就可以通过其返回的画布对象 Canvas，在其上面进行画图等操作了。
- public abstract Canvas lockCanvas(Rect dirty)：锁定画布的某个区域进行画图。相对部分内存要求比较高的游戏来说，可以不用重画 dirty 外的其他区域的像素，这样一来就可以提高速度了。
- public abstract unlockCanvasAndPost(Canvas canvas)：结束锁定画图，并提交改变。

下面通过实例介绍如何利用 MediaPlayer 结合 SurfaceView 来播放视频文件。

(1) 新建工程 chapter14_4，编辑 res/layout/activity_main.xml 文件，源代码如下：

```
<? xml version = "1.0" encoding = "utf-8"? >
<LinearLayout xmlns:android = "http://schemas.android.com/apk/res/android"
    android:layout_width = "fill_parent"
    android:layout_height = "fill_parent"
    android:orientation = "vertical" >
    <SurfaceView
        android:id = "@ + id/surfaceView"
        android:layout_width = "fill_parent"
        android:layout_height = "360px" />
    <LinearLayout
        android:layout_width = "fill_parent"
        android:layout_height = "wrap_content"
        android:gravity = "center_horizontal"
        android:orientation = "horizontal" >
   <!-- 播放按钮 -->
        <Button
            android:id = "@ + id/btnplay"
            android:layout_width = "wrap_content"
            android:layout_height = "wrap_content"
    android:text = "播放"/>
```

```
<!-- 暂停按钮 -->
    <Button
        android:id = "@ + id/btnpause"
        android:layout_width = "wrap_content"
        android:layout_height = "wrap_content"
        android:text = "暂停"/>
  <!-- 停止按钮 -->
    <Button
        android:id = "@ + id/btnstop"
        android:layout_width = "wrap_content"
        android:layout_height = "wrap_content"
        android:text = "停止"/>
  </LinearLayout>
</LinearLayout>
```

(2) 将需要测试的 test. mp4 文件复制到 SDCard 根目录下。

(3) 编辑 MainActivity. java 文件,源代码如下:

```
public class MainActivity extends Activity  implements OnClickListener
{
  private static final String TAG = "MainActivity";
  Button btnplay, btnstop, btnpause;  //声明播放,暂停和停止按钮
  SurfaceView surfaceView;
  MediaPlayer mediaPlayer;
  int position;
  public void onCreate(Bundle savedInstanceState)
  {
    super.onCreate(savedInstanceState);
    setContentView(R.layout.activity_main);
    btnplay = (Button)findViewById(R.id.btnplay);//找到 id 为 btnplay 的播放按钮
    btnpause = (Button)findViewById(R.id.btnpause);//找到 id 为 btnpause 的暂停按钮
    btnstop = (Button)findViewById(R.id.btnstop);//找到 id 为 btnstop 的停止按钮
    btnplay.setOnClickListener(this);//绑定单击事件监听器
    btnpause.setOnClickListener(this);
    btnstop.setOnClickListener(this);
    mediaPlayer = new MediaPlayer();
    surfaceView = (SurfaceView) this.findViewById(R.id.surfaceView);
    //设置 SurfaceView 自己不管理的缓冲区
    surfaceView.getHolder().setType(SurfaceHolder.SURFACE_TYPE_PUSH_BUFFERS);
    surfaceView.getHolder().addCallback(new Callback()
    {
      @Override
      public void surfaceDestroyed(SurfaceHolder holder) //surface 销毁时回调的方法
```

```
        {
          Log.d(TAG, "surfaceDestroyed method was invoked");
        }
        @Override
        public void surfaceCreated(SurfaceHolder holder) //surface 创建时回调的方法
        {
          if (position>0) {  //如果位置大于 0
            try
            {
              //开始播放
              play();
              //并直接从指定位置开始播放
              mediaPlayer.seekTo(position);
              position = 0;
            }
            catch (Exception e)
            {
              Log.d(TAG, "surfaceCreated occurs errors:" + e.getMessage());
            }
          }
        }
        @Override
        public void surfaceChanged(SurfaceHolder holder, int format, int width,
              int height)
        {
        Log.d(TAG, "surfaceChanged method was invoked");
        }
      });
    }
    @Override
    public void onClick(View v)
    {
      switch (v.getId())
      {
      case R.id.btnplay:              //播放按钮
          play();
          break;
      case R.id.btnpause:             //暂停播放
          if (mediaPlayer.isPlaying())
          {//如果正在播放
              mediaPlayer.pause();
          }
```

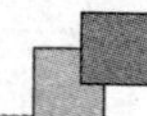

```
        else
        { //如果没有在播放
            mediaPlayer.start();
        }
        break;
    case R.id.btnstop:          //停止播放
        if (mediaPlayer.isPlaying())
        {
            mediaPlayer.stop();
        }
        break;
    default:
        break;
        }
    }
    @Override
    protected void onPause()
    {
        //先判断是否正在播放
        if (mediaPlayer.isPlaying())
        { //如果正在播放
            //如果正在播放就先保存这个播放位置
            position = mediaPlayer.getCurrentPosition();
            mediaPlayer.stop();
        }
        super.onPause();
    }
    private void play()
    {
      try {
          mediaPlayer.reset();
          mediaPlayer
          .setAudioStreamType(AudioManager.STREAM_MUSIC);
          //设置需要播放的视频
          mediaPlayer.setDataSource(Environment.getExternalStorageDirectory() + "/
test.mp4");
          //把视频画面输出到 SurfaceView
          mediaPlayer.setDisplay(surfaceView.getHolder());
          mediaPlayer.prepare();
          //播放
          mediaPlayer.start();
      }
```

```
            catch (Exception e)
            {
             Log.e(TAG, "play occurs errors:" + e.getMessage());
            }
        }
    }
```

(4) 右击工程 chapter14_4,在弹出的快捷菜单中选择 Run As|Android Application 命令,运行效果笔者在这里就不贴图展示了。

14.3 音频录制

可以使用 Android SDK 提供的 MediaRecorder 类来完成音频的录制。下面笔者通过实例介绍如何利用 MediaRecorder 类来完成音频的录制。

(1) 新建工程 chapter14_5,编辑 res/layout/activity_main.xml 文件,源代码如下:

```
<?xml version="1.0" encoding="utf-8"?>
<LinearLayout xmlns:android="http://schemas.android.com/apk/res/android"
    android:layout_width="fill_parent"
    android:layout_height="fill_parent"
    android:orientation="vertical" >
    <!-- 状态提示 -->
     <TextView
            android:id="@+id/state_info"
            android:layout_width="fill_parent"
            android:layout_height="wrap_content"
            android:text=""/>
  <!--开始按钮-->
        <Button
            android:id="@+id/btn_start"
            android:layout_width="fill_parent"
            android:layout_height="wrap_content"
            android:text="开始"/>
  <!-- 停止按钮 -->
        <Button
            android:id="@+id/btn_stop"
            android:layout_width="fill_parent"
            android:layout_height="wrap_content"
            android:text="停止"/>
    <!-- 播放按钮 -->
```

```
    <Button
        android:id = "@ + id/btn_play"
        android:layout_width = "fill_parent"
        android:layout_height = "wrap_content"
        android:text = "播放"/>
  <!-- 结束按钮 -->
    <Button
        android:id = "@ + id/btn_finish"
        android:layout_width = "fill_parent"
        android:layout_height = "wrap_content"
        android:text = "结束"/>
</LinearLayout>
```

(2) 由于需要录制音频,将需要录制好的音频文件保存在 SDCard 卡当中,所以需要添加相关权限。编辑 AndroidManifest. xml 文件,源代码如下:

```
<manifest xmlns:android = "http://schemas.android.com/apk/res/android"
    package = "com.example.chapter14_5"
    android:versionCode = "1"
    android:versionName = "1.0" >
    <uses-sdk  android:minSdkVersion = "8"  android:targetSdkVersion = "15" />
    <!-- 往 SDCard 中写入数据权限 -->
    <uses-permission android:name = "android.permission.WRITE_EXTERNAL_STORAGE"/>
    <!-- 在 SDCard 中创建与删除文件权限 -->
     < uses-permission  android: name  =  " android.  permission.  MOUNT _ UNMOUNT _
FILESYSTEMS"/>
    <!-- 录制声音权限 -->
    <uses-permission  android:name = "android.permission.RECORD_AUDIO"/>
    <application
        android:icon = "@drawable/ic_launcher"
        android:label = "@string/app_name"
        android:theme = "@style/AppTheme" >
        <activity
            android:name = ".MainActivity"
            android:label = "@string/title_activity_main" >
            <intent-filter>
                <action android:name = "android.intent.action.MAIN" />
                <category android:name = "android.intent.category.LAUNCHER" />
            </intent-filter>
        </activity>
    </application>
</manifest>
```

(3) 编辑 MainActivity.java 文件,源代码如下:

```
public class MainActivity extends Activity implements OnClickListener
{
    private static final String TAG = "MainActivity";
    Button startButton, stopButton, playButton, finishButton;  //声明开始录音、结
                                              //束录音、播放录音以及结束 Activity 按钮
    TextView stateView;//显示录音状态文本
    private MediaRecorder recorder;
    private MediaPlayer player;
    private File audioFile;
    private Uri fileUri;
    @Override
    public void onCreate(Bundle savedInstanceState)
    {
        super.onCreate(savedInstanceState);
        setContentView(R.layout.activity_main);
        startButton = (Button) findViewById(R.id.btn_start);
                                             //找到 id 为 btn_start 的开始录音按钮
        stopButton = (Button) findViewById(R.id.btn_stop);
                                              //找到 id 为 btn_stop 的停止录音按钮
        playButton = (Button) findViewById(R.id.btn_play);
                                              //找到 id 为 btn_play 的播放录音按钮
        finishButton = (Button) findViewById(R.id.btn_finish);
                                         //找到 id 为 btn_finish 的结束 Activity 按钮
        startButton.setOnClickListener(this);//绑定单击事件监听器
        stopButton.setOnClickListener(this);
        playButton.setOnClickListener(this);
        finishButton.setOnClickListener(this);
        stateView = (TextView)findViewById(R.id.state_info);
                                            //找到 id 为 state_info 的显示状态文本
        stateView.setText("准备开始");
    }
    @Override
    public void onClick(View v)
    {
        int id = v.getId();
        switch (id)
        {
        case R.id.btn_start: //开始录音按钮
          //开始录制
          //我们需要实例化一个 MediaRecorder 对象,然后进行相应的设置
```

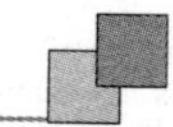

```
            recorder = new MediaRecorder();
            //指定 AudioSource 为 MIC(Microphone audio source ),这是最常用的
            recorder.setAudioSource(MediaRecorder.AudioSource.MIC);
            //指定 OutputFormat,我们选择 3gp 格式
            //其他格式,MPEG-4:这将指定录制的文件为 mpeg-4 格式,可以保护 Audio 和 Video
            //RAW_AMR:录制原始文件,这只支持音频录制,同时要求音频编码为 AMR_NB
            //THREE_GPP:录制后文件是一个 3gp 文件,支持音频和视频录制
            recorder.setOutputFormat(MediaRecorder.OutputFormat.THREE_GPP);
            //指定 Audio 编码方式,目前只有 AMR_NB 格式
            recorder.setAudioEncoder(MediaRecorder.AudioEncoder.AMR_NB);
            //接下来我们需要指定录制后文件的存储路径
            File fpath = new File(Environment.getExternalStorageDirectory().getAbso-
lutePath() + "/data/files/");
            fpath.mkdirs();//创建文件夹
            try
            {
              //创建临时文件
              audioFile = File.createTempFile("recording", ".3gp", fpath);
              recorder.setOutputFile(audioFile.getAbsolutePath());
              recorder.prepare();
            }
            catch (Exception e)
            {
              Log.e(TAG, "Occurs errors:" + e.getMessage());
            }
            //下面就开始录制了
            recorder.start();
            stateView.setText("正在录制");
            startButton.setEnabled(false);
            playButton.setEnabled(false);
            stopButton.setEnabled(true);
            break;
        case R.id.btn_stop://结束录音按钮
              recorder.stop();
              recorder.release();
              //然后可以将录制文件存储到 MediaStore 中
              ContentValues values = new ContentValues();
              values.put(MediaStore.Audio.Media.TITLE, "this is my first record-audio");
             values.put(MediaStore.Audio.Media.DATE_ADDED, System.currentTimeMillis());
             values.put(MediaStore.Audio.Media.DATA , audioFile.getAbsolutePath());
              fileUri = this.getContentResolver().insert(MediaStore.Audio.Media.
EXTERNAL_CONTENT_URI, values);
```

```
        //录制结束后,我们实例化一个 MediaPlayer 对象,然后准备播放
        player = new MediaPlayer();
        player.setOnCompletionListener(new MediaPlayer.OnCompletionListener()
        {
            @Override
            public void onCompletion(MediaPlayer arg0)
            {
                //更新状态
                stateView.setText("准备录制");
                startButton.setEnabled(true);
                playButton.setEnabled(true);
                stopButton.setEnabled(false);
            }
        });
        //准备播放
        try
        {
            player.setDataSource(audioFile.getAbsolutePath());
            player.prepare();
        }
        catch (Exception e)
        {
            Log.e(TAG, "Prepare for playing occurs errors:" + e.getMessage());
        }
        //更新状态
        stateView.setText("准备播放");
        playButton.setEnabled(true);
        startButton.setEnabled(true);
        stopButton.setEnabled(false);
    break;
case R.id.btn_play://播放录音按钮
        //播放录音
    //注意,在录音结束的时候,已经实例化了 MediaPlayer,做好了播放的准备
    player.start();
    //更新状态
    stateView.setText("正在播放");
    startButton.setEnabled(false);
    stopButton.setEnabled(false);
    playButton.setEnabled(false);
      break;
case R.id.btn_finish://结束 Activity 按钮
      break;
```

```
        default:
            finish();
            break;
        }
    }
    @Override
    public boolean onCreateOptionsMenu(Menu menu) {
        getMenuInflater().inflate(R.menu.activity_main, menu);
        return true;
    }
}
```

(4) 右击工程 chapter14_5，在弹出的快捷菜单中选择 Run As|Android Application 命令，运行效果笔者在这里就不贴图展示了。

14.4　视频录制

同样也可以使用 Android SDK 提供的 MediaRecorder 类来完成视频的录制。下面就通过实例介绍如何利用 MediaRecorder 类来完成视频的录制。

(1) 新建工程 chapter14_6，编辑 res/layout/activity_main.xml 文件，源代码参考配套资料。

(2) 由于需要录制视频，还需要将录制好的视频文件保存在 SDCard 卡当中，所以需要添加相关权限。编辑 AndroidManifest.xml 文件，源代码如下：

```
<manifest xmlns:android="http://schemas.android.com/apk/res/android"
    package="com.example.chapter14_6"
    android:versionCode="1"
    android:versionName="1.0" >
    <uses-sdk  android:minSdkVersion="8"    android:targetSdkVersion="15" />
    <!-- 使用照相机权限 -->
    <uses-permission android:name="android.permission.CAMERA" />
    <!-- 录制音视频权限 -->
    <uses-permission android:name="android.permission.RECORD_AUDIO" />
    <!-- 在 SDCARD 写入数据权限 -->
    <uses-permission android:name="android.permission.WRITE_EXTERNAL_STORAGE" />
    <!-- 在 SDCard 中创建与删除文件权限 -->
    <uses-permission android:name="android.permission.MOUNT_UNMOUNT_FILESYSTEMS"/>
    <application
        android:icon="@drawable/ic_launcher"
```

```
        android:label = "@string/app_name"
        android:theme = "@style/AppTheme" >
        <activity
            android:name = ".MainActivity"
            android:label = "@string/title_activity_main" >
            <intent-filter>
                <action android:name = "android.intent.action.MAIN" />
                <category android:name = "android.intent.category.LAUNCHER" />
            </intent-filter>
        </activity>
    </application>
</manifest>
```

(3) 编辑 MainActivity.java 文件,源代码如下:

```
public class MainActivity extends Activity implements SurfaceHolder.Callback
{
    private Button start;//开始录制按钮
    private Button stop;//停止录制按钮
    private MediaRecorder mediarecorder;//录制视频的类
    private SurfaceView surfaceview;//显示视频的控件
    private SurfaceHolder surfaceHolder;
    public void onCreate(Bundle savedInstanceState)
    {
        super.onCreate(savedInstanceState);
        requestWindowFeature(Window.FEATURE_NO_TITLE);//去掉标题栏
        getWindow().setFlags(WindowManager.LayoutParams.FLAG_FULLSCREEN, WindowManager.LayoutParams.FLAG_FULLSCREEN);//设置全屏
        //设置横屏显示
        setRequestedOrientation(ActivityInfo.SCREEN_ORIENTATION_LANDSCAPE);
        //选择支持半透明模式,在有 surfaceview 的 activity 中使用
        getWindow().setFormat(PixelFormat.TRANSLUCENT);
        setContentView(R.layout.activity_main);
        init();
    }
    private void init()
    {
        start = (Button) this.findViewById(R.id.start);
        stop = (Button) this.findViewById(R.id.stop);
        start.setOnClickListener(new TestVideoListener());
        stop.setOnClickListener(new TestVideoListener());
        surfaceview = (SurfaceView) this.findViewById(R.id.surfaceview);
        SurfaceHolder holder = surfaceview.getHolder();//取得 holder
```

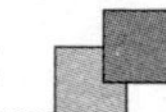

```
        holder.addCallback(this); //holder 加入回调接口
        //setType 必须设置,否则会出错
        holder.setType(SurfaceHolder.SURFACE_TYPE_PUSH_BUFFERS);
    }
    class TestVideoListener implements OnClickListener
    {
        @Override
        public void onClick(View v)
        {
            if (v == start) { //开始录制按钮
              mediarecorder = new MediaRecorder();//创建 mediarecorder 对象
              //设置录制视频源为 Camera(相机)
              mediarecorder.setVideoSource(MediaRecorder.VideoSource.CAMERA);
              //设置录制完成后视频的封装格式 THREE_GPP 为 3gp.MPEG_4 为 mp4
              mediarecorder.setOutputFormat(MediaRecorder.OutputFormat.THREE_GPP);
              //设置录制的视频编码 h263、h264
              mediarecorder.setVideoEncoder(MediaRecorder.VideoEncoder.H264);
              //设置视频录制的分辨率。必须放在设置编码和格式的后面,否则报错
              mediarecorder.setVideoSize(176, 144);
              //设置录制的视频帧率。必须放在设置编码和格式的后面,否则报错
              mediarecorder.setVideoFrameRate(20);
              mediarecorder.setPreviewDisplay(surfaceHolder.getSurface());
              //设置视频文件输出的路径
              mediarecorder.setOutputFile("/sdcard/love.3gp");
              try
              {
                //准备录制
                mediarecorder.prepare();
                //开始录制
                mediarecorder.start();
            }
            catch (IllegalStateException e)
            {
                //TODO Auto-generated catch block
                e.printStackTrace();
            } catch (IOException e)
            {
                //TODO Auto-generated catch block
                e.printStackTrace();
            }
        }
        if (v == stop)
```

```
            {  //停止录制按钮
                if (mediarecorder != null)
                {//如果 mediarecorder 对象不为 null
                    //停止录制
                    mediarecorder.stop();
                    //释放资源
                    mediarecorder.release();
                    mediarecorder = null;
                }
            }
        }
    }
    @Override
    public void surfaceChanged(SurfaceHolder holder, int format, int width,
            int height)
    {
        //将 holder(这个 holder 为开始在 oncreat 里面取得的 holder)赋给 surfaceHolder
        surfaceHolder = holder;
    }
    @Override
    public void surfaceCreated(SurfaceHolder holder)
    {
        //将 holder(这个 holder 为开始在 oncreat 里面取得的 holder)赋给 surfaceHolder
        surfaceHolder = holder;
    }
    @Override
    public void surfaceDestroyed(SurfaceHolder holder) {
        //surfaceDestroyed 方法被回调时将对象设置为空,回收内存
        surfaceview = null;
        surfaceHolder = null;
        mediarecorder = null;
    }
}
```

(4) 右击工程 chapter14_6,在弹出的快捷菜单中选择 Run As|Android Application 命令,运行效果笔者在这里就不贴图展示了。

14.5 TTS 的使用

TextToSpeech 简称 TTS,称为语音合成。是 Android 从版本 1.6 开始支持的新功能,能将所指定的文本转换成不同语言音频输出。TTS 功能需要有 TTS En-

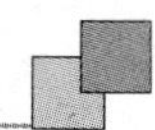

gine 的支持，下面首先来了解一下 Android 提供的 TTS Engine。Android 使用了叫 Pico 支持多种语言的语音合成引擎，负责在后台分析输入的文本，把文本分解它能识别的各个片段，再把合成的各个语音片段以听起来比较自然的方式连接在一起。TTS engine 依托于当前 Android Platform 所支持的几种主要的言：English、French、German、Italian、和 Spanish 共 5 大语言（暂时没有对中文提供支持）。TTS 可以将文本随意地转换成以上任意 5 种语言的语音输出。与此同时，对于个别的语言版本将取决于不同的时区，例如：对于 English，在 TTS 中可以分别输出美式和英式两种不同的版本。

下面就通过实例介绍如何在 Android 应用程序当中使用 TTS 相关的 API。

(1) 新建工程 chapter14_7，编辑 res/layout/activity_main.xml 文件，源代码参考配套资料。

(2) 编辑 MainActivity.java 文件，源代码如下：

```
public class MainActivity extends Activity implements OnInitListener
{
    /** Called when the activity is first created. */
    private EditText inputText = null;
    private Button speakBtn = null;
    private static final int REQ_TTS_STATUS_CHECK = 0;
    private static final String TAG = "TTS Demo";
    private TextToSpeech mTts;
    @Override
    public void onCreate(Bundle savedInstanceState)
    {
        super.onCreate(savedInstanceState);
        setContentView(R.layout.activity_main);
        //检查 TTS 数据是否已经安装并且可用
        Intent checkIntent = new Intent();
        checkIntent.setAction(TextToSpeech.Engine.ACTION_CHECK_TTS_DATA);
        startActivityForResult(checkIntent, REQ_TTS_STATUS_CHECK);
        inputText = (EditText) findViewById(R.id.inputText);
        speakBtn = (Button) findViewById(R.id.speakBtn);
        inputText.setText("This is an example of speech synthesis.");
        speakBtn.setOnClickListener(new OnClickListener()
        {
            public void onClick(View v)
            {
                //TODO Auto-generated method stub
                mTts.speak(inputText.getText().toString(), TextToSpeech.QUEUE_ADD,
null);
                //朗读输入框里的内容
            }
        });
    }
```

```
        //实现 TTS 初始化接口
        @Override
        public void onInit(int status)
        {
            //TODO Auto-generated method stub
            //TTS Engine 初始化完成
            if (status == TextToSpeech.SUCCESS) {//如果 TTS Engine 初始化成功
                int result = mTts.setLanguage(Locale.US);
                //设置发音语言
                  if (result == TextToSpeech.LANG_MISSING_DATA || result == Text-
ToSpeech.LANG_NOT_SUPPORTED)
                //判断语言是否可用
                {//如果不支持发音语言
                    Log.v(TAG, "Language is not available");
                    speakBtn.setEnabled(false);
                }
                else
                {//如果支持发音语言
                    mTts.speak("This is an example of speech synthesis.", TextToSpeech.
QUEUE_ADD, null);
                    speakBtn.setEnabled(true);
                }
            }
        }
        protected void onActivityResult(int requestCode, int resultCode, Intent data)
        {
            if (requestCode == REQ_TTS_STATUS_CHECK)
            {
                switch (resultCode)
                {
                case TextToSpeech.Engine.CHECK_VOICE_DATA_PASS:
                //这个返回结果表明 TTS Engine 可以用
                {
                    mTts = new TextToSpeech(this, this);
                    Log.v(TAG, "TTS Engine is installed!");
                }
                    break;
                case TextToSpeech.Engine.CHECK_VOICE_DATA_BAD_DATA:
                    //需要的语音数据已损坏
                case TextToSpeech.Engine.CHECK_VOICE_DATA_MISSING_DATA:
                    //缺少需要语言的语音数据
                case TextToSpeech.Engine.CHECK_VOICE_DATA_MISSING_VOLUME:
                //缺少需要语言的发音数据
                {
                    //这 3 种情况都表明数据有错,重新下载安装需要的数据
                    Log.v(TAG, "Need language stuff:" + resultCode);
```

```
                Intent dataIntent = new Intent();
                dataIntent.setAction(TextToSpeech.Engine.ACTION_INSTALL_TTS_DATA);
                startActivity(dataIntent);
            }
                break;
            case TextToSpeech.Engine.CHECK_VOICE_DATA_FAIL:
                //检查失败
            default:
                Log.v(TAG, "Got a failure. TTS apparently not available");
                break;
            }
        }
        else
        {
            //其他 Intent 返回的结果
        }
    }
    @Override
    protected void onPause()
    {
        //TODO Auto-generated method stub
        super.onPause();
        if (mTts != null)
        //activity 暂停时也停止 TTS
        {
            mTts.stop();
        }
    }
    @Override
    protected void onDestroy()
    {
        //TODO Auto-generated method stub
        super.onDestroy();
        //释放 TTS 的资源
        mTts.shutdown();
    }
}
```

(3) 右击工程 chapter14_7，在弹出的快捷菜单中选择 Run As|Android Application 命令，运行效果笔者在这里就不贴图展示了。

第 15 章 传感器

截止到最新版本的 Android SDK，系统已经内置了对多达 13 种传感器的支持，它们分别是：加速度传感器（accelerometer），环境温度传感器（ambient temperature），重力传感器（gravity），陀螺仪（gyroscope），环境光照传感器（light），线性加速传感器（linear accelerometer），磁力传感器（magnetic field），方向传感器（orientation），压力传感器（pressure），距离传感器（proximity），相对湿度传感器（relative humidity），旋转向量传感器（rotation vector）和温度传感器（temperature）。

15.1 传感器入门

在本小节当中笔者将会为读者介绍如何获取真机设备上的传感器的类别以及如何对特定传感器事件进行监听。

15.1.1 获取传感器类别

(1) Android 设备中所有的传感器都归传感器管理器 SensorManager 管理，获取传感器管理器的方法很简单，如以下代码：

```
//声明并实例化传感器管理器
SensorManager sensorManager = (SensorManager) getSystemService(Context.SENSOR_
SERVICE);
```

(2) Android SDK 各版本所支持的传感器类别如图 15－1 所示。

(3) 从传感器管理器获取其中某个或某些传感器的方法有如下 3 种：

❑ 获取某种传感器的默认传感器，如以下代码：

```
Sensor defaultGyroscope = sensorManager.getDefaultSensor(Sensor.TYPE_GYROSCOPE);
```

❑ 获取某种传感器的列表，如以下代码：

Sensor	Android 4.0 (API Level 14)	Android 2.3 (API Level 9)	Android 2.2 (API Level 8)	Android 1.5 (API Level 3)
TYPE_ACCELEROMETER	Yes	Yes	Yes	Yes
TYPE_AMBIENT_TEMPERATURE	Yes	n/a	n/a	n/a
TYPE_GRAVITY	Yes	Yes	n/a	n/a
TYPE_GYROSCOPE	Yes	Yes	n/a[1]	n/a[1]
TYPE_LIGHT	Yes	Yes	Yes	Yes
TYPE_LINEAR_ACCELERATION	Yes	Yes	n/a	n/a
TYPE_MAGNETIC_FIELD	Yes	Yes	Yes	Yes
TYPE_ORIENTATION	Yes[2]	Yes[2]	Yes[2]	Yes
TYPE_PRESSURE	Yes	Yes	n/a[1]	n/a[1]
TYPE_PROXIMITY	Yes	Yes	Yes	Yes
TYPE_RELATIVE_HUMIDITY	Yes	n/a	n/a	n/a
TYPE_ROTATION_VECTOR	Yes	Yes	n/a	n/a
TYPE_TEMPERATURE	Yes[2]	Yes	Yes	Yes

[1] This sensor type was added in Android 1.5 (API Level 3), but it was not available for use until Android 2.3 (API Level 9).

[2] This sensor is available, but it has been deprecated.

图 15-1 各 SDK 版本支持的传感器类别

```
List<Sensor> pressureSensors = sensorManager.getSensorList(Sensor.TYPE_PRESSURE);
```

- 获取所有传感器的列表，如以下代码：

```
List<Sensor> allSensors = sensorManager.getSensorList(Sensor.TYPE_ALL);
```

(4) 对于某一个具体的传感器，有以下几个比较常用的方法：

- getMaxiumRange()：获取最大取值范围。
- getName()：获取设备名称。
- getPower()：获取功率。
- getResolution()：获取精度。
- getType()：获取传感器类型。
- getVentor()：获取设备供应商。
- getVersion()：获取设备版本号。

下面通过实例介绍如何获取真机设备上的各种传感器的信息。

(5) 新建工程 chapter15_1，编辑 res/layout/activity_main.xml 文件，源代码参考配套资料。

(6) 编辑 MainActivity.java 文件，源代码如下：

```
public class MainActivity extends Activity
```

```
{
    TextView sensorInfoTextView;//声明显示传感器信息文本
    @Override
    public void onCreate(Bundle savedInstanceState)
    {
        super.onCreate(savedInstanceState);
        setContentView(R.layout.activity_main);
        sensorInfoTextView = (TextView) findViewById(R.id.sensor_info);
                                        //找到 id 为 sensor_info 的 TextView 控件
        //声明并实例化传感器管理器
        SensorManager sensorManager = (SensorManager) getSystemService(Context.SEN-
SOR_SERVICE);
        //获得全部的传感器列表
        List<Sensor> allSensors = sensorManager.getSensorList(Sensor.TYPE_ALL);
        StringBuffer infoBuffer = new StringBuffer();//声明显示文本
        infoBuffer.append("经检测该手机有" + allSensors.size() + "个传感器,它们
分别是:\n");
        //迭代传感器列表
        for(Sensor sensor : allSensors)
        {
            switch(sensor.getType())
            {
            case Sensor.TYPE_ACCELEROMETER:
                    infoBuffer.append(sensor.getType() + " 加速度传感器 acceler-
omter");
                break;
            case Sensor.TYPE_GYROSCOPE:
                infoBuffer.append(sensor.getType() + " 陀螺仪传感器 gyroscope");
                break;
            case Sensor.TYPE_LIGHT:
                infoBuffer.append(sensor.getType() + " 环境光线传感器 light");
                break;
            case Sensor.TYPE_MAGNETIC_FIELD:
                    infoBuffer.append(sensor.getType() + " 电磁场传感器 magnetic
field");
                break;
            case Sensor.TYPE_ORIENTATION:
                infoBuffer.append(sensor.getType() + " 方向传感器 orientation");
                break;
            case Sensor.TYPE_PRESSURE:
                infoBuffer.append(sensor.getType() + " 压力传感器 pressure");
                break;
```

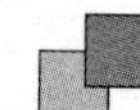

```
                case Sensor.TYPE_PROXIMITY:
                    infoBuffer.append(sensor.getType() + " 距离传感器 proximity");
                    break;
                case Sensor.TYPE_TEMPERATURE:
                    infoBuffer.append(sensor.getType() + " 温度传感器 temperature");
                    break;
                default:
                    infoBuffer.append(sensor.getType() + " android2.2 以下系统不支持
触感器");
                    break;
                }
                infoBuffer.append("\n" + "
设备名称:" + sensor.getName() + "\n" + " 设
备版本:" + sensor.getVersion() + "\n" + " 供
应商:" + sensor.getVendor() + "\n");
            }
            sensorInfoTextView.setText(in-
foBuffer);
        }
        @Override
        public boolean onCreateOptionsMenu
(Menu menu)
        {
            getMenuInflater().inflate(R.men-
u.activity_main, menu);
            return true;
        }
    }
```

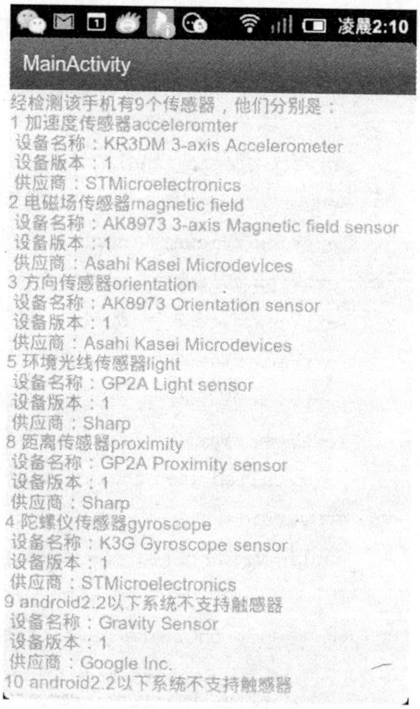

图 15 - 2　chapter15_1 运行效果

(7) 右击工程 chapter15_1,在弹出的快捷菜单中选择 Run As|Android Application 命令,运行效果如图 15 - 2 所示。

15.1.2　监听传感器事件

程序当中如果想要对特定的传感器事件进行监听,则需要去实现 android.hardware.SensorListener(或者是 android.hardware.SensorEventListener)这个接口。SensorListener 接口包含以下两个方法:

- onSensorChanged(int sensor,float values[]) 方法在传感器值更改时调用。该方法只对受此应用程序监视的传感器调用。该方法的参数包括:一个整数,指示更改的传感器;一个浮点值数组,表示传感器数据本身。有些传感器只提供一个数据值,另一些则提供 3 个浮点值。方向和加速表传感器都提供

3 个数据值。

- 当传感器的精度值更改时，将调用 onAccuracyChanged(int sensor,int accuracy) 方法。参数包括两个整数：一个表示传感器，另一个表示该传感器新的精度值。

下面就通过实例演示如何在 Android 应用程序当中监听加速度传感器以及方向传感器的数值变化。

(1) 新建工程 chapter15_2，编辑 res/layout/activity_main. xml 文件，源代码参考配套资料。

(2) 编辑 MainActivity. java 文件，源代码如下：

```
public class MainActivity extends Activity implements SensorListener
{
    private static final String TAG = "MainActivity";
    SensorManager sm = null;//声明传感器管理器
    TextView xViewA = null;//声明显示加速度传感器 X 轴值的文本
    TextView yViewA = null;//声明显示加速度传感器 Y 轴值的文本
    TextView zViewA = null;//声明显示加速度传感器 Z 轴值的文本
    TextView xViewO = null;//声明显示方向传感器 X 轴值的文本
    TextView yViewO = null;//声明显示方向传感器 Y 轴值的文本
    TextView zViewO = null;//声明显示方向传感器 Z 轴值的文本
    /** Called when the activity is first created. */
    @Override
    public void onCreate(Bundle savedInstanceState)
    {
        super.onCreate(savedInstanceState);
        sm = (SensorManager) getSystemService(SENSOR_SERVICE);
                                                        //实例化传感器管理器
        setContentView(R.layout.activity_main);
        xViewA = (TextView) findViewById(R.id.x_accelerometer_value);
        yViewA = (TextView) findViewById(R.id.y_accelerometer_value);
        zViewA = (TextView) findViewById(R.id.z_accelerometer_value);
        xViewO = (TextView) findViewById(R.id.x_orientation_value);
        yViewO = (TextView) findViewById(R.id.y_orientation_value);
        zViewO = (TextView) findViewById(R.id.z_orientation_value);
    }
    //传感器值发生变化时调用
    public void onSensorChanged(int sensor, float[] values)
    {
        synchronized (this) //同步代码段
          {
            Log.d(TAG, "onSensorChanged: " + sensor + ", x: " + values[0] + ", y: "
```

```
+ values[1] + ", z: " + values[2]);
                if (sensor == SensorManager.SENSOR_ORIENTATION)
                {  //方向传感器
                    xViewO.setText("Orientation X: " + values[0]);
                    yViewO.setText("Orientation Y: " + values[1]);
                    zViewO.setText("Orientation Z: " + values[2]);
                }
                if (sensor == SensorManager.SENSOR_ACCELEROMETER)
                {//加速度传感器
                    xViewA.setText("Accel X: " + values[0]);
                    yViewA.setText("Accel Y: " + values[1]);
                    zViewA.setText("Accel Z: " + values[2]);
                }
            }
        }
        //传感器准确性更改时调用
        public void onAccuracyChanged(int sensor, int accuracy)
        {
        Log.d(TAG,"onAccuracyChanged: " + sensor + ", accuracy: " + accuracy);
        }
        @Override
        protected void onResume()
        {
            super.onResume();
             sm.registerListener(this, SensorManager.SENSOR_ORIENTATION SensorManager.
SENSOR_ACCELEROMETER, SensorManager.SENSOR_DELAY_NORMAL);
        }
        @Override
        protected void onStop()
        {
            sm.unregisterListener(this);
            super.onStop();
        }
        @Override
        public boolean onCreateOptionsMenu(Menu menu)
        {
            getMenuInflater().inflate(R.menu.activity_main, menu);
            return true;
        }
    }
```

上述代码当中在 MainActivity 活动的 onResume()回调方法当中为系统的加速

度传感器以及方向传感器(这两种传感器也是项目当中比较常用的两种传感器类型)注册了监听器。当系统的加速度传感器或者方向传感器的值发生变化时,就会调用 MainActivity 当中的 onSensorChanged(int sensor, float[] values)这个方法;当系统的加速度传感器或者是方向传感器的精度值发生变化时,就会调用 MainActivity 活动的 onAccuracyChanged(int sensor, int accuracy)这个方法。

上述代码当中不关注传感器精度的变化,但关注传感器当前的 X、Y 和 Z 值。onAccuracyChanged 方法实质上不执行任何操作,它只在每次调用时添加一个日志项。

onSensorChanged(int sensor, float[] values)方法会被频繁地调用。方法的第一个参数确定哪个传感器的值发生了变化,确定了哪个传感器的值正在发生变化过后,就可以使用方法的第二个参数传递的浮点数数组中所包含的数据去更新相应的 UI 界面元素。

上述代码在 MainActivity 活动的 onStop()方法当中取消了对系统的加速度传感器以及方向传感器的注册。

图 15-3　chapter15_2 运行效果

(3) 右击工程 chapter15_2,在弹出的快捷菜单中选择 Run As|Android Application 命令,运行效果如图 15-3 所示。

15.2　仿微信摇一摇功能

微信是腾讯公司于 2011 年 1 月 21 日推出的一款通过网络快速发送语音短信、视频、图片和文字,支持多人群聊的手机聊天软件。用户可以通过微信与好友进行形式上更加丰富的类似于短信、彩信等方式的联系。微信软件本身完全免费,使用任何功能都不会收取费用,微信时产生的上网流量费由网络运营商收取。

微信自从 3.0 版本开始,加入"摇一摇"功能,支持通过摇一摇手机找到附近同时也在摇手机的朋友。进入"摇一摇"界面,轻摇手机,微信会帮您搜寻同一时刻摇晃手机的用户。聚会上一起摇,会快速帮您列出一起摇的朋友;千里摇一摇,可以为您匹配这个世界上同时也在摇手机的朋友。摇到的朋友,直接单击就可以开始聊天。

实现这个功能的核心就是手机设备里面集成的加速度传感器。它能够分别测量 X、Y、Z 三个方向的加速度值,X 方向值的大小代表手机水平方向的加速度值,Y 方向值的大小代表垂直于手机水平方向的加速度的值,Z 方向值的大小代表手机空间垂直方向的加速度的值,天空的方向为正,地球的方向为负。通过判断这三个方向的加速度的值,就能判断用户是否在"摇一摇"手机设备了。微信"摇一摇"功能截图如图 15-4 和图 15-5 所示。

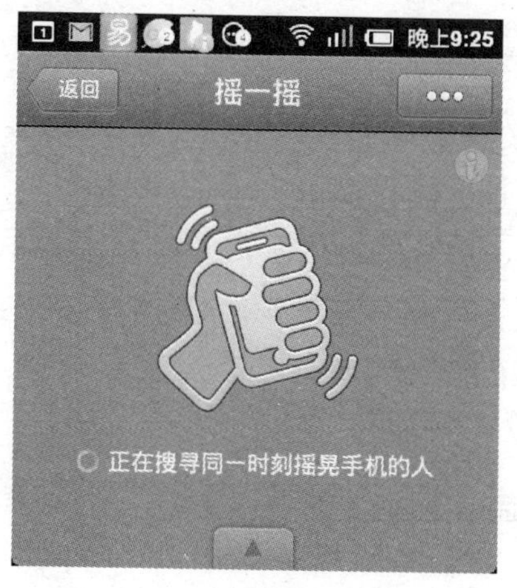

图 15-4　微信摇一摇功能截图

图 15-5　微信摇一摇功能截图 2

下面就通过利用 Android 真机设备内置的传感器的支持来完成仿微信“摇一摇”的功能，当应用检测到用户正在“摇一摇”手机设备的时候，使手机设备振动。

(1) 新建工程 chapter15_3，编辑 res/layout/activity_main. xml 文件，源代码参考配套资料。

(2) 由于当检测到用户正在“摇一摇”的时候，需要使手机设备振动起来。所以需要添加相应的权限。编辑 AndroidManifest. xml 文件，源代码如下：

```
<manifest xmlns:android = "http://schemas.android.com/apk/res/android"
    package = "com.example.chapter15_3"
    android:versionCode = "1"
    android:versionName = "1.0" >
    <!-- 震动权限 -->
    <uses-permission  android:name = "android.permission.VIBRATE"/>
    <uses-sdk   android:minSdkVersion = "8"    android:targetSdkVersion = "15" />
    <application
        android:icon = "@drawable/ic_launcher"
        android:label = "@string/app_name"
        android:theme = "@style/AppTheme" >
        <activity
            android:name = ".MainActivity"
            android:label = "@string/title_activity_main" >
            <intent-filter>
                <action android:name = "android.intent.action.MAIN" />
                <category android:name = "android.intent.category.LAUNCHER" />
            </intent-filter>
        </activity>
```

```
    </application>
</manifest>
```

(3) 编辑 MainActivity.java 文件,源代码如下:

```
public class MainActivity extends Activity implements SensorEventListener
{
    Button myButton;//声明复原文本按钮
    TextView statusTextView;//摇动发生时,显示文本
    SensorManager sensorManager = null;//传感器管理器
    Vibrator vibrator = null;     //震动管理器
    @Override
    public void onCreate(Bundle savedInstanceState)
    {
        super.onCreate(savedInstanceState);
        setContentView(R.layout.activity_main);
        myButton = (Button)findViewById(R.id.my_button);
                                                    //找到 id 为 my_button 的按钮
        statusTextView = (TextView)findViewById(R.id.show_status);
                                                    //找到 id 为 show_status 的文本
        myButton.setOnClickListener(new OnClickListener()
        {//为按钮绑定单击事件监听器
            @Override
            public void onClick(View v)
            {
                statusTextView.setText("");//文本清空
            }
        });
        sensorManager = (SensorManager)getSystemService(Context.SENSOR_SERVICE);
                                                    //实例化传感器管理器
        vibrator = (Vibrator)getSystemService(Context.VIBRATOR_SERVICE);
                                                    //实例化振动管理器
    }
    //传感器值发生变化时调用
    @Override
    public void onSensorChanged(SensorEvent event)
    {
        int sensorType = event.sensor.getType();
          //values[0]:X 轴,values[1]:Y 轴,values[2]:Z 轴
          float[] values = event.values;
          if(sensorType == Sensor.TYPE_ACCELEROMETER)
          { //加速度传感器
              if((Math.abs(values[0])>17||Math.abs(values[1])>17||Math.abs(val-
```

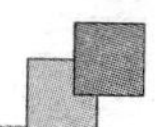

```
ues[2])>17))
                {
                    Log.d("sensor ", "============ values[0] = " + values[0]);
                    Log.d("sensor ", "============ values[1] = " + values[1]);
                    Log.d("sensor ", "============ values[2] = " + values[2]);
                    statusTextView.setText("摇一摇");
                    //摇动手机后,再伴随震动提示~~
                    vibrator.vibrate(500);
                }
            }
        }
        //在传感器精度发生变化时调用
        @Override
        public void onAccuracyChanged(Sensor sensor, int accuracy)
        {
        //do nothing
        }
        @Override
        protected void onResume()
        {
            super.onResume();
            sensorManager.registerListener(this, sensorManager.getDefaultSensor(Sensor.
TYPE_ACCELEROMETER), SensorManager.SENSOR_DELAY_NORMAL);
        }
        @Override
        protected void onPause()
        {
            super.onPause();
            sensorManager.unregisterListener(this);
        }
        @Override
        public boolean onCreateOptionsMenu(Menu menu)
        {
            getMenuInflater().inflate(R.menu.activity_main, menu);
            return true;
        }
    }
```

上述代码当中在 MainActivity 活动的 onResume()回调方法当中为系统的加速度传感器注册了监听器。当加速度传感器的 X、Y 以及 Z 轴的加速度都达到 17 时，笔者就认为用户正在“摇一摇”手机，并更新相应的文本控件的内容。当然这个临近值可以根据开发者的需要去自定义。

(4) 右击工程 chapter15_3，在弹出的快捷菜单中选择 Run As | Android Application 命令，“摇一摇”手机。当程序检测到用户正在“摇一摇”手机的时候，程序会使设备振动起来，并且更新相关文本内容，运行效果如图 15－6 所示。

图 15－6　chapter15_3 运行效果

15.3　方向传感器

方向传感器用于感应手机设备的摆放状态。方向传感器可以返回 3 个角度，这 3 个角度即可确定手机的摆放状态。这 3 个角度分别是：方位角、倾斜角以及旋转角。要想弄清楚这 3 个角度代表的意思，首先要来看看 Android 设备传感器的坐标系。如图 15－7 所示。

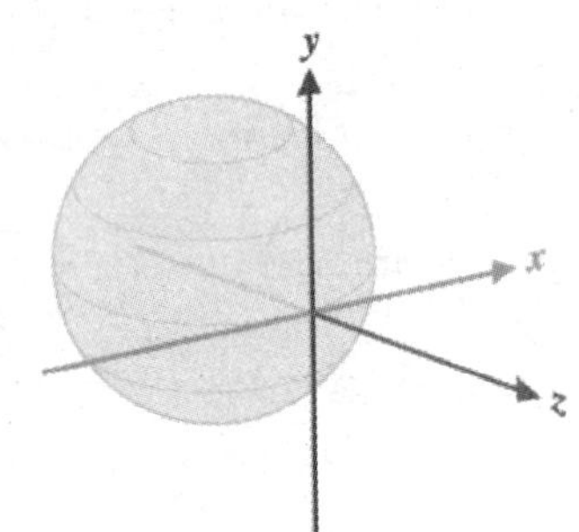

图 15－7　Android 传感器坐标

其中：

- X 方向就是手机的水平方向，右为正。
- Y 方向就是手机的水平垂直方向，前为正。
- Z 方向手机空间的垂直方向，天空的方向为正，地球的方向为负。

那么：

- 方位角的定义是手机 Y 轴水平面上的投影与正北方向的夹角。(取值范围是：0°～359°，其中 0°＝North，90°＝East，180°＝South，270°＝West)
- 倾斜角的定义是手机 X 轴与水平面的夹角。(手机 Z 轴向 Y 轴方向移动为正，取值范围是－180°～180°)
- 旋转角的定义是手机 Y 轴与水平面的夹角。(手机 X 轴离开 Z 轴方向为正，取值范围是－90°～90°)

下面笔者就通过测试真机设备内置的方向传感器的支持来判断手机的方向。

(1) 新建工程 chapter15_4，编辑 res/layout/activity_main.xml 文件，源代码参考配套资料。

(2) 编辑 MainActivity.java 文件，源代码如下：

```
public class MainActivity extends Activity
{
    TextView orientationTextView;//声明显示方向文本
    SensorManager sensorManager;//声明传感器管理器
    Sensor orientaionSensor;//声明方向传感器
```

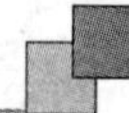

```
private float decDegree = 0;
private String message = "正北 0°";
private SensorEventListener orientationListener = new SensorEventListener()
{
    //当传感器值发生变化时调用
    @Override
    public void onSensorChanged(SensorEvent sensorEvent)
    {
        if (sensorEvent.sensor.getType() == Sensor.TYPE_ORIENTATION)
         {
            float x = sensorEvent.values[SensorManager.DATA_X];//方位角
            //设置灵敏度
            if(Math.abs(decDegree - x) >= 2 )
            {
                decDegree = x;
                int range = 22;
                String degreeStr = String.valueOf(decDegree);
                //指向正北
                if(decDegree > 360 - range && decDegree < 360 + range)
                {
                    message = "正北 " + degreeStr + "°";
                }
                //指向正东
                if(decDegree > 90 - range && decDegree < 90 + range)
                {
                    message = "正东 " + degreeStr + "°";
                }
                //指向正南
                if(decDegree > 180 - range && decDegree < 180 + range)
                {
                    message = "正南 " + degreeStr + "°";
                }
                //指向正西
                if(decDegree > 270 - range && decDegree < 270 + range)
                {
                    message = "正西 " + degreeStr + "°";
                }
                //指向东北
                if(decDegree > 45 - range && decDegree < 45 + range)
                {
                    message = "东北 " + degreeStr + "°";
                }
                //指向东南
```

```
                    if(decDegree > 135 - range && decDegree < 135 + range)
                    {
                        message = "东南 " + degreeStr + "°";
                    }
                    //指向西南
                    if(decDegree > 225 - range && decDegree < 225 + range)
                    {
                        message = "西南 " + degreeStr + "°";
                    }
                    //指向西北
                    if(decDegree > 315 - range && decDegree < 315 + range)
                    {
                        message = "西北 " + degreeStr + "°";
                    }
                    orientationTextView.setText(message);
                    }
                }
            }
            //当传感器精度变化时调用
            @Override
            public void onAccuracyChanged(Sensor sensor, int accuracy)
            {
                //do nothing
            }
        };

    @Override
    public void onCreate(Bundle savedInstanceState)
      {
        super.onCreate(savedInstanceState);
        setContentView(R.layout.activity_main);
        orientationTextView = (TextView)findViewById(R.id.show_orientation);
                                        //找到 id 为 show_orientation 的 TextView 控件
        sensorManager = (SensorManager)getSystemService(Context.SENSOR_SERVICE);
                                                                //实例化传感器管理器
         List<android.hardware.Sensor> sensorList = sensorManager.getSensorList
(Sensor.TYPE_ORIENTATION);
         if (sensorList.size() > 0)
         {
           orientaionSensor = sensorList.get(0);
         }
         else
         {
```

```
            orientationTextView.setText("Orientation sensor not present");
        }
    }
    @Override
    protected void onResume()
    {
    super.onResume();
    //注册对方向传感器的监听
     sensorManager.registerListener(orientationListener, orientaionSensor, Sensor-
Manager.SENSOR_DELAY_NORMAL);
    }
    @Override
    protected void onStop()
    {
    super.onStop();
    sensorManager.unregisterListener(orientationListener);
    }
}
```

上述程序代码在 MainActivity 活动的 onResume()方法当中，为真机设备的方向传感器注册了监听器。这个监听器是以内部类的方式存在的。当方向传感器的值发生变化的时候，此内部类当中的 onSensorChanged()方法就会被回调。

通过对方位角的度数的判断，就可以判断出手机设备当前的方向是怎么样的。

上述程序代码在 MainActivity 活动的 onStop()方法当中，为真机设备的方向传感器取消了监听器。

(3) 右击工程 chapter15_4，在弹出的快捷菜单中选择 Run As|Android Application 命令，运行效果如图 15－8 所示。

图 15－8　chapter15_4 运行效果

第 16 章

Android 图形和图像

Android 处理图形的能力非常强大。在前面的章节当中，笔者已经为读者介绍过如何使用 ImageView 控件来显示单张图片。在本章当中，笔者将会为读者介绍如何显示一组图片（图片浏览器），除此之外，笔者还会为读者介绍使用简单图片的几种方式、如何进行 2D 绘图以及 Android 动画的相关知识。

16.1 图片浏览器

在本节当中，笔者将会通过两种方式来实现图片浏览器的功能，分别是单独使用 Gallery 控件的方式以及使用 Gallery 控件和 ImageSwitcher 控件组合的方式。

16.1.1 Gallery

Android 的 Gallery 控件是一个水平的列表选择框，一般用来展示一组图片。Gallery 控件的水平列表可以让用户以滑动的方式来切换列表项，所以用户体验也比较好。下面就通过实例演示如何通过使用 Gallery 控件来实现图片浏览器的功能。

（1）新建工程 chapter16_1，编辑 res/layout/activity_main.xml 文件，源代码参考配套资料。

（2）将事先准备好的几张图片复制到 res/drawable-hdpi/子目录当中。

（3）编辑 MainActivity.java 文件，源代码如下：

```
public class MainActivity extends Activity
{
    private static final String TAG = "MainActivity";
    private Gallery mGallery;  //声明 Gallery 控件
    @Override
    public void onCreate(Bundle savedInstanceState)
```

```
{
    super.onCreate(savedInstanceState);
    setContentView(R.layout.activity_main);
    mGallery = (Gallery)findViewById(R.id.gallery);//找到 id 为 gallery 的控件
    try
    {
        mGallery.setAdapter(new ImageAdapter(this));
    }
    catch (Exception e)
    {
        Log.e(TAG, "set adpater occurs errors:" + e.toString());
    }
    //匿名内部类形式绑定 Gallery 控件的各选项单击事件监听器
    mGallery.setOnItemClickListener(new OnItemClickListener()
    {
    //各选项单击时回调
        public void onItemClick(AdapterView parent, View v, int position, long id)
        {
          MainActivity.this.setTitle(String.valueOf(position));
                                                        //设置应用程序标题
        }
    });
}
/*
 * class ImageAdapter is used to control gallery source and operation.
 */
private class ImageAdapter extends BaseAdapter
{
    private Context mContext;
    private ArrayList<Integer> imgList = new ArrayList<Integer>();
    private ArrayList<Object> imgSizes = new ArrayList<Object>();
    public ImageAdapter(Context c) throws IllegalArgumentException, IllegalAc-
cessException
      {
        mContext = c;
        //用反射机制来获取资源中的图片 ID 和尺寸
      Field[] fields = R.drawable.class.getDeclaredFields();
      for (Field field : fields)//迭代 fields 数组
      {
          if (!"ic_action_search".equals(field.getName()) && ! "ic_launcher".e-
quals(field.getName()))              //除了 ic_action_search 和 ic_launcher 之外的图片
          {
```

```
                int index = field.getInt(R.drawable.class);
                //保存图片 ID
                imgList.add(index);
                //保存图片大小
                int size[] = new int[2];
                Bitmap bmImg = BitmapFactory.decodeResource(getResources(),index);
                size[0] = bmImg.getWidth();
                size[1] = bmImg.getHeight();
                imgSizes.add(size);
            }
        }
    }
    @Override
    public int getCount()
    {
        return imgList.size();
    }
    @Override
    public Object getItem(int position)
    {
        return position;
    }
    @Override
    public long getItemId(int position)
    {
        return position;
    }
    @Override
    public View getView(int position, View convertView, ViewGroup parent)
    {
      //加入 Adapter 优化策略
        ImageView imageView = null;
        if(convertView == null)
        {//如果可复用旧视图为 null
          imageView = new ImageView(mContext);
          //从 imgList 取得图片 ID
          imageView.setImageResource(imgList.get(position).intValue());
          imageView.setScaleType(ImageView.ScaleType.FIT_XY);
          //从 imgSizes 取得图片大小
          int size[] = new int[2];
          size = (int[]) imgSizes.get(position);
          imageView.setLayoutParams(new Gallery.LayoutParams(size[0], size[1]));
```

```
            }
            else
            {   //如果可复用旧视图不为 null
                imageView = (ImageView)convertView;
            }
            return imageView;
        }
    };
}
```

填充 Gallery 控件数据的方式也是通过调用相应 Gallery 对象的 setAdapter()方法来实现的。所以在上述代码当中，定义了一个继承自 android. widget. BaseAdapter 父类的 ImageAdapter 子类，并实现了相应需要实现的方法。

(4) 右击工程 chapter16_1，在弹出的快捷菜单中选择 Run As|Android Application 命令，运行效果如图 16－1 所示，向右滑动 Gallery 控件效果如图 16－2 所示。

图 16－1 chapter16－1 运行效果

图 16－2 从右向左滑动 Gallery 的效果

16.1.2 ImageSwitcher

ImageSwitcher 控件继承于 android. widget. FrameLayout 父类。ImageSwitcher 控件的功能与 ImageView 控件的功能类似，都可以显示单张图片。不过 ImageSwicher 控件在切换图片显示的时候可以添加相应的动画效果。

下面就通过实例介绍如何通过 Gallery 控件结合 ImageSwitcher 控件来完成一个用户体验更好的图片浏览器。

(1) 新建工程 chapter16_2，编辑 res/layout/activity_main. xml 文件，源代码参考配套资料。

(2) 将事先准备好的几张图片复制到 res/drawable-hdpi/子目录当中。

(3) 编辑 MainActivity. java 文件，源代码如下：

```
public class MainActivity extends Activity implements ViewFactory
{
    private static final String TAG = "MainActivity";
```

```
private ImageSwitcher is; //声明图像切换控件
private Gallery gallery; //声明 Gallery 控件
private int downX, upX;
private ArrayList<Integer> imgList = new ArrayList<Integer>();//图像 ID
@Override
protected void onCreate(Bundle savedInstanceState)
  {
    //TODO Auto-generated method stub
    super.onCreate(savedInstanceState);
    setContentView(R.layout.activity_main);
    //用反射机制来获取资源中的图片 ID
    Field[] fields = R.drawable.class.getDeclaredFields();
    for (Field field : fields)
    { //迭代 fields 数组
        if (!"ic_action_search".equals(field.getName()) && !"ic_launcher".e-
quals(field.getName()))  //除了 ic_action_search 和 ic_launcher 之外的图片
        {
            int index = 0;
            try
            {
                index = field.getInt(R.drawable.class);
            }
            catch (Exception e)
            {
                Log.e(TAG, "occurs errors:" + e.getMessage());
            }
            //保存图片 ID
            imgList.add(index);
        }
    }
    //设置 ImageSwitcher 控件
    is = (ImageSwitcher) findViewById(R.id.switcher);
    is.setFactory(this);//显示 makeView()返回的 ImageView 控件
    is.setInAnimation(AnimationUtils.loadAnimation(this, android.R.anim.fade_
in));//设置切入动画
    is.setOutAnimation(AnimationUtils.loadAnimation(this, android.R.anim.fade_
out));//设置切出动画
    //为 ImageSwitcher 控件添加触摸事件监听器
    is.setOnTouchListener(new OnTouchListener()
    {
        /*
         * 在 ImageSwitcher 控件上滑动可以切换图片
```

```
 */
@Override
public boolean onTouch(View v, MotionEvent event)
{
    if (event.getAction() == MotionEvent.ACTION_DOWN)
    {//如果为触摸按下事件
        downX = (int) event.getX();//取得按下时的坐标
        return true;
    }
    else if (event.getAction() == MotionEvent.ACTION_UP)
    {//如果为触摸释放事件
        upX = (int) event.getX();//取得松开时的坐标
        int index = 0;
        if (upX - downX > 100)//从左拖到右,即看前一张
        {
            //如果当前是第一张,则从左向右拖,回到最后一张
            if (gallery.getSelectedItemPosition() == 0)
            {
                index = gallery.getCount() - 1;
            }
            else
            {
                index = gallery.getSelectedItemPosition() - 1;
            }
        } else if (downX - upX > 100)//从右拖到左,即看后一张
        {
            //如果当前是最后一张,从右向左拖,回到第一张
             if (gallery.getSelectedItemPosition() == (gallery.get-
Count() - 1))

            {
                index = 0;
            }
            else
            {
                index = gallery.getSelectedItemPosition() + 1;
            }
        }
        //改变 gallery 图片所选,自动触发 Gallery 控件的
        //setOnItemSelectedListener 的回调 onItemSelected 方法
        gallery.setSelection(index, true);
        return true;
    }
    return false;
}
```

```
            });
            //设置 gallery 控件
            gallery = (Gallery) findViewById(R.id.gallery);
            gallery.setAdapter(new ImageAdapter(this));//绑定 Gallery 数据源
            //绑定 Gallery 选项选中监听器
            gallery.setOnItemSelectedListener(new OnItemSelectedListener()
              {
                @Override
                 public void onItemSelected(AdapterView<?> arg0, View arg1, int posi-
tion, long arg3)
                {
                    is.setImageResource(imgList.get(position));
                }
                @Override
                public void onNothingSelected(AdapterView<?> arg0)
                {
                    //TODO Auto-generated method stub
                }
            });
        }
        //设置 ImgaeSwitcher
        @Override
        public View makeView()
        {
            ImageView i = new ImageView(this);
            i.setBackgroundColor(0xFF000000);
            i.setScaleType(ImageView.ScaleType.CENTER);//居中
            i.setLayoutParams(new ImageSwitcher.LayoutParams(//自适应图片大小
                LayoutParams.FILL_PARENT, LayoutParams.FILL_PARENT));
            return i;
        }
        public class ImageAdapter extends BaseAdapter
        {
            private Context mContext;
            public ImageAdapter(Context c)
            {
                mContext = c;
            }
            public int getCount()
            {
                return imgList.size();
            }
            public Object getItem(int position)
            {
```

```
            return position;
        }
        public long getItemId(int position)
        {
            return position;
        }
        public View getView(int position, View convertView, ViewGroup parent)
        {
            ImageView imageView = null;
              if (convertView == null)
                {//如果可复用旧视图对象为 null
                imageView = new ImageView(mContext);
                imageView.setImageResource(imgList.get(position));
                imageView.setAdjustViewBounds(true);
                 imageView.setLayoutParams(new Gallery.LayoutParams(LayoutParams.
WRAP_CONTENT, LayoutParams.WRAP_CONTENT));
              }
              else
              {  //如果可复用旧视图对象不为 null
                 imageView = (ImageView) convertView;
              }
              return imageView;
        }
    }
}
```

(4) 右击工程 chapter16_2,在弹出的快捷菜单中选择 Run As|Android Application 命令,运行效果如图 16 - 3 所示,从右向左方向触摸 ImageSwitcher 控件,运行效果如图 16 - 4 所示。

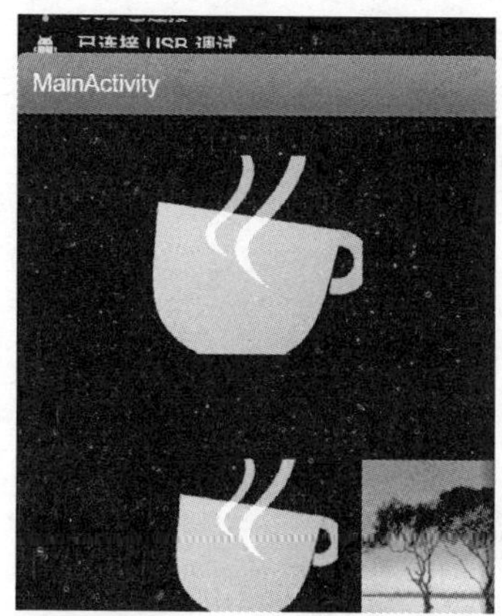

图 16 - 3　chapter16 - 2 运行效果

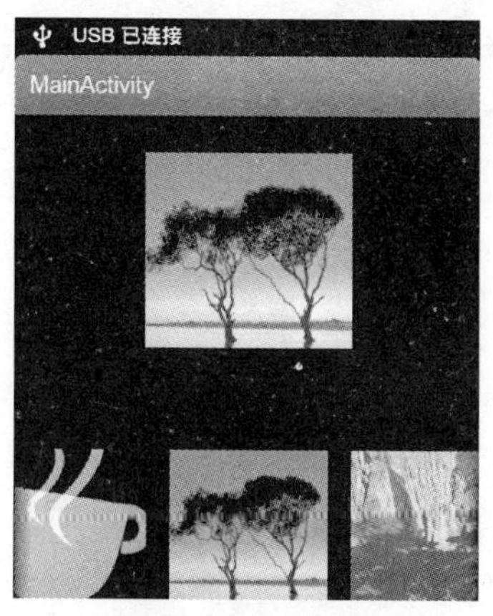

图 16 - 4　从右向左滑动 ImageSwitcher 运行效果

16.2 访问图片

Android 对于静态图形的处理，也就是不经常变化的图片，如 Icon 和 Logo 等，一般是通过各种 Drawable 类来进行处理。在本小节当中，笔者还会为读者介绍 Bitmap 类和 BitmapFactory 类的使用，以及如何避免在使用图片的时候，出现内存溢出的情况。

16.2.1 Drawable

Android 访问图片的时候，使用 Drawable 类及其子类 BitmapDrawable、LayerDrawable、ShapeDrawable 等类处理。下面笔者分别通过使用 Java 代码的方式以及 XML 文件引用的方式来介绍如何使用 Drawable 对象访问图片。

1. Java 代码

(1) 新建工程 chapter16_3，将事先准备好的 images.jpg 文件复制到 res/drawable-hdpi/子目录下。

(2) 编辑 MainActivity.java 文件，源代码如下：

```
public class MainActivity extends Activity
{
    @Override
    public void onCreate(Bundle savedInstanceState)
    {
        super.onCreate(savedInstanceState);
        LinearLayout mLinearLayout = new LinearLayout(this);//界面整体线性布局
        ImageView mImageView = new ImageView(this);
        //通过 R.drawable.XXX 的方式在 Java 代码当中引用图片资源文件
        mImageView.setImageResource(R.drawable.images);//设置 ImageView 显示的资源文件
        mImageView.setLayoutParams(new Gallery.LayoutParams(LayoutParams.WRAP_CON-
TENT, LayoutParams.WRAP_CONTENT));
        mLinearLayout.addView(mImageView);  //将 ImageView 控件加入到整体线性布局当中
        setContentView(mLinearLayout);
    }
    @Override
    public boolean onCreateOptionsMenu(Menu menu)
    {
        getMenuInflater().inflate(R.menu.activity_main, menu);
        return true;
    }
}
```

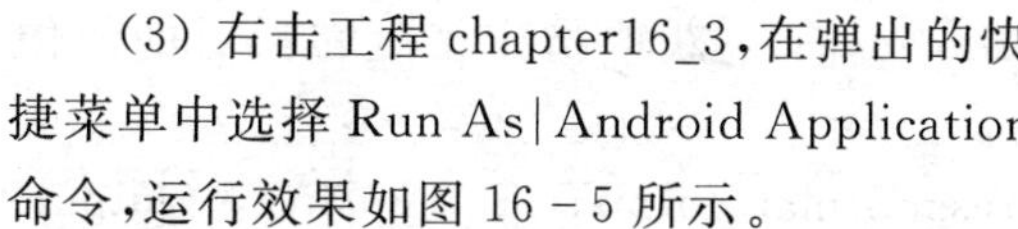

(3) 右击工程 chapter16_3,在弹出的快捷菜单中选择 Run As|Android Application 命令,运行效果如图 16-5 所示。

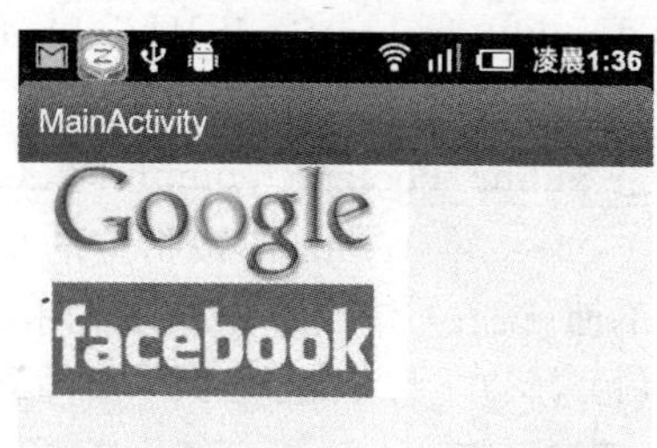

图 16-5　chapter16_3 运行效果

2. XML 文件引用

(1) 新建工程 chapter16_4,将事先准备好的 images.jpg 文件复制到 res/drawable-hdpi/子目录下。

(2) 编辑 res/layout/activity_main.xml 文件,源代码如下:

```
<LinearLayout xmlns:android = "http://schemas.android.com/apk/res/android"
    xmlns:tools = "http://schemas.android.com/tools"
    android:layout_width = "fill_parent"
    android:layout_height = "fill_parent"
    android:orientation = "vertical">
    <!-- 通过@drawable 的方式来引用图片资源文件 -->
    <ImageView
        android:layout_width = "wrap_content"
        android:layout_height = "wrap_content"
        android:src = "@drawable/images" />
</LinearLayout>
```

(3) 工程 chapter16_4 与工程 chapter16_3 运行效果一致。

16.2.2　Bitmap 和 BitmapFactory

Bitmap 是 Android 系统中图像处理的最重要类之一。用它可以获取图像文件信息,进行图像剪裁、旋转、缩放等操作,并可以指定格式保存图像文件。Bitmap 位于 android.graphics 包中。由于 Bitmap 类的构造函数是私有的,所以在类的外面并不能对其实例化。

利用 BitmapFactory 可以从一个指定文件中,调用 decodeFile()方法返回 Bitmap 对象;也可以调用 decodeResource()从工程的资源文件中返回 Bitmap 对象。

Bitmap 类有以下常用方法:

- public void recycle():回收位图占用的内存空间。
- public final boolean isRecycled():判断位图内存是否已释放。
- public final int getWidth():获取位图的宽度。
- public final int getHeight():获取位图的高度。
- public final boolean isMutable():判断图片是否可修改。
- public int getScaledWidth(Canvas canvas):获取指定密度转换后的图像宽度。

- public int getScaledHeight(Canvas canvas)：获取指定密度转换后的图像高度。
- public boolean compress (CompressFormat format, int quality, OutputStream stream)：按指定的图片格式以及画质,将图片转换为输出流。

下面就通过实例演示如何使用 Bitmap 类以及 BitmapFactory 类来访问图片。

(1) 新建工程 chapter16_5,将事先准备好的 qq. jpg 文件复制到 res/drawable-hdpi/子目录下。

(2) 编辑 MainActivity. java 文件,源代码如下：

```
public class MainActivity extends Activity
{
    @Override
    public void onCreate(Bundle savedInstanceState)
    {
        super.onCreate(savedInstanceState);
        setContentView(R.layout.activity_main);
        //读取一个资源文件得到一个位图
        Bitmap bm = BitmapFactory.decodeResource(getResources(), R.drawable.qq);
        ImageView imageView = new ImageView(this);
        imageView.setImageBitmap(bm);
        this.setContentView(imageView);
    }
    @Override
    public boolean onCreateOptionsMenu(Menu menu)
    {
        getMenuInflater().inflate(R.menu.activi-
ty_main, menu);
        return true;
    }
}
```

(3) 右击工程 chapter16_5,在弹出的快捷菜单中选择 Run As|Android Application 命令,运行效果如图 16-6 所示。

图 16-6　chapter16_5 运行效果

16.3 内存优化

由于移动设备内存的大小限制,使得在开发 Android 应用程序的时候,内存的优化始终是贯穿于整个开发过程的一条线索。下面就通过比较 Drawable 对象与 Bitmap 对象占用内存的大小以及比较 BitmapFactory 类的 decodeResource 方法与 decodeStream 方法的效率来进行一些简单的 Android 内存优化的测试。

16.3.1　Drawable 与 Bitmap 占用内存比较

下面通过实例比较 Drawable 对象与 Bitmap 对象占用内存的大小。

(1) 新建工程 chapter16_6，将事先准备好的 qq.jpg 文件复制到 res/drawable-hdpi/子目录下。

(2) 编辑 MainActivity.java 文件，源代码如下：

```
public class MainActivity extends Activity
{
    private static final String TAG = "MainActivity";
    int number = 1000;
    Drawable[] array ;
    @Override
    public void onCreate(Bundle savedInstanceState)
    {
        super.onCreate(savedInstanceState);
        setContentView(R.layout.activity_main);
        array = new BitmapDrawable[number];
        for(int i = 0; i < number; i++)//初始化 array 数组
        {
            Log.d(TAG, "测试第" + (i+1) + "张图片");
            array[i] = getResources().getDrawable(R.drawable.qq);
        }
    }
    @Override
    public boolean onCreateOptionsMenu(Menu menu)
    {
        getMenuInflater().inflate(R.menu.activity_main, menu);
        return true;
    }
}
```

(3) 右击工程 chapter16_6，在弹出的快捷菜单中选择 Run As|Android Application 命令，Logcat 控制台输出效果如图 16-7 所示。

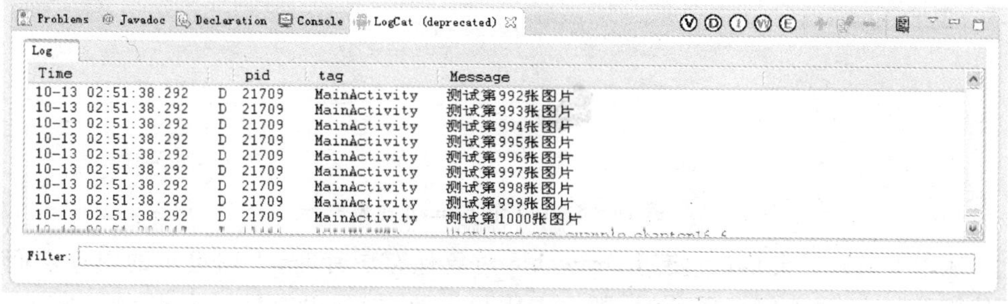

Problems
Javadoc
Declaration
Console
LogCat (deprecated)
Log
Time
pid
tag
Message
10-13 02:51:38.292 D 21709 MainActivity 测试第992张图片
10-13 02:51:38.292 D 21709 MainActivity 测试第993张图片
10-13 02:51:38.292 D 21709 MainActivity 测试第994张图片
10-13 02:51:38.292 D 21709 MainActivity 测试第995张图片
10-13 02:51:38.292 D 21709 MainActivity 测试第996张图片
10-13 02:51:38.292 D 21709 MainActivity 测试第997张图片
10-13 02:51:38.292 D 21709 MainActivity 测试第998张图片
10-13 02:51:38.292 D 21709 MainActivity 测试第999张图片
10-13 02:51:38.292 D 21709 MainActivity 测试第1000张图片
Filter:

图 16-7　1000 个 Drawable 对象

(4) 重新编辑 MainActivity.java 文件,源代码如下:

```
public class MainActivity extends Activity
{
    private static final String TAG = "MainActivity";
    int number = 1000;
    Bitmap[] array ;
    @Override
    public void onCreate(Bundle savedInstanceState)
    {
        super.onCreate(savedInstanceState);
        setContentView(R.layout.activity_main);
        array = new Bitmap[number];
        for(int i = 0; i < number; i++)//初始化 array 数组
        {
          Log.d(TAG, "测试第" + (i+1) + "张图片");
          array[i] = BitmapFactory.decodeResource(getResources(), R.drawable.qq);
        }
}
    @Override
    public boolean onCreateOptionsMenu(Menu menu)
    {
        getMenuInflater().inflate(R.menu.activity_main, menu);
        return true;
    }
}
```

(5) 右击工程 chapter16_6,在弹出的快捷菜单中选择 Run As|Android Application 命令,Logcat 控制台输出效果如图 16-8 所示。

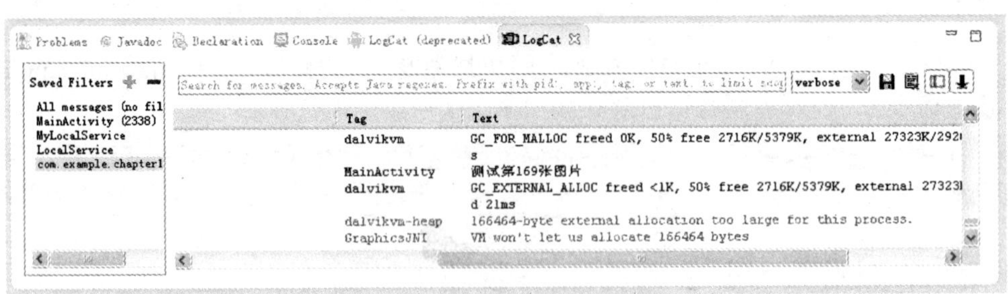

图 16-8　169 个 Bitmap 对象

从以上可以得出结论:使用 Drawable 对象保存图片时,占用更小的内存空间。而使用 Bitmap 对象保存图片时,则会占用很大的空间,很容易就出现 OOM(java.lang.OutOfMemoryError)了。

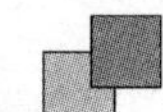

16.3.2　decodeResource 方法与 decodeStream 效率

下面笔者就通过实例比较 BitmapFactory 类的 decodeResource 方法与 decodeStream 方法的效率。

(1) 新建工程 chapter16_7，将事先准备好的 qq.jpg 文件复制到 res/drawable-hdpi/子目录下。

(2) 编辑 MainActivity.java 文件，源代码如下：

```
public class MainActivity extends Activity
{
    private static final String TAG = "MainActivity";
    int number = 1000;
    Bitmap bitmap[];
    @Override
    public void onCreate(Bundle savedInstanceState)
     {
        super.onCreate(savedInstanceState);
        setContentView(R.layout.activity_main);
        bitmap = new Bitmap[number];
        for (int i = 0; i < number; i++)
        {//初始化 bitmap 数组
            Log.d(TAG, "测试第" + (i+1) + "张图片");
            bitmap[i] = BitmapFactory.decodeStream(getResources().openRawResource
(R.drawable.qq));
        }
    }
}
```

(3)右击工程 chapter16_7，在弹出的快捷菜单中选择 Run As|Android Application 命令，Logcat 控制台输出效果如图 16 - 9 所示。

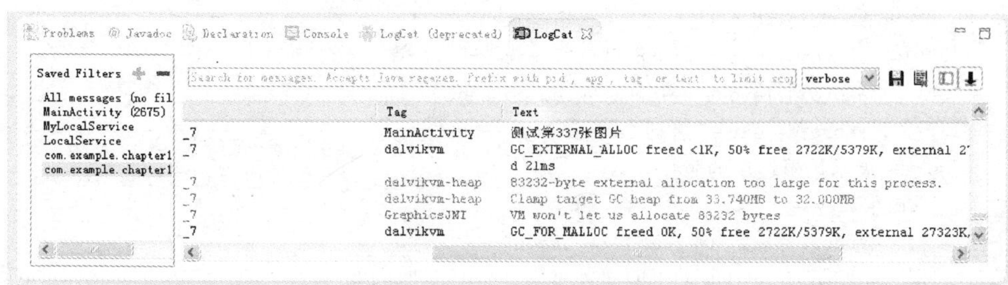

图 16 - 9　337 个 Bitmap 对象

从以上可以得出结论：使用 BitmapFactory 类的 decodeStream 方法会比使用

BitmapFactory 类的 decodeResource 方法占据更小的内存。

16.3.3 防止内存溢出

通过上面章节使用 Bitmap 的情况来看，经常会发生内存溢出的错误，特别是在处理比较大的图片的时候，发生 OOM 错误的几率会更高。在本节当中，笔者就会为读者介绍在使用 Bitmap 对象来保存图片，或者对图片进行处理的时候，如何避免内存溢出问题的发生。

首先看看 BitmapFactory 的一个静态内部类：BitmapFactory.Options，此静态内部类有以下一些常用的属性：

- public boolean inJustDecodeBounds：如果设置为 true，不获取图片，不分配内存，但会返回图片的宽度和高度信息。
- public int inSampleSize：图片缩放的倍数。如果设为 4，则宽和高都为原来的 1/4，而图则是原来的 1/16。
- public int outWidth：获取图片的宽度值。
- public int outHeight：获取图片的高度值。
- public int inDensity：用于位图的像素比。
- public int inTargetDensity：用于目标位图的像素压缩比(要生成的位图)。

下面就通过实例介绍如何在 Bitmap 的时候，防止内存溢出的发生。

(1) 新建工程 chapter16_8，编辑 res/layout/activity_main.xml 文件，源代码如下：

```
<LinearLayout xmlns:android = "http://schemas.android.com/apk/res/android"
    xmlns:tools = "http://schemas.android.com/tools"
    android:layout_width = "fill_parent"
    android:layout_height = "fill_parent"
    android:orientation = "vertical" >
   <!-- 启动 Camera 按钮 -->
    <Button
        android:id = "@ + id/launch_camera"
        android:layout_width = "fill_parent"
        android:layout_height = "wrap_content"
        android:text = "启动 Camera"></Button>
   <!-- 展示拍摄过后的照片 -->
   <ImageView
        android:id = "@ + id/show_image"
        android:layout_width = "wrap_content"
        android:layout_height = "wrap_content"></ImageView>
</LinearLayout>
```

(2) 由于本实例需要调用系统 Camera，需要在 SDCard 卡创建文件以及写入数

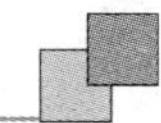

据，所以编辑 AndroidManifest.xml 文件，添加相应权限，源代码如下：

```
<manifest xmlns:android = "http://schemas.android.com/apk/res/android"
    package = "com.example.chapter16_8"
    android:versionCode = "1"
    android:versionName = "1.0" >
      <!-- Camera 权限 -->
      <uses-permission  android:name = "android.permission.CAMERA"/>
      <!-- 在 SDCard 中创建与删除文件权限 -->
    <uses-permission android:name = "android.permission.MOUNT_UNMOUNT_FILESYSTEMS"/>
    <!-- 往 SDCard 写入数据权限 -->
    <uses-permission android:name = "android.permission.WRITE_EXTERNAL_STORAGE"/>
    <uses-sdk  android:minSdkVersion = "8"    android:targetSdkVersion = "15" />
    <application
        android:icon = "@drawable/ic_launcher"
        android:label = "@string/app_name"
        android:theme = "@style/AppTheme" >
        <activity
            android:name = ".MainActivity"
            android:label = "@string/title_activity_main" >
            <intent-filter>
                <action android:name = "android.intent.action.MAIN" />
                <category android:name = "android.intent.category.LAUNCHER" />
            </intent-filter>
        </activity>
    </application>
</manifest>
```

(3) 由于调用系统 Camera 拍摄的照片会自动旋转 90°，所以需要对拍摄后的照片进行旋转处理。除此之外，还需要防止处理过大图片引起的内存溢出问题。新建一个工具类：ImageUtil，编辑 ImageUtil.java 文件，源代码如下：

```
public final class ImageUtil
{
    private static final String TAG  =  "ImageUtil";
    public static Bitmap getBitmap(String url)
    {
        Log.d(TAG, "Get Image Url:"  +  url);
        Bitmap bm  =  null;
        File file  =  new File(url);
        FileInputStream fs  =  null;
        try
        {
```

```
            fs = new FileInputStream(file);
        }
        catch (FileNotFoundException e)
        {
            Log.d(TAG, "download image occurs errors:" + e);
        }
        BitmapFactory.Options bfOptions = new BitmapFactory.Options();
        bfOptions.inDither = false;
        bfOptions.inPurgeable = true; //告诉 GC,当 GC 需要内存的时候,随时都可以释放
        bfOptions.inInputShareable = true;
        bfOptions.inTempStorage = new byte[32 * 1024];
        try
        {
            //bm = BitmapFactory.decodeFile(path, bfOptions); This one causes
            //解决 java.lang.OutOfMemoryError: bitmap size exceeds VM budget 异常
            if (fs != null)
              bm = BitmapFactory.decodeFileDescriptor(fs.getFD(), null, bfOptions);
        }
        catch (IOException e)
        {
            e.printStackTrace();
        } finally
        {
            if (fs != null)
            {
                try
                {
                    fs.close();
                }
                catch (IOException e)
                {
                    e.printStackTrace();
                }
            }
        }
        return bm;
    }
    /**
    * 对图片旋转 90°
    * @param url
    * @return
    */
```

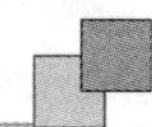

```
        public static Bitmap getRotatedBitmap(String url)
        {
            Bitmap bm = getBitmap(url);
            Matrix matrix = new Matrix();
            matrix.postRotate(90);//旋转 90°
            bm = Bitmap.createBitmap(bm, 0, 0, bm.getWidth(), bm.getHeight(), matrix, true);
            return bm;
        }
        public static Bitmap rotateMatrixBitmap(String fromUrl)
        {
            Bitmap bm = getRotatedBitmap(fromUrl);
            return bm;
        }
    }
```

(4) 由于对拍摄后的照片进行处理的这些工作,相对来讲比较耗时。所以新建一个 AsyncTask：CameraAsyncTask,编辑 CameraAsyncTask.java 文件,源代码如下：

```
    public class CameraAsyncTask extends AsyncTask<Void, Void, Bitmap>
    {
        private Context context;
        private String url;
        private ImageView imageView;
        public CameraAsyncTask(Context context, String url, ImageView imageView)
        {
            this.context = context;
            this.url = url;
            this.imageView = imageView;
        }
        @Override
        protected void onPreExecute()
        {
            super.onPreExecute();
            Toast.makeText(context, "图片正在进行压缩,请您耐心等待", Toast.LENGTH_
SHORT).show();
        }
        @Override
        protected Bitmap doInBackground(Void... params)//子线程当中执行
        {
            return ImageUtil.rotateMatrixBitmap(url);
        }
```

```
    @Override
    protected void onPostExecute(Bitmap result)
    {
        super.onPostExecute(result);
        imageView.setImageBitmap(result);
    }
}
```

(5) 编辑 MainActivity.java 文件,源代码如下:

```
public class MainActivity extends Activity
{
    private static final String TAG = "MainActivity";
    //打开 Camera 请求码
    public static final int  OPEN_CAMERA_REQUEST_CODE = 0x0001;
    //拍摄照片保存路径
    public static String CAMERA_CAPTURE_SAVE_PATH = "";
    Button launchCameraButton;//声明打开摄像头的按钮
    ImageView showImage;//声明显示摄像头拍过的照片的图像显示控件
    @Override
    public void onCreate(Bundle savedInstanceState)
      {
        super.onCreate(savedInstanceState);
        setContentView(R.layout.activity_main);
        launchCameraButton = (Button)findViewById(R.id.launch_camera);
                                                  //找到 id 为 launch_camera 的按钮
        showImage = (ImageView)findViewById(R.id.show_image);
                                              //找到 id 为 show_image 的图像显示控件
        //绑定按钮的单击事件监听器
        launchCameraButton.setOnClickListener(new OnClickListener()
         {
            @Override
            public void onClick(View v)
            {//按钮单击事件回调方法
                open();
            }
        });
    }
    @Override
    protected void onActivityResult(int requestCode, int resultCode, Intent data)
    {
        super.onActivityResult(requestCode, resultCode, data);
        //如果发送意图请求码为 OPEN_CAMERA_REQUEST_CODE 且结果返回码为 RESULT_OK
```

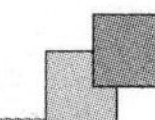

```
        if (requestCode == OPEN_CAMERA_REQUEST_CODE && resultCode == RESULT_OK)
        {
            AsyncTask<Void, Void, Bitmap> cameraAsyncTask = new CameraAsyncTask
(this, CAMERA_CAPTURE_SAVE_PATH, showImage);
           cameraAsyncTask.execute((Void[])null);
        }
    }
    //打开系统照相机
    private void open()
    {
    if (Environment.MEDIA_MOUNTED.equals(Environment.getExternalStorageState()))
      {//判断是否有 SD 卡
          File dir = new File(Environment.getExternalStorageDirectory() + "/DCIM/
Camera");     //拍摄照片保存路径
          File destFile = new File(dir, generatePhotoFileName()); //拍摄照片名称
          if (! dir.exists())
          {//如果 DCIM/Camera 目录不存在
              dir.mkdirs();                       //创建 DCIM/Camera 目录
          }
          if (! destFile.exists())
          {//如果目标文件不存在
              try
              {
                  destFile.createNewFile();   //创建文件
              }
              catch (IOException e)
              {
                  Log.e(TAG,"创建文件抛出异常:" + e);
              }
          }
          CAMERA_CAPTURE_SAVE_PATH = destFile.getPath();
          //调用系统 Camera
          Intent cameraIntent = new Intent();
          cameraIntent.putExtra(MediaStore.EXTRA_OUTPUT, Uri.fromFile(destFile));
          cameraIntent.setAction(MediaStore.ACTION_IMAGE_CAPTURE);
          startActivityForResult(cameraIntent, OPEN_CAMERA_REQUEST_CODE);
        }
        else
        {//如果设备上没有 SDCARD 卡
            Toast.makeText(this, "没有找到 SDCard 卡", Toast.LENGTH_LONG).show();
        }
    }
```

```
        //生成拍摄照片的名称
        private String generatePhotoFileName()
        {
        Date date = new Date(System.currentTimeMillis());
            SimpleDateFormat dateFormat = new SimpleDateFormat("'IMG'_yyyyMMdd_
HHmmss");
            return dateFormat.format(date) + ".JPG";
        }
        @Override
        public boolean onCreateOptionsMenu(Menu menu)
        {
            getMenuInflater().inflate(R.menu.activity_main, menu);
            return true;
        }
    }
```

(6) 右击工程 chapter16_8,在弹出的快捷菜单中选择 Run As|Android Application 命令,单击"启动 Camera"按钮过后,进入到系统的 Camera 当中,拍摄完成过后,运行效果如图 16-10 所示。

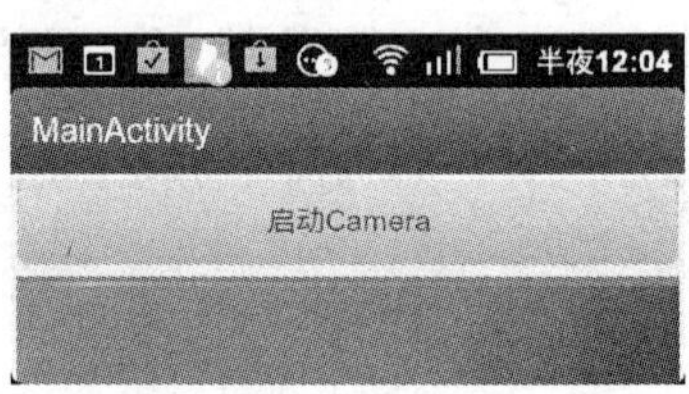

图 16-10 chapter16_8 运行效果

16.4 2D 绘图

2D 图形的接口实际上是 Android 图形系统的基础,GUI 的各种可见元素也是基于 2D 图形接口构建的,各种控件实际上是基于图形 API 绘制出来的。Android 系统提供的 UI 控件通过继承 android.view.View 类,并实现其中的 onDraw()函数来实现绘制的工作,绘制的工作主要是由 android.graphics 包来实现。android.graphics 包中的内容是 Android 系统的 2D 图形 API,其中包含以下一些主要类:

- Point、Rect 和 Color 等:一些基础类、分别定义点、矩形、颜色等基础信息元素。
- Paint:画笔,用于控制绘制的样式和颜色等信息。
- Canvas:画布,2D 图形系统最核心的一个类,处理 onDraw()时调用。

主要绘制的设置和操作在 Paint(画笔)和 Canvas(画布)两个类当中,使用这两个类就可以完成所有的绘制。

16.4.1 View 类

任何自定义的控件都需要继承 android.view.View 类,通过重写 android.view.View 父类的 onDraw()函数来完成绘制的工作。自定义控件的 onDraw()方法不能

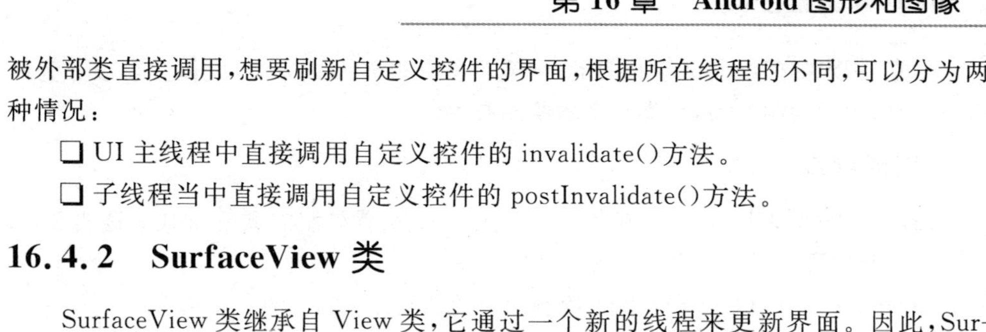

被外部类直接调用，想要刷新自定义控件的界面，根据所在线程的不同，可以分为两种情况：

- UI 主线程中直接调用自定义控件的 invalidate()方法。
- 子线程当中直接调用自定义控件的 postInvalidate()方法。

16.4.2 SurfaceView 类

SurfaceView 类继承自 View 类，它通过一个新的线程来更新界面。因此，SurfaceView 类更适合需要快速加载 UI，或渲染代码阻塞 UI 主线程的时间过长的情形。SurfaceView 封装了一个 Surface 对象，而不是 Canvas 对象，这一点对于那些资源敏感的操作特别有用。SurfaceView 一般通过使用 SurfaceHolder 类来控制 Canvas 在其 surface 上的操作，SurfaceHolder 类的实例可以通过 SurfaceHolder 对象的 getHolder()方法来获得。SurfaceHolder 类有以下几个比较常用的方法：

- lockCanvas()：用于锁定画布，这样 surface 中就可以指定画布了，之后就可以在画布上进行绘画。
- unlockCanvasAndPost(Canvas canvas)：用于释放处于锁定状态的 Canvas。

surface 的状态发生变化时，可以通过 SurfaceHolder. CallBack 来获取这些信息。SurfaceHolder. callback 有 3 个重要的方法：surfaceCreated()、surfaceChanged()和 surfaceDestory()分别在 surface 创建、改变和销毁时调用。

16.4.3 Canvas(画布)和 Paint(画笔)

Canvas 类主要实现屏幕的绘制过程，Canvas 类有以下常用方法：

- void drawRect(RectF rect, Paint paint)：绘制区域，第一个参数 RectF 为一个区域，第二个参数为 Paint 画笔对象。
- void drawPath(Path path, Paint paint)：绘制一个路径，第一个参数 Path 为一个路径对象，第二个参数为 Paint 画笔对象。
- void drawLine(float startX, float startY, float stopX, float stopY, Paint paint)：画线，第 1 个参数为起始点的 *X* 轴坐标的值，第 2 个参数为起始点的 *Y* 轴坐标的值，第 3 个参数为终点的 *X* 轴坐标位置的值，第 4 个参数为终点的 *Y* 轴坐标的值，最后一个参数为 Paint 画笔对象。
- void drawPoint(float x, float Y, Paint paint)：画点，第 1 个参数是点的 X 轴坐标的值，第 2 个参数是点的 Y 轴坐标的值，第 3 个参数为 Paint 画笔对象。
- void drawText(String text, float x, float y, Paint paint)：绘制文本，第 1 个参数是要绘制的 String 类型的文本，第 2 个参数是绘制的文本所在的起始位置的 X 轴坐标的值，第 3 个参数是绘制文本所在的起始位置的 Y 轴坐标的值，第 4 个参数为 Paint 画笔对象。

Paint 即画笔，在绘图过程中起到了非常重要的作用，画笔主要保存颜色、样式等绘制信息，指定如何绘制文本和图形，画笔对象有很多设置方法，大体上可以分为两类，一类与图形绘制有关，一类与文本绘制有关。

1. 图形绘制

- void setARGB(int a, int r, int g, int b)：设置绘制的颜色，a 代表透明度，r、g、b 代表颜色值。
- void setAlpha(int a)：设置绘制图形的透明度。
- void setColor(int color)：设置绘制的颜色，使用颜色值来表示，该颜色值包括透明度和 RGB 颜色。
- void setAntiAlias(boolean aa)：设置是否使用抗锯齿功能，会消耗较大资源，绘制图形速度会变慢。
- void setDither(boolean dither)：设定是否使用图像抖动处理，会使绘制出来的图片颜色更加平滑和饱满，图像更加清晰。
- void setFilterBitmap(bolean filter)：如果该项设置为 true，则图像在动画进行中会滤掉对 Bitmap 图像的优化操作，加快显示速度，本设置项依赖于 dither 和 xfermode 的设置。
- MaskFilter setMaskFilter(MaskFilter maskFilter)：设置 MaskFilter，可以用不同的 MaskFilter 实现滤镜的效果，如滤化、立体等。
- ColorFilter setColorFilter(ColorFilter filter)：设置颜色过滤器，可以在绘制颜色时实现不用颜色的变换效果。
- PathEffect setPathEffect(PathEffect effect)：设置绘制路径的效果，如点画线等。
- Shader setShader(Shader shader)：设置图像效果，使用 Shader 可以绘制出各种渐变效果。
- void setShadowLayer(float radius, float dx, float dy, int color)：在图形下面设置阴影层，产生阴影效果，radius 为阴影的角度，dx 和 dy 为阴影在 X 轴和 Y 轴上的距离，color 为阴影的颜色。
- void setStyle(Paint. Style style)：设置画笔的样式，为 FILL、FILL_OR_STROKE 或 STROKE 。
- void setStrokeCap(Paint. Cap cap)：当画笔样式为 STROKE 或 FILL_OR_STROKE 时，设置笔刷的图形样式，如圆形样式 Cap. ROUND 或方形样式 Cap. SQUARE。
- void setStrokeJoin(Paint. Join join)：设置绘制时各图形的结合方式，如平滑效果等。
- void setStrokeWidth(float width)：当画笔样式为 STROKE 或 FILL_OR_STROKE 时，设置笔刷的粗细度 。

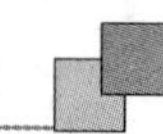

- Xfermode setXfermode(Xfermode xfermode)：设置图形重叠时的处理方式，如合并、取交集或并集，经常用来制作橡皮的擦除效果。

2. 文本绘制

- void setFakeBoldText(boolean fakeBoldText)：模拟实现粗体文字，设置在小字体上效果会非常差。
- void setSubpixelText(boolean subpixelText)：设置该项为 true，将有助于文本在 LCD 屏幕上的显示效果。
- void setTextAlign(Paint. Align align)：设置绘制文字的对齐方向。
- void setTextScaleX(float scaleX)：设置绘制文字 X 轴的缩放比例，可以实现文字的拉伸效果。
- void setTextSize(float textSize)：设置绘制文字的字号大小。
- void setTextSkewX(float skewX)：设置斜体文字，skewX 为倾斜弧度 。
- Typeface setTypeface(Typeface typeface)：设置 Typeface 对象，即字体风格，包括粗体、斜体以及衬线体、非衬线体等 。
- void setUnderlineText(boolean underlineText)：设置带下划线的文字效果。
- void setStrikeThruText(boolean strikeThruText)：设置带有删除线的效果。

下面就通过实例介绍如何使用 Canvas(画布)以及 Paint(画笔)。

(1) 新建工程 chapter16_9，新建一个自定义 View：DrawView，编辑 DrawView.java 文件，源代码如下：

```
public class DrawView extends View
{
    public DrawView(Context context)
    {
        super(context);
    }
    @Override
    protected void onDraw(Canvas canvas)
    {
        super.onDraw(canvas);
        /*
         * 方法 说明 drawRect 绘制矩形 drawCircle 绘制圆形 drawOval 绘制椭圆 draw-
Path 绘制任意多边形
         * drawLine 绘制直线 drawPoin 绘制点
         */
        //创建画笔
        Paint p = new Paint();
        p.setColor(Color.RED);//设置红色
        canvas.drawText("画圆：", 10, 20, p);          //画文本
```

```
canvas.drawCircle(60, 20, 10, p);                    //小圆
p.setAntiAlias(true);//设置画笔的锯齿效果。true 是去除,大家一看效果就明白了
canvas.drawCircle(120, 20, 20, p);                   //大圆
canvas.drawText("画线及弧线:", 10, 60, p);
p.setColor(Color.GREEN);                             //设置绿色
canvas.drawLine(60, 40, 100, 40, p);                 //画线
canvas.drawLine(110, 40, 190, 80, p);                //斜线
//画笑脸弧线
p.setStyle(Paint.Style.STROKE);                      //设置空心
RectF oval1 = new RectF(150,20,180,40);
canvas.drawArc(oval1, 180, 180, false, p);           //小弧形
oval1.set(190, 20, 220, 40);
canvas.drawArc(oval1, 180, 180, false, p);           //小弧形
oval1.set(160, 30, 210, 60);
canvas.drawArc(oval1, 0, 180, false, p);             //小弧形
canvas.drawText("画矩形:", 10, 80, p);
p.setColor(Color.GRAY);                              //设置灰色
p.setStyle(Paint.Style.FILL);                        //设置填满
canvas.drawRect(60, 60, 80, 80, p);                  //正方形
canvas.drawRect(60, 90, 160, 100, p);                //长方形
canvas.drawText("画扇形和椭圆:", 10, 120, p);
/* 设置渐变色 这个正方形的颜色是改变的 */
Shader mShader = new LinearGradient(0, 0, 100, 100,
    new int[] { Color.RED, Color.GREEN, Color.BLUE, Color.YELLOW,
        Color.LTGRAY }, null, Shader.TileMode.REPEAT);
                                        //一个材质,打造出一个线性梯度沿着一条线
p.setShader(mShader);
//p.setColor(Color.BLUE);
RectF oval2 = new RectF(60, 100, 200, 240); //设置个新的长方形,扫描测量
canvas.drawArc(oval2, 200, 130, true, p);
//画弧,第 1 个参数是 RectF:该类是第 2 个参数是角度的开始,第 3 个参数是
//多少度,第 4 个参数是真的时候画扇形,是假的时候画弧线
//画椭圆,把 oval 改一下
oval2.set(210,100,250,130);
canvas.drawOval(oval2, p);
canvas.drawText("画三角形:", 10, 200, p);
//绘制这个三角形,可以绘制任意多边形
Path path = new Path();
path.moveTo(80, 200);                                //此点为多边形的起点
path.lineTo(120, 250);
path.lineTo(80, 250);
path.close();                                        //使这些点构成封闭的多边形
```

```
canvas.drawPath(path, p);
//可以绘制很多任意多边形,比如下面画六连形
p.reset();//重置
p.setColor(Color.LTGRAY);
p.setStyle(Paint.Style.STROKE);                    //设置空心
Path path1 = new Path();
path1.moveTo(180, 200);
path1.lineTo(200, 200);
path1.lineTo(210, 210);
path1.lineTo(200, 220);
path1.lineTo(180, 220);
path1.lineTo(170, 210);
path1.close();//封闭
canvas.drawPath(path1, p);
/*
 * Path 类封装了一组区域的描述,包括直线、二次曲线和三次方曲线。可以通过
调用 Canvas 或 drawPath 将 Path 定义的区域画到画布上或用来填充图形
 */
//画圆角矩形
p.setStyle(Paint.Style.FILL);                      //充满
p.setColor(Color.LTGRAY);
p.setAntiAlias(true);                              //设置画笔的锯齿效果
canvas.drawText("画圆角矩形:", 10, 260, p);
RectF oval3 = new RectF(80, 260, 200, 300); //设置个新的长方形
canvas.drawRoundRect(oval3, 20, 15, p);//第二个参数是 x 半径,第三个参数是 y 半径
//画贝塞尔曲线
canvas.drawText("画贝塞尔曲线:", 10, 310, p);
p.reset();
p.setStyle(Paint.Style.STROKE);
p.setColor(Color.GREEN);
Path path2 = new Path();
path2.moveTo(100, 320);                            //设置 Path 的起点
path2.quadTo(150, 310, 170, 400); //设置贝塞尔曲线的控制点坐标和终点坐标
canvas.drawPath(path2, p);                         //画出贝塞尔曲线
//画点
p.setStyle(Paint.Style.FILL);
canvas.drawText("画点:", 10, 390, p);
canvas.drawPoint(60, 390, p); //画一个点
canvas.drawPoints(new float[]{60,400,65,400,70,400}, p);//画多个点
//画图片,就是贴图
Bitmap bitmap = BitmapFactory.decodeResource(getResources(), R.drawable.ic_
launcher);
```

```
            canvas.drawBitmap(bitmap, 250,360, p);
        }
    }
```

(2) 编辑 MainActivity.java 文件,源代码如下:

```
public class MainActivity extends Activity
{
    @Override
    public void onCreate(Bundle savedInstanceState)
    {
        super.onCreate(savedInstanceState);
        init();
    }
    private void init()
    {
        LinearLayout layout = new LinearLayout(this);
        final DrawView view = new DrawView(this);
        view.setMinimumHeight(500);
        view.setMinimumWidth(300);
        //通知 view 组件重绘
        view.invalidate();
        layout.addView(view);
        setContentView(layout);
    }
}
```

(3) 右击工程 chapter16_9,在弹出的快捷菜单中选择 Run As|Android Application 命令,运行效果如图 16-11 所示。

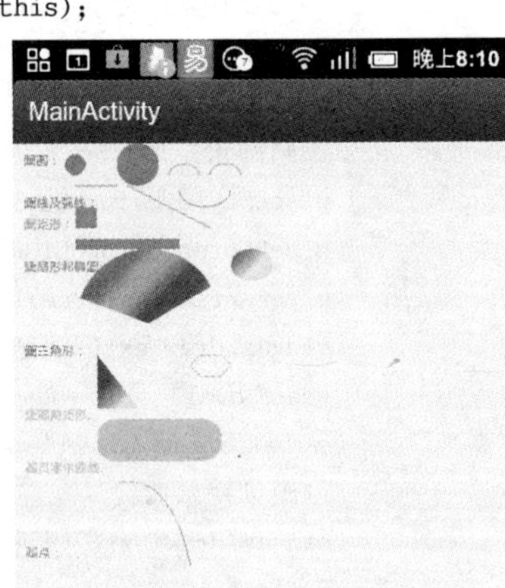

图 16-11 chapter16_9 运行效果

16.5 Android 动画

Android 的动画实现分两种方式,一种方式是补间动画(Tween Animation),开发人员只需指定动画开始和动画结束的"关键帧",而动画变化的"中间帧"由系统计算并补齐,所以被称为补间动画;另外一种方式是逐帧动画(Frame Animation),就是通过播放预先排序好的图片来实现动态的画面。

16.5.1 补间动画

补间动画可以使视图组件移动、放大、缩小以及产生透明的变化。例如在一个 ImageView 组件当中,通过补间动画可以使该视图实现放大、缩小、旋转及渐变等效

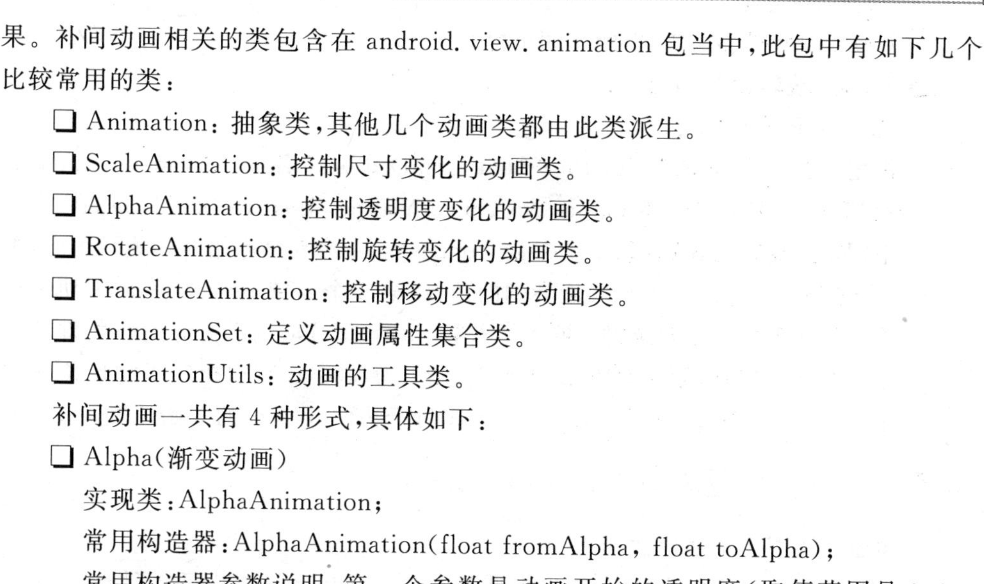

果。补间动画相关的类包含在 android. view. animation 包当中，此包中有如下几个比较常用的类：

- ❑ Animation：抽象类，其他几个动画类都由此类派生。
- ❑ ScaleAnimation：控制尺寸变化的动画类。
- ❑ AlphaAnimation：控制透明度变化的动画类。
- ❑ RotateAnimation：控制旋转变化的动画类。
- ❑ TranslateAnimation：控制移动变化的动画类。
- ❑ AnimationSet：定义动画属性集合类。
- ❑ AnimationUtils：动画的工具类。

补间动画一共有4种形式，具体如下：

- ❑ Alpha(渐变动画)

  实现类：AlphaAnimation；

  常用构造器：AlphaAnimation(float fromAlpha, float toAlpha)；

  常用构造器参数说明：第一个参数是动画开始的透明度(取值范围是0.0～1.0)，第二个参数动画结束的透明度(取值范围是0.0～1.0)。

- ❑ Scale(尺寸变化动画)

  实现类：ScaleAnimation；

  常用构造器：ScaleAnimation(float fromX, float toX, float fromY, float toY, int pivotXType, float pivotXValue, int pivotYType, float pivotYValue)；

  常用构造器参数说明：第1个参数 fromX 为动画起始时 X 坐标上的伸缩尺寸，第2个参数 toX 为动画结束时 X 坐标上的伸缩尺寸，第3个参数 fromY 为动画初始时 Y 坐标上的伸缩尺寸，第4个参数 toY 为动画结束时 Y 坐标上的伸缩尺寸。其中当取值为0.0时表示收缩到没有，当取值为1.0时表示正常无伸缩，当取值小于1.0时表示收缩，当取值大于1.0时表示放大。第5个参数 pivotXType 为动画在 X 轴相对于物件的位置类型，第6个参数 pivotXValue 为动画相对于物件的 X 坐标的开始位置，第7个参数 pivotYType 为动画在 Y 轴相对于物件位置类型，第8个参数 pivotYValue 为动画相对于物件的 Y 坐标的开始位置。

- ❑ Translate(位置变化动画)

  实现类：TranslateAnimation；

  常用构造器：TranslateAnimation(float fromXDelta, float toXDelta, float fromYDelta, float toYDelta)；

  常用构造器参数说明：第1个参数 fromXDelta 为动画起始时 X 坐标上的移动位置，第2个参数 toXDelta 为动画结束时 X 坐标上的移动位置，第3个参数 fromYDelta 为动画开始时 Y 坐标上的移动位置，第4个参数为 toYDelta

为动画结束时 Y 坐标上的移动位置。

- Rotate(旋转变化动画)

  实现类:RotateAnimation;

  常用构造器:RotateAnimation(float fromDegress, float toDegress, int pivotXType, float pivotXValue, float pivotYType, float pivotYValue);

  常用构造器参数说明:第 1 个参数 fromDegrees 为动画起始时的旋转角度,第 2 个参数 toDegrees 为动画旋转到的角度,第 3 个参数 pivotXType 为动画在 X 轴相对于物件位置类型 ,第 4 个参数 pivotXValue 为动画相对于物件的 X 坐标的开始位置,第 5 个参数 pivotXType 为动画在 Y 轴相对于物件位置类型 ,第 6 个参数 pivotYValue 为动画相对于物件 Y 坐标的开始位置。

16.5.2 渐变动画(AlphaAnimation)

下面通过实例演示如何使用渐变动画。

(1) 新建工程 chapter16_10,将事先准备好的 splash.jpg 图片复制到 res/drawable-hdpi 目录,编辑 res/layout/activity_main.xml 文件,源代码如下:

```
<?xml version="1.0" encoding="utf-8"?>
<FrameLayout
  xmlns:android="http://schemas.android.com/apk/res/android"
  android:layout_width="fill_parent"
  android:layout_height="fill_parent"
  android:background="#FFFFFF">
  <ImageView
    android:id="@+id/splash"
    android:layout_width="fill_parent"
    android:layout_height="fill_parent"
    android:layout_gravity="center"
    android:src="@drawable/splash"/>
  <Button
    android:layout_width="fill_parent"
    android:layout_height="wrap_content"
    android:layout_gravity="bottom"
    android:text="alpha"
    android:onClick="alpha"/>
</FrameLayout>
```

(2) 在 res 目录下新建 anim 子目录,并在 res/anim/目录下新建 alpha.xml 文件,编辑 alpha.xml 文件,源代码如下:

```
<?xml version="1.0" encoding="utf-8"?>
<set xmlns:android="http://schemas.android.com/apk/res/android">
  <alpha
```

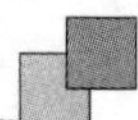

```
    android:fromAlpha = "0.0"
    android:toAlpha = "1.0"
    android:duration = "3000"/>
</set>
```

(3) 编辑 MainActivity.java 文件，源代码如下：

```
public class MainActivity extends Activity implements AnimationListener
{
    private ImageView splash;
    @Override
    protected void onCreate(Bundle savedInstanceState)
    {
        super.onCreate(savedInstanceState);
        setContentView(R.layout.activity_main);
        splash = (ImageView) findViewById(R.id.splash);
        Animation anim = AnimationUtils.loadAnimation(this, R.anim.alpha);
        anim.setAnimationListener(this);
        splash.startAnimation(anim);
    }

    public void alpha(View view)
    {
        Animation anim = new AlphaAnimation(1.0f, 0.0f);
        anim.setDuration(3000);
        anim.setFillAfter(true);
        splash.startAnimation(anim);
    }
    @Override
    public void onAnimationStart(Animation animation)
    {
        Log.i("alpha", "onAnimationStart called.");
    }
    @Override
    public void onAnimationEnd(Animation animation)
    {
        Log.i("alpha", "onAnimationEnd called");
    }
    @Override
    public void onAnimationRepeat(Animation animation)
    {
        Log.i("alpha", "onAnimationRepeat called");
    }

}
```

(4) 右击工程 chapter16_10,在弹出的快捷菜单中选择 Run As|Android Application 命令,运行效果如图 16-12 所示,单击界面上的 alpha 按钮过后,运行效果如图 16-13 所示。

图 16-12　chapter16_10 运行效果

图 16-13　chapter16_10 单击 Alpha 按钮的效果

16.5.3　尺寸变化动画(ScaleAnimation)

下面通过实例演示如何使用尺寸变化动画。

(1) 新建工程 chapter16_11,将事先准备好的 person.png 图片复制到 res/drawable-hdpi 目录,编辑 res/layout/activity_main.xml 文件,源代码参考配套资料。

(2) 在 res 目录下新建 anim 子目录,并在 res/anim/目录下新建 scale.xml 文件,编辑 scale.xml 文件,源代码如下:

```
<?xml version="1.0" encoding="utf-8"?>
<set xmlns:android="http://schemas.android.com/apk/res/android"
    android:interpolator="@android:anim/bounce_interpolator">
    <scale
        android:fromXScale="1.0"
        android:toXScale="2.0"
        android:fromYScale="1.0"
        android:toYScale="2.0"
        android:pivotX="0.5"
        android:pivotY="50%"
        android:duration="2000"/>
    <alpha
        android:fromAlpha="0.0"
        android:toAlpha="1.0"
        android:duration="3000"/>
</set>
```

(3) 编辑 MainActivity.java 文件,源代码如下:

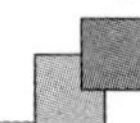

```
public class MainActivity extends Activity
{
    private ImageView scale_iamge;
    @Override
    protected void onCreate(Bundle savedInstanceState)
  {
        super.onCreate(savedInstanceState);
        setContentView(R.layout.activity_main);
        scale_iamge = (ImageView) findViewById(R.id.scale_image);
        Animation anim = AnimationUtils.loadAnimation(this, R.anim.scale);
        anim.setFillAfter(true);
        scale_iamge.startAnimation(anim);
    }
    public void sclae(View view)
     {
        Animation anim = new ScaleAnimation(2.0f, 1.0f, 2.0f, 1.0f,
            Animation.RELATIVE_TO_SELF, 0.5f,
            Animation.RELATIVE_TO_SELF, 0.5f);
        anim.setDuration(2000);
        anim.setFillAfter(true);
        BounceInterpolator bounce = new BounceInterpolator();
        anim.setInterpolator(bounce);
        scale_iamge.startAnimation(anim);
    }
}
```

(4) 右击工程 chapter16_11，在弹出的快捷菜单中选择 Run As|Android Application 命令，运行效果如图 16-14 所示，单击界面上的 scale 按钮，运行效果如图 16-15所示。

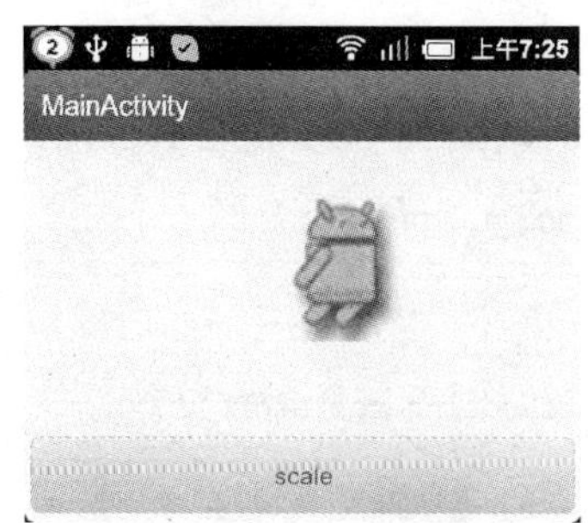

图 16-14　chapter16_11 运行效果

图 16-15　chapter16_11 单击 scale 按钮的效果

16.5.4 位置变化动画(TranslateAnimation)

下面通过实例演示如何使用位置变化动画。

(1) 新建工程 chapter16_12,将事先准备好的 person.png 图片复制到 res/drawable-hdpi 目录,编辑 res/layout/activity_main.xml 文件,源代码参考配套资料。

(2) 在 res 目录下新建 anim 子目录,并在 res/anim/目录下新建 translate.xml 文件,编辑 translate.xml 文件,源代码如下:

```
<?xml version = "1.0" encoding = "utf - 8"?>
<set xmlns:android = "http://schemas.android.com/apk/res/android"
    android:interpolator = "@android:anim/bounce_interpolator">
    <translate
        android:fromXDelta = "0"
        android:fromYDelta = "0"
        android:toXDelta = "200"
        android:toYDelta = "300"
        android:duration = "2000"/>
</set>
```

(3) 编辑 MainActivity.java 文件,源代码如下:

```
public class MainActivity extends Activity {
    private ImageView trans_iamge;
    @Override
    protected void onCreate(Bundle savedInstanceState)
    {
        super.onCreate(savedInstanceState);
        setContentView(R.layout.activity_main);
        trans_iamge = (ImageView) findViewById(R.id.trans_image);
        Animation anim = AnimationUtils.loadAnimation(this, R.anim.translate);
        anim.setFillAfter(true);
        trans_iamge.startAnimation(anim);
    }
        public void translate(View view)
    {
        Animation anim = new TranslateAnimation(200, 0, 300, 0);
        anim.setDuration(2000);
        anim.setFillAfter(true);
        OvershootInterpolator overshoot = new OvershootInterpolator();
        anim.setInterpolator(overshoot);
        trans_iamge.startAnimation(anim);
    }
}
```

(4) 右击工程 chapter16_12,在弹出的快捷菜单中选择 Run As|Android Application 命令,运行效果如图 16－16 所示,单击界面上的 translate 按钮,运行效果如图 16－17 所示。

图 16－16　chapter16_12 运行效果

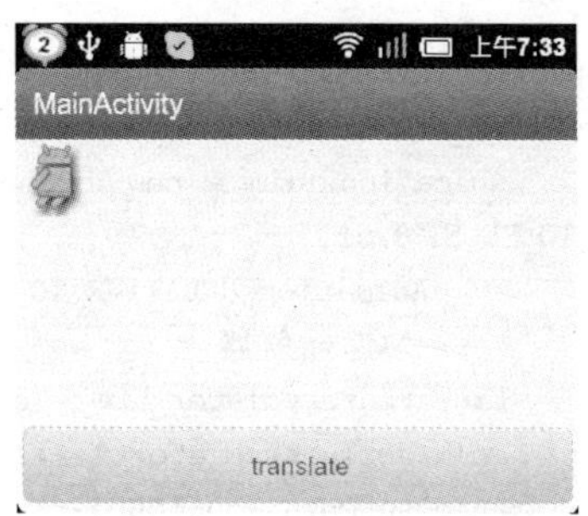

图 16－17　chapter16_12 单击 translate 按钮的效果

16.5.5　旋转变化动画(RotateAnimation)

下面通过实例演示如何使用旋转变化动画。

(1) 新建工程 chapter16_13,将事先准备好的 piechart.png 图片复制到 res/drawable-hdpi 目录,编辑 res/layout/activity_main.xml 文件,源代码参考配套资料。

(2) 在 res 目录下新建 anim 子目录,并在 res/anim/目录下新建 rotate.xml 文件,编辑 rotate.xml 文件,源代码如下:

```
<?xml version = "1.0" encoding = "utf - 8"?>
<set xmlns:android = "http://schemas.android.com/apk/res/android"
    android:interpolator = "@android:anim/accelerate_decelerate_interpolator">
    <rotate
        android:fromDegrees = "0"
        android:toDegrees = " + 360"
        android:pivotX = "50 % "
        android:pivotY = "50 % "
        android:duration = "5000"/>
</set>
```

(3) 编辑 MainActivity.java 文件,源代码如下:

```
public class MainActivity extends Activity
{
    private int currAngle;
    private View piechart;
    @Override
    public void onCreate(Bundle savedInstanceState)
  {
        super.onCreate(savedInstanceState);
```

```
            setContentView(R.layout.activity_main);
            piechart = findViewById(R.id.piechart);
            Animation animation = AnimationUtils.loadAnimation(this, R.anim.rotate);
            piechart.startAnimation(animation);
        }
        public void positive(View v)
        {
            Animation anim = new RotateAnimation(currAngle, currAngle + 180, Animation.
RELATIVE_TO_SELF, 0.5f,
                Animation.RELATIVE_TO_SELF, 0.5f);
            /** 匀速插值器 */
            LinearInterpolator lir = new LinearInterpolator();
            anim.setInterpolator(lir);
            anim.setDuration(1000);
            /** 动画完成后不恢复原状 */
            anim.setFillAfter(true);
            currAngle += 180;
            if (currAngle > 360)
            {
                currAngle = currAngle - 360;
            }
            piechart.startAnimation(anim);
        }
        public void negative(View v)
        {
            Animation anim = new RotateAnimation(currAngle, currAngle - 180, Animation.
RELATIVE_TO_SELF, 0.5f,Animation.RELATIVE_TO_SELF, 0.5f);
            /** 匀速插值器 */
            LinearInterpolator lir = new LinearInterpolator();
            anim.setInterpolator(lir);
            anim.setDuration(1000);
            /** 动画完成后不恢复原状 */
            anim.setFillAfter(true);
            currAngle -= 180;
            if (currAngle < -360)
            {
                currAngle = currAngle + 360;
            }
            piechart.startAnimation(anim);
        }
        @Override
        public boolean onCreateOptionsMenu(Menu menu)
        {
            getMenuInflater().inflate(R.menu.activity_main, menu);
            return true;
        }
    }
```

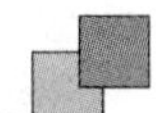

（4）右击工程 chapter16_13，在弹出的快捷菜单中选择 Run As|Android Application 命令，运行效果如图 16 - 18 所示，单击界面上的“逆时针”按钮，运行效果如图 16 - 19 所示。

图 16 - 18　chapter16_13 运行效果

图 16 - 19　chapter16_13 单击逆时针按钮的效果

16.5.6　逐帧动画(Frame Animation)

下面通过实例演示如何使用逐帧动画。

（1）新建工程 chaper16_14，将事先准备好的图片都复制到 res/drawable-hdpi 目录，编辑 res/layout/activity_main.xml 文件，源代码如下：

```
<?xml version = "1.0" encoding = "utf - 8"?>
<LinearLayout xmlns:android = "http://schemas.android.com/apk/res/android"
    android:layout_width = "wrap_content" android:layout_height = "wrap_content"
    android:orientation = "vertical">
    <Button
        android:id = "@ + id/button_animation1"
        android:text = "动画开始"
        android:layout_gravity = "center_horizontal"
        android:layout_width = "wrap_content"
        android:layout_height = "wrap_content"
        android:onClick = "myClickHandler"></Button>
    <ImageView
        android:id = "@ + id/imageView_animation1"
        android:layout_width = "wrap_content"
        android:layout_height = "wrap_content"
        android:layout_weight = "1"></ImageView>
    <ImageView
        android:id  = "@ + id/imageView_role"
        android:layout_width = "wrap_content"
        android:layout_height = "wrap_content"
        android:layout_marginTop = "50dp"
        />
```

```
</LinearLayout>
```

(2) 在 res/drawable-hdpi 目录中新建 number_animation_drawable.xml 文件，编辑 number_animation_drawable.xml 文件，源代码如下：

```
    <?xml version="1.0" encoding="utf-8"?>
<!--
    根标签为 animation-list，其中 oneshot 代表着是否只展示一遍，设置为 false 则不停地循环播放动画
    根标签下，通过 item 标签对动画中的每一个图片进行声明
    android:duration 表示展示所用的该图片的时间长度
    每个 item 就是一帧，drawable 是该帧显示的图片，duration 是显示时间
    android:oneshot="false"表示动画一直进行，若为 true 则表示只播放一次动画
-->
<animation-list
  xmlns:android="http://schemas.android.com/apk/res/android"
  android:oneshot="false"
  >
    <item android:drawable="@drawable/number0" android:duration="50"></item>
    <item android:drawable="@drawable/number1" android:duration="50"></item>
    <item android:drawable="@drawable/number2" android:duration="50"></item>
    <item android:drawable="@drawable/number3" android:duration="50"></item>
    <item android:drawable="@drawable/number4" android:duration="50"></item>
    <item android:drawable="@drawable/number5" android:duration="50"></item>
</animation-list>
```

(3) 在 res/drawable-hdpi 目录中新建 role_animation_drawable.xml 文件，编辑 role_animation_drawable.xml 文件，源代码如下：

```
    <?xml version="1.0" encoding="utf-8"?>
<!--
    根标签为 animation-list，其中 oneshot 代表是否只展示一遍，设置为 false 会不停地循环播放动画
    根标签下，通过 item 标签对动画中的每一个图片进行声明
    android:duration 表示展示所用的该图片的时间长度

    每个 item 就是一帧，drawable 是该帧显示的图片，duration 是显示时间
    android:oneshot="false"表示动画一直进行，若为 true 则表示只播放一次动画
-->
<animation-list
  xmlns:android="http://schemas.android.com/apk/res/android"
  android:oneshot="false"
  >
    <item android:drawable="@drawable/role1" android:duration="60"></item>
    <item android:drawable="@drawable/role2" android:duration="60"></item>
    <item android:drawable="@drawable/role3" android:duration="60"></item>
    <item android:drawable="@drawable/role4" android:duration="60"></item>
```

```
</animation-list>
```

(4) 编辑 MainActivity.java 文件,源代码如下:

```
/*
 * @description android 中的逐帧动画。
 * 逐帧动画的原理很简单,跟电影的播放一样,一张张类似的图片不停地切换,当切换速度达到一定值时,我们的视觉就会出现残影,残影的出现保证了视觉上变化的连续性,这时候图片的切换看在我们眼中就跟真实的一样了。
 * 想使用逐帧动画需要以下几步:
 * 第一步:需要在 res/drawable 文件夹下新建一个 xml 文件,该文件详细定义了动画播放时所用的图片、切换每张图片所用的时间、是否为连续播放等。
 * 第二步:在代码中,将该动画布局文件,赋值给特定的图片展示控件,如本例子中的 ImageView。
 * 第三步:通过 imageView.getBackGround()获取相应的 AnimationDrawable 对象,然后通过该对象的方法进行控制动画。
 */
public class MainActivity extends Activity
{
    private ImageView imageViewNumber,imageViewRole;
    @Override
    public void onCreate(Bundle savedInstanceState)
    {
        super.onCreate(savedInstanceState);
        setContentView(R.layout.activity_main);
        imageViewNumber = (ImageView) findViewById(R.id.imageView_animation1);
        imageViewNumber.setBackgroundResource(R.drawable.number_animation_drawable);
        imageViewRole = (ImageView) findViewById(R.id.imageView_role);
        imageViewRole.setBackgroundResource(R.drawable.role_animation_drawable);
    }
    public void myClickHandler(View targetButton)
   {
        //获取 AnimationDrawable 对象
        AnimationDrawable animationDrawable = (AnimationDrawable) imageViewNumber.
getBackground();
        AnimationDrawable animationDrawable2 = (AnimationDrawable) imageViewRole.
getBackground();
        //动画是否正在运行
        if(animationDrawable.isRunning())
        {   //如果动画正在运行,则停止动画播放
            animationDrawable.stop();
        }
        else if(! animationDrawable.isRunning())
        {   //如果动画不在运行,则开始或者继续动画播放
            animationDrawable.start();
        }
```

```
        //动画是否正在运行
        if(animationDrawable2.isRunning())
        {  //如果动画正在运行,则停止动画播放
            animationDrawable2.stop();
        }
        else if(! animationDrawable2.isRunning())
         {  //如果动画不在运行,则开始或者继续动画播放
          animationDrawable2.start();
        }
    }
    /***
    * 一开始就运动
    */
    public void onWindowFocusChanged(boolean hasFocus)
    {
      //获取 AnimationDrawable 对象
        AnimationDrawable animation1 = (AnimationDrawable)imageViewNumber.getBackground();
        AnimationDrawable animation2 = (AnimationDrawable)imageViewRole.getBackground();
      super.onWindowFocusChanged(hasFocus);
      if(hasFocus)
      {//得到焦点
        animation1.start();
        animation2.start();
      }
      else
      {  //失去焦点
        animation1.stop();
        animation2.stop();
      }
    }
    @Override
      public boolean onCreateOptionsMenu
(Menu menu) {
            getMenuInflater().inflate(R.
menu.activity_main, menu);
            return true;
    }
}
```

(5) 右击工程 chapter16_14,在弹出的快捷菜单中选择 Run As|Android Application 命令,运行效果如图 16 - 20 所示。

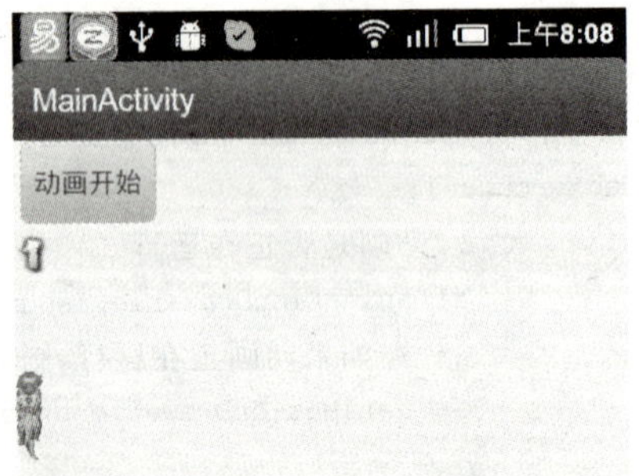

图 16 - 20　chapter16_14 运行效果

第 17 章 Android 硬件接口

在本章当中笔者将会介绍有关如何使用 Bluetooth(蓝牙),如何使用 Telephony(电话)以及一些获取系统信息和对设备进行相应控制的方法。

17.1 蓝牙基本介绍

蓝牙(Bluetooth)是目前广泛应用的无线通信协议,近距离无线通信的标准,主要针对短距离设备通信(10 m 之内)。传说瑞典有个国王特别爱吃蓝莓导致自己的牙齿天天都是蓝色的,在他执政期间这位国王非常善于交际,能说会道,和邻国的关系非常好,这个 Bluetooth 的发明者觉得蓝牙它的作用就是在近距离沟通周围的设备,跟这个国王很类似,于是起名叫蓝牙。

Android SDK 直到 2.0 以后的版本才开始支持蓝牙编程。蓝牙是一项无线电连接系统,它可以将不同的电子器材连接起来,常用于连接耳机、鼠标和移动通信设备等。蓝牙不仅仅是一项技术,而是一种概念。

17.1.1 蓝牙工作流程

两个设备要想通过蓝牙协议进行数据通信的前提是:两个设备上都要有蓝牙设备或者专业一点叫蓝牙适配器,以手机和计算机为例笔者画了如图 17-1 所示的蓝牙工作流程图。手机开始扫描周围的蓝牙设备,扫描发现附近的计算机有蓝牙设备时,给它发出一个信号需要进行蓝牙配对并设置一个临时密钥,计算机收到请求信息,输入临时密钥,配对成功。

配对成功后手机和计算机就可以进行文件传输了,这是一个最基本的一个流程,如图 17-1 所示。

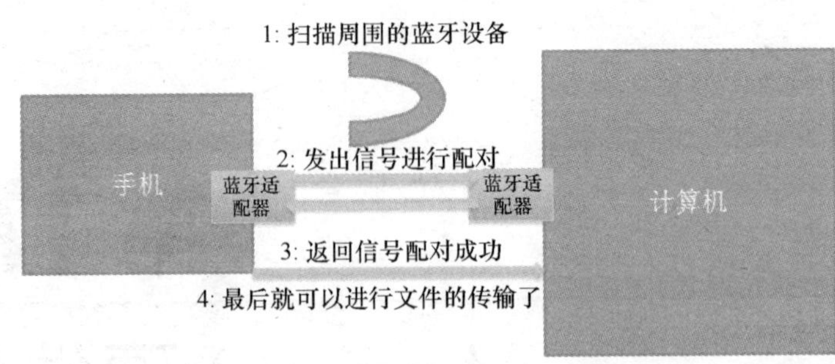

图 17－1　蓝牙工作流程图

每一个蓝牙设备都会有一个可见性的设置，如果将蓝牙设备设置为可见，那么其他近距离的蓝牙设备就可以搜索到这个蓝牙设备，如果将蓝牙设备设置为不可见，那么其他近距离的蓝牙设备就无法扫描到这个蓝牙设备。如图 17－2 所示。

图 17－2　蓝牙设置界面

如图 17－2 所示，如果将"可检测性"打开，则其他近距离蓝牙设备可以搜索到此蓝牙设备，如果将"可检测性"关闭，则其他近距离设备无法搜索到此蓝牙设备。

17.1.2　蓝牙编程核心类

有关蓝牙编程的核心类都位于 android. bluetooth 包当中。android. bluetooth 包当中有如下几个比较常用的核心类：

- BluetoothAdapter：这个类的对象代表了本地的蓝牙适配器，相当于图 17－1 蓝牙工作流程图当中手机里的蓝牙适配器，例如这个应用程序运行在手机上，那么手机上的蓝牙适配器就是本地的蓝牙适配器。
- BluetoothDevice：这个类的对象代表了远程的蓝牙设备，相当于图 17－1 蓝牙工作流程图当中的计算机里的蓝牙适配器，例如这个应用程序运行在手机上，那么 BluetoothDevice 就代表了手机要连接的远程设备上的蓝牙适配器。
- BluetoothSocket：这个类的对象代表了一个蓝牙 Socket 的接口(类似于 TCP Socket)。这是应用程序通过 InputStream 或者 OutputStream 与其他蓝牙设备进行数据交换的连接点。
- BluetoothServerSocket：这个类的对象代表了一个服务器 Socket，监听进入的连接请求(类似于 TCP 的 ServerSocket)。为了连接两台 Android 设备，其中一台必须打开一个 ServerSocket。当一台远程蓝牙设备发出一个连接请求并被接受时，BluetoothServerSocket 将返回一个已连接的 BluetoothSocket。

17.1.3　蓝牙权限

为了在应用程序当中使用蓝牙的相关功能 API，在项目清单文件当中至少需要声明两方面的权限：BLUETOOTH 权限和 BLUETOOTH_ADMIN 权限。只有在项目清单当中声明了 BLUETOOTH 权限，才能够实现蓝牙设备之间的互相通信，例如请求一个连接、接受一个连接以及数据的传输。只有在项目清单当中声明了 BLUETOOTH_ADMIN 这个权限，才能够发现近距离的蓝牙设备以及对蓝牙设备进行管理设置。

要想使用 BLUETOOTH_ADMIN 这个权限，必须首先声明 BLUETOOTH 这个权限。

17.1.4　找寻周围蓝牙设备

下面就通过实例介绍如何用一个蓝牙设备找到周围临近的蓝牙设备，并打印出这些蓝牙设备的 MAC 地址。

(1) 新建工程 chapter17_1，编辑 res/layout/activity_main. xml 文件，源代码参考配套资料。

(2) 编辑 AndroidManifest. xml 文件，添加相关权限，源代码如下：

```
<manifest xmlns:android = "http://schemas.android.com/apk/res/android"
    package = "com.example.chapter17_1"
    android:versionCode = "1"
    android:versionName = "1.0" >
    <!-- 蓝牙设备间数据传输权限 -->
    <uses-permission android:name = "android.permission.BLUETOOTH"/>
    <!-- 设置管理蓝牙设备权限 -->
    <uses-permission android:name = "android.permission.BLUETOOTH_ADMIN"/>
    <uses-sdk  android:minSdkVersion = "8"  android:targetSdkVersion = "15" />
    <application
        android:icon = "@drawable/ic_launcher"
        android:label = "@string/app_name"
        android:theme = "@style/AppTheme" >
        <activity
            android:name = ".MainActivity"
            android:label = "@string/title_activity_main" >
            <intent-filter>
                <action android:name = "android.intent.action.MAIN" />
                <category android:name = "android.intent.category.LAUNCHER" />
            </intent-filter>
        </activity>
    </application>
```

```
</manifest>
```

(3) 编辑 MainActivity.java 文件，源代码如下：

```
public class MainActivity extends Activity
{
    private static final String TAG = "MainActivity";
    Button searchBTDevices;//声明搜索临近蓝牙设备按钮
    TextView showDevices;
    BluetoothReceiver bluetoothReceiver;
    StringBuffer stringBuffer;
    @Override
    public void onCreate(Bundle savedInstanceState)
    {
        super.onCreate(savedInstanceState);
        setContentView(R.layout.activity_main);
        showDevices = (TextView)findViewById(R.id.show_devices);
        searchBTDevices = (Button)findViewById(R.id.search_bt_devices);
        searchBTDevices.setOnClickListener(new SearchBtListenner());
                                        //为按钮绑定内部类形式的单击事件监听器
        stringBuffer = new StringBuffer();
    }
    private class SearchBtListenner implements OnClickListener
    {
    //处理单击事件回调
    @Override
    public void onClick(View v)
    {
        //获取运行当前程序设备的蓝牙适配器
        BluetoothAdapter localAdapter = BluetoothAdapter.getDefaultAdapter();
        //设备不支持蓝牙
        if(localAdapter == null)
        {//如果 localAdapter 对象为 null
         Toast.makeText(MainActivity.this, "设备不支持蓝牙", Toast.LENGTH_LONG).show();
        }
        //打开蓝牙，并设置蓝牙可见性
        if(!localAdapter.isEnabled())
        {//如果蓝牙设备不可见
            Intent intent = new Intent(BluetoothAdapter.ACTION_REQUEST_ENABLE);
            // 设置蓝牙可见性，最多 300 s，当值大于 300 时默认为 300
            intent.putExtra(BluetoothAdapter.EXTRA_DISCOVERABLE_DURATION, 300);
            MainActivity.this.startActivity(intent);
        }
```

```
            localAdapter.startDiscovery();
        }
    }
    //onResume 注册发现相关蓝牙设备广播接收者
    @Override
    protected void onResume()
    {
        super.onResume();
        //设定广播接收的 filter
        IntentFilter intentFilter = new IntentFilter(BluetoothDevice.ACTION_FOUND);

        //创建蓝牙广播信息的 receiver
        bluetoothReceiver = new BluetoothReceiver ();
        //注册广播接收器
        registerReceiver(bluetoothReceiver,intentFilter);
    }
    @Override
    protected void onStop()
    {
        super.onStop();
        unregisterReceiver(bluetoothReceiver);
    }
    /**
     * 搜到到蓝牙设备的广播接收者
     *
     */
    private class BluetoothReceiver extends BroadcastReceiver
    {
        @Override
        public void onReceive(Context context, Intent intent)
        {
            //获得扫描到的远程蓝牙设备
            BluetoothDevice device = intent.getParcelableExtra(BluetoothDevice.EX-
TRA_DEVICE);
            Log.d(TAG, "new device address:" + device.getAddress());
            stringBuffer.append(device.getAddress() + "\n");
            showDevices.setText(stringBuffer);
        }
    }
    @Override
    public boolean onCreateOptionsMenu(Menu menu) {
        getMenuInflater().inflate(R.menu.activity_main, menu);
```

```
            return true;
        }
    }
```

（4）右击工程 chapter17_1，在弹出的快捷菜单中选择 Run As|Android Application 命令，运行效果如图 17-3 所示。

图 17-3　chapter17_1 运行效果

17.2　Telephony 介绍

Android Telephony API 提供了一种对基本电话信息进行监听的方式，不但可以获取网络类型，连接状态等基本信息，还可以用来对电话号码等字符串进行相应的操作。

17.2.1　使用 Telephony Manager

Android Telephony API 当中有一个 TelephonyManager 类，这个类事实上代表的是 Android 的一种系统服务，用来访问设备上有关 Telephony Service 的信息。通过 Telephony API 提供的 TelephonyManager 这个类，在程序当中可以用来获取 SIM 卡的相关信息，除此之外，还可以用来获取电信网络的一些相关信息，例如电信网络国家、代码、名称和网络类型等。其中有一些 Telephony Service 的信息是有权限保护的，所以在访问这些信息的时候必须声明相应的权限。

下面就通过实例介绍如何使用 TelephonyManager 这个类。

（1）新建工程 chapter17_2，编辑 res/layout/activity_main.xml 文件，源代码参考配套资料。

（2）由于需要添加相关获取 Telephony Services 的权限，所以编辑 AndroidManifest.xml 文件，源代码如下：

```
<manifest xmlns:android = "http://schemas.android.com/apk/res/android"
    package = "com.example.chapter17_2"
    android:versionCode = "1"
    android:versionName = "1.0" >
    <!-- 读取相关 Telephony Services 权限 -->
   <uses-permission android:name = "android.permission.READ_PHONE_STATE"></uses-permission>
    <uses-sdk   android:minSdkVersion = "8"   android:targetSdkVersion = "15" />
    <application
        android:icon = "@drawable/ic_launcher"
        android:label = "@string/app_name"
        android:theme = "@style/AppTheme" >
```

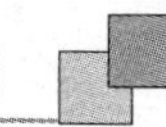

```
        <activity
            android:name=".MainActivity"
            android:label="@string/title_activity_main" >
            <intent-filter>
                <action android:name="android.intent.action.MAIN" />
                <category android:name="android.intent.category.LAUNCHER" />
            </intent-filter>
        </activity>
    </application>
</manifest>
```

(3) 编辑 MainActivity.java 文件，源代码如下：

```
public class MainActivity extends Activity
{
    TextView showInfo;//声明显示相关信息的文本控件
    TelephonyManager telManager;//声明电话管理器
    @Override
    public void onCreate(Bundle savedInstanceState)
    {
      super.onCreate(savedInstanceState);
      setContentView(R.layout.activity_main);
      showInfo = (TextView)findViewById(R.id.telephony_info);//找到 id 为 telephony
                                                              //_info 的文本控件
       telManager = (TelephonyManager)getSystemService(Context.TELEPHONY_SERV-
ICE);//实例化电话管理器

      StringBuilder sb = new StringBuilder();
      sb.append("设备 ID:").append(telManager.getDeviceId()).append("\n");
      sb.append("IMSI:").append(telManager.getSubscriberId()).append("\n");
      sb.append("网络运营商国家:").append(telManager.getNetworkCountryIso()).ap-
pend("\n");
      sb.append("网络运营商代码:").append(telManager.getNetworkOperator()).append("\n");
      sb.append("网络运营商名称:").append(telManager.getNetworkOperatorName()).
append("\n");
      sb.append("SIM 卡运营商国家:").append(telManager.getSimCountryIso()).append("\n");
      sb.append("SIM 运营商代码:").append(telManager.getSimOperator()).append("\n");
      sb.append("SIM 卡运营商名称:").append(telManager.getSimOperatorName()).ap-
pend("\n");
      sb.append("IMEI:").append(telManager.getSimSerialNumber()).append("\n");
      showInfo.setText(sb);
    }
    @Override
```

```
        public boolean onCreateOptionsMenu(Menu menu) {
            getMenuInflater().inflate(R.menu.activity_main, menu);
            return true;
        }
    }
```

(4) 右击工程 chapter17_2,在弹出的快捷菜单中选择 Run As|Android Application 命令,运行效果如图 17-4 所示。

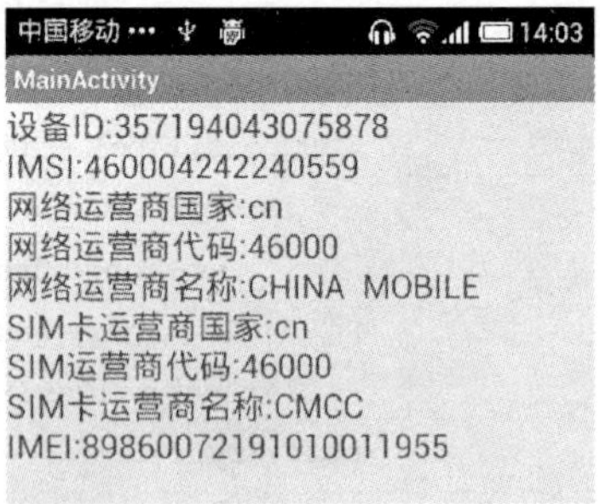

设备ID:357194043075878
IMSI:460004242240559
网络运营商国家:cn
网络运营商代码:46000
网络运营商名称:CHINA MOBILE
SIM卡运营商国家:cn
SIM运营商代码:46000
SIM卡运营商名称:CMCC
IMEI:89860072191010011955

图 17-4 chapter17_2 运行效果

17.2.2 广播接收者监听来电信息

下面通过实例介绍如何采用广播接收者的方式监听来电信息。

(1) 新建工程 chapter17_3,编辑 res/layout/activity_main.xml 文件,源代码参考配套资料。

(2) 编辑 AndroidManifest.xml 文件,添加相关权限,源代码如下:

```
<manifest xmlns:android = "http://schemas.android.com/apk/res/android"
    package = "com.example.chapter17_3"
    android:versionCode = "1"
    android:versionName = "1.0" >
    <!-- 读取电话状态 -->
    <uses-permission android:name = "android.permission.READ_PHONE_STATE" />
    <uses-sdk android:minSdkVersion = "8"   android:targetSdkVersion = "15" />
    <application
        android:icon = "@drawable/ic_launcher"
        android:label = "@string/app_name"
        android:theme = "@style/AppTheme" >
        <activity
            android:name = ".MainActivity"
            android:label = "@string/title_activity_main" >
            <intent-filter>
                <action android:name = "android.intent.action.MAIN" />
                <category android:name = "android.intent.category.LAUNCHER" />
            </intent-filter>
        </activity>
    </application>
</manifest>
```

(3) 新建一个广播接收者:MyBroadcastReceiver,用于对系统发出的电话状态改变广播进行监听,编辑 MyBroadcastReceiver.java 文件,源代码如下:

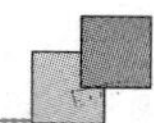

```
public class MyBroadcastReceiver extends BroadcastReceiver
{
    private static final String TAG = "MyBroadcastReceiver";
    @Override
    public void onReceive(Context context, Intent intent)
    {
        String action = intent.getAction();
        Log.i(TAG, "[Broadcast]" + action);
        //如果是呼入电话意图
        if(action.equals(TelephonyManager.ACTION_PHONE_STATE_CHANGED))
        {
            Log.i(TAG, "[Broadcast]PHONE_STATE");
            doReceivePhone(context,intent);
        }
    }
    /**
     * 处理电话广播
     * @param context
     * @param intent
     */
    public void doReceivePhone(Context context, Intent intent)
    {
        String phoneNumber = intent.getStringExtra(TelephonyManager.EXTRA_INCOMING
_NUMBER);
        TelephonyManager telephony = (TelephonyManager) context.getSystemService
(Context.TELEPHONY_SERVICE);
        int state = telephony.getCallState();
        switch (state)
        {
        case TelephonyManager.CALL_STATE_RINGING://等待接电话
            Log.i(TAG, "[Broadcast]等待接电话 = " + phoneNumber);
            break;
        case TelephonyManager.CALL_STATE_IDLE:  //电话挂断
            Log.i(TAG, "[Broadcast]电话挂断 = " + phoneNumber);
            break;
        case TelephonyManager.CALL_STATE_OFFHOOK://通话中
            Log.i(TAG, "[Broadcast]通话中 = " + phoneNumber);
            break;
        }
    }
}
```

(4) 编辑 MainActivity.java 文件,源代码如下:

```
public class MainActivity extends Activity  implements OnClickListener
{
    Button registerBtn;           //声明注册广播接收者按钮
    Button unregisterBtn;         //声明取消注册广播接收者按钮
    private MyBroadcastReceiver mBroadcastReceiver;
    @Override
    public void onCreate(Bundle savedInstanceState)
    {
        super.onCreate(savedInstanceState);
        setContentView(R.layout.activity_main);
        registerBtn = (Button)findViewById(R.id.register);//找到 id 为 register 的按钮
        unregisterBtn = (Button)findViewById(R.id.unregister);
                                                      //找到 id 为 unregister 的按钮
        registerBtn.setOnClickListener(this);
        unregisterBtn.setOnClickListener(this);
        mBroadcastReceiver = new MyBroadcastReceiver();
    }
    @Override
    public void onClick(View v)
    {
        int viewId = v.getId();
        if(R.id.register == viewId)
        {//注册广播接收者
            IntentFilter intentFilter = new IntentFilter();
            intentFilter.addAction(TelephonyManager.ACTION_PHONE_STATE_CHANGED);
            intentFilter.setPriority(Integer.MAX_VALUE);
            registerReceiver(mBroadcastReceiver, intentFilter);
        }
        else if(R.id.unregister == viewId)
          {//取消注册广播接收者
            unregisterReceiver(mBroadcastReceiver);
        }
    }
    @Override
    public boolean onCreateOptionsMenu(Menu menu) {
        getMenuInflater().inflate(R.menu.activity_main, menu);
        return true;
    }
}
```

(5) 右击工程 chapter17_3,在弹出的快捷菜单中选择 Run As|Android Appli-

cation 命令,运行效果如图 17-5 所示。

(6)单击界面上的“来电监听器”按钮,用另外一台 Android 手机往测试设备上拨打电话,测试设备响起电话铃声后,直接挂掉电话,Logcat 输出日志效果如图 17-6 所示。

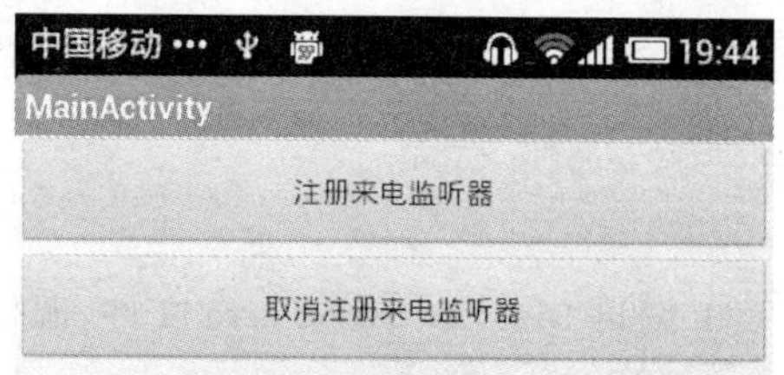

图 17-5 chapter17_3 运行效果

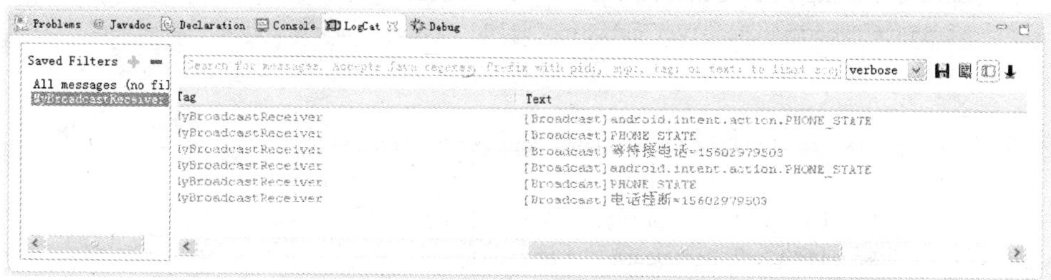

图 17-6 挂断电话 Logcat 输出日志

17.2.3 广播接收者监听去电信息

下面通过实例介绍如何采用广播接收者的方式监听去电信息。

(1) 新建工程 chapter17_4,由于需要对用户拨出的电话进行监听,所以需要添加相应权限。编辑 AndroidManifest.xml 文件,源代码如下:

```
<manifest xmlns:android = "http://schemas.android.com/apk/res/android"
    package = "com.example.chapter17_4"
    android:versionCode = "1"
    android:versionName = "1.0" >
<!-- 允许程序监视、修改有关播出电话权限 -->
<uses-permission android:name = "android.permission.PROCESS_OUTGOING_CALLS" ></u-
ses-permission>
<uses-sdk  android:minSdkVersion = "8" android:targetSdkVersion = "15" />
    <application
        android:icon = "@drawable/ic_launcher"
        android:label = "@string/app_name"
        android:theme = "@style/AppTheme" >
        <activity
            android:name = ".MainActivity"
            android:label = "@string/title_activity_main" >
            <intent-filter>
                <action android:name = "android.intent.action.MAIN" />
                <category android:name = "android.intent.category.LAUNCHER" />
```

```
            </intent-filter>
        </activity>
    </application>
</manifest>
```

(2) 编辑 MainActivity.java 文件，源代码如下：

```
public class MainActivity extends Activity
{
    private static final String TAG = "MainActivity";
    private BroadcastReceiver outgoingCallReceiver = new BroadcastReceiver()
    {
        @Override
        public void onReceive(Context context, Intent intent)
        {
          String action = intent.getAction();
          Log.i(TAG, "[Broadcast]" + action);
          //如果是呼出电话意图
          if(action.equals(Intent.ACTION_NEW_OUTGOING_CALL))
          {
            String outPhoneNumber = intent.getStringExtra(Intent.EXTRA_PHONE_NUMBER);
            Log.i(TAG, "[Broadcast]ACTION_NEW_OUTGOING_CALL:" + outPhoneNumber);
            //this.setResultData(null);
            //这里可以更改呼出电话号码。如果设置为 null,电话就永远不会拨出了
          }
        }
    };
    @Override
    public void onCreate(Bundle savedInstanceState)
    {
        super.onCreate(savedInstanceState);
        setContentView(R.layout.activity_main);
        IntentFilter intentFilter = new IntentFilter();
        intentFilter.addAction(Intent.ACTION_NEW_OUTGOING_CALL);
        intentFilter.setPriority(Integer.MAX_VALUE);
        registerReceiver(outgoingCallReceiver, intentFilter);
    }

    @Override
    public boolean onCreateOptionsMenu(Menu menu)
    {
        getMenuInflater().inflate(R.menu.activity_main, menu);
        return true;
    }
}
```

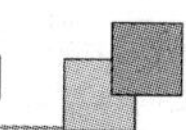

上述代码在 onCreate()方法当中注册了一个对拨出电话广播进行监听的广播接收者,笔者在 MainActivity 当中是以内部类的形式来定义这个广播接收者的。当监听到用户拨出电话过后,广播接收者的 onReceive 方法自动会被回调,可以通过获取 Intent 对象的相关参数来获取用户拨出的具体电话号码。

(3)右击工程 chapter17_4,在弹出的快捷菜单中选择 Run As|Android Application 命令,运行程序 chapter17_4 过后。当用户拨打电话时,Logcat 日志输出效果如图 17-7 所示。

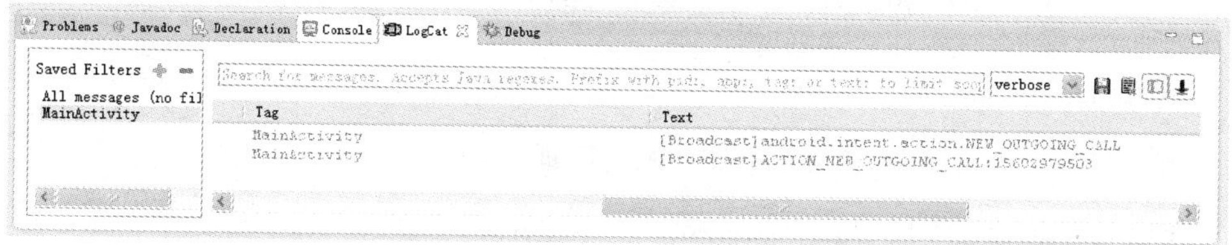

图 17-7 拨打电话 Logcat 输出日志

17.3 系统和控制设备

17.3.1 设置声音模式

手机都有声音模式,声音、静音还有震动,甚至震动加声音兼备,这些都是手机的基本功能。在 Android 手机中,同样可以通过 Android 的 SDK 提供的声音管理接口来管理手机声音模式以及调整声音大小,这就是 Android 中 AudioManager 的使用。

下面通过实例介绍如何设置手机的声音模式。

(1)新建工程 chapter17_5,编辑 res/layout/activity_main.xml 文件,源代码如下:

```
<LinearLayout xmlns:android = "http://schemas.android.com/apk/res/android"
    xmlns:tools = "http://schemas.android.com/tools"
    android:layout_width = "fill_parent"
    android:layout_height = "fill_parent"
    android:orientation = "vertical" >
    <!-- 静音模式按钮 -->
    <Button
        android:id = "@ + id/slient"
        android:layout_width = "fill_parent"
        android:layout_height = "wrap_content"
        android:text = "静音模式" />
    <!-- 声音模式按钮 -->
    <Button
```

```
            android:id = "@ + id/normal"
            android:layout_width = "fill_parent"
            android:layout_height = "wrap_content"
            android:text = "声音模式" />
        <!-- 震动模式按钮 -->
        <Button
            android:id = "@ + id/vibrate"
            android:layout_width = "fill_parent"
            android:layout_height = "wrap_content"
            android:text = "震动模式" />
</LinearLayout>
```

(2) 编辑 MainActivity.java 文件，源代码如下：

```
public class MainActivity extends Activity implements OnClickListener
{
    Button silentBtn;              // 静音模式按钮
    Button normalBtn;              // 声音模式按钮
    Button vibrateBtn;             // 震动模式按钮
    private AudioManager am;       // 声音管理器
    @Override
    public void onCreate(Bundle savedInstanceState)
    {
        super.onCreate(savedInstanceState);
        setContentView(R.layout.activity_main);
        silentBtn = (Button) findViewById(R.id.slient);
        normalBtn = (Button) findViewById(R.id.normal);
        vibrateBtn = (Button) findViewById(R.id.vibrate);
        silentBtn.setOnClickListener(this);
        normalBtn.setOnClickListener(this);
        vibrateBtn.setOnClickListener(this);
        am = (AudioManager) getSystemService(Context.AUDIO_SERVICE);// 实例化声音管理器
    }
    @Override
    public void onClick(View v)
    {
        int viewId = v.getId();
        switch (viewId)
        {
        case R.id.slient:          // 静音模式
            am.setRingerMode(AudioManager.RINGER_MODE_SILENT);
            Toast.makeText(getApplicationContext(), "Silent Mode Activated.", Toast.LENGTH
_LONG).show();
```

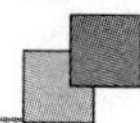

```
                break;
            case R.id.normal:        // 声音模式
                am.setRingerMode(AudioManager.RINGER_MODE_NORMAL);
                Toast.makeText(getApplicationContext(), "Normal Mode Activated", Toast.
LENGTH_LONG).show();
                break;
            case R.id.vibrate:       // 震动模式
                am.setRingerMode(AudioManager.RINGER_MODE_VIBRATE);
                  Toast.makeText(getApplicationContext(), "Vibrate Mode Activated",
Toast.LENGTH_LONG).show();
                break;
            default:
                break;
            }
        }
        @Override
        public boolean onCreateOptionsMenu(Menu menu) {
            getMenuInflater().inflate(R.menu.activity_main, menu);
            return true;
        }
    }
```

(3) 右击工程 chapter17_5，在弹出的快捷菜单中选择 Run As | Android Application 命令，运行效果如图 17－8 所示。

图 17－8　chapter17_5 运行效果

17.3.2　获取安装程序列表

下面通过实例介绍如何获取正在运行的应用程序列表。

(1) 新建工程 chapter17_6，编辑 res/layout/activity_main.xml 文件，源代码如下：

```
<?xml version="1.0" encoding="UTF-8"?>
<LinearLayout xmlns:android="http://schemas.android.com/apk/res/android"
    android:layout_width="fill_parent"
    android:layout_height="fill_parent"
    android:orientation="vertical" >
    <ListView
        android:id="@+id/program_list"
        android:layout_width="fill_parent"
        android:layout_height="fill_parent" />
</LinearLayout>
```

(2) 新建应用程序数据模型类:Program,编辑 Program.java 文件,源代码如下:

```
public class Program
{
    //图标
    private Drawable icon;
    //程序名
    private String name;
    public Drawable getIcon()
    {
        return icon;
    }
    public void setIcon(Drawable icon)
    {
        this.icon = icon;
    }
    public String getName()
    {
        return name;
    }
    public void setName(String name)
    {
        this.name = name;
    }
}
```

(3) 新建显示列表每一行的布局文件:list_item.xml,编辑 list_item.xml 文件,源代码如下:

```
<?xml version = "1.0" encoding = "UTF-8"? >
<LinearLayout xmlns:android = "http://schemas.android.com/apk/res/android"
    android:layout_width = "wrap_content"
    android:layout_height = "wrap_content"
    android:orientation = "horizontal" >
    <!-- 显示应用程序图标 -->
    <ImageView
        android:id = "@ + id/image"
        android:layout_width = "wrap_content"
        android:layout_height = "wrap_content"
        android:layout_marginRight = "10dip" />
    <!-- 显示应用程序名称 -->
    <TextView
        android:id = "@ + id/text"
```

```
        android:layout_width = "wrap_content"
        android:layout_height = "wrap_content" />
</LinearLayout>
```

(4) 新建 ListView 数据适配器:ProgramAdapter,编辑 ProgramAdapter.java 文件,源代码如下:

```
public class ProgramAdapter extends BaseAdapter
{
    List<Program> list = new ArrayList<Program>();
    LayoutInflater la;
    Context context;
    public ProgramAdapter(List<Program> list,Context context)
    {
        this.list = list;
        this.context = context;
    }
    @Override
    public int getCount()
    {
        return list.size();
    }
    @Override
    public Object getItem(int position)
    {
        return list.get(position);
    }
    @Override
    public long getItemId(int position)
    {
        return position;
    }
    @Override
    public View getView(int position, View convertView, ViewGroup parent)
    {
        ViewHolder holder;
        if(convertView == null)//如果可复用旧视图为 null
        {
            la = LayoutInflater.from(context);
            convertView = la.inflate(R.layout.list_item, null);
            holder = new ViewHolder();
            holder.image = (ImageView) convertView.findViewById(R.id.image);
            holder.text = (TextView) convertView.findViewById(R.id.text);
            convertView.setTag(holder);
```

```
        }
        else
        {//如果可复用旧视图不为 null
            holder = (ViewHolder) convertView.getTag();
        }
        final Program pr = (Program) list.get(position);
        // 设置图标
        holder.imgage.setImageDrawable(pr.getIcon());
        // 设置程序名
        holder.text.setText(pr.getName());
        return convertView;
    }
}
class ViewHolder
{
    TextView text;
    ImageView imgage;
}
```

(5) 编辑 MainActivity.java 文件,源代码如下:

```
public class MainActivity extends Activity
{
    ListView programListView = null;// 声明显示应用程序列表控件
    @Override
    public void onCreate(Bundle savedInstanceState)
    {
        super.onCreate(savedInstanceState);
        setContentView(R.layout.activity_main);
        programListView = (ListView) findViewById(R.id.program_list);
        List<Program> list = getAllInstalled();  //获取所有安装的非系统应用
        ProgramAdapter adapter = new ProgramAdapter(list, this);
        programListView.setAdapter(adapter);  //填充 ListView 数据
    }
    // 所有安装的应用程序
    public List<Program> getAllInstalled()
    {
        ArrayList<Program> appList = new ArrayList<Program>();
                                            // 用来存储获取的应用信息数据
        List<PackageInfo> packages = getPackageManager().getInstalledPackages(0);
        for (int i = 0; i < packages.size(); i++) //迭代 packages 列表
        {
            PackageInfo packageInfo = packages.get(i);
            Program appInfo = new Program();
            appInfo.setName(packageInfo.applicationInfo.loadLabel(getPackageManag-
```

```
er()).toString();
                appInfo.setIcon(packageInfo.applicationInfo.loadIcon(getPackageManager()));
                //如果是非系统应用
                if ((packageInfo.applicationInfo.flags & ApplicationInfo.FLAG_SYSTEM) == 0)
                {
                    appList.add(appInfo);// 如果非系统应用,则添加至 appList
                }
            }
            return appList;
        }
    }
```

(6) 右击工程 chapter17_6,在弹出的快捷菜单中选择 Run As | Android Application 命令,运行效果如图 17－9 所示。

图 17－9　所有已安装的非系统应用列表

17.3.3　控制设备振动

手机设备振动是向用户反馈信息的一种很好的方法,特别是在游戏应用中得到了极大的发挥。

下面就通过实例介绍如何控制设备的振动。

(1) 新建工程 chapter17_7,编辑 res/layout/activity_main.xml 文件,源代码参考配套资料。

(2) 由于需要使设备振动起来,因此需要相应权限。编辑 AndroidManifest.xml 文件,源代码如下:

```
<manifest xmlns:android = "http://schemas.android.com/apk/res/android"
    package = "com.example.chapter17_7" android:versionCode = "1" android:versionName
= "1.0" >
    <!-- 设备振动权限 -->
  <uses-permission android:name = "android.permission.VIBRATE"/>
    <uses-sdk    android:minSdkVersion = "8"    android:targetSdkVersion = "15" />
    <application
        android:icon = "@drawable/ic_launcher"
        android:label = "@string/app_name"
        android:theme = "@style/AppTheme" >
        <activity
            android:name = ".MainActivity"
            android:label = "@string/title_activity_main" >
            <intent-filter>
                <action android:name = "android.intent.action.MAIN" />
```

```
                <category android:name = "android.intent.category.LAUNCHER" />
            </intent-filter>
        </activity>
    </application>
</manifest>
```

(3) 编辑 MainActivity.java 文件,源代码如下:

```
public class MainActivity extends Activity implements OnClickListener
{
    Button vibrateButton;           //声明振动设备按钮
    Vibrator vibrator;              //振动管理器
    @Override
    public void onCreate(Bundle savedInstanceState)
    {
        super.onCreate(savedInstanceState);
        setContentView(R.layout.activity_main);
        vibrateButton = (Button)findViewById(R.id.vibrate_device);
        vibrator = (Vibrator)getSystemService(Context.VIBRATOR_SERVICE);
                                    //实例化振动管理器
        vibrateButton.setOnClickListener(this);
    }
    @Override
    public void onClick(View v)
    {
        long[] pattern = {1000, 2000, 4000, 8000, 16000};
        vibrator.vibrate(pattern, 0);
    }
    @Override
    protected void onDestroy()
    {
        super.onDestroy();
        //调用 cancel 函数可以取消振动,注意,退出振动程序后,如果不手动调用 cancel
//函数取消振动,该振动将会持续下去(如果是持续振动的话),而不管应用程序是否退出
        vibrator.cancel();
    }
}
```

(4) 右击工程 chapter17_7,在弹出的快捷菜单中选择 Run As|Android Application 命令,运行效果如图 17-10 所示。(单击"振动设备"按钮过后 1 s,振动 2 s,停止振动 4 s,接着振动 8 s,停止振动 16 s,周而复始。)

图 17-10　chapter17_10 运行效果

17.3.4 管理网络和 WIFI 连接

在 Android 中，ConnectivityManager 类代表网络连接服务，它被用来监控网络连接状态，配置失效重连，并控制网络天线等。

1. ConnectivityManager

获取 Connectivity Manager 实例的方法是使用 Context 对象的 getSystemService(String name)方法，并指定 Context. CONNECTIVITY_SERVICE 作为参数，代码片段如下：

```
String cserviceName = Context.CONNECTIVITY_SERVICE;
ConnectivityManager cm = (ConnectivityManager) getSystemService(cserviceName);
```

要使用 Connectivity Manager 来读写网络状态时，需要在 AndroidManifest. xml 文件中加入如下的权限：

```
<!-- 访问网络状态 -->
<uses-permission android:name = "android.permission.ACCESS_NETWORK_STATE" />
<!-- 更改网络状态 -->
<uses-permission android:name = "android.permission.CHANGE_NETWORK_STATE" />
```

Connectivity Manager 在较高层面提供了管理可用网络连接的接口，使用 getActiveNetworkInfo 和 getNetworkInfo 方法可以查询获取 NetworkInfo 对象，该对象包含了当前活动网络连接或者指定类型的不可用网络连接的详细信息。

2. 设置首选网络

当任何认证的应用程序请求网络连接时，Android 都会优先尝试使用首选网络连接。设置首选网络连接的函数是 setNetworkPreference，代码片段如下：

```
cm.setNetworkPreference(ConnectivityManager.TYPE_WIFI);
```

当首选网络连接不可用或者连接丢失时，Android 将自动尝试使用第二优先连接类型。

3. 监控网络连接

ConnectivityManager 最常用的一个功能就是当网络连接状态改变时通知应用程序，这是通过应用程序实现 Broadcast Receiver 来监听 ConnectivityManager. CONNECTIVITY_ACTION 类型的 Intent 来实现的。这个 Intent 提供了以下几种 extra 来进一步明确发生改变的网络状态。

- ❑ ConnectivityManager. EXTRA_IS_FAILOVER：值为 true 说明当前的连接是首选网络失效重连后的连接。
- ❑ ConnectivityManager. EXTRA_NO_CONNECTIVITY：值为 true 表示当前

设备没有连接到网络。

- ConnectivityManager. EXTRA_REASON：如果当前的广播代表网络失效，则这个值包含了连接失效的原因描述。
- ConnectivityManager. EXTRA_NETWORK_INFO：返回 NetworkInfo 对象，包含了当前连接事件相关的网络详细信息。
- ConnectivityManager. EXTRA_OTHER_NETWORK_INFO：在一个网络断开连接时，这个值返回 NetworkInfo 对象，包含了可能的网络失效重连的详细信息。
- ConnectivityManager. EXTRA_EXTRA_INFO：包含可选的网络连接的额外信息。

4. WIFI 连接

使用 WiFi Manager 可以进行网络配置，控制连接到哪个网络。当连接建立后，可以进一步获取活动网络连接的额外配置信息。

下面通过实例介绍如何获取当前使用的 WIFI 连接信息以及如何扫描 WIFI 热点。

(1) 新建工程 chapter17_8，编辑 res/layout/activity_main. xml 文件，源代码参考配套资料。

(2) 由于需要获取 WIFI 连接信息，所以需要添加相关权限。编辑 AndroidManifest. xml 文件，源代码如下：

```
<manifest xmlns:android = "http://schemas.android.com/apk/res/android"
    package = "com.example.chapter17_8"
    android:versionCode = "1"
    android:versionName = "1.0" >
    <!-- 访问 WIFI 状态权限 -->
    <uses-permission android:name = "android.permission.ACCESS_WIFI_STATE"/>
    <!-- 扫描 WIFI 热点所需权限 -->
    <uses-permission android:name = "android.permission.CHANGE_WIFI_STATE"/>
    <!-- 访问网络权限 -->
    <uses-permission android:name = "android.permission.INTERNET"/>
    <uses-sdk   android:minSdkVersion = "8"   android:targetSdkVersion = "15" />
    <application
        android:icon = "@drawable/ic_launcher"
        android:label = "@string/app_name"
        android:theme = "@style/AppTheme" >
        <activity
            android:name = ".MainActivity"
            android:label = "@string/title_activity_main" >
            <intent-filter>
```

```
            <action android:name = "android.intent.action.MAIN" />
            <category android:name = "android.intent.category.LAUNCHER" />
        </intent-filter>
    </activity>
</application>
</manifest>
```

(3) 编辑 MainActivity.java 文件,源代码如下:

```
public class MainActivity extends Activity implements OnClickListener
{
    private static final String TAG = "MainActivity";
    TextView showCurrentWIFI; // 声明显示当前 WIFI 状态信息文本
    Button getCurrentWIFIBtn;// 获取当前 WIFI 状态按钮
    Button scanWIFIBtn; // 扫描 WIFI 热点按钮
    WifiManager wifiManager = null;// WIFI 管理器
    @Override
    public void onCreate(Bundle savedInstanceState)
    {
        super.onCreate(savedInstanceState);
        setContentView(R.layout.activity_main);
        showCurrentWIFI = (TextView) findViewById(R.id.show_current_wifi);
        getCurrentWIFIBtn = (Button) findViewById(R.id.get_current_wifi);
        getCurrentWIFIBtn.setOnClickListener(this);
        scanWIFIBtn = (Button) findViewById(R.id.scan_wifi);
        scanWIFIBtn.setOnClickListener(this);
        wifiManager = (WifiManager) getSystemService(Context.WIFI_SERVICE);// 实例
// 化 WIFI 管理器
    }
    @Override
    public void onClick(View v)
    {
        int viewId = v.getId();
        switch (viewId)
        {
            case R.id.get_current_wifi: // 获取当前 WIFI 状态按钮
                WifiInfo info = wifiManager.getConnectionInfo();
                String maxText = info.getMacAddress();
                String ipText = intToIp(info.getIpAddress());
                String status = "";
                if (wifiManager.getWifiState() == WifiManager.WIFI_STATE_ENABLED)
                { //如果 WIFI 可用
                    status = "WIFI_STATE_ENABLED";
```

```
                }
                String ssid = info.getSSID();
                int networkID = info.getNetworkId();
                int speed = info.getLinkSpeed();
                showCurrentWIFI.setText("mac:" + maxText + "\n\r" + "ip:" + ip-
Text + "\n\r" + "wifi status :" + status + "\n\r" + "ssid :" + ssid + "\n\r" + "net
work id :" + networkID + "\n\r"
                    + "connection speed:" + speed + "\n\r");
                break;

        case R.id.scan_wifi: // 扫描 WIFI 热点按钮
            registerReceiver(new BroadcastReceiver()
            {
                @Override
                public void onReceive(Context context, Intent intent)
                {
                  List<ScanResult> results = wifiManager.getScanResults();
                  ScanResult bestSignal = null;
                  for (ScanResult result : results) {//迭代 results 列表
                     if (null == bestSignal || WifiManager.compareSignalLevel(be-
                        stSignal.level, result.level) < 0)
                     {//如果 bestSignal 对象为 null 或者是 result.level 信号强度千古
                        bestSignal.level
                        bestSignal = result;
                     }
                  }
                  Log.d(TAG, results.size() + " networks found, " + bestSignal.
                        SSID + " is the strongest");
                }
            }, new IntentFilter(WifiManager.SCAN_RESULTS_AVAILABLE_ACTION));
            wifiManager.startScan();
            break;
        }
    }
    private String intToIp(int ip)
    {
        return (ip & 0xFF) + "." + ((ip >> 8) & 0xFF) + "." + ((ip >> 16) &
0xFF) + "." + ((ip >> 24) & 0xFF);
    }
    @Override
    public boolean onCreateOptionsMenu(Menu menu) {
      getMenuInflater().inflate(R.menu.activity_main, menu);
```

```
        return true;
    }
}
```

(4)右击工程 chapter17_8，在弹出的快捷菜单中选择 Run As|Android Application 命令，单击界面上的“获取当前 WIFI 连接”按钮，运行效果如图 17－11 所示，单击界面上的“扫描 WIFI 热点”按钮，Logcat 输出日志信息如图 17－12 所示。

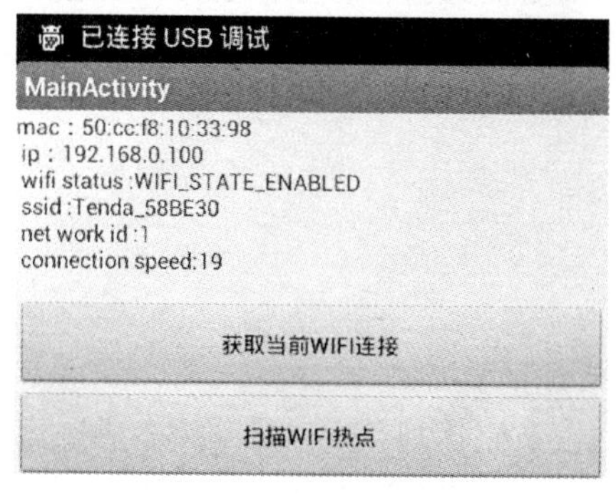

图 17－11　当前连接 WIFI 信息

图 17－12　扫描 WIFI 信息

5. 监控 WIFI 连接

在 WIFI 网络连接状态改变时，WIFI Manager 将广播相应的 intent，有如下几种：

- ❑ WifiManager. WIFI_STATE_CHANGED_ACTION：标识 WiFi 硬件状态改变，可能在 enabling、enabled、disabling、disabled 和 unknown 之间改变。它包含了两个额外的键值 EXTRA_WIFI_STATE 和 EXTRA_PREVIOUS_STATE，分别表示新的和前一个 WiFi 状态。
- ❑ WifiManager. SUPPLICANT_CONNECTION_CHANGE_ACTION：WiFi 硬件和当前接入点之间的连接状态改变时，这个 intent 将被广播。额外的键值 EXTRA_NEW_STATE 表示是新的连接建立事件还是已存在的连接中断事件，为 true 表示新的连接建立。
- ❑ WifiManager. NETWORK_STATE_CHANGED_ACTION：WiFi 连接状态改变时广播这个 intent，包含两个额外的键值：一是 EXTRA_NETWORK_INFO，它包含表示当前网络状态的 NetworkInfo 对象；一是 EXTRA_BSSID，包含连接到的接入点的 BSSID 值。
- ❑ WifiManager. RSSI CHANGED ACTION：监听这个 intent 可以使应用程序监控当前 WIFI 连接的信号强度。包含一个额外键值 EXTRA_NEW_RSSI，包含了当前信号强度。使用这个信号强度，需要使用静态函数 calculateSignalLevel 将这个值按指定的缩放转换为整型值。

第 18 章

Android 桌面组件

在本章中，笔者将会介绍如何使用 Android 系统的桌面组件，包括 Live Folder（实时文件夹）的使用、Shortcuts（快捷方式的使用）以及桌面插件（Widget）的使用。

18.1 实时文件夹

Android 的实时文件夹是在 SDK—1.5 版本当中引入的，支持开发人员在设备的默认打开屏幕（也就是设备的主页）上显示 ContentProvider 提供的数据，如联系人信息。将联系人信息在主页上设置为实时文件夹之后，在联系人数据库中添加、删除或修改联系人时，此实时文件夹能够刷新自身所包含的内容。

Android 中的实时文件夹对 ContentProvider 的作用就相当于 RSS 阅读器对发布网站的作用。Content Provider 也是类似于根据 URI 提供信息的网站。随着网站的迅速增加，每个网站会以独特的方式发布自己的信息，这就需要集中多个网站的信息，以便用户可以通过单一阅读器了解网站的最新动态。为此，RSS 应运而生。RSS 强制在不同的信息集之间提供一种通用的使用模式。有了通用模式，只需设计一次阅读器，就可以使用它阅读任何内容，只需该内容具有统一的结构即可。

实时文件夹在概念上也没有什么不同。就像 RSS 阅读器为所发布的网站内容提供通用的接口一样，实时文件夹也为 Android 中的 ContentProvider 定义了一种通用接口。只要 ContentProvider 遵守此协议，Android 就能够在设备的主页上创建活动文件夹图标来表示该 ContentProvider。当用户单击此活动文件夹图标时，系统将联系相应的 ContentProvider。相应的 ContentProvider 应该会返回一个游标。根据活动文件夹的契约，此游标必须具有一组预定义的列。此游标通过 ListView 或 GridView 直观地显示出来。

Android 实时文件夹，为什么说它是实时的，因为它可以根据设备后台数据库的变化更新自身的 UI 界面，这样无论什么时候显示的内容都是最新的。例如，用户删除了一条联系人信息，相应的 ContentProvider 所对应的实时文件夹马上也会更新。

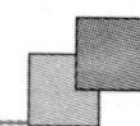

18.1.1 使用实时文件夹

(1) 新建一个 Android—2.2 模拟器,打开模拟器,来到模拟器主页(默认屏幕)。

(2) 在主页的空白处长按(大约 2 秒钟左右),就可以看到一个可供用户单击的上下文菜单了,如图 18-1 所示。

(3) 找到一个名为 Folders(实时文件夹)的上下文菜单选项,单击此选项就可以查看所有可用的实时文件夹的一个列表,如图 18-2 所示。

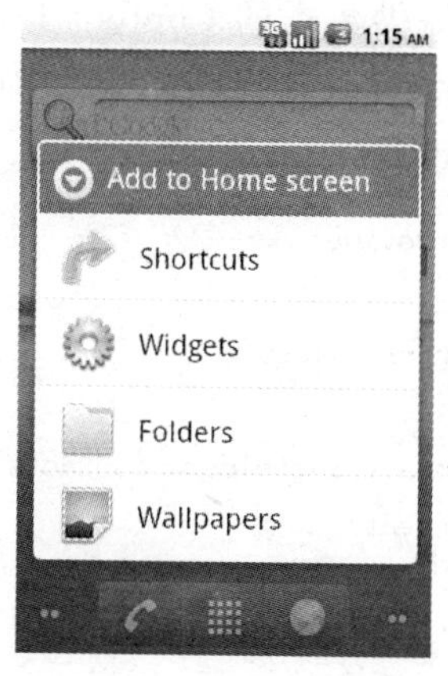

图 18-1 Add to Home Screen 效果

图 18-2 单击 Folders 选项效果

(4) 从列表中选择并单击希望在主页上公开的实时文件夹名称,这会在主页上创建一个图标来表示所选的实时文件夹。

(5) 笔者选中并单击的是 All contacts 这个选项,单击过后,主页会生成一个 All contacts 图标,如图 18-3 所示。

(6) 单击桌面主页新生成的 All contacts 图标,就可以查看到所有联系人的一个列表,如图 18-4 所示。

图 18-3 选中 All contacts 主页效果

图 18-4 单击主页 All contacts 图标运行效果

(7) 单击这个列表中的任何一项，就可以进入到相应的联系人的信息编辑页面。

18.1.2 实时文件夹实例

创建一个自定义的实时文件夹，需要一个 Activity 以及一个提供数据的 Content Provider。Activity 的标签会作为类似 All contacts 这样的实时文件夹的名称出现在用户单击 Folders 上下文菜单选项所看到的那个所有可用的实时文件夹的列表当中。下面就通过实例介绍如何创建一个自定义的实时文件夹。

(1) 新建工程 chapter18_1，新建一个用来提供数据的 Content Provider：MyContactsProvider，当用户单击实时文件夹的图标时，系统将调用此 URL 来检索数据。编辑 MyContactsProvider.java 文件，源代码如下：

```
public class MyContactsProvider extends ContentProvider
{
    private static final String TAG = "MyContactsProvider";
    //内容提供者的名称
    public static final String AUTHORITY = "com.ai.livefolders.contacts";
    //Uri that goes as input to the live-folder creation
    public static final Uri CONTACTS_URI = Uri.parse("content://" + AUTHORITY + "/
contacts");
    //区分 URI 类型
    private static final int TYPE_MY_URI = 0;
    private static final UriMatcher URI_MATCHER;
    static
    {
        URI_MATCHER = new UriMatcher(UriMatcher.NO_MATCH);
        URI_MATCHER.addURI(AUTHORITY, "contacts", TYPE_MY_URI);
    }
    @Override
    public boolean onCreate()
    {
        return true;
    }
    @Override
    public int bulkInsert(Uri arg0, ContentValues[] values)
    {
        return 0; //nothing to insert
    }
    //实时文件夹需要的列
    private static final String[] CURSOR_COLUMNS = new String[] { BaseColumns._ID,
LiveFolders.NAME, LiveFolders.DESCRIPTION, LiveFolders.INTENT, LiveFolders.ICON_PACKAGE,
LiveFolders.ICON_RESOURCE };
```

```
        //没有数据显示时的错误信息列
        private static final String[] CURSOR_ERROR_COLUMNS = new String[] { BaseColumns._
ID, LiveFolders.NAME, LiveFolders.DESCRIPTION };
        //没有数据显示时的错误信息
        private static final Object[] ERROR_MESSAGE_ROW = new Object[] { -1, //id
            "No contacts found", //名称
            "Check your contacts database" //描述
        };
        //没有数据显示时返回的 Cursor 对象
        private static MatrixCursor sErrorCursor = new MatrixCursor(CURSOR_ERROR_COLUMNS);
        static
        {
            sErrorCursor.addRow(ERROR_MESSAGE_ROW);
        }
        //需要从数据库获取的联系人的字段
        private static final String[] CONTACTS_COLUMN_NAMES = new String[]
        {
            People._ID,
            People.DISPLAY_NAME,
            People.TIMES_CONTACTED,
            People.STARRED
        };
        @Override
        public Cursor query(Uri uri, String[] projection, String selection, String[] se-
lectionArgs, String sortOrder)
        {
            //进行 URI 的匹配
            int type = URI_MATCHER.match(uri);
            if(UriMatcher.NO_MATCH == type)
            {
                Log.d(TAG, "查询取消");
                return sErrorCursor;
            }
            try
            {
                //调用 loadNewData 方法进行查询,返回匹配好的 MatrixCursor 对象
                MatrixCursor mc = loadNewData(this);
                //setNotificationUri 方法用来指定一个 Uri 用来观察它的变化
                //参数一:需要一个 ContentResolver 对象,从上下文对象获得
                //参数二:要监视的 Uri
                mc.setNotificationUri(getContext().getContentResolver(), Uri.parse("
content://contacts/people/"));
```

```
                MyCursor wmc = new MyCursor(mc,this);
                return wmc;
            }
            catch(Throwable e)
            {
                //返回定义的错误对象
                return sErrorCursor;
            }
        }
        public static MatrixCursor loadNewData(ContentProvider cp)
        {
            MatrixCursor mc = new MatrixCursor(CURSOR_COLUMNS);
            Cursor allContacts = null;
            try
            {
                //query方法的参数
                /*
                 * 参数1:要查询的URI,我们查询的完整URI为"content://com.android.
contacts/contacts"
                 * 也就是所有的联系人数;
                 * 参数2:都要查询那些列,上边已经定义好;
                 * 参数3:本质上就是一个WHERE子句,它以SQL WHERE子句(不包含WHERE本
身)的格式声明要返回的行,传递null将返回给定URI的所有行
                 * 参数4:用来替换where子句中的?号
                 * 参数5:指定排序的方式
                 */
                allContacts = cp.getContext().getContentResolver().query(People.CON-
TENT_URI, CONTACTS_COLUMN_NAMES, null, null, People.DISPLAY_NAME); //按照名称排序
                while (allContacts.moveToNext())
                {
                  String timesContacted = "Times contacted: " + allContacts.getInt(2);
                  Object[] rowObject = new Object[]
                    { allContacts.getLong(0), //id
                      allContacts.getString(1), //名称
                      timesContacted, //描述
                        Uri.parse("content://contacts/people/" + allContacts.getLong
(0)), //intent
                      cp.getContext().getPackageName(), //package
                      android.R.drawable.ic_dialog_dialer //返回列表每一行的图标
                  };
                  mc.addRow(rowObject);
                }
```

```
            return mc;
        }
        finally
        {
            //关闭 Cursor 对象,释放所有资源并使其无效
            allContacts.close();
        }
    }
    //根据指定的 URI 返回相应的 MIME 类型,例如:"vnd.android.cursor.dir/vnd.google.note"
    public String getType(Uri uri)
    {
        return People.CONTENT_TYPE;
    }
    //不开放 Insert 接口
    public Uri insert(Uri uri, ContentValues initialValues)
    {
        throw new UnsupportedOperationException("no insert as this is just a wrapper");
    }
    //不开放 delete 接口
    @Override
    public int delete(Uri uri, String selection, String[] selectionArgs)
    {
        throw new UnsupportedOperationException("no delete as this is just a wrapper");
    }
    //不开放 update 接口
    public int update(Uri uri, ContentValues values, String selection, String[] selectionArgs)
    {
        throw new UnsupportedOperationException("no update as this is just a wrapper");
    }
}
```

首先来解释一下 timesContacted 所代表的意义,例如读者手机里存着“张三”的号码,那么 timeContacted 就代表“张三”这个人呼叫了读者的次数。

loadNewData()方法读取所有联系人,并返回 MatrixCursor 对象,MatrixCursor 对象包含实时文件夹所要显示的字段。之后,MatrixCursor 对象指定了一个 Uri 对象来观察它的变化,当要监视的 Uri 数据发生变化的时候,ContentResolver 对象会主动更新 MatrixCursor 对象。

MyCursor 是笔者自定义的 个 Cursor 类,这里要理解为什么需要对 MatrixCursor 对象进行封装,以及需要了解视图如何更新更改的内容。由于 MatrixCursor 对象注册了对指定 Uri 对象的监听,所以当指定 Uri 的数据发生变化的时候,会调用 MatrixCursor 对象的 requery()方法。

由于 MatrixCursor 自身的一些限制，所以笔者将其包装到一个游标包装器中，并重写 requery()方法用来丢弃内部的 MatrixCursor 对象，并使用更新的数据创建一个新游标。

(2) 新建自定义 Cursor：MyCursor，编辑 MyCursor.java 文件，源代码如下：

```
public class MyCursor extends BetterCursorWrapper
{
    private ContentProvider mcp = null;
    public MyCursor(MatrixCursor mc, ContentProvider inCp)
    {
        super(mc);
        mcp = inCp;
    }
    public boolean requery()
    {
        //每次 URI 的数据更新时，调用 requery()方法返回新的 MatrixCursor 对象
        MatrixCursor mc = MyContactsProvider.loadNewData(mcp);
        this.setInternalCursor(mc);
        return super.requery();
    }
}
```

(3) 编辑 BetterCursorWrapper.java 文件，源代码如下：

```
//BetterCursorWrapper 用来装饰包装 CrosssProcessCursor
public class BetterCursorWrapper implements CrossProcessCursor
{
  protected CrossProcessCursor internalCursor;
  public BetterCursorWrapper(CrossProcessCursor inCursor)
    {
        this.setInternalCursor(inCursor);
    }
    public void setInternalCursor(CrossProcessCursor inCursor)
    {
    internalCursor = inCursor;
    }
    @Override
    public void close()
    {
        internalCursor.close();
    }
    @Override
    public void copyStringToBuffer(int columnIndex, CharArrayBuffer buffer)
```

```
{
    internalCursor.copyStringToBuffer(columnIndex, buffer);
}
@Override
public void deactivate()
{
    internalCursor.deactivate();
}
@Override
public byte[] getBlob(int columnIndex)
{
    return internalCursor.getBlob(columnIndex);
}
@Override
public int getColumnCount()
{
    return internalCursor.getColumnCount();
}
@Override
public int getColumnIndex(String columnName)
{
    return internalCursor.getColumnIndex(columnName);
}
@Override
public int getColumnIndexOrThrow(String columnName) throws IllegalArgumentException
{
    return internalCursor.getColumnIndexOrThrow(columnName);
}
@Override
public String getColumnName(int columnIndex)
{
    return internalCursor.getColumnName(columnIndex);
}
@Override
public String[] getColumnNames()
{
    return internalCursor.getColumnNames();
}
@Override
public int getCount()
{
    return internalCursor.getCount();
```

```
}
@Override
public double getDouble(int columnIndex)
{
    return internalCursor.getDouble(columnIndex);
}
@Override
public Bundle getExtras()
{
    return internalCursor.getExtras();
}
@Override
public float getFloat(int columnIndex)
{
    return internalCursor.getFloat(columnIndex);
}
@Override
public int getInt(int columnIndex)
{
    return internalCursor.getInt(columnIndex);
}
@Override
public long getLong(int columnIndex)
{
    return internalCursor.getLong(columnIndex);
}
@Override
public int getPosition()
{
    return internalCursor.getPosition();
}
@Override
public short getShort(int columnIndex)
{
    return internalCursor.getShort(columnIndex);
}
@Override
public String getString(int columnIndex)
{
    return internalCursor.getString(columnIndex);
}
@Override
```

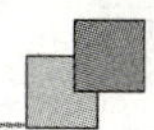

```
public boolean getWantsAllOnMoveCalls()
{
    return internalCursor.getWantsAllOnMoveCalls();
}
@Override
public boolean isAfterLast()
{
    return internalCursor.isAfterLast();
}
@Override
public boolean isBeforeFirst()
{
    return internalCursor.isBeforeFirst();
}
@Override
public boolean isClosed()
{
    return internalCursor.isClosed();
}
@Override
public boolean isFirst()
{
    return internalCursor.isFirst();
}
@Override
public boolean isLast()
{
    return internalCursor.isLast();
}
@Override
public boolean isNull(int columnIndex)
{
    return internalCursor.isNull(columnIndex);
}
@Override
public boolean move(int offset)
{
    return internalCursor.move(offset);
}
@Override
public boolean moveToFirst()
{
```

```
    return internalCursor.moveToFirst();
}
@Override
public boolean moveToLast()
{
    return internalCursor.moveToLast();
}
@Override
public boolean moveToNext()
{
    return internalCursor.moveToNext();
}
@Override
public boolean moveToPosition(int position)
{
    return internalCursor.moveToPosition(position);
}
@Override
public boolean moveToPrevious()
{
    return internalCursor.moveToPrevious();
}
@Override
public void registerContentObserver(ContentObserver observer)
{
    internalCursor.registerContentObserver(observer);
}
@Override
public void registerDataSetObserver(DataSetObserver observer)
{
    internalCursor.registerDataSetObserver(observer);
}
@Override
public boolean requery()
{
    return internalCursor.requery();
}
@Override
public Bundle respond(Bundle extras)
{
    return internalCursor.respond(extras);
}
```

```
    @Override
    public void setNotificationUri(ContentResolver cr, Uri uri)
    {
        internalCursor.setNotificationUri(cr, uri);
    }
    @Override
    public void unregisterContentObserver(ContentObserver observer)
    {
        internalCursor.unregisterContentObserver(observer);
    }

    @Override
    public void unregisterDataSetObserver(DataSetObserver observer) {
        internalCursor.unregisterDataSetObserver(observer);
    }
    @Override
    public void fillWindow(int position, CursorWindow window)
    {
        internalCursor.fillWindow(position, window);
    }
    @Override
    public CursorWindow getWindow()
    {
        return internalCursor.getWindow();
    }
    @Override
    public boolean onMove(int oldPosition, int newPosition)
    {
        return internalCursor.onMove(oldPosition, newPosition);
    }
}
```

BetterCursorWrapper 类非常类似于 Android 数据库框架中的 CursorWrapper 类。但是实时文件夹还需要 CursorWrapper 所缺少的另外两项功能。首先，它没有提供 set 方法来替换 requery 方法中的内部游标。其次，CursorWrapper 不是 CrossProcessCursor。活动文件夹需要 CrossProcessCursor，而不是普通游标，因为实时文件夹会跨进程进行工作。

（4）新建活动：AllContactsLiveFolderCreatorActivity，编辑 AllContactsLiveFolderCreatorActivity.java 文件，源代码如下：

```
public class AllContactsLiveFolderCreatorActivity extends Activity
{
```

```
    @Override
    protected void onCreate(Bundle savedInstanceState)
    {
        super.onCreate(savedInstanceState);
        final Intent intent = getIntent();
        final String action = intent.getAction();
         if (LiveFolders.ACTION_CREATE_LIVE_FOLDER.equals(action))
        {//如果是创建实时文件夹意图,返回一个创建实时文件夹的意图对象
            setResult(RESULT_OK, createLiveFolder(MyContactsProvider.CONTACTS_URI,
"Contacts LF",    android.   R.drawable.ic_dialog_dialer));
        }
         else
         {//如果不是创建实时文件夹意图
            setResult(RESULT_CANCELED);
        }
        finish();
    }
    private Intent createLiveFolder(Uri uri, String name, int icon)
    {
        final Intent intent = new Intent();
        intent.setData(uri);
        intent.putExtra(LiveFolders.EXTRA_LIVE_FOLDER_NAME, name);
        intent.putExtra(LiveFolders.EXTRA_LIVE_FOLDER_ICON,
                        Intent.ShortcutIconResource.fromContext(this, icon));
        intent.putExtra(LiveFolders.EXTRA_LIVE_FOLDER_DISPLAY_MODE,
                        LiveFolders.DISPLAY_MODE_LIST);
        return intent;
    }
}
```

AllContactsLiveFolderCreatorActivity 类只有一个功能,担当实时文件夹的创建程序。Activity 的标签会作为类似 All contacts 的实时文件夹的名称出现在用户单击 Folders 上下文菜单选项所看到的那个所有可用的实时文件夹的列表当中。

此活动的 createLiveFolder()方法将会告诉 Android 系统:

- ❑ 实时文件夹的名称。
- ❑ 实时文件夹所使用的图标。
- ❑ 实时文件夹的显示模式:列表或网络。
- ❑ 实时文件夹的数据来源 Uri。

(5) 编辑 AndroidManifest.xml 文件,源代码如下:

```
<manifest xmlns:android="http://schemas.android.com/apk/res/android"
```

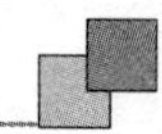

```
    package = "com.example.chapter18_1"
    android:versionCode = "1"
    android:versionName = "1.0" >
    <!-- 读取联系人权限 -->
    <uses-permission android:name = "android.permission.READ_CONTACTS" ></uses-permission>
    <uses-sdk android:minSdkVersion = "8"   android:targetSdkVersion = "15" />
    <application
        android:icon = "@drawable/ic_launcher"
        android:label = "@string/app_name"
        android:theme = "@style/AppTheme" >
        <activity
            android:name = ".MainActivity"
            android:label = "@string/title_activity_main" >
            <intent-filter>
                <action android:name = "android.intent.action.MAIN" />
                <category android:name = "android.intent.category.LAUNCHER" />
            </intent-filter>
        </activity>
       <!-- 实时文件夹 Activity -->
       <activity
            android:name = ".AllContactsLiveFolderCreatorActivity"
            android:icon = "@drawable/ic_launcher"
            android:label = "New live folder " >
            <intent-filter>
                <action android:name = "android.intent.action.CREATE_LIVE_FOLDER" />
                <category android:name = "android.intent.category.DEFAULT" />
            </intent-filter>
        </activity>
         <!-- 实时文件夹 Content Provider -->
        <provider
            android:name = ".MyContactsProvider"
            android:authorities = "com.ai.livefolders.contacts"
            android:multiprocess = "true" />
    </application>
</manifest>
```

(6) 右击工程 chapter18_1，在弹出的快捷菜单中选择 Run As|Android Application 命令进入模拟器主页长按，选择并单击弹出的上下文菜单的 Folders 选项，运行效果如图 18-5 所示，单击主页生成的 Contacts LF 图标，运行效果如图 18-6 所示。

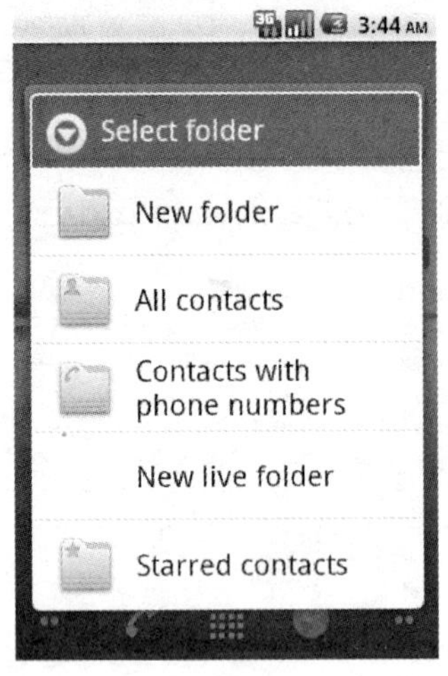

图 18-5 New live folder

图 18-6 选中 Contacts LF 效果

18.2 快捷方式

Android 快捷方式也是桌面的一种组件，它是用于直接启动某一应用程序的某个组件。一般情况下，可以在 Launcher 的应用程序列表上，通过长按某一个应用程序的图标，可以在主页上创建该应用程序的快捷方式。下面就通过实例介绍如何添加应用程序的快捷方式，如何查询应用程序是否创建了快捷方式以及如何删除应用程序的快捷方式。

(1) 新建工程 chapter18_2，编辑 res/layout/activity_main.xml 文件，源代码如下：

```
<LinearLayout xmlns:android = "http://schemas.android.com/apk/res/android"
    xmlns:tools = "http://schemas.android.com/tools"
    android:layout_width = "fill_parent"
    android:layout_height = "fill_parent"
    android:orientation = "vertical">
    <!-- 创建应用程序的快捷方式 -->
    <Button
          android:id = "@ + id/create_shortcut"
          android:layout_width = "fill_parent"
          android:layout_height = "wrap_content"
          android:text = "创建快捷方式"/>
    <Button
          android:id = "@ + id/query_shortcut"
          android:layout_width = "fill_parent"
```

```
        android:layout_height = "wrap_content"
        android:text = "查询快捷方式"/>
    <!-- 删除应用程序的快捷方式 -->
        <Button
        android:id = "@ + id/delete_shortcut"
        android:layout_width = "fill_parent"
        android:layout_height = "wrap_content"
        android:text = "删除快捷方式"/>
</LinearLayout>
```

(2) 由于添加快捷方式需要使用相关权限，所以编辑 AndroidManifest.xml 文件，源代码如下：

```
<manifest xmlns:android = "http://schemas.android.com/apk/res/android"
    package = "com.example.chapter18_2"
    android:versionCode = "1"
    android:versionName = "1.0" >
    <!-- Android 创建快捷方式权限 -->
    <uses-permission android:name = "com.android.launcher.permission.INSTALL_SHORTCUT"/>
    <!-- Android 删除快捷方式权限 -->
    <uses-permission android:name = "com.android.launcher.permission.UNINSTALL_
SHORTCUT"/>
    <uses-permission android:name = "com.android.launcher.permission.READ_SETTINGS" />
    <uses-permission android:name = "com.android.launcher.permission.WRITE_SETTINGS" />
    <uses-sdk    android:minSdkVersion = "8"    android:targetSdkVersion = "15" />
    <application
        android:icon = "@drawable/ic_launcher"
        android:label = "@string/app_name"
        android:theme = "@style/AppTheme" >
        <activity
            android:name = ".MainActivity"
            android:label = "@string/title_activity_main" >
            <intent-filter>
                <action android:name = "android.intent.action.MAIN" />
                <category android:name = "android.intent.category.LAUNCHER" />
            </intent-filter>
        </activity>
    </application>
</manifest>
```

(3) 编辑 MainActivity.java 文件，源代码如下：

```
public class MainActivity extends Activity implements OnClickListener ,
{
    //创建快捷方式意图 Action
```

```
        private static final String ACTION_INSTALL_SHORTCUT = "com.android.launcher.action.INSTALL_SHORTCUT";
        //删除快捷方式意图 Action
        private static final String ACTION_UNINSTALL_SHORTCUT = "com.android.launcher.action.UNINSTALL_SHORTCUT";
        Button createShortCutBtn;                       //声明创建快捷方式按钮
        Button queryShortCutBtn;                        //声明查询快捷方式按钮
        Button deleteShortCutBtn;                       //声明删除快捷方式按钮
        @Override
        public void onCreate(Bundle savedInstanceState)
        {
            super.onCreate(savedInstanceState);
            setContentView(R.layout.activity_main);
            createShortCutBtn = (Button)findViewById(R.id.create_shortcut);
            queryShortCutBtn = (Button)findViewById(R.id.query_shortcut);
            deleteShortCutBtn = (Button)findViewById(R.id.delete_shortcut);
            createShortCutBtn.setOnClickListener(this);//绑定单击事件监听器
            queryShortCutBtn.setOnClickListener(this);
            deleteShortCutBtn.setOnClickListener(this);
        }
        @Override
        public void onClick(View v)
        {
        int viewId = v.getId();
        switch(viewId)
        {
            case R.id.create_shortcut:                  //创建快捷方式按钮
                createShortCut();
                break;
            case R.id.query_shortcut:                   //查询快捷方式是否创建按钮
                queryShortCut();
                break;
            case R.id.delete_shortcut:                  //删除快捷方式按钮
                deleteShortCut();
                break;
            default:
                break;
        }
        }
        //创建快捷方式
        private void createShortCut()
        {
            Intent shortcutIntent = new Intent(ACTION_INSTALL_SHORTCUT);
            //快捷方式的名称
```

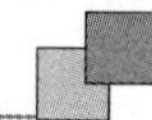

```
        shortcutIntent.putExtra(Intent.EXTRA_SHORTCUT_NAME, getString(R.string.app_name));
        //不允许重复创建
        shortcutIntent.putExtra("duplicate", false);
        ComponentName comp = new ComponentName(this.getPackageName(), ".MainActivity");
        shortcutIntent.putExtra(Intent.EXTRA_SHORTCUT_INTENT, new Intent(
            Intent.ACTION_MAIN).setComponent(comp));
        //快捷方式图片
        shortcutIntent.putExtra(Intent.EXTRA_SHORTCUT_ICON_RESOURCE, Intent.Short-
cutIconResource.fromContext(this, android.R.drawable.ic_menu_call));
        sendBroadcast(shortcutIntent);
    }
    //查询快捷方式按钮
    private void queryShortCut()
    {
        boolean isInstallShortcut = false;
        final ContentResolver cr = this.getContentResolver();
        final String AUTHORITY ;
        //在 andriod 2.1 即 SDK7 以上,是读取 launcher.settings 中的 favorites 表的数据
        //在 andriod 2.2 即 SDK8 以上,是读取 launcher2.settings 中的 favorites 表的数据
        if(getSystemVersion() < 8)
      { //如果 android2.2 以下系统
          AUTHORITY = "com.android.launcher.settings";
      }
      else
      { //android2.2 以上系统(包括 2.2)
        AUTHORITY = "com.android.launcher2.settings";
      }
        final Uri CONTENT_URI = Uri.parse("content://" + AUTHORITY + "/favorites?
notify=true");
        Cursor c = cr.query(CONTENT_URI, new String[] { "title", "iconResource" }, "
title=?", new String[] { getString(R.string.app_name).trim() }, null);
        if (c != null && c.getCount() > 0)
        {//如果游标对象不为 null 且 getCount()大于 0
            isInstallShortcut = true;
        }
         Toast.makeText(this, "hasInstallShortCut:" + isInstallShortcut, Toast.
LENGTH_LONG).show();
    }
    /**
     * 返回系统 SDK 版本号
     * @return
     */
    public static int getSystemVersion()
    {
```

```
            return android.os.Build.VERSION.SDK_INT;
        }
        //删除快捷方式
        private void deleteShortCut()
        {
            Intent shortcut = new Intent(ACTION_UNINSTALL_SHORTCUT);
            //快捷方式的名称
            shortcut.putExtra(Intent.EXTRA_SHORTCUT_NAME, getString(R.string.app_name));
            //指定当前的 Activity 为快捷方式启动的对象:如 com.everest.video.VideoPlayer
            //注意: ComponentName 的第二个参数必须是完整的类名(包名 + 类名),否则无法
            //删除快捷方式
            String appClass = this.getPackageName() + ".MainActivity";
            ComponentName comp = new ComponentName(this.getPackageName(), appClass);
            shortcut.putExtra(Intent.EXTRA_SHORTCUT_INTENT,  new Intent(Intent.ACTION_
MAIN).setComponent(comp));
            sendBroadcast(shortcut);
        }
        @Override
        public boolean onCreateOptionsMenu(Menu menu) {
            getMenuInflater().inflate(R.menu.activity_main, menu);
            return true;
        }
    }
```

(4) 右击工程 chapter18_2,在弹出的快捷菜单中选择 Run As|Android Application 命令,单击界面上的“创建快捷方式”按钮,运行效果如图 18-7 所示(主页上多了一个名为 chapter18_2 的快捷方式),单击界面上的“删除快捷方式”按钮,运行效果如图 18-8 所示。

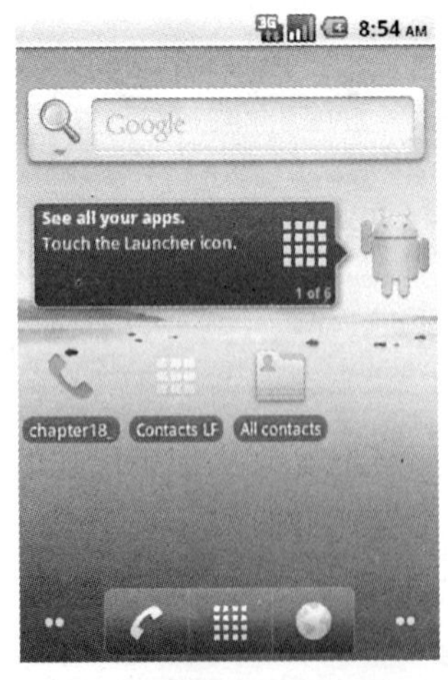

图 18-7 chapter18_2 应用程序快捷方式

图 18-8 chapter18_2 删除快捷方式按钮

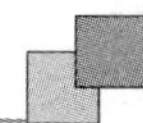

18.3 桌面插件(Widget)

Widget 是实现一个特定功能的一个视图部件(小插件),例如实时更新沪深股市指数信息,例如显示今天待办日程等功能。

18.3.1 使用 Widget

- 打开模拟器,进入模拟器主页(默认屏幕)。
- 在主页的空白处长按(大约 2 s),可以看到一个可供用户单击的上下文菜单。
- 找到一个名为 Widgets(桌面小插件)的上下文菜单选项,单击此选项可以查看所有可用的 Widget 的一个列表,如图 18-9 所示。
- 从列表中选择并单击希望在主页上显示的 Widget 名称,这会在主页上创建一块区域来显示这个桌面插件,如图 18-10 所示(笔者选择 Music)。

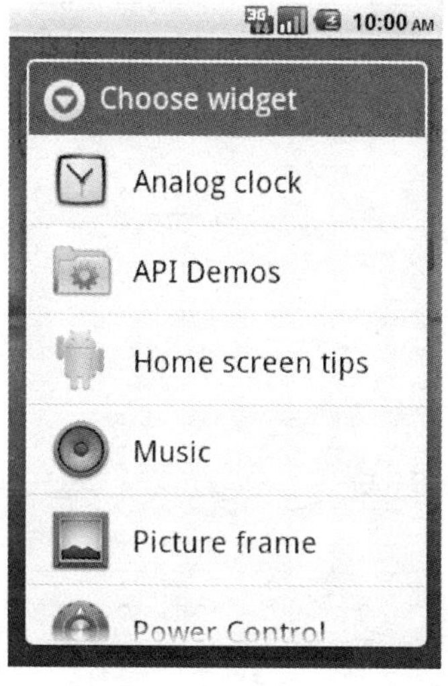

图 18-9 Widget 列表

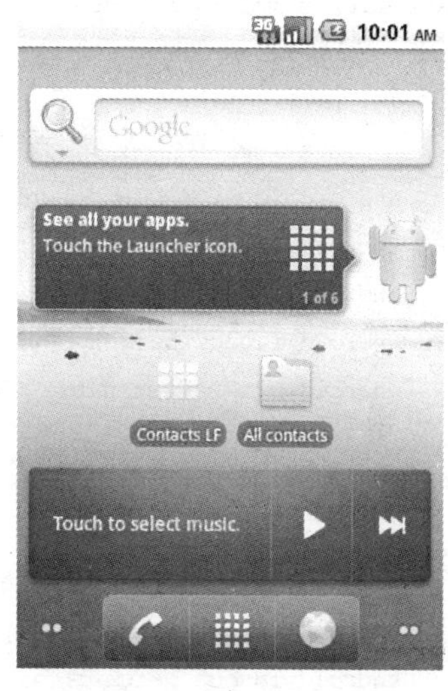

图 18-10 Music Widget 效果

18.3.2 AppWidget 框架类

Android 提供的 AppWidget 框架当中有如下几个重要的类:

- AppWidgetProvider:继承自 BroadcastRecevier,在 AppWidget 应用 update、enable、disable 和 delete 时接收通知。其中,onUpdate 和 onReceive 是最常用到的方法,它们接收更新通知。
- AppWidgetProviderInfo:描述 AppWidget 的大小、更新频率和初始界面等信息,以 XML 文件形式存在于应用的 res/xml/目录下。

- ❑ AppWidgetManager：负责管理 AppWidget ，向 AppwidgetProvider 发送通知。
- ❑ RemoteViews：一个可以在其他应用进程中运行的类，向 AppWidgetProvider 发送通知。

18.3.3 桌面插件(Widget)实例

下面就通过实例介绍如何创建一个自定义的桌面插件(Widget)。

(1) 新建工程 chapter18_3，将准备好的图片复制到 res/drawable-hdpi/目录下。

(2) 整个工程主要实现两个部分，一个是 AppWidget 部分，实现桌面 Widget 的显示、更新等，另一个部分就是单击 Widget 后出现的显示电池详细信息的 Activity 的实现。对于 Widget，首先需要编辑它的显示布局，所以新建一个布局文件 widget_layout. xml，编辑 widget_layout. xml 文件，源代码如下：

```
<?xml version="1.0" encoding="utf-8"?>
<RelativeLayout xmlns:android="http://schemas.android.com/apk/res/android"
    android:layout_width="fill_parent"
    android:layout_height="fill_parent">
        <ImageView
        android:id="@+id/imageView"
        android:layout_centerInParent="true"
        android:layout_width="wrap_content"
        android:layout_height="wrap_content"
        android:src="@drawable/j"
        />
        <TextView
        android:id="@+id/tv"
        android:layout_centerInParent="true"
        android:layout_width="wrap_content"
        android:layout_height="wrap_content"
        android:textColor="#000000"
        android:textStyle="bold"
        android:textSize="14sp"
        />
</RelativeLayout>
```

(3) 除了设置 Widget 的布局以外，还需要设置 Widget 所占区域的大小。在 res 目录下新建一个 xml 子目录用来存放 Widget 的大小配置文件。新建 Widget 的大小配置文件：battery_widget. xml。编辑 battery_widget. xml 文件，源代码如下：

```
<?xml version="1.0" encoding="utf-8"?>
<appwidget-provider xmlns:android="http://schemas.android.com/apk/res/android"
```

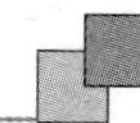

```
    android:minHeight = "72dip"
    android:minWidth = "72dip"
    android:updatePeriodMillis = "1000000"
    android:initialLayout = "@layout/newrelativelayout"    >
</appwidget-provider>
```

（4）新建一个继承自 android. appwidget. AppWidgetProvider 的子类：BatteryAppWidgetProvider。编辑 BatteryAppWidgetProvider. java 文件，源代码如下：

```
public class BatteryAppWidgetProvider extends AppWidgetProvider
{
    private static int currentBatteryLevel;
    private static int currentBatteryStatus;
    public void onUpdate(Context context,AppWidgetManager appWidgetManager,int[] app-
WidgetIds)
    {
        super.onUpdate(context, appWidgetManager, appWidgetIds);
        /** 启动自动更新电池信息的 service */
        context.startService(new Intent(context,updateService.class));
        /** 为 AppWidget 设置单击事件的响应，启动显示电池信息详情的 activity */
        Intent startActivityIntent = new Intent(context,MainActivity.class);
        PendingIntent intent = PendingIntent.getActivity(context,0,startActivity-
Intent,0);
        RemoteViews views = new RemoteViews(context.getPackageName(),R.layout.wid-
get_layout);
        //单击 widget_layout 布局文件的 id 为 imageview 的控件时，启动 MainActivity
        views.setOnClickPendingIntent(R.id.imageView,intent);
        appWidgetManager.updateAppWidget(appWidgetIds,views);
    }
    /** 自动更新电池信息的 service,通过 AlarmManager 实现定时不间断地发送电池信息 */
    public static class updateService extends Service
    {
        Bitmap bmp;    //定义机器人图片
        @Override
        public IBinder onBind(Intent intent)
        {
            //TODO Auto-generated method stub
            return null;
        }
        /** 定义一个接收电池信息的 broascastReceiver */
        private BroadcastReceiver batteryReceiver = new BroadcastReceiver()
        {
            @Override
```

```
            public void onReceive(Context context, Intent intent)
            {
                //TODO Auto-generated method stub
                currentBatteryLevel = intent.getIntExtra("level", 0);
                currentBatteryStatus = intent.getIntExtra("status", 0);
            }
        };
        public void onStart(Intent intent,int startId)
        {
            super.onStart(intent, startId);
            /** 注册接收器 */
             registerReceiver(batteryReceiver,new IntentFilter(Intent.ACTION_BAT-
TERY_CHANGED));
            /** 定义一个 AppWidgetManager */
            AppWidgetManager manager = AppWidgetManager.getInstance(this);
            /** 定义一个 RemoteViews,实现对 AppWidget 界面的控制 */
            RemoteViews views = new RemoteViews(getPackageName(),R.layout.widget_layout);
            if(currentBatteryStatus == 2||currentBatteryStatus == 5)
                                              //当正在充电或充满电时,显示充电的图片
            {
              if(currentBatteryLevel>= 95) //如果 currentBatteryLevel 的值大于等于 95
              {
              bmp = BitmapFactory.decodeResource(getResources(),R.drawable.jcharge);
              }
              else if(currentBatteryLevel>= 85&&currentBatteryLevel<95)
                                    //如果 currentBatteryLevel 的值大于等于 85 且小于 95
              {
              bmp = BitmapFactory.decodeResource(getResources(),R.drawable.icharge);
              }
              else if(currentBatteryLevel>= 75&&currentBatteryLevel<85)
                                    //如果 currentBatteryLevel 的值大于等于 75 且小于 85
              {
              bmp = BitmapFactory.decodeResource(getResources(),R.drawable.hcharge);
              }
              else if(currentBatteryLevel>= 65&&currentBatteryLevel<75)
                                    //如果 currentBatteryLevel 的值大于等于 65 且小于 75
              {
              bmp = BitmapFactory.decodeResource(getResources(),R.drawable.gcharge);
              }
              else if(currentBatteryLevel>= 55&&currentBatteryLevel<65)
                                    //如果 currentBatteryLevel 的值大于等于 55 且小于 65
              {
```

```
        bmp = BitmapFactory.decodeResource(getResources(),R.drawable.fcharge);
        }
        else if(currentBatteryLevel> = 45&&currentBatteryLevel<55)
                        //如果 currentBatteryLevel 的值大于等于 45 且小于 55
        {
        bmp = BitmapFactory.decodeResource(getResources(),R.drawable.echarge);
        }
        else if(currentBatteryLevel> = 35&&currentBatteryLevel<45)
                        //如果 currentBatteryLevel 的值大于等于 35 且小于 45
        {
        bmp = BitmapFactory.decodeResource(getResources(),R.drawable.dcharge);
        }
        else if(currentBatteryLevel> = 25&&currentBatteryLevel<35)
                        //如果 currentBatteryLevel 的值大于等于 25 且小于 35
        {
        bmp = BitmapFactory.decodeResource(getResources(),R.drawable.ccharge);
        }
        else if(currentBatteryLevel> = 15&&currentBatteryLevel<25)
                        //如果 currentBatteryLevel 的值大于等于 15 且小于 25
        {
        bmp = BitmapFactory.decodeResource(getResources(),R.drawable.bcharge);
        }
        else
        {
        bmp = BitmapFactory.decodeResource(getResources(),R.drawable.acharge);
        }
    }
    else                            //未在充电时,显示不在充电状态的系列图片
    {
        if(currentBatteryLevel> = 95)//如果 currentBatteryLevel 的值大于等于 95
        {
        bmp = BitmapFactory.decodeResource(getResources(),R.drawable.j);
        }
        else if(currentBatteryLevel> = 85&&currentBatteryLevel<95)
                        //如果 currentBatteryLevel 的值大于等于 85 且小于 95
        {
        bmp = BitmapFactory.decodeResource(getResources(),R.drawable.i);
        }
        else if(currentBatteryLevel> = 75&&currentBatteryLevel<85)
                        //如果 currentBatteryLevel 的值大于等于 75 且小于 85
        {
        bmp = BitmapFactory.decodeResource(getResources(),R.drawable.h);
```

```
            }
            else if(currentBatteryLevel>=65&&currentBatteryLevel<75)
                            //如果 currentBatteryLevel 的值大于等于 65 且小于 75
            {
            bmp = BitmapFactory.decodeResource(getResources(),R.drawable.g);
            }
            else if(currentBatteryLevel>=55&&currentBatteryLevel<65)
                            //如果 currentBatteryLevel 的值大于等于 55 且小于 65
            {
            bmp = BitmapFactory.decodeResource(getResources(),R.drawable.f);
            }
            else if(currentBatteryLevel>=45&&currentBatteryLevel<55)
                            //如果 currentBatteryLevel 的值大于等于 45 且小于 55
            {
            bmp = BitmapFactory.decodeResource(getResources(),R.drawable.e);
            }
            else if(currentBatteryLevel>=35&&currentBatteryLevel<45)
                            //如果 currentBatteryLevel 的值大于等于 35 且小于 45
            {
            bmp = BitmapFactory.decodeResource(getResources(),R.drawable.d);
            }
            else if(currentBatteryLevel>=25&&currentBatteryLevel<35)
                            //如果 currentBatteryLevel 的值大于等于 25 且小于 35
            {
            bmp = BitmapFactory.decodeResource(getResources(),R.drawable.c);
            }
            else if(currentBatteryLevel>=15&&currentBatteryLevel<25)
                            //如果 currentBatteryLevel 的值大于等于 15 且小于 25
            {
            bmp = BitmapFactory.decodeResource(getResources(),R.drawable.b);
            }
            else
            {
            bmp = BitmapFactory.decodeResource(getResources(),R.drawable.a);
            }
        }
        /** 设置 AppWidget 上显示的图片和文字的内容 */
        views.setImageViewBitmap(R.id.imageView,bmp);
        views.setTextViewText(R.id.tv,currentBatteryLevel + "%");
         ComponentName thisWidget = new ComponentName(this,BatteryAppWidgetPro-
vider.class);
        //android:updatePeriodMillis 的设置没有效果
```

```
            /** 使用 AlarmManager 实现每隔一秒发送一次更新提示信息,实现信息实时
动态变化 */
            long now = System.currentTimeMillis();
            long pause = 1000;
            Intent alarmIntent = new Intent();
            alarmIntent = intent;
            PendingIntent pendingIntent = PendingIntent.getService(this, 0, alarmIntent, 0);
            AlarmManager alarm = (AlarmManager) getSystemService(Context.ALARM_SERVICE);
            alarm.set(AlarmManager.RTC_WAKEUP, now + pause, pendingIntent);
            /** 更新 AppWidget */
            manager.updateAppWidget(thisWidget, views);
        }
    }
}
```

(5)编辑 res/layout/activity_main.xml 文件,源代码如下:

```
<?xml version = "1.0" encoding = "utf-8"?>
<LinearLayout xmlns:android = "http://schemas.android.com/apk/res/android"
    android:layout_width = "fill_parent"
    android:layout_height = "fill_parent"
    android:orientation = "vertical" >
    <TextView
        android:id = "@+id/tvBatteryStatus"
        android:layout_marginLeft = "3sp"
        android:layout_width = "wrap_content"
        android:layout_height = "wrap_content"
        android:text = "电池状态:"
        android:textSize = "18dp"
        android:textColor = "#FFFFFF"
        />
    <TextView
        android:id = "@+id/tvBatteryLevel"
        android:layout_marginLeft = "3sp"
        android:layout_width = "wrap_content"
        android:layout_height = "wrap_content"
        android:text = "电池电量:"
        android:textSize = "18dp"
        android:textColor = "#FFFFFF"
        />
        <TextView
        android:id = "@+id/tvBatteryHealth"
        android:layout_marginLeft = "3sp"
```

```
android:layout_width = "wrap_content"
android:layout_height = "wrap_content"
android:text = "电池健康:"
android:textSize = "18dp"
android:textColor = "#FFFFFF"
/>
<TextView
android:id = "@ + id/tvBatteryTemperature"
android:layout_marginLeft = "3sp"
android:layout_width = "wrap_content"
android:layout_height = "wrap_content"
android:text = "电池温度:"
android:textSize = "18dp"
android:textColor = "#FFFFFF"
/>
<TextView
android:id = "@ + id/tvBatteryVoltage"
android:layout_marginLeft = "3sp"
android:layout_width = "wrap_content"
android:layout_height = "wrap_content"
android:text = "电池电压:"
android:textSize = "18dp"
android:textColor = "#FFFFFF"
/>
<TextView
android:id = "@ + id/tvBatteryTechnology"
android:layout_marginLeft = "3sp"
android:layout_width = "wrap_content"
android:layout_height = "wrap_content"
android:text = "电池技术:"
android:textSize = "18dp"
android:textColor = "#FFFFFF"
/>
<TextView
android:id = "@ + id/tvInfo"
android:layout_marginLeft = "3sp"
android:layout_width = "wrap_content"
android:layout_height = "wrap_content"
android:text = "http://www.qinjianping.com"
android:textSize = "15dp"
android:textColor = "#FFFFFF"
/>
```

```
</LinearLayout>
```

(6) 编辑 MainActivity.java 文件，源代码如下：

```
public class MainActivity extends Activity {
  /** 定义电池信息变量   */
    private static int currentBatteryPlugged = 0;
    private static int currentBatteryStatus = 0;
    private static int currentBatteryLevel = 0;
    private static int currentBatteryHealth = 0;
    private static int currentBatteryTemperature = 0;
    private static int currentBatteryVoltage = 0;
    private static String currentBatteryTechnology = "";
    /** TextView 声明 */
    private static TextView tvBatteryStatus;
    private static TextView tvBatteryLevel;
    private static TextView tvBatteryHealth;
    private static TextView tvBatteryTemperature;
    private static TextView tvBatteryVoltage;
    private static TextView tvBatteryTechnology;
    /** 定义好字符串以备使用 */
    private static String batteryStatus = "电池状态：";
    private static String batteryLevel = "电池电量：";
    private static String batteryHealth = "电池健康：";
    private static String batteryTemperature = "电池温度：";
    private static String batteryVoltage = "电池电压：";
    private static String  batteryTechnology = "电池技术：";
    private static String  batteryStatusCharging = "正在充电";
    private static String  batteryStatusDischarging = "正在放电";
    private static String  batteryStatusFull = "已充满";
    private static String  batteryStatusNotCharging = "未在充电";
    private static String  batteryStatusUnknown = "状态未知";
    private static String  batteryPluggedAC = "(AC)";
    private static String  batteryPluggedUSB = "(USB)";
    private static String  batteryHealthCold = "过冷";
    private static String  batteryHealthDead = "损坏";
    private static String  batteryHealthGood = "良好";
    private static String  batteryHealthOverheat = "过热";
    private static String  batteryHealthOverVoltage = "过压";
    private static String  batteryHealthUnknown = "未知";
    private static String  batteryHealthUnspecifiedFailure = "未知的故障";
  /** 提示 Service 启动标志位 */
  private static boolean flag;
```

```
/** 提示信息接收器 */
BroadcastReceiver infoReceiver;
public void onCreate(Bundle savedInstanceState)
{
    super.onCreate(savedInstanceState);
    this.requestWindowFeature(Window.FEATURE_NO_TITLE);//设置 activity 无标题
    setContentView(R.layout.activity_main);  //使用 newlayout 的布局
    tvBatteryStatus = (TextView)findViewById(R.id.tvBatteryStatus);
                                    //找到 id 为 tvBatteryStatus 的文本
    tvBatteryLevel = (TextView)findViewById(R.id.tvBatteryLevel);
                                    //找到 id 为 tvBatteryLevel 的文本
    tvBatteryHealth = (TextView)findViewById(R.id.tvBatteryHealth);
                                    //找到 id 为 tvBatteryHealth 的文本
    tvBatteryTemperature = (TextView)findViewById(R.id.tvBatteryTemperature);
                                    //找到 id 为 tvBatteryTemperature 的文本
    tvBatteryVoltage = (TextView)findViewById(R.id.tvBatteryVoltage);
                                    //找到 id 为 tvBatteryVoltage 的文本
    tvBatteryTechnology = (TextView)findViewById(R.id.tvBatteryTechnology);
                                    //找到 id 为 tvBatteryTechnology 的文本
    flag = true;                    //提示 service 的标志位置为 true
    infoReceiver = new BroadcastReceiver()  //提示信息接收器的定义
    {
        @Override
        public void onReceive(Context context, Intent intent)
        {
            //TODO Auto-generated method stub
            setText();
//收到 intent,就及时修改 TextView 信息,使得 Activity 显示时,电池信息也能动态显示
        }
    };
    /** 注册提示信息的 intentFilter */
    IntentFilter filter = new IntentFilter();
    filter.addAction("com.ritterliu.newBatteryWidget");
    registerReceiver(infoReceiver,filter);
    /** 启动提示 service */
    Intent startService = new Intent(this,updateService.class);
    startService(startService);
}
/** 单击屏幕任意位置,关闭电池信息 Activity */
public boolean onTouchEvent(MotionEvent event)
{
    this.finish();//结束当前 Activity
```

```
    return true;
}
@Override
protected void onDestroy()
{
    //TODO Auto-generated method stub
    flag = false;
    unregisterReceiver(infoReceiver);
    super.onDestroy();
}
/** 及时动态修改 Activity 上文字信息的函数 */
public static void setText()
{
    String plugState = "";
    switch(currentBatteryPlugged)
    {
    case 0:   //无类型充电
        plugState = "";
        break;
    case 1:   //AC 充电
        plugState = batteryPluggedAC;
        break;
    case 2:  //USB 充电
        plugState = batteryPluggedUSB;
        break;
        default:
        plugState = "";
    }
    switch(currentBatteryStatus)
    {
    case 1://状态未知
        tvBatteryStatus.setText(batteryStatus + batteryStatusUnknown);
        break;
    case 2://正在充电
        tvBatteryStatus.setText(batteryStatus + batteryStatusCharging + plugState);
        break;
    case 3://正在放电
        tvBatteryStatus.setText(batteryStatus + batteryStatusDischarging);
        break;
    case 4://未在充电
        tvBatteryStatus.setText(batteryStatus + batteryStatusNotCharging);
        break;
```

```
            case 5://已经充满
                tvBatteryStatus.setText(batteryStatus + batteryStatusFull + plugState);
                break;
            default://默认状态未知
                tvBatteryStatus.setText(batteryStatus + batteryStatusUnknown);
            }

            tvBatteryLevel.setText(batteryLevel + String.valueOf(currentBatteryLevel) + "%");
            switch(currentBatteryHealth)
            {
            case 1://健康未知
                tvBatteryHealth.setText(batteryHealth + batteryHealthUnknown);
                break;
            case 2://良好
                tvBatteryHealth.setText(batteryHealth + batteryHealthGood);
                break;
            case 3://过热
                tvBatteryHealth.setText(batteryHealth + batteryHealthOverheat);
                break;
            case 4://损坏
                tvBatteryHealth.setText(batteryHealth + batteryHealthDead);
                break;
            case 5://过压
                tvBatteryHealth.setText(batteryHealth + batteryHealthOverVoltage);
                break;
            case 6://未知的故障
                tvBatteryHealth.setText(batteryHealth + batteryHealthUnspecifiedFailure);
                break;
            case 7://过冷
                tvBatteryHealth.setText(batteryHealth + batteryHealthCold);
                break;
                default://默认未知
                tvBatteryHealth.setText(batteryHealth + batteryHealthUnknown);
            }
              tvBatteryTemperature.setText(batteryTemperature + currentBatteryTempera-
ture/10f + "℃");
            tvBatteryVoltage.setText(batteryVoltage + currentBatteryVoltage + "mv");
            tvBatteryTechnology.setText(batteryTechnology + currentBatteryTechnology);
        }

        /** 提示信息变化的 service,约每隔一秒,就发送 intent
         * 提醒 activity 更新电池信息,主要为了检测电池状态的变化
```

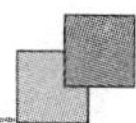

```
     * 例如连上充电时,状态会从"未在充电"变为"正在充电"
     * 通过调用 plugged 方式,还能判断是 AC 方式充电还是 USB 方式充电
     */
    public static class updateService extends Service
    {
        @Override
        public IBinder onBind(Intent intent)
        {
            //TODO Auto-generated method stub
            return null;
        }
        /** 定义得到电池信息的 BroadcastReceiver,提取出关键信息,存入变量中 */
        private BroadcastReceiver batteryReceiver = new BroadcastReceiver()
        {
            @Override
            public void onReceive(Context context, Intent intent)
            {
                //TODO Auto-generated method stub
                currentBatteryStatus = intent.getIntExtra("status", 0);
                currentBatteryLevel = intent.getIntExtra("level", 0);
                currentBatteryHealth = intent.getIntExtra("health", 0);
                currentBatteryTemperature = intent.getIntExtra("temperature", 0);
                currentBatteryVoltage = intent.getIntExtra("voltage", 0);
                currentBatteryTechnology = intent.getStringExtra("technology");
                currentBatteryPlugged = intent.getIntExtra("plugged", 0);
            }
        };
        public void onStart(Intent intent,int startId)
        {
             registerReceiver(batteryReceiver,new IntentFilter(Intent.ACTION_BAT-
TERY_CHANGED));//注册 BroadcastReceiver
            /** 启动一个线程,约每隔一秒就发送 intent 提醒 Activity 更新电池信息 */
            new Thread()
            {
                public void run()
                {
                    while(flag)//如果 flag 为 true
                    {
                        Intent sendInfoToActivity = new Intent();
                        sendInfoToActivity.setAction("com.ritterliu.newBatteryWidget");
                        sendBroadcast(sendInfoToActivity);
                          try
```

```
                        {
                          Thread.sleep(1000);//休眠 1s
                        }
                        catch(Exception ex)
                        {//捕获异常
                          ex.printStackTrace();
                        }
                    }
                }
            }.start();
            super.onStart(intent, startId);
        }
    }
}
```

(7) 编辑 AndroidManifest.xml 文件,源代码如下:

```
<manifest xmlns:android = "http://schemas.android.com/apk/res/android"
    package = "com.example.chapter18_3"
    android:versionCode = "1"
    android:versionName = "1.0" >
    <uses-sdk  android:minSdkVersion = "8"   android:targetSdkVersion = "15" />
    <application
        android:icon = "@drawable/ic_launcher"
        android:label = "@string/app_name"
        android:theme = "@style/AppTheme" >
        <activity
            android:name = ".MainActivity"
            android:theme = "@android:style/Theme.Dialog"
            android:label = "@string/title_activity_main" >
            <intent-filter>
                <action android:name = "android.intent.action.MAIN" />
                <category android:name = "android.intent.category.LAUNCHER" />
            </intent-filter>
        </activity>
        <receiver
            android:name = ".BatteryAppWidgetProvider"
            android:label = "@string/app_name" >
            <intent-filter>
              <action android:name = "android.appwidget.action.APPWIDGET_UPDATE" />
            </intent-filter>
            <meta-data
                android:name = "android.appwidget.provider"
```

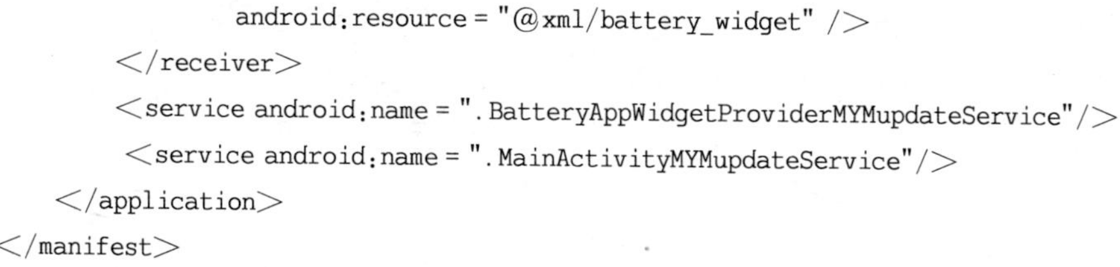

```
            android:resource = "@xml/battery_widget" />
        </receiver>
        <service android:name = ".BatteryAppWidgetProviderMYMupdateService"/>
        <service android:name = ".MainActivityMYMupdateService"/>
    </application>
</manifest>
```

(8) 右击工程 chapter18_3，在弹出的快捷菜单中选择 Run As|Android Application 命令，回到模拟器主页长按，选中并单击 Widgets 上下文菜单选项，运行效果如图 18－11 所示，单击“Android 机器人电池插件”选项，桌面主页运行效果如图 18－12 所示。

图 18－11　Android 机器人电池插件

图 18－12　添加电池 Widget 后主页效果

第 3 篇　Android 实战应用

第 19 章

电子订餐系统

本章将会介绍一个完整的 Android 应用，包括客户端的开发以及服务端的开发，是一个 Android+Servlet+JDBC+JSP 整合的应用。Servlet+JDBC+JSP 提供了一个基于 B/S 结构的电子订餐系统。对于使用 PC 的用户而言，他们可以使用浏览器来访问该系统；对于使用 Android 手机的用户而言，可以选择安装 Android 客户端程序，这样即可通过手机来使用该订餐系统。

本应用的服务端采用了完整的 JAVA EE 应用架构：技术实现上依赖于适合中小型项目开发的 Servlet+JSP+JDBC 的组合，应用架构采用了具有高度可扩展性的控制器层(Servlet)+视图层(JSP)+数据访问层(DAO)的分层架构。Android 客户端通过网络与服务器的控制器组件(Servlet)交互，整个应用具有极好的可扩展性和示范性。

19.1　系统功能简介和架构设计

本章介绍的应用属于具备实战意义的整合应用，应用的服务端由良好的 JAVA EE 架构实现，由控制器层(Servlet)+ 视图层(JSP)+ 数据访问层(DAO)实现了扩展性比较良好的分层架构。客户端则使用 Android 程序。

19.1.1　系统功能简介

现代生活紧张，外卖业务急剧增长。而利用手机进行网上订餐方便快捷，而且功能服务齐全。越来越多的人接受了这种业务方式。本系统主要包括 3 个模块的功能，分别是会员模块、菜品展示模块以及购物车模块。

1. 会员模块

会员模块主要包括以下两个子模块。

- ❑ 会员登录:游客填写完会员账号和会员密码后,立即登录。登录成功,则进入到菜品展示页面;登录失败,则提示相应的失败信息。
- ❑ 会员注册:游客进入主页页面后,填写注册信息后单击注册按钮。如果注册成功,则提示成功信息并跳转至会员登录模块;如果注册失败,则提示相应的失败信息。

2. 菜品展示模块

会员登录成功后,进入到菜品展示模块。菜品展示模块会根据商品的种类进行分类展示,对于每一个具体的菜品,则会显示菜品的图片、名称以及价格等信息。单击 Android 客户端程序的选项菜单,则可以进入到每一个菜品的详细信息,包括图片、编号、名称、类型、价格以及描述等信息,同时也可以将相应菜品加入到购物车当中。

3. 购物车模块

购物车模块显示用户选购的菜品的一个列表,以及所要支付的总金额。对于列表的每一项,则会显示这一项菜品的编号、菜品名称、单价、数量以及金额信息。

19.1.2　系统架构设计

本系统的服务器采用 JAVA EE 比较经典的分层结构,分为视图层(JSP)、控制器层(Servlet)和数据访问(DAO)层。分层结构将数据访问逻辑以及业务处理逻辑与视图展示逻辑分离开来,使得程序的耦合性大大降低。客户端不直接与数据库交互,而是通过控制器与中间层建立连接,再由中间层与数据库进行交互。

- ❑ 控制器层,就是 MVC 模式里面的 C,负责表现层与业务逻辑层的交互,调用业务逻辑层,并将业务数据返回给表现层进行展示。
- ❑ 数据访问层,也就是 DAO 层,负责调用 JDBC 的 API 与数据库进行交互,对数据库的表进行增删改查等操作。
- ❑ 视图层,也就是 JSP 页面展示,负责显示相应的页面。

本系统使用 MYSQL 数据库存放数据。

当使用 Android 应用程序作为客户端时,Android 应用可以通过网络与服务器端进行交互,Android 应用将会通过 Apache HttpClient 向服务器的控制器发送请求(此处的控制器直接采用 Servlet 充当),并获取服务端的响应。整体系统架构如图 19-1 所示。

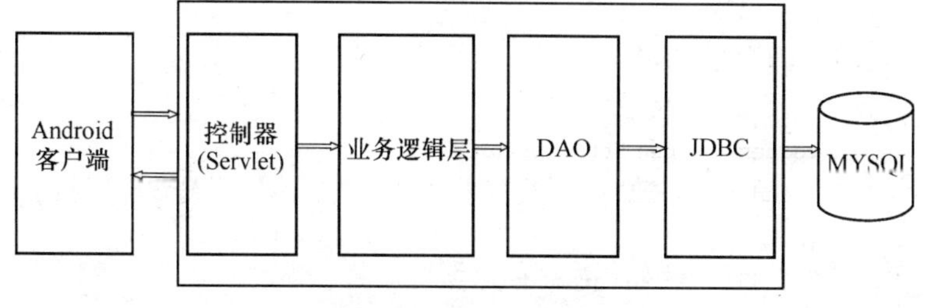

图 19-1　系统架构图

19.2 发送 Http 请求的工具类

在本应用当中，Android 客户端程序与服务端程序之间的通信通过使用 Apache HttpClient 类库来完成，在客户端程序当中为了达到代码复用的目的，所以对 HttpClient 类请求数据进行了相应的封装。新建了一个工具类：HttpCallUtil，此工具类当中主要有两个方法：

- ❑ getConnectionPost：发送 POST 请求。
- ❑ getConnectionGet：发送 GET 请求。

该工具类的源代码如下：

```
/**
 * 封装了 HTTP 的连接
 *
 * @author Administrator
 *
 */
public class HttpCallUtil
{
  /**
   * post 请求
   * @param url
   * @param name:请求的参数,map:请求的参数
   * @return
   */
  public String getConntionPost(String url, String name,String psw,List data)
   {
   //Log.v("map2 == == ", data.size() + "");
   String sb = new String();
   HttpPost request = null;
   HttpResponse response = null;
   List<NameValuePair> list = null;
   try
   {
       if (url != null)
        {
           request = new HttpPost(url);
           if(data != null)
            {
               list = putParam(data);
            }
```

```
            else if(name != null && psw != null)
            {
                list = new ArrayList<NameValuePair>();
                list.add(new BasicNameValuePair("username", name));
                list.add(new BasicNameValuePair("password", psw));
            }
            request.setEntity(new UrlEncodedFormEntity(list,HTTP.UTF_8));
            response = new DefaultHttpClient().execute(request);
            if(response.getStatusLine().getStatusCode() == 200)
            {
                String temp = EntityUtils.toString(response.getEntity());
                if(temp.length() > 0)
                {
                    sb = temp.trim().toString();
                }
                else
                {
                    sb = "error response data length";
                }
            }
            else
            {
                sb = "error response code:" + response.getStatusLine().getStatusCode();
            }
        }
        else
        {
            return null;
            }
        }
        catch (Exception e)
        {
            e.printStackTrace();
        }
        return sb;
    }
    /**
     * 传递给服务端的数据,用 Map 进行封装
     * @param data
     * @return
     */
    public List putParam(List data)
```

```
        {
            Log.v("map 3 == == ", data.size() + "");
            List<NameValuePair> list = new ArrayList<NameValuePair>();
            if(data != null)
            {
                for(int i = 0;i<data.size();i++)
                {
                    Log.v(" == " + data.get(i).toString(), data.get(i).toString());
                    list.add(new BasicNameValuePair(data.get(i).toString(), data.get
(i).toString()));
                }
            }
            return list;
        }
        /**
         * get 请求
         * @param url
         * @param
         * @return
         */
        public String getConntionGet(String url)
        {
            String str = new String();
            HttpGet request = null;
            HttpResponse response = null;
            try
            {
                if(url != null)
                    {
                    request = new HttpGet(url);
                        response = new DefaultHttpClient().execute(request);
                        if(response.getStatusLine().getStatusCode() == 200)
                        {
                            String temp = EntityUtils.toString(response.getEntity());
                            if(temp.length() >= 0)
                            {
                                str = temp.substring(0, temp.length() - 1);
                            }
                            else
                            {
                                str = "error response data length " + response.getSta-
tusLine().getStatusCode();
```

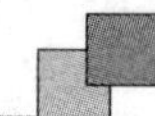

```
                }
            }
            else
            {
                str = "error response :" + response.getStatusLine().getStatusCode();
            }
        }
    }
    catch (Exception e)
    {
        e.printStackTrace();
    }
    return str;
}
/**
 * 取得图片
 * @param url
 * @param iv
 */
public void getConntionImage(String url,ImageView iv)
{
    URL imageUrl = null;
    HttpURLConnection conn = null;
    try
    {
        if(url != null)
        {
            imageUrl = new URL(url);
            conn = (HttpURLConnection) imageUrl.openConnection();
            conn.setDoInput(true);
            conn.connect();
            InputStream in = conn.getInputStream();
            Bitmap bp = BitmapFactory.decodeStream(in);
            if(iv != null)
            {
                iv.setImageBitmap(bp);
            }
        }
        else
        {
            ;
        }
```

```
            }
            catch (Exception e)
            {
                e.printStackTrace();
            }
        }
    }
```

19.3 用户注册

只有注册的用户才能使用本订餐系统进行订餐。游客启动应用程序后，单击登录界面的“注册”按钮，进入到用户注册界面，输入相应的用户名及密码，单击注册界面上的“注册”按钮，则 Android 客户端程序会通过 HttpCallUtil 向服务端发送用户注册的请求，服务端接收到请求后，会将相应的信息增加到数据库相关用户表当中，如果增加用户成功，则返回相关用户注册成功的信息；如果增加用户失败，则返回相关用户注册失败的信息。

19.3.1 用户注册 Servlet

处理用户注册的 Servlet 只是前端控制器，它的作用只有 3 个。

- 获取请求参数。
- 根据相应业务逻辑处理用户请求，调用 DAO 层操作数据。
- 根据处理结果生成输出。

处理用户注册请求的 Servlet：RegisterServlet，源代码如下：

```
public class RegisterServlet extends HttpServlet
{
    /** 执行注册用户的 Servlet */
    public void doGet(HttpServletRequest request, HttpServletResponse response)
            throws ServletException, IOException
    {
        response.setContentType("text/html");
        PrintWriter out = response.getWriter();
        UserDao dao = new UserDaoImpl();
        String username = request.getParameter("username");
        String password = request.getParameter("password");
        Users user = new Users();
        user.setUserName(username);
        user.setUserPassword(password);
        int a = dao.addUser(user);
        if(a == 0)
```

```
            {
                out.println("success");
            }
            else
            {
                out.print("add failue");
            }
            out.flush();
            out.close();
        }
        public void doPost(HttpServletRequest request, HttpServletResponse response)
            throws ServletException, IOException
        {
            doGet(request,response);
        }
    }
```

19.3.2 用户模型

M 实际上就是 MVC 模型当中的数据模型的概念。用户模型 Users 用来代替现实生活中的客户对象，包括客户 ID 属性、客户名称属性、客户密码属性以及客户是否是会员属性。Users 源代码如下：

```
public class Users
{
    private int userId;//客户的 Id
    private String userName;//客户的 name
    private String userPassword;
    private int isMember;//是否是会员
    public Users()
    {
        super();
        // TODO Auto-generated constructor stub
    }
    public Users(int userId, String userName, String userPassword, int isMember)
    {
        super();
        this.userId = userId;
        this.userName = userName;
        this.userPassword = userPassword;
        this.isMember = isMember;
    }
    public int getUserId()
    {
```

```
        return userId;
    }
    public void setUserId(int userId)
    {
        this.userId = userId;
    }
    public String getUserName()
    {
        return userName;
    }
    public void setUserName(String userName)
    {
        this.userName = userName;
    }
    public String getUserPassword()
    {
        return userPassword;
    }
    public void setUserPassword(String userPassword)
    {
        this.userPassword = userPassword;
    }
    public int getIsMember()
    {
        return isMember;
    }
    public void setIsMember(int isMember)
    {
        this.isMember = isMember;
    }
    @Override
    public String toString()
    {
        return "Users [isMember=" + isMember + ", userId=" + userId
            + ", userName=" + userName + ", userPassword=" + userPassword
            + "]";
    }
}
```

19.3.3 用户 DAO

由于控制器不负责直接和数据库进行交互，而是与 DAO 层先进行交互，DAO 层再调用 JDBC 提供的 API 对相关数据库表做增删改查等操作。UserDaoImpl 中主

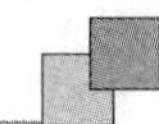

要实现了两个具体的方法，一个是添加用户的方法，另外一个是进行登录判断的方法。访问用户表的 UserDaoImpl 源代码如下：

```
public class UserDaoImpl implements UserDao
{
    //添加用户的方法
    public int addUser(Users user)
    {
        Connection conn = null;
        PreparedStatement pstmt  =  null;
        String sql  =  "insert into t_user(username,userpassword) values(?,?)";
                                                              //增加用户 SQL 语句
        try
        {
            conn  =  DButil2.getConnection();
            pstmt  =  DButil2.getPreparedStatement(conn, sql);
            pstmt.setString(1, user.getUserName());
            pstmt.setString(2, user.getUserPassword());
            pstmt.executeUpdate();
        }
        catch (SQLException e)
        {
            e.printStackTrace();
        }
        return 0;
    }
    //进行登录判断方法
    public int login(String user, String psw)
    {
        DButil2 dbutil  =  new DButil2();
        String sql  = "select *  from t_user where username = ? and userpassword = ?";
        PreparedStatement pstmt  =    null;
        Connection conn  = dbutil.getConnection();
        try
        {
            pstmt  =  conn.prepareStatement(sql);
            pstmt.setString(1, user);
            pstmt.setString(2, psw);
            System.out.println("query the table");
            ResultSet rs  =  pstmt.executeQuery();
            if(rs.next())
            {
```

```
                System.out.println("query from users success ");
                Users u = new Users();
                u.setUserName(user);
                u.setUserPassword(psw);
                return 1;
            }
        }
        catch (Exception e)
        {
            e.printStackTrace();
        }
        return 0;
    }
}
```

UserDaoImpl 通过调用 JDBC 提供的 API 主要实现了两个方法，用户注册 Servlet 对 addUser 方法进行调用，用户登录 Servlet 对 login 方法进行调用。

19.3.4 用户注册

Android 客户端的用户注册界面，需要游客填写用户名和密码信息，填好过后，单击界面上的“注册”按钮，会通过 HttpCallUtil 类向服务端发送相应的请求，服务端收到请求后，根据相应的 URL 地址，找到需要处理用户注册请求的 RegisterServlet，对用户的请求进行处理，处理完成后给 Android 客户端程序输出相应的结果。

用户注册界面的布局文件：register.xml，源代码如下：

```
<?xml version = "1.0" encoding = "utf - 8"? >
<RelativeLayout
android:id = "@ + id/widget33"
android:layout_width = "fill_parent"
android:layout_height = "fill_parent"
xmlns:android = "http://schemas.android.com/apk/res/android"
>
<Button
android:id = "@ + id/brevert"
android:layout_width = "140px"
android:layout_height = "70px"
android:text = "返回登录"
android:textSize = "18sp"
android:layout_below = "@ + id/etcomment"
android:layout_alignParentBottom = "true"
android:layout_alignParentRight = "true"
```

```
android:layout_alignTop = "@ + id/bdone"
>
</Button>
<Button
android:id = "@ + id/bdone"
android:layout_width = "140px"
android:layout_height = "wrap_content"
android:text = "注册"
android:textSize = "18sp"
android:layout_alignParentBottom = "true"
>
</Button>
<EditText
android:id = "@ + id/etconfirmpwd"
android:layout_width = "250px"
android:layout_height = "wrap_content"
android:text = ""
android:textSize = "18sp"
android:password = "true"
android:layout_alignTop = "@ + id/tvconfirmpwd"
android:layout_alignLeft = "@ + id/etpassword"
>
</EditText>
<TextView
android:id = "@ + id/tvconfirmpwd"
android:layout_width = "wrap_content"
android:layout_height = "wrap_content"
android:text = "确认密码:"
android:textSize = "18sp"
android:layout_below = "@ + id/etpassword"
android:layout_alignParentLeft = "true"
>
</TextView>
<EditText
android:id = "@ + id/etpassword"
android:layout_width = "250px"
android:layout_height = "wrap_content"
android:text = ""
android:textSize = "18sp"
android:password = "true"
android:layout_alignTop = "@ + id/tvpassword"
android:layout_alignLeft = "@ + id/etname"
```

```
>
</EditText>
<TextView
android:id = "@ + id/tvpassword"
android:layout_width = "wrap_content"
android:layout_height = "wrap_content"
android:text = "密码:"
android:textSize = "18sp"
android:layout_below = "@ + id/etname"
android:layout_alignParentLeft = "true"
>
</TextView>
<EditText
android:id = "@ + id/etname"
android:layout_width = "249px"
android:layout_height = "wrap_content"
android:text = ""
android:textSize = "18sp"
android:layout_alignParentTop = "true"
android:layout_toRightOf = "@ + id/tvname"
>
</EditText>
<TextView
android:id = "@ + id/tvname"
android:layout_width = "100px"
android:layout_height = "wrap_content"
android:textSize = "18sp"
android:text = "用户名:"
android:layout_alignParentTop = "true"
android:layout_alignParentLeft = "true"
>
</TextView>
</RelativeLayout>
```

用户注册界面需要用户输入用户名、密码以及确认密码信息，在用户输入完这些信息后，会调用服务端的 RegisterServlet 进行注册的处理。用户注册的 Activity：ActivityRegister，源代码如下：

```
public class ActivityRegister extends Activity
{
    private Button bdone,brevert;
    private EditText pwd,cfpwd,username;
```

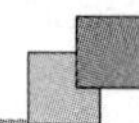

```
        private HttpCallUtil call;
        public void onCreate(Bundle savedInstanceState)
        {
            super.onCreate(savedInstanceState);
            setContentView(R.layout.register);
            call = new HttpCallUtil();
            bdone = (Button) findViewById(R.id.bdone);
            brevert = (Button) findViewById(R.id.brevert);
            bdone.setOnClickListener(new OnClickListener()
            {
            @Override
            public void onClick(View v)
            {
                //用户名和密码不能为空
                username = (EditText) findViewById(R.id.etname);
                pwd = (EditText) findViewById(R.id.etpassword);
                cfpwd = (EditText) findViewById(R.id.etconfirmpwd);
                if(username.getText().toString().length() == 0 || pwd.getText().toString().length() == 0)
                {
                    Toast.makeText(ActivityRegister.this, "username and password can't be blank,please input again!", Toast.LENGTH_LONG).show();
                        return;
                    }
                    //判断密码和确认密码是否相等
                    if(pwd.getText().toString().trim().equals(cfpwd.getText().toString().trim())){
                        Toast.makeText(ActivityRegister.this, "register succeed, please revert to the login interface to login!", Toast.LENGTH_LONG).show();
                        String url = "http://211.155.227.204:8080/DestineFoodServer/RegisterServlet";
                      System.out.println("---------register--------");
                    String resultData = call.getConntionPost(url, username.getText().toString(), pwd.getText().toString(),null);
                      System.out.println(resultData);
                      if (resultData.trim().equals("success")) {
                         setTitle("register succeed,info:" + username.getText() + ":" + pwd.getText());
                         Intent intent = new Intent(ActivityRegister.this,OrderFoodLogin.class);
                         startActivity(intent);
                      }
```

```
                }
                else
                {
                    Toast.makeText(ActivityRegister.this, "the pwd is not the same with cfpwd,please input again!", Toast.LENGTH_LONG).show();
                }
            }
        });
        brevert.setOnClickListener(new OnClickListener() {
            @Override
            public void onClick(View v) {
                //结束当前应用程序,停止歌曲的播放
                System.exit(0);
                Intent intent = new Intent(ActivityRegister.this, OrderFoodLogin.class);
                startActivity(intent);
            }
        });
    }
}
```

上述程序首先对用户输入的用户名以及密码信息是否为空进行校验,紧接着又对用户输入的密码以及确认密码是否匹配进行判断,这样基本可以保证用户输入的注册信息的有效性以及完整性。

在确认数据没有问题后,通过调用 HttpCallUtil 工具类的 getConnectionPost 方法向服务端发送请求。如果服务端返回 success,则表明用户注册成功。如图 19-2 所示。

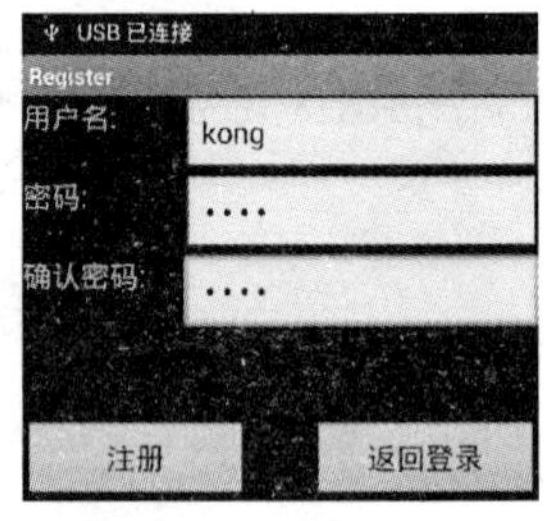

图 19-2 注册页面

19.4 用户登录

只有登录成功的会员才能进行菜品的预定。用户注册成功后,会跳转到用户登录界面,这时需要用户输入相应的用户名以及密码。如果用户名和密码输入正确,则跳转至菜品展示界面;如果用户名或者密码输入错误,会进行相应错误信息的提示。

19.4.1 用户登录 Servlet

LoginServlet 是用来处理用户登录请求的控制器,在 LoginServlet 的 doGet()方法当中,会对客户端传过来的 username 参数以及 password 参数进行校验,如果调用 UserDaoImpl 的 login()方法后,得到的 userId 大于 0,则说明用户登录成功,否则说

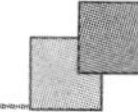

明用户登录失败。LoginServlet的源代码如下：

```
public class LoginServlet extends HttpServlet
{
    public void doGet(HttpServletRequest request, HttpServletResponse response)
        throws ServletException, IOException
    {
        response.setContentType("text/html");
        PrintWriter out = response.getWriter();
        UserDao dao = new UserDaoImpl();
        String username = request.getParameter("username");
        String password = request.getParameter("password");
        int userId = dao.login(username, password);
        if(userId>0)
        {
            //将信息传给客户端
            out.println("success");
            out.println(userId);
            System.out.println("success!!! the response");
        }
        System.out.println("username = " + username);
        out.flush();
        out.close();
    }
    public void doPost(HttpServletRequest request, HttpServletResponse response)
            throws ServletException, IOException
    {
        doGet(request,response);
    }
}
```

上述代码中首先获取Android客户端程序传过来的username以及password参数，之后调用UserDao的login方法查询用户信息表对用户名和密码进行校验。

19.4.2 用户登录

Android客户端的用户登录界面，需要用户填写用户名和密码信息，填好过后，单击界面上的“登录”按钮，会通过HttpCallUtil类向服务端发送相应的请求，服务端收到请求后，根据相应的URL地址，找到需要处理用户登录请求的LoginServlet，对用户的请求进行处理，如果登录成功则会跳转到菜品展示界面；如果登录失败，则会提示相应的错误信息。

用户登录活动OrderFoodLogin启动时，会首先创建一个对话框，对话框的内容布局文件是 dialogitem1. xml,源代码如下：

```
<?xml version = "1.0" encoding = "utf - 8"? >
<RelativeLayout
android:id = "@ + id/widget30"
android:layout_width = "fill_parent"
android:layout_height = "fill_parent"
xmlns:android = "http://schemas.android.com/apk/res/android"
>
<ImageView
android:id = "@ + id/ivpausesong"
android:layout_width = "30px"
android:layout_height = "30px"
android:src = "@drawable/pause1"
android:layout_alignTop = "@ + id/ivregister"
android:layout_alignParentRight = "true"
>
</ImageView>
<TextView
android:id = "@ + id/tvregister"
android:layout_width = "wrap_content"
android:layout_height = "wrap_content"
android:text = "注册"
android:textSize = "18sp"
android:gravity = "center_vertical"
android:layout_below = "@ + id/etpassword"
android:layout_toRightOf = "@ + id/ivregister"
>
</TextView>
<ImageView
android:id = "@ + id/ivregister"
android:layout_width = "wrap_content"
android:layout_height = "wrap_content"
android:src = "@drawable/register4"
android:layout_below = "@ + id/etpassword"
android:layout_alignParentLeft = "true"
>
</ImageView>
<EditText
android:id = "@ + id/etpassword"
android:layout_width = "250px"
android:layout_height = "wrap_content"
```

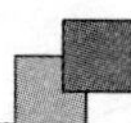

```
android:text = ""
android:textSize = "18sp"
android:password = "true"
android:layout_alignTop = "@ + id/tvpassword"
android:layout_alignLeft = "@ + id/etuser"
>
</EditText>
<TextView
android:id = "@ + id/tvpassword"
android:layout_width = "wrap_content"
android:layout_height = "wrap_content"
android:text = "密码:"
android:textSize = "18sp"
android:layout_below = "@ + id/etuser"
android:layout_alignParentLeft = "true"
android:layout_alignRight = "@ + id/tvuser"
>
</TextView>
<EditText
android:id = "@ + id/etuser"
android:layout_width = "249px"
android:layout_height = "wrap_content"
android:text = ""
android:textSize = "18sp"
android:layout_alignParentTop = "true"
android:layout_toRightOf = "@ + id/tvuser"
>
</EditText>
<TextView
android:id = "@ + id/tvuser"
android:layout_width = "wrap_content"
android:layout_height = "wrap_content"
android:text = "用户名:"
android:textSize = "18sp"
android:layout_alignParentTop = "true"
android:layout_alignParentLeft = "true"
>
</TextView>
</RelativeLayout>
```

OrderFoodLogin 活动对用户输入的用户名和密码等信息是否为空进行校验，如果用户输入的用户名以及密码都不为空，则程序紧接着会调用 HttpCallUtil 工具类的 getConnectionPost 方法向指定 URL 地址发送 Http 请求，Android 客户端程序接收到服务端的响应信息过后，会根据信息作出相应的判断，如果登录成功，则会跳转到菜品展示界面。OrderFoodLogin 源代码如下：

```
public class OrderFoodLogin extends Activity {
    private static final int DIALOG = 1;
    private View dialogitem;
    //播放音乐的对象
    private MediaPlayer mp;
    private Button bpause;
    private TextView tvsongname;
    private HttpCallUtil call;
    //暂停图片和注册图片
    private ImageView ivpause,ivregister;
    private boolean imagebflag = true;
    private SharedPreferences sp = null;
    public void onCreate(Bundle savedInstanceState)
    {
        super.onCreate(savedInstanceState);
        call = new HttpCallUtil();
        showDialog(DIALOG);
        try {
            mp = MediaPlayer.create(this, R.raw.higher);
        } catch (IllegalArgumentException e) {
            e.printStackTrace();
        } catch (IllegalStateException e) {
            e.printStackTrace();
        }
    }
    protected Dialog onCreateDialog(int id)
    {
        super.onCreateDialog(id);
        dialogitem = LayoutInflater.from(this).inflate(R.layout.dialogitem1,null);
        final EditText etname = ((EditText) dialogitem.findViewById(R.id.etuser));
        final EditText etpwd = ((EditText) dialogitem.findViewById(R.id.etpassword));
        AlertDialog.Builder builder = new AlertDialog.Builder(this);
        // 给 Dialog 设置用户自定义的界面
        builder.setView(dialogitem);
        // 设置 Dialog 的标题
        builder.setTitle("Welcome to order food system!");
        //builder.set
        // 设置 Dialog 的图标
```

```
            builder.setIcon(R.drawable.loginimage1);
            // 设置 Dialog 的显示信息
            // builder.setMessage("请输入用户名和密码:");
            // 设置 Dialog 上的一个按钮和按钮的单击事件
            builder.setPositiveButton("Enter", new DialogInterface.OnClickListener() {
              public void onClick(DialogInterface dialog, int which) {
                if(etname.getText().toString().equals("") || etpwd.getText().toString
().equals("")){
                    Toast.makeText(OrderFoodLogin.this, "请输入信息", Toast.LENGTH_
SHORT).show();
                  removeDialog(DIALOG);// 空格 +"/":快捷键出来相关的函数
                  showDialog(DIALOG);
                }else{
                  Map map = new HashMap();
                  map.put("name", etname.getText().toString());
                  map.put("psw", etpwd.getText().toString());
                  Log.v("map == == ", map.size() + "");
                  List list = new ArrayList();
                  list.add(etname.getText().toString());
                  list.add(etpwd.getText().toString());
                  String url = "http://211.155.227.204:8080/DestineFoodServer/Login-
Servlet";
                  String resultData = call.getConntionPost(url, etname.getText().toS-
tring(), etpwd.getText().toString(),null);
                  String sb[] = resultData.split("\n");
                  //System.out.println("sb.length:" + sb.length + ",sb[0]:" + sb[0] + ",
sb[1]:" + sb[1]);
                  //Log.e("-----------------", "sb.length:" + sb.length + ",sb[0]:" + sb[0]
+ ",sb[1]:" + sb[1]);
                  // 调用另外一个 activity,sb[0]是字符"success",sb[1]是 userId
                  if (sb.length == 2 && sb[0].trim().equals("success")) {
                  setTitle("login succeed,info:" + etname.getText() + ":" + etpwd.
getText());
                  Intent intent = new Intent(OrderFoodLogin.this,AndroidFoodMain.class);
                  //将用户 id 和用户姓名传给主界面
                  //intent.putExtra("userId", sb[1]);
                  //intent.putExtra("userName", etname.getText().toString());
                   sp = OrderFoodLogin.this.getSharedPreferences("android", Context.
MODE_WORLD_WRITEABLE);
                  Editor e = sp.edit();
                  e.putString("userName", etname.getText().toString());
                  e.putString("userId", sb[1]);
```

```
            e.commit();
            startActivity(intent);
            finish();
        } else {
            removeDialog(DIALOG);// 空格 + "/":快捷键出来相关的函数
            showDialog(DIALOG);
              Toast.makeText(OrderFoodLogin.this," your username or password is
                            wrong,please input again!" , Toast.LENGTH_LONG).
                            show();
        }
        }
    }
});
    builder.setNegativeButton("Exit", new DialogInterface.OnClickListener() {
      public void onClick(DialogInterface dialog, int which) {
        setTitle("Exit");
        OrderFoodLogin.this.finish();
      }
    });
    //播放音乐
    mp = new MediaPlayer();
    ivpause = (ImageView) dialogitem.findViewById(R.id.ivpausesong);
    ivregister = (ImageView) dialogitem.findViewById(R.id.ivregister);
    ivpause.setOnClickListener(new OnClickListener() {
        @Override
        public void onClick(View v) {
            if(imagebflag)
            {
                ivpause.setImageResource(R.drawable.start1);
                imagebflag = false;
            }
            else
            {
                ivpause.setImageResource(R.drawable.pause1);
                imagebflag = true;
            }
            if(mp.isPlaying())
            {
                mp.pause();
            }
        else
        {
```

```
                mp.start();
            }
        }
    });
    ivregister.setOnClickListener(new OnClickListener() {
        @Override
        public void onClick(View v) {
            Intent intent = new Intent(OrderFoodLogin.this, ActivityRegister.class);
              startActivity(intent);
            }
        });
        return builder.create();
    }
    @Override
    protected void onDestroy() {
        super.onDestroy();
        mp.release();
    }
}
```

图 19-3　登录界面

登录界面上还有一个播放音乐的按钮，单击此按钮后，会开始播放音乐。登录界面效果如图 19-3 所示。

19.5　菜品展示

用户登录完成后，程序就会跳转到菜品的分类展示界面。菜品展示模块会根据商品的种类进行分类展示，对于每一个具体的菜品，则会显示菜品的图片、名称以及价格等信息。

19.5.1　菜品展示 Servlet

GetAllFoodServlet 是处理获取所有菜品请求的控制器，在 GetAllFoodServelt 的 doGet()方法当中会调用 FoodDaoImpl 对象的 getAllFood()方法来获取所有的菜品，如果 getAllFood()返回的列表不为空，则 GetAllFoodServlet 会向客户端直接输出这个列表；如果 getAllFood()返回的列表为空，则控制器会向客户端输出错误信息。GetAllFoodServlet 的源代码如下：

```
public class GetAllFoodServlet extends HttpServlet
{
    public void doGet(HttpServletRequest request, HttpServletResponse response)
```

```
        throws ServletException, IOException
    {
        response.setContentType("text/html;charset = GBK");
        PrintWriter out  =  response.getWriter();
        FoodDao dao  =  new FoodDaoImpl();
        List foodlist  =  dao.getAllFood();
        if (! foodlist.isEmpty())
        {
            out.print(foodlist);
        else
        {
            out.print("find failue");
        }
        out.flush();
        out.close();
    }
    public void doPost(HttpServletRequest request, HttpServletResponse response)
        throws ServletException, IOException
    {
        doGet(request, response);
    }
}
```

19.5.2 菜品模型

菜品数据模型包括菜品 ID 属性、菜品名称属性、菜品价格属性、菜品类型属性、菜品描述属性以及菜品图片属性。菜品模型 Foods 的源代码如下：

```
public class Foods {
    private int foodId;            //菜品 ID
    private String foodName;       //菜品名称
    private float foodPrice;       //菜品价格
    private int foodType;          //菜的类型
    private String foodDescri;     //菜品描述
    private String foodImage;      //菜品图片
    public Foods() {
        super();
        // TODO Auto-generated constructor stub
    }
    public Foods(int foodId, String foodName, int foodPrice, int foodType,
            String foodDescri, String foodImage) {
        super();
```

```
        this.foodId = foodId;
        this.foodName = foodName;
        this.foodPrice = foodPrice;
        this.foodType = foodType;
        this.foodDescri = foodDescri;
        this.foodImage = foodImage;
    }
    public int getFoodId() {
        return foodId;
    }
    public void setFoodId(int foodId) {
        this.foodId = foodId;
    }
    public String getFoodName() {
        return foodName;
    }
    public void setFoodName(String foodName) {
        this.foodName = foodName;
    }
    public float getFoodPrice() {
        return foodPrice;
    }
    public void setFoodPrice(float foodPrice) {
        this.foodPrice = foodPrice;
    }
    public int getFoodType() {
        return foodType;
    }
    public void setFoodType(int foodType) {
        this.foodType = foodType;
    }
    public String getFoodDescri() {
        return foodDescri;
    }
    public void setFoodDescri(String foodDescri) {
        this.foodDescri = foodDescri;
    }
    public String getFoodImage() {
        return foodImage;
    }
    public void setFoodImage(String foodImage) {
        this.foodImage = foodImage;
```

```
    }
    @Override
    public String toString() {

        return "foodDescri=" + foodDescri + ";foodId=" + foodId
             + ";foodImage=" + foodImage + ";foodName=" + foodName
             + ";foodPrice=" + foodPrice + ";foodType=" + foodType + "]";
    }
}
```

19.5.3 菜品 DAO

实现菜品 DAO 接口的具体实现类是 FoodDaoImpl，主要实现了 FoodDao 接口当中定义的几个方法，分别是添加菜品的方法、删除菜品的方法、更新菜品的方法、显示所有菜品的方法、添加订单信息的方法以及查询所有订单信息等方法。FoodDaoImpl 源码如下：

```
public class FoodDaoImpl implements FoodDao
{
    public int addFood(Foods foods)
    {
        Connection conn = null;
        PreparedStatement pstmt = null;
        String sql = "insert into t_food(foodname,foodprice,foodtype,fooddescri,
foodimage) values(?,?,?,?,?)";
        try {
            conn = DButil2.getConnection();
            pstmt = DButil2.getPreparedStatement(conn, sql);
            pstmt.setString(1, foods.getFoodName());
            pstmt.setFloat(2, foods.getFoodPrice());
            pstmt.setInt(3, foods.getFoodType());
            pstmt.setString(4, foods.getFoodDescri());
            pstmt.setString(5, foods.getFoodImage());
            pstmt.executeUpdate();
        } catch (SQLException e) {
            e.printStackTrace();
        }
        return 0;
    }
    //添加订单用户的信息。但是要返回 orderId，以便稍后插入到另外一张表中当外键
    public int addOrder(OrderFoods order) {
        Connection conn = null;
        ResultSet rs;
```

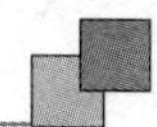

```
        PreparedStatement pstmt = null;
        int orderId = 0;
        String sql = "insert into t_order(userid,username,address,telephone,email,
orderSuggest) values(?,?,?,?,?,?)";
        try {
            conn = DButil2.getConnection();
            pstmt = conn.prepareStatement(sql, Statement.RETURN_GENERATED_KEYS);
            pstmt.setInt(1, order.getUserId());
            pstmt.setString(2, order.getUserName());
            pstmt.setString(3, order.getAddress());
            pstmt.setString(4, order.getTelephone());
            pstmt.setString(5, order.getEmail());
            pstmt.setString(6, order.getOrderSuggest());
            pstmt.executeUpdate();
            rs = pstmt.getGeneratedKeys();
            rs.next();
            orderId = rs.getInt(1);
        } catch (SQLException e) {
            e.printStackTrace();
        }finally{
            DButil2.closeStatement(pstmt);
            DButil2.closeConnection(conn);
        }
        return orderId;
    }
    //添加购物车中菜的数量和订单的 id
    public boolean addCartInformation(int orderId,int foodId,float quantity){
        Connection conn = null;
        PreparedStatement pstmt = null;
          String sql = " insert into t _ food _ order ( foodId, orderId, quantity)
values(?,?,?)";
        try {
            conn = DButil2.getConnection();
            pstmt = DButil2.getPreparedStatement(conn, sql);
            pstmt.setInt(1, foodId);
            pstmt.setInt(2,orderId);
            pstmt.setFloat(3, quantity);
            pstmt.executeUpdate();
            return true;
        } catch (SQLException e) {
            e.printStackTrace();
        }finally{
            DButil2.closeStatement(pstmt);
```

```
            DButil2.closeConnection(conn);
        }
        return false;
    }
    //显示所有订单,多表查询
    public ArrayList getFoodOrder() {
        Connection conn = null;
        PreparedStatement pstmt = null;
        ResultSet rs = null;
        ArrayList orderList = new ArrayList();
        String sql = "select t_order.userName '姓名'," +
            "t_order.address '客户地址'," +
            "t_order.telephone '客户电话', " +
            "t_food.foodName '菜名'," +
            "t_food.foodPrice '菜的价格(圆/份)'," +
            "t_food_order.quantity '份数'," +
            "t_order.orderSuggest '客户的要求'," +
            "from t_food,t_order,t_food_order " +
            "where t_food.foodId = " +
            "t_food_order.foodId and t_food_order.orderId = t_order.orderId";
        try {
            conn = DButil2.getConnection();
            pstmt = DButil2.getPreparedStatement(conn, sql);
            rs = pstmt.executeQuery();
        while(rs.next()){
            orderList.add(rs.getString(1));//userName
            orderList.add(rs.getString(2));//address
            orderList.add(rs.getString(3));//telephone
            orderList.add(rs.getString(7)); //suggest
            orderList.add(rs.getString(4));//foodName
            orderList.add(rs.getFloat(5));//foodPrice
            System.out.println(rs.getFloat(5));
            orderList.add(rs.getInt(6));//quantity
            }
        } catch (SQLException e) {
            e.printStackTrace();
        } finally {
            DButil2.closeStatement(pstmt);
            DButil2.closeConnection(conn);
        }
        return orderList;
    }
    public int deleteFood(String id, String name) {
```

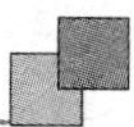

```
    Connection conn = null;
    PreparedStatement pstmt = null;
    int foodid = Integer.valueOf(id);
    String sql = "delete from t_food where foodid = ? and foodname = ?";
    try {
        conn = DButil2.getConnection();
        pstmt = DButil2.getPreparedStatement(conn, sql);
        pstmt.setInt(1, foodid);
        pstmt.setString(2, name);
        pstmt.executeUpdate();
    } catch (SQLException e) {
        e.printStackTrace();
    } finally {
        DButil2.closeStatement(pstmt);
        DButil2.closeConnection(conn);
    }
    return 0;
}
public int deleteOrder(String id) {
    Connection conn = null;
    PreparedStatement pstmt = null;
    int orderid = Integer.valueOf(id);
    String sql = "delete from t_order where orderid = ?";
    try {
        conn = DButil2.getConnection();
        pstmt = DButil2.getPreparedStatement(conn, sql);
        pstmt.setInt(1, orderid);
        pstmt.executeUpdate();
    } catch (SQLException e) {
        e.printStackTrace();
    } finally {
        DButil2.closeStatement(pstmt);
        DButil2.closeConnection(conn);
    }
    return 0;
}
public List getFoodMes(String name, String type) {
    Connection conn = null;
    PreparedStatement pstmt = null;
    ResultSet rs = null;
    int foodtype = Integer.valueOf(type);
    List<Foods> foodsList = new ArrayList<Foods>();
    String sql = "select * from t_food where foodname = ? and foodtype = ?";
```

```
        try {
            conn = DButil2.getConnection();
            pstmt = DButil2.getPreparedStatement(conn, sql);
            pstmt.setString(1, name);
            pstmt.setInt(2, foodtype);
            rs = pstmt.executeQuery();
            while (rs.next()) {
                Foods foods = new Foods();
                foods.setFoodId(rs.getInt(1));
                foods.setFoodName(rs.getString(2));
                foods.setFoodPrice(rs.getInt(3));
                foods.setFoodType(rs.getInt(4));
                foods.setFoodDescri(rs.getString(5));
                foods.setFoodImage(rs.getString(6));
                foodsList.add(foods);
            }
        } catch (SQLException e) {
            e.printStackTrace();
        } finally {
            DButil2.closeStatement(pstmt);
            DButil2.closeConnection(conn);
        }
        return foodsList;
    }
    public List getAllFood() {//显示所有菜品
        Connection conn = null;
        PreparedStatement pstmt = null;
        ResultSet rs = null;
        List<Foods> foodList = new ArrayList<Foods>();
        String sql = "select * from t_food";
        try {
            conn = DButil2.getConnection();
            pstmt = DButil2.getPreparedStatement(conn, sql);
            rs = pstmt.executeQuery();
            while (rs.next()) {
                Foods foods = new Foods();
                foods.setFoodId(rs.getInt(1));
                foods.setFoodName(rs.getString(2));
                foods.setFoodPrice(rs.getFloat(3));
                foods.setFoodType(rs.getInt(4));
                foods.setFoodDescri(rs.getString(5));
                foods.setFoodImage(rs.getString(6));
                foodList.add(foods);
```

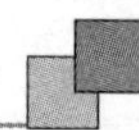

```
            }
        } catch (SQLException e) {
            e.printStackTrace();
        } finally {
            DButil2.closeStatement(pstmt);
            DButil2.closeConnection(conn);
        }
        return foodList;
    }
    public OrderFoods updateOrder(String id) {//未实现
        Connection conn = null;
        PreparedStatement pstmt = null;
        String sql = "update t_order set user_name = ?,password = ? where user_id = ?";
        try {
            conn = DButil2.getConnection();
            pstmt = DButil2.getPreparedStatement(conn, sql);
            pstmt.executeUpdate();
        } catch (SQLException e) {
            e.printStackTrace();
        } finally {
            DButil2.closeStatement(pstmt);
            DButil2.closeConnection(conn);
        }
        return null;
    }
    public ArrayList findrOrderByOrderId(int orderId) {
        ArrayList list = new ArrayList();
        Connection conn = null;
        PreparedStatement pstmt = null;
        ResultSet rs = null;
        String sql = "select t_food.foodName '菜名'," +
                    "t_food.foodPrice '菜的价格(圆/份)'," +
                    "t_food_order.quantity '份数'" +
                    "from t_food,t_food_order where t_food.foodId = t_food_order.
foodId and t_food_order.orderId = " + orderId;
        conn = DButil2.getConnection();
        try {
            pstmt = conn.prepareStatement(sql);
            rs = pstmt.executeQuery();
            while(rs.next()){
            list.add(rs.getString(1));
            list.add(rs.getFloat(2));
            list.add(rs.getInt(3));
```

```
            }
        } catch (SQLException e) {
            // TODO Auto-generated catch block
            e.printStackTrace();
        }finally{
            DButil2.closeStatement(pstmt);
            DButil2.closeConnection(conn);
        }
        return list;
    }
}
```

19.5.4 菜品展示

菜品展示页面 AndroidFoodMain 是一个继承自 android.app.TabActivity 的活动，此页面总共有 3 个标签页，分别是：肉食类、素食类以及鲜汤类，源代码如下：

```
public class AndroidFoodMain extends TabActivity {
    public void onCreate(Bundle savedInstanceState) {
        super.onCreate(savedInstanceState);
        final TabHost tabHost = getTabHost();
        Drawable meat = getResources().getDrawable(R.drawable.last_3);
        Drawable vegetarian = getResources().getDrawable(R.drawable.last_2);
        Drawable soup = getResources().getDrawable(R.drawable.last_1);
        Intent intent1 = new Intent(this, ActivityShowFoodByType.class);
        intent1.setFlags(FoodType.FOOD_MEAT);
        Intent intent2 = new Intent(this, ActivityShowFoodByType.class);
        intent2.setFlags(FoodType.FOOD_VEGETARIAN);
        Intent intent3 = new Intent(this, ActivityShowFoodByType.class);
        intent3.setFlags(FoodType.FOOD_SOUP);
        tabHost.addTab(tabHost.newTabSpec("tab1").setIndicator("肉食类",meat)
                .setContent(intent1));
        tabHost.addTab(tabHost.newTabSpec("tab2").setIndicator("素食类",vegetarian)
                .setContent(intent2));
        tabHost.addTab(tabHost.newTabSpec("tab3").setIndicator("鲜汤类",soup)
                .setContent(intent3));
    }
}
```

ActivityShowFoodByType 活动是每一个具体 Tab 页所要展示的页面，在向 ActivityShowFoodByType 发送意图的时候，AndroidFoodMain 活动会将菜品的类型传过去，这样 AndroidShowFoodByType 活动就可以根据具体的菜品类型，展示某一具体类型下面的所有菜品。

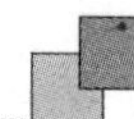

AndroidShowFoodByType 活动布局文件：show.xml，源代码如下：

```
<?xml version = "1.0" encoding = "utf - 8"? >
<LinearLayout xmlns:android = "http://schemas.android.com/apk/res/android"
    android:orientation = "vertical"
    android:layout_width = "fill_parent"
    android:layout_height = "fill_parent"
    >
<GridView
    android:id = "@ + id/grid"
    android:layout_width = "fill_parent"
    android:layout_height = "fill_parent"
    android:columnWidth = "90dp"
    android:numColumns = "auto_fit"
    android:verticalSpacing = "10dp"
    android:horizontalSpacing = "10dp">
</GridView>
</LinearLayout>
```

AndroidShowFoodByType 活动源代码如下：

```
public class ActivityShowFoodByType extends Activity {
    // 用于存储需要显示的数据
    private ArrayList<HashMap<String, Object>> data = new ArrayList<HashMap<
String, Object>>();
    // 用于接收服务端
    private ArrayList<Food> serverFoods = new ArrayList<Food>();
    private GridView gv = null;
    public final static int SHOW_FOOD_DETAIL = 0;
    public final static int CONFIRM_PURCHASE = 1;
    public final static int LOOK_UP_SHOPPING = 2;
    private DButil db;
    // 用户在界面上选中的所有商品信息
    public Food selectedFood = new Food();
    /**
     * 根据菜品类型,获取服务器端的菜品信息
     *
     * @return
     */
    private ArrayList<Food> getServerFoodInfoByType(int flagType) {
        serverFoods.clear();
        HashMap<String, String> request = new HashMap<String, String>();
        // 获得服务端数据"根据约定的字符串格式"
        String resData = GetNetWorkData.getConnectionPost(GetNetWorkData.URL_All-
```

```
Food, request).toString();
            // String resData =
            // "[foodName = test;foodType = 0;foodPrice = 33.3f;foodImage = 500008.jpg],
[foodPrice = 44;foodType = 0;foodName = 红烧肉;foodImage = 500022.jpg]";
            ArrayList<Food> serverAllFoods = GetNetWorkData.decodeResponseData(resData);
            Food oneFood;
            Iterator<Food> it = serverAllFoods.iterator();
            while (it.hasNext()) {
                oneFood = it.next();
                if (oneFood.foodType == flagType) {
                    serverFoods.add(oneFood);
                }
            }
            return serverFoods;
        }// end of getServerData()

        /**
         * 准备显示在界面的数据
         */
        private void prepareView() {
            data.clear();
            // 获取要显示菜品的类型
            Intent intent = this.getIntent();
            int showType = intent.getFlags();
            // 获取指定类型的菜品
            ArrayList<Food> assignedFoods = getServerFoodInfoByType(showType);
            // 设定界面显示数据
            HashMap<String, Object> item;
            for (Food food : assignedFoods) {
                item = new HashMap<String, Object>();
                item.put("price", "价格:" + food.foodPrice);
                item.put("name", "菜名:" + food.foodName);
                item.put("icon", food.foodImage);
                data.add(item);
            }
        }// end of prepareView()
        public void onCreate(Bundle saved) {
            super.onCreate(saved);
            setContentView(R.layout.show);
            try {
                gv = (GridView) this.findViewById(R.id.grid);
                prepareView();
                MyGridViewAdapter mgva = new MyGridViewAdapter(this);
```

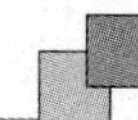

```
            gv.setAdapter(mgva);
            // 监听 GridView 单击事件
            gv.setOnItemClickListener(new OnItemClickListener() {
                public void onItemClick(AdapterView<?> parent, View view, int po-
sition, long id) {
                    selectedFood = serverFoods.get(position);
                    setTitle("测试选中事件---" + selectedFood.foodId);
                    Log.e("----test selected---", "" + selectedFood.foodId);
                    ActivityShowFoodByType.this.openOptionsMenu();
                    Log.e("test GridView onClickListener", "------");
                }
            });
        } catch (Exception e) {
            e.printStackTrace();
        }
    }
    protected void onRestart() {
        super.onRestart();
        Log.e("TabIntent------", "test--onRestart()" + getIntent().getFlags());
    }
    protected void onResume() {
        super.onResume();
        prepareView();
        Log.e("TabIntent------", "test--onResume()" + getIntent().getFlags());
    }
    protected void onStart() {
        super.onStart();
        Log.e("TabIntent------", "test--onStart()" + getIntent().getFlags());
    }
    protected void onStop() {
        super.onStop();
        Log.e("TabIntent------", "test--onStop()" + getIntent().getFlags());
    }
    protected void onPause() {
        super.onPause();
        Log.e("TabIntent------", "test--onPause()" + getIntent().getFlags());
    }
    /**
     * 添加功能 Menu 键
     */
    public boolean onCreateOptionsMenu(Menu menu) {

        super.onCreateOptionsMenu(menu);
```

```
        menu.add(1, SHOW_FOOD_DETAIL, Menu.CATEGORY_SYSTEM, "菜品详情").setIcon(R.
drawable.zqz_menu_lookdetail);
        menu.add(1, CONFIRM_PURCHASE, Menu.CATEGORY_SYSTEM, "加入购物车").setIcon
(R.drawable.zqz_menu_purchase);
        return true;
    }

    /**
     * 监听用户选中事件
     */
    public boolean onOptionsItemSelected(MenuItem item) {
        super.onOptionsItemSelected(item);
        switch (item.getItemId()) {
        case SHOW_FOOD_DETAIL:
            // 跳转到详情显示页面
            Intent intent = new Intent(ActivityShowFoodByType.this, FoodDetail.class);
            intent.putExtra("foodId", selectedFood.foodId);
            intent.putExtra("foodName", selectedFood.foodName);
            intent.putExtra("foodDescri", selectedFood.foodDescri);
            intent.putExtra("foodType", selectedFood.foodType);
            intent.putExtra("foodPrice", selectedFood.foodPrice);
            intent.putExtra("foodImage", selectedFood.foodImage);
            startActivity(intent);
            break;
        case CONFIRM_PURCHASE:
            Log.e("test==", "" + CONFIRM_PURCHASE);
            if (selectedFood.foodId != 0) {
                // 跳转到购物车界面,并且提示用户输入要几份菜品
                final View zqz_dialog = LayoutInflater.from(this).inflate(R.lay-
out.zqz_dialog, null);
                // 提示用户输入数量
                new AlertDialog.Builder(this).setTitle("选购").setIcon(R.drawable.
tap_vegetarian2).setView(zqz_dialog).setPositiveButton("订购", new OnClickListener() {
                public void onClick(DialogInterface dialog, int which) {
                  EditText etnumber = (EditText) zqz_dialog.findViewById(R.id.zd_
etnumber);
                  String etnumberStr = etnumber.getText().toString().trim();
                  if (! etnumberStr.equals("") && etnumberStr != null) {
                    db = new DButil(ActivityShowFoodByType.this);
                    ShopCart shopcart = new ShopCart();
                    shopcart.setFoodId(selectedFood.foodId);
                    shopcart.setFoodName(selectedFood.foodName);
                    shopcart.setFoodPrice(selectedFood.foodPrice);
```

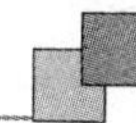

```
                    int foodnum = Integer.parseInt(etnumberStr);
                    shopcart.setFoodNum(foodnum);
                    // float sumprices = selectedFood.foodPrice
                    // * foodnum;
                    float sumprices = getTotalMoney(selectedFood.foodPrice, foodnum);
                    shopcart.setSumPrices(sumprices);
                    db.addFood(shopcart);
                    db.close();
                    Log.e("-------- shopcart", "" + shopcart);
                    Intent intent = new Intent(ActivityShowFoodByType.this, ShopC-
artShowAll.class);

                    Bundle extras = new Bundle();
                    extras.putInt("foodId", selectedFood.foodId);
                    extras.putFloat("foodPrice", selectedFood.foodPrice);
                    extras.putInt("foodNumber", Integer.parseInt(etnumberStr));
                    extras.putString("foodName", selectedFood.foodName);
                    extras.putString("用户名", "");
                    extras.putInt("用户 Id", 0);
                    intent.putExtras(extras);
                    startActivity(intent);
                } else {
                    Toast.makeText(ActivityShowFoodByType.this, "请输入数量",
Toast.LENGTH_SHORT).show();
                }
            }
        }).setNegativeButton("取消", new OnClickListener() {
            public void onClick(DialogInterface dialog, int which) {
            }
        }).create().show();
    } else {
        Toast.makeText(ActivityShowFoodByType.this, "请选中菜品", Toast.LENGTH_
LONG).show();
    }
    break;
  }// end of switch
  return true;
}// end of onOptionsItemSelected()
/**
 * 计算(可能含小数位)总金额
 */
private float getTotalMoney(float price, int number) {
  BigDecimal bd_price = new BigDecimal(price);
  BigDecimal bd_number = new BigDecimal(number);
```

```
    return bd_price.multiply(bd_number).floatValue();
  }// end of getTotalMoney()
  /**
   * Bean 对象,用来封装 GridView 中的一个数据
   *
   * @author zhuqiuzhu
   *
   */
  public final class ViewHolder {
    public ImageView foodImagView;
    public TextView foodNameView;
    public TextView foodPriceView;
  }
  class MyGridViewAdapter extends BaseAdapter {
    LayoutInflater mInflater = null;
    public MyGridViewAdapter(Context ctx) {
      mInflater = LayoutInflater.from(ctx);
    }
    // 返回需要显示的数据集合的 size
    public int getCount() {
      return data.size();
    }
    // 返回指定集合位置的子项数据
    public Object getItem(int position) {
        return data.get(position);
      }

    // 返回指定集合子项的 ID 号
    public long getItemId(int position) {
      return position;
    }
    // 返回某一条数据的视图
    public View getView(int position, View convertView, ViewGroup parent) {
      ViewHolder holder = null;
      // 若 GridView 传进来的代表上一行数据的视图参数 arg1 为空,代表当前是第一行数据
      // 这个时候则解析代表一行数据的视图展现的 xml 文件 zqz_gridview.xml
      // 并且把视图里的控件和 bean 对象里的控件关联起来
      if (convertView == null) {
        holder = new ViewHolder();
        convertView = mInflater.inflate(R.layout.zqz_gridview, null);
        holder.foodImagView = (ImageView) convertView.findViewById(R.id.zg_gridicon);
        holder.foodNameView = (TextView) convertView.findViewById(R.id.zg_gridname);
```

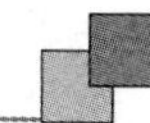

```
                holder.foodPriceView = (TextView) convertView.findViewById(R.id.zg_gridprice);
                // 设置标识
                convertView.setTag(holder);
            } else {
                // 如果传进来的 arg1 不为空,则可以沿用上一行数据的视图
                holder = (ViewHolder) convertView.getTag();
            }
            // 只是需要将数据替换成本行的数据即可
            /*------------图片的显示实现------------- */
            String fi = ((HashMap<String, Object>) data.get(position)).get("icon").
toString();
            String mUrl = GetNetWorkData.getAssignedFoodBitmapUrl(GetNetWorkData.URL, fi);
            GetNetWorkData.setAssignedFoodBimapView(mUrl, holder.foodImagView);
            /*-------------------------------------- */
            holder.foodNameView.setText(((HashMap<String, Object>) data.get(posi-
tion)).get("name").toString());
            holder.foodPriceView.setText(((HashMap<String, Object>) data.get(posi-
tion)).get("price").toString());
            return convertView;
        }
    }// end of MyGridViewAdapter
}
```

上述代码首先根据 Intent 对象,获取 AndroidFoodMain 活动传过来的菜品种类,然后 Android 客户端程序通过 Apache HttpClient 去请求 GetAllFoodServlet 这个控制器,GetAllFoodServlet 控制器会调用数据访问层的 FoodDaoImpl 的 getAllFood 方法获取数据库当中所有的菜品,然后将所有菜品返回给 Android 客户端程序,接着 Android 客户端程序会根据此前获取到的菜品种类,对服务端返回的菜品的列表进行一个迭代,将菜品种类匹配的菜品通过 GridView 控件显示出来,包括菜品的图片、菜品的价格以及菜品的名称等信息。

单击每一个菜品,ActivityShowFoodByType 活动的选项菜单就会弹出来,选项菜单当中包含两个选项,一个是菜品详情,一个是加入购物车,如图 19-4 所示。

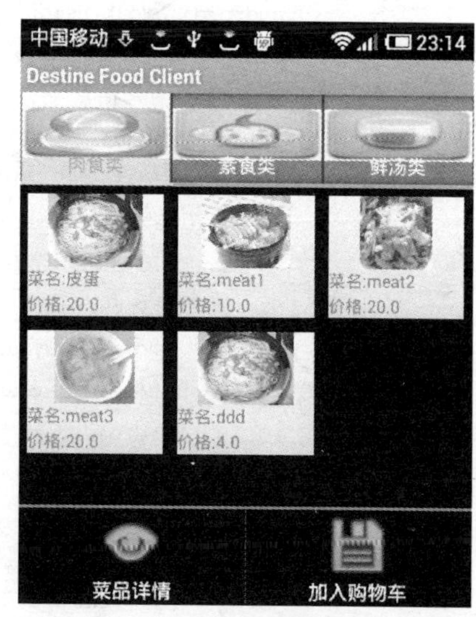

图 19-4　菜品展示界面

19.6 菜品详情

单击 ActivityShowFoodByType 活动的菜品详情选项菜单，则会进入到每一个菜品的菜品详情展示界面：FoodDetail 活动，菜品详情界面布局文件：zqz_food_detail1.xml 文件的源代码如下：

```
<?xml version = "1.0" encoding = "utf - 8"? >
<LinearLayout xmlns:android = "http://schemas.android.com/apk/res/android"
    android:layout_width = "fill_parent"
    android:layout_height = "fill_parent"
    android:orientation = "vertical">
    <LinearLayout
        android:layout_width = "fill_parent"
        android:layout_height = "wrap_content"
        android:orientation = "horizontal"
        >
        <!-- 菜品图片 -->
        <ImageView
                android:id = "@ + id/iv_zfd_image"
            android:layout_width = "100dip"
            android:layout_height = "100dip"></ImageView>
        <LinearLayout
            android:layout_width = "wrap_content"
            android:layout_height = "wrap_content"
            android:orientation = "vertical">
           <!-- 显示编号信息 -->
            <LinearLayout
                android:layout_width = "wrap_content"
                android:layout_height = "wrap_content"
                android:orientation = "horizontal" >
                <TextView
                    android:layout_width = "wrap_content"
                    android:layout_height = "wrap_content"
                    android:text = "编号:" >
                </TextView>
                <TextView
                    android:id = "@ + id/tv_zfd_id"
                        android:layout_marginLeft = "10dip"
                    android:layout_width = "wrap_content"
                    android:layout_height = "wrap_content" >
```

```
        </TextView></LinearLayout>
      <!-- 显示名称信息 -->
    <LinearLayout
        android:layout_width = "wrap_content"
        android:layout_height = "wrap_content"
            android:layout_marginTop = "10dip"
        android:orientation = "horizontal" >
        <TextView
            android:layout_width = "wrap_content"
            android:layout_height = "wrap_content"
            android:text = "名称:" >
        </TextView>
        <TextView
            android:id = "@ + id/tv_zfd_name"
                android:layout_marginLeft = "10dip"
            android:layout_width = "wrap_content"
            android:layout_height = "wrap_content" >
        </TextView></LinearLayout>
    <!-- 显示类型信息 -->
<LinearLayout
    android:layout_width = "wrap_content"
    android:layout_height = "wrap_content"
            android:layout_marginTop = "10dip"
    android:orientation = "horizontal" >
     <TextView
        android:layout_width = "wrap_content"
        android:layout_height = "wrap_content"
            android:text = "类型:" >
        </TextView>
        <TextView
            android:id = "@ + id/tv_zft_type"
                android:layout_marginLeft = "10dip"
            android:layout_width = "wrap_content"
            android:layout_height = "wrap_content" >
        </TextView></LinearLayout>
    <!-- 显示价格信息 -->
<LinearLayout
        android:layout_width = "wrap_content"
        android:layout_height = "wrap_content"
            android:layout_marginTop = "10dip"
        android:orientation = "horizontal" >
        <TextView
```

```
                    android:layout_width = "wrap_content"
                    android:layout_height = "wrap_content"
                    android:text = "价格:" >
                </TextView>
                <TextView
                    android:id = "@ + id/tv_zft_price"
                    android:layout_marginLeft = "10dip"
                    android:layout_width = "wrap_content"
                    android:layout_height = "wrap_content" >
                </TextView>   </LinearLayout>
        </LinearLayout>
    </LinearLayout>
    <TextView
        android:layout_width = "wrap_content"
        android:layout_height = "wrap_content"
        android:text = "描述"/>
    <EditText
        android:id = "@ + id/tv_zft_descri"
        android:layout_width = "fill_parent"
        android:layout_height = "wrap_content"
        android:lines = "8"/>
    <ImageButton
        android:id = "@ + id/ib_zfd_toback"
        android:layout_width = "wrap_content"
        android:layout_height = "wrap_content"
        android:background = "@drawable/to_black"
        android:layout_gravity = "center_horizontal"/>
</LinearLayout>
```

FoodDetail 活动的源代码如下：

```
public class FoodDetail extends Activity {
    private ImageView iv;
    private ImageButton ib;
    private TextView tvId,tvName,tvType,tvMes,tvPrice;
    @Override
    protected void onCreate(Bundle savedInstanceState) {
        super.onCreate(savedInstanceState);
        setContentView(R.layout.zqz_food_detail1);
        Intent intent = this.getIntent();
        int foodId = intent.getIntExtra("foodId", 0);
        int foodType = intent.getIntExtra("foodType", 0);
        String foodImage = intent.getStringExtra("foodImage");
```

```
        float foodPrice = intent.getFloatExtra("foodPrice", 0.0f);
        String foodDescri = intent.getStringExtra("foodDescri");
        String foodName = intent.getStringExtra("foodName");
        iv = (ImageView) findViewById(R.id.iv_zfd_image);
        ib = (ImageButton) findViewById(R.id.ib_zfd_toback);
        tvId = (TextView) findViewById(R.id.tv_zfd_id);
        tvName = (TextView) findViewById(R.id.tv_zfd_name);
        tvPrice = (TextView) findViewById(R.id.tv_zft_price);
        tvType = (TextView) findViewById(R.id.tv_zft_type);
        tvMes = (TextView) findViewById(R.id.tv_zft_descri);
        tvId.setText("" + foodId);
        tvName.setText("" + foodName);
        tvType.setText("" + foodTypeTransform(foodType));
        tvPrice.setText("" + foodPrice);
        tvMes.setText("" + foodDescri);
        iv.setBackgroundResource(R.drawable.yellowbackground);
        ib.setBackgroundResource(R.drawable.to_black);
        ib.setOnClickListener(new OnClickListener() {
            @Override
            public void onClick(View v) {
                finish();
            }
        });
        String mUrl = GetNetWorkData.getAssignedFoodBitmapUrl(GetNetWorkData.URL,
foodImage);
        GetNetWorkData.setAssignedFoodBimapView(mUrl, iv);
        }// end of onCreate()
        public String foodTypeTransform(int type){
        String transform = "";
        if(type == 0){
            transform = "肉食类";
        }else if(type == 1){
            transform = "素食类";
        }else if(type == 2){
            transform = "汤品类";
        }else{
            transform new IllegalArgumentException();
        }
        return transform;
    }
}
```

当单击 ActivityShowFoodByType 活动的“菜品详情”选项菜单的时候，会将选中菜品的编号、菜品的名称、菜品的描述信息、菜品的种类、菜品的价格以及菜品的图片 URL 地址通过 Intent 对象传给菜品详情活动。

FoodDetail 活动拿到这些信息后，除了图片之外，其余信息直接找到相应的控件，然后显示出来即可，而对于图片的显示，则需要从网络上去下载，所以在 FoodDetail 的 onCreate 方法中调用了 GetNetworkData 类的 setAssignedFoodBitmapView 方法将相应的菜品图片显示出来，如图 19－5 所示。

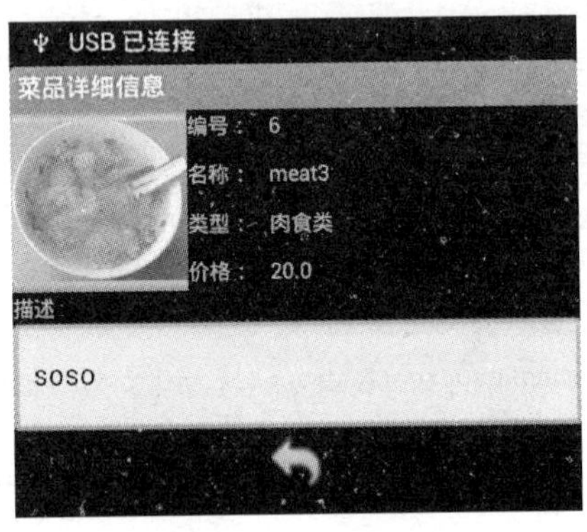

图 19－5　菜品详情页面

19.7　购物车

在菜品展示界面，单击某一个菜品的时候，会弹出 ActivityShowFoodByType 活动的选项菜单，其中有一个选项叫做“加入购物车”。当用户单击这个选项的时候，会弹出一个订购数量的对话框，在这个对话框里面，用户可以输入想要预定这款菜品的数量，用户输入数量后，单击对话框的“订购”按钮，会进入到购物车总计界面。

19.7.1　购物车总计

单击 ActivityShowFoodByType 活动的“加入购物车”选项菜单后，会弹出一个输入数量的对话框，输入想要预定的数量后，单击“订购”按钮，则会调用操作本地 SQLite 数据库的工具类：DBUtil 的 addFood 方法，DBUtil 工具类的源代码如下：

```
public class DButil {
    //列出购物车中的所有字段
    private static final String[] COLS = new String[]{"foodid","foodname","foodprice",
        "foodnum","foodsumprices","foodimage"};
    //数据库名字
    public static String DB = "shopcartDB";
    //购物车的表名
```

```
    public static String TABLE_SHOP_CART = "shopcart";
    //创建购物车表
    public static String create_table_shopcard = "create table " + TABLE_SHOP_CART
        + "(foodid integer,foodname text,foodprice float,foodnum int,foodsumprices
float,foodimage int)";
    public int version = 7;
    private Context ctx;
    private SQLiteDatabase sqllite;
    private SQLiteOpenHelper helper;
    //new 出数据库并初始化
    public DButil(Context context){
        this.ctx = context;
        init();
    }
    //数据库初始化
    private void init(){
        if(helper == null){
            helper = new MySQLiteHelper(ctx, DB, null, version);
            sqllite = helper.getWritableDatabase();
        }
    }
    // 新增一条购买商品记录
    public long addFood(ShopCart food) {
        //用来装数据
        try {
            Cursor cur = sqllite.query(TABLE_SHOP_CART, COLS, null, null, null, null, null);
            cur.moveToFirst();
            int foodId;
            int count = cur.getCount();
            for (int i = 0; i < count; i++ ) {
                foodId = cur.getInt(0);
                //若存在此 foodid 则只增加数量和总价
                    if(food.getFoodId() == foodId){
                    ContentValues value = new ContentValues();
                      value.put("foodnum", food.getFoodNum() + cur.getInt(3));
                      value.put("foodsumprices", food.getSumPrices() + cur.getFloat(4));
                       return sqllite.update(TABLE_SHOP_CART, value, "foodid = " +
foodId, null);
                    }
                cur.moveToNext();
```

```
            }
        } catch (Exception e) {
            e.printStackTrace();
        }
        ContentValues value = new ContentValues();
        value.put("foodid", food.getFoodId());
        value.put("foodname", food.getFoodName());
        value.put("foodprice", food.getFoodPrice());
        value.put("foodnum", food.getFoodNum());
        value.put("foodsumprices", food.getSumPrices());
        value.put("foodimage", food.getImage());
        return sqllite.insert(TABLE_SHOP_CART, null, value);
    }
    //查询所有的游标
    public Cursor getAllCartCursor(){
        return   sqllite.query(TABLE_SHOP_CART, COLS, null, null, null, null, null);
    }
    //得到购物车表里的所有数据
    public ArrayList getAllShopCart(){
        ArrayList rs = new ArrayList();
        try {
            Cursor cur = sqllite.query(TABLE_SHOP_CART, COLS, null, null, null, null, null);
            cur.moveToFirst();
            int count = cur.getCount();
            for (int i = 0; i < count; i++) {
                HashMap item = new HashMap();
                item.put("foodid", cur.getInt(0));
                item.put("foodname", cur.getString(1));
                item.put("foodprice", cur.getFloat(2));
                item.put("foodnum", cur.getInt(3));
                item.put("foodsumprices", cur.getFloat(4));
                item.put("foodimage", cur.getInt(5));
                rs.add(item);
                cur.moveToNext();
            }
        } catch (Exception e) {
            e.printStackTrace();
        }
        return rs;
    }
```

```
    //返回所有 shopCart 类型的 ArrayList
    public ArrayList<ShopCart> getFoodFromCart(){
        ArrayList rs = new ArrayList();
        try {
             Cursor cur = sqllite.query(TABLE_SHOP_CART, COLS, null, null, null,
null, null);
            cur.moveToFirst();
            int count = cur.getCount();
            for (int i = 0; i < count; i++) {
                ShopCart shopcart = new ShopCart();
                shopcart.setFoodId(cur.getInt(0));
                shopcart.setFoodName(cur.getString(1));
                shopcart.setFoodPrice(cur.getFloat(2));
                shopcart.setFoodNum(cur.getInt(3));
                shopcart.setSumPrices(cur.getFloat(4));
                rs.add(shopcart);
                cur.moveToNext();
            }
        } catch (Exception e) {
            e.printStackTrace();
        }
        return rs;
    }
    //修改购物车中所有的数据
    public int ModifyShopcart(int foodId,int foodnum,float foodsumprices ){
        ContentValues value = new ContentValues();
        value.put("foodnum", foodnum);
        value.put("foodsumprices",foodsumprices);
        return sqllite.update(TABLE_SHOP_CART, value, "foodid=" + foodId, null);
    }
    //删除购物车中所有的数据
    public int deleteAllShopcart(){
        return sqllite.delete(TABLE_SHOP_CART, null, null);
    }
    //删除不需要的商品
    public int deleteFood(String whereparam){
        return sqllite.delete(TABLE_SHOP_CART, whereparam, null);
    }
    //查找需要的商品
    public ShopCart findFood(int foodid){
```

```
            ShopCart shopcart = new ShopCart();
            Cursor cur = sqllite.query(TABLE_SHOP_CART,COLS, "foodid = " + foodid, null,
null, null, null);
            cur.moveToFirst();
            shopcart.setFoodId(cur.getInt(0));
            shopcart.setFoodName(cur.getString(1));
            shopcart.setFoodPrice(cur.getFloat(2));
            shopcart.setFoodNum(cur.getInt(3));
            shopcart.setSumPrices(cur.getFloat(4));
            shopcart.setImage(cur.getInt(5));
            return shopcart;
        }
        /**
         * 获取数据库连接
         * @return
         */
        public SQLiteDatabase getDB(){
            return this.sqllite;
        }
        /**
         * 关闭数据库
         */
        public void close(){
            this.sqllite.close();
        }
        class MySQLiteHelper extends SQLiteOpenHelper {
            public MySQLiteHelper(Context context, String name,
                    CursorFactory factory, int version) {
                super(context, name, factory, version);
            }
            @Override
            public void onCreate(SQLiteDatabase db) {
                db.execSQL(create_table_shopcard);
            }
            @Override
            public void onUpgrade(SQLiteDatabase db, int oldVersion, int newVersion) {
                db.execSQL("drop table if exists " + TABLE_SHOP_CART);
                onCreate(db);
            }
        }
    }
```

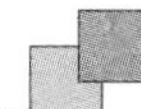

运行效果如图 19－6 所示。

显示购物车里面所有菜品的活动是 ShopCartShowAll，此活动当中用到两个布局文件，其中 wjf_shopcarthead.xml 布局文件的源代码如下：

图 19－6 加入购物车对话框

```
<? xml version = "1.0" encoding = "utf-8"? >
<LinearLayout
android:id = "@ + id/widget35"
android:layout_width = "fill_parent"
android:layout_height = "fill_parent"
xmlns: android = " http://schemas. android.com/apk/res/android"
>
<TextView
android:id = "@ + id/foodid"
android:layout_width = "40px"
android:layout_height = "wrap_content"
android:text = "&#32534;&#21495;"
>
</TextView>
<TextView
android:id = "@ + id/foodname"
android:layout_width = "100px"
android:layout_height = "wrap_content"
android:text = "&#21830;&#21697;&#21517;&#31216;"
>
</TextView>
<TextView
android:id = "@ + id/foodprice"
android:layout_width = "50px"
android:layout_height = "wrap_content"
android:text = "&#21333;&#20215;"
>
</TextView>
<TextView
android:id = "@ + id/foodnumber"
android:layout width = "50px"
android:layout_height = "wrap_content"
android:text = "  &#25968;&#37327;"
>
```

```
</TextView>
<TextView
android:id="@+id/sumpirces"
android:layout_width="50px"
android:layout_height="wrap_content"
android:text="  &#37329;&#39069;"
>
</TextView>
</LinearLayout>
```

ShopCartShowAll 活动使用到的另外一个布局文件是 wjf_shopcartlast.xml，其源代码如下：

```
<?xml version="1.0" encoding="utf-8"?>
<LinearLayout
android:id="@+id/widget35"
android:layout_width="fill_parent"
android:layout_height="fill_parent"
xmlns:android="http://schemas.android.com/apk/res/android"
>
<TextView
android:id="@+id/allfood"
android:layout_width="100px"
android:layout_marginLeft="100px"
android:layout_height="wrap_content"
android:text="总金额"
>
</TextView>
<TextView
android:id="@+id/toatalmoney"
android:layout_width="100px"
android:layout_height="wrap_content"
android:text=""
android:textColor="#FFFFFF"
>
</TextView>
</LinearLayout>
```

ShopCartShowAll 活动的源代码如下：

```
public class ShopCartShowAll extends Activity{
    //定义一个数据库类的引用
    private DButil db;
    //定义 ArrayList 包含所有的餐品
```

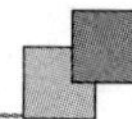

```
private ArrayList shopcartList = new ArrayList();
//定义购物车的一个类
private ShopCart shopCart;
//弹出删除餐品的标志 ,1 表示删除或修改选中的餐品
private static final int DIALOG1  =  1;
//2 表示清空购物车中的食品
private static final int DIALOG2  =  2;
private   HashMap item;
//定义餐品的一个 hashmap food
private   HashMap food;
//表示选中 listview 的那一行的商品 id
private int food_id ;
public int foodID;
private Cursor shopcur;
//设置总金额
public float totalprices;
//表示选中 listview 的那一行
public int lvpostion;
//设置头视图
private View headview;
//设置尾视图
private View lastview;
private TextView foodId;
private TextView foodName;
private TextView foodPrice;
private TextView foodNum;
//表示每一种餐品的总价
private TextView foodSumPrices;
//表示所有餐品的总价
private TextView foodtotalmoney;
//定义传递的参数
public Intent param;
public int userId;
public String username;
public String foodname;
public float foodprice;
private int foodnum;
public void onCreate(Bundle savedInstanceState) {
    super.onCreate(savedInstanceState);
    //初始化数据库
    db = new DButil(this);
    //获得购物车中所有的游标
```

```
shopcur = db.getAllCartCursor();
//获取数据库中所有的数据
shopcartList = db.getAllShopCart();
//总价初始化为 0
totalprices = 0;
param = this.getIntent();
// 解析 wjf_shopcartthead 文件
headview = LayoutInflater.from(this).inflate(R.layout.wjf_shopcarthead, null);
// 解析 wjf_shopcartlast 文件
lastview = LayoutInflater.from(this).inflate(R.layout.wjf_shopcartlast, null);
foodId = (TextView) headview.findViewById(R.id.tvid);
foodName = (TextView) headview.findViewById(R.id.tvname);
foodPrice = (TextView) headview.findViewById(R.id.tvprice);
foodNum = (TextView) headview.findViewById(R.id.tvnum);
foodSumPrices = (TextView) headview.findViewById(R.id.tvsumprices);
foodtotalmoney = (TextView) lastview.findViewById(R.id.toatalmoney);
for(int i = 0;i<shopcartList.size();i++){
    food = (HashMap) shopcartList.get(i);
    Float prices = (Float) food.get("foodsumprices");
    //得到所有餐品总的金额
    totalprices += prices;
}
//显示所有餐品的总金额
foodtotalmoney.setText(totalprices + "");
ListView lv = new ListView(this);
SimpleAdapter sa = new SimpleAdapter(this,shopcartList, R.layout.wjf_shopc-
artshowall,
    new String[]{"foodid","foodname","foodprice","foodnum","foodsumprices"},
    new int[]{R.id.foodid,R.id.foodname,R.id.foodprice, R.id.foodnumber,R.
id.sumpirces});
sa.notifyDataSetChanged();
//加载头文件视图
lv.addHeaderView(headview);
//加载尾文件视图
lv.addFooterView(lastview);
lv.setAdapter(sa);
//设置整个界面文件
setContentView(lv);
lv.setOnItemClickListener(new OnItemClickListener() {
  public void onItemClick(AdapterView<?> arg0, View arg1, int arg2,
      long arg3) {
    if(0<arg2&&arg2<=shopcartList.size()){
```

```
            //移动游标位置
            shopcur.moveToPosition(arg2 - 1);
            //获取选中的 listview 的某一行的 foodid
            foodID = shopcur.getInt(shopcur.getColumnIndex("foodid"));
            foodname = (String) shopcur.getString(shopcur.getColumnIndex("foodname"));
            foodprice = shopcur.getFloat(shopcur.getColumnIndex("foodprice"));
            foodnum = shopcur.getInt(shopcur.getColumnIndex("foodnum"));
            lvpostion = arg2 - 1;
            Log.e("---------foodId", "" + foodID);
            //表示 listview 的位置
          showDialog(DIALOG1);
            }
            else{
              foodID = - 1;
            }

        }
    });
      lv.setOnItemSelectedListener(new OnItemSelectedListener() {
        public void onItemSelected(AdapterView<?> arg0, View arg1, int arg2,
            long arg3) {
          if(0<arg2&&arg2< = shopcartList.size()){
          //移动游标位置
          shopcur.moveToPosition(arg2 - 1);
          //获取选中的 listview 的某一行的 foodid
          food_id = shopcur.getInt(shopcur.getColumnIndex("foodid"));
          Log.e("---------Id", "" + food_id);
          //表示 listview 的位置
        lvpostion = arg2 - 1;
          }
          else{
              food_id = - 1;
          }
        }
        public void onNothingSelected(AdapterView<?> arg0) {
        }
    });
    }
    protected Dialog onCreateDialog(int id) {
        super.onCreateDialog(id);
        switch (id) {
        case DIALOG1: {
```

```
            return builder1(this);
        }
        case DIALOG2: {
            return builder2(this);
        }
        }
        return null;
    }
    private Dialog builder1(Context ctx) {
        AlertDialog.Builder builder = new AlertDialog.Builder(ctx);
        // 设置 Dialog 的标题
        builder.setTitle("修改或删除");
        // 设置 Dialog 的图标
        builder.setIcon(R.drawable.modify);
        // 设置 Dialog 的显示信息
        builder.setMessage("修改还是删除");
        // 设置 Dialog 上的一个按钮和按钮的单击事件
        builder.setPositiveButton("修改", new DialogInterface.OnClickListener() {
            public void onClick(DialogInterface dialog, int which) {
                Intent tent = new Intent(ShopCartShowAll.this,ModifyListCart.class);
                tent.putExtra("foodname", foodname);
                tent.putExtra("foodprice", foodprice);
                tent.putExtra("foodid", foodID);
                tent.putExtra("foodnum", foodnum);
                //关闭数据库
                db.close();
                startActivity(tent);
                finish();
            }
        });
        builder.setNeutralButton("删除", new DialogInterface.OnClickListener() {
            public void onClick(DialogInterface dialog, int which) {
                if(foodID>0){
                    HashMap map = (HashMap) shopcartList.get(lvpostion);
                    Float price = (Float) map.get("foodsumprices");
                    totalprices - = price;
                    db.deleteFood("foodid = " + foodID);
                    Intent tent = new Intent(ShopCartShowAll.this,ShopCartShowAll.class);
                    tent.addFlags(Intent.FLAG_ACTIVITY_NEW_TASK);
                    //关闭数据库
                    db.close();
                    startActivity(tent);
```

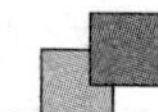

```
                finish();
            }
        }
    });
    builder.setNegativeButton("取消", new DialogInterface.OnClickListener() {
        public void onClick(DialogInterface dialog, int which) {
            }
    });
    return builder.create();
}
private Dialog builder2(Context ctx) {
    AlertDialog.Builder builder = new AlertDialog.Builder(ctx);
    // 设置 Dialog 的标题
    builder.setTitle("删除确认");
    // 设置 Dialog 的图标
    builder.setIcon(R.drawable.delete);
    // 设置 Dialog 的显示信息
    builder.setMessage("真的要删除吗?");
    // 设置 Dialog 上的一个按钮和按钮的单击事件
    builder.setPositiveButton("确认", new DialogInterface.OnClickListener() {
        public void onClick(DialogInterface dialog, int which) {
            if(db.getAllShopCart().size()>0){
                Log.e("-----------get shopcart", "" + db.getAllShopCart().size());
                db.deleteAllShopcart();
                Intent tent = new Intent(ShopCartShowAll.this,ShopCartShowAll.class);
                tent.addFlags(Intent.FLAG_ACTIVITY_NEW_TASK);
                //关闭数据库
                db.close();
                startActivity(tent);
                finish();
            }
        }
    });
    builder.setNegativeButton("取消", new DialogInterface.OnClickListener() {
        public void onClick(DialogInterface dialog, int which) {
            setTitle("取消按钮按下");
        }
    });
    return builder.create();
}
public boolean onCreateOptionsMenu(Menu menu) {
    menu.add(2, Menu.FIRST + 1, 2, "清空购物车").setIcon(R.drawable.delete);
```

```
        menu.add(3, Menu.FIRST + 2, 3, "继续购物").setIcon(R.drawable.goshop);
        menu.add(4, Menu.FIRST + 3, 4, "下订单").setIcon(R.drawable.lastlist);
        return super.onCreateOptionsMenu(menu);
    }
    public boolean onOptionsItemSelected(MenuItem item) {
        super.onOptionsItemSelected(item);
        switch (item.getItemId()) {
        //删除所有的餐品
        case Menu.FIRST + 1: {
            showDialog(DIALOG2);
            break;
        }
        //继续购物
        case Menu.FIRST + 2: {
            Intent intent = new Intent(ShopCartShowAll.this,AndroidFoodMain.class);
            db.close();
            startActivity(intent);
            finish();
            break;
        }
        //生产订单
        case Menu.FIRST + 3: {
            Intent intent = new Intent(ShopCartShowAll.this,OrderListActivity.class);
            intent.putExtra("totalprices", totalprices);
            db.close();
            startActivity(intent);
            finish();
            break;
        }
        }
        return true;
    }
    protected void onStop() {
        super.onStop();
        db.close();
    }
}
```

购物车显示所有菜品运行效果如图 19-7 所示。

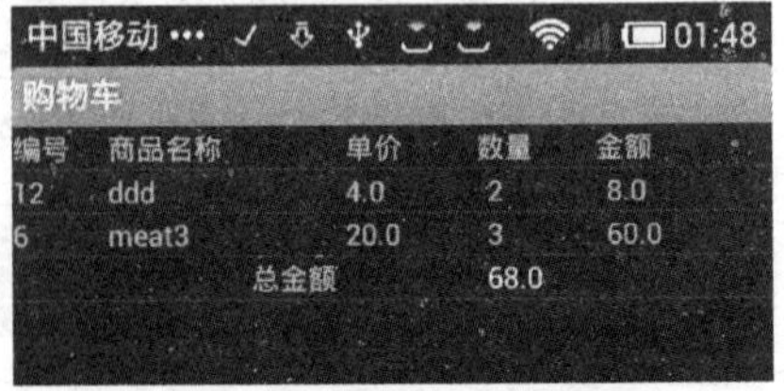

图 19-7　购物车显示所有菜品

19.7.2　修改购物车

图 19-8　单击购物车列表选项

单击购物车当中的每一行，会弹出一个对话框。对话框的底部会有两个按钮，一个是“修改”按钮，一个是“删除”按钮。单击“修改”按钮的时候，会跳转到对当前选中项菜品进行编辑的页面；单击“删除”按钮的时候，会将当前选中项从购物车当中删除，同时更新购物车的总结界面。如图 19-8 所示。

当用户单击“修改”按钮的时候，会跳转到相应菜品的编辑页面：ModifyListCart，此活动使用的布局文件是 wjf_updatelistcart.xml，其源代码如下：

```
<?xml version = "1.0" encoding = "utf-8"?>
<LinearLayout
android:id = "@ + id/widget52"
android:layout_width = "fill_parent"
android:layout_height = "fill_parent"
android:orientation = "vertical"
xmlns:android = "http://schemas.android.com/apk/res/android"
>
<TextView
android:id = "@ + id/foodname"
android:layout_width = "121px"
android:layout_height = "wrap_content"
android:text = "餐品名称"
>
</TextView>
<TextView
android:id = "@ + id/foodprice"
android:layout_width = "121px"
android:layout_height = "18px"
android:text = "餐品价格"
>
</TextView>
<TextView
android:id = "@ + id/foodnum"
android:layout_width = "122px"
```

```
android:layout_height = "21px"
android:text = "餐品数量"
>
</TextView>
<EditText
android:id = "@ + id/etfoodnum"
android:layout_width = "160px"
android:layout_height = "wrap_content"
android:textSize = "18sp"
>
</EditText>
<Button
android:id = "@ + id/bmodify"
android:layout_width = "160px"
android:layout_height = "wrap_content"
android:text = "修改"
>
</Button>
</LinearLayout>
```

ModifyCartList.java 文件源代码如下：

```
public class ModifyListCart extends Activity {
    private Button bmodify;
    private DButil db;
    private TextView tvname;
    private TextView tvprice;
    private EditText etnum;
    public Intent intent;
    //食品编号
    int foodid;
    //食品单价
    float foodprice;
    //食品名字
    String foodname;
    int foodnum;
    float sumfoodprices;
    public void onCreate(Bundle savedInstanceState) {
        super.onCreate(savedInstanceState);
        setContentView(R.layout.wjf_updatelistcart);
        db = new DButil(this);
        Intent intent = getIntent();
        //拿到购物车中传过来的食品编号 foodID
```

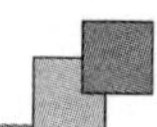

```
        foodid = intent.getIntExtra("foodid", 0);
        //拿到购物车中传过来的食品名称 foodname
        foodname = intent.getStringExtra("foodname");
        //拿到购物车中传过来的食品单价 foodprice
        foodprice = intent.getFloatExtra("foodprice", 1.0f);
        foodnum = intent.getIntExtra("foodnum", 1);
        int lvpostion = intent.getIntExtra("lvpostion", 0);
        tvname = (TextView) this.findViewById(R.id.foodname);
        tvprice = (TextView) this.findViewById(R.id.foodprice);
        etnum = (EditText) this.findViewById(R.id.etfoodnum);
        etnum.setText(foodnum + "");
        tvname.setText("餐品名称:  " + foodname);
        tvprice.setText("餐品单价:  " + foodprice + "");
        bmodify = (Button) this.findViewById(R.id.bmodify);
        bmodify.setOnClickListener(new OnClickListener() {
            public void onClick(View v) {
                foodnum = Integer.parseInt(etnum.getText().toString());
                sumfoodprices = foodnum * foodprice;
                foodnum = Integer.parseInt(etnum.getText().toString());
                //对购物车中的食品进行修改
                db.ModifyShopcart(foodid, foodnum, sumfoodprices);
                //回到购物车界面
                Intent intent = new Intent(ModifyListCart.this,ShopCartShowAll.class);
                startActivity(intent);
                db.close();
                finish();
            }
        });
    }
}
```

19.7.3 下　单

当用户已经预定完所需的菜品后，就可以通过单击 ShopCartShowAll 活动的“下订单“选项菜单(按住设备的 Menu 键弹出)进行确认购物的操作。

用户通过 Android 客户端发起确认购物操作的时候，会向服务端的 AddOrderServlet 控制器发起 Http 请求，并得到 AddOrderServlet 的响应。

AddOrderServlet 控制器的源代码如下：

```
public class AddOrderServlet extends HttpServlet {
    public void doGet(HttpServletRequest request, HttpServletResponse response)
            throws ServletException, IOException {
```

```
response.setContentType("text/html");
PrintWriter out = response.getWriter();
//订单的用户信息
int userId = Integer.parseInt(request.getParameter("userId"));
String userName = request.getParameter("userName");
String address = request.getParameter("address");
String email = request.getParameter("email");
String telephone = request.getParameter("telephone");
String suggest = request.getParameter("suggest");
OrderFoods orderfoods = new OrderFoods();//保存订单用户的所有信息。
orderfoods.setUserId(userId);
orderfoods.setUserName(userName);
orderfoods.setAddress(address);
orderfoods.setEmail(email);
orderfoods.setTelephone(telephone);
orderfoods.setOrderSuggest(suggest);
FoodDao dao = new FoodDaoImpl();
int orderId = dao.addOrder(orderfoods);//添加订单用户的信息和 OrderId
//得到的是购物车中的菜的种数
int foodsNum = Integer.parseInt( request.getParameter("foodsNum"));
//在这里可以将菜的种类打印一下
//System.out.println(foodsNum);
ArrayList<ShopCart> list = new ArrayList<ShopCart>();
  for(int i = 0;i<foodsNum;i ++ ){
    ShopCart cart = new ShopCart();
    int foodId = Integer.parseInt(request.getParameter("foodId" + i));
    int foodNum = Integer.parseInt(request.getParameter("foodNum" + i));
    cart.setFoodId(foodId);//购物车中第一个菜的 id
    cart.setFoodName(request.getParameter("foodName" + i));
                          //购物车中第一个菜的 name
    cart.setFoodNum(foodNum);
                          //购物车中第一个菜的订购数量
    cart.setFoodPrice(Float.parseFloat(request.getParameter("foodPrice" + i)));
                          //购物车中第一个菜的价格
    cart.setSumPrices(Float.parseFloat(request.getParameter("sumPrices" + i)));
                          //购物车中的第一个菜订购的总金额
    list.add(cart);
    //将购物车中的菜的信息添加到数据库中
    boolean success = dao.addCartInformation(orderId, foodId, foodNum);
                          //插入到 t_food_order 表中去
}
    ArrayList dingdan = new ArrayList();
```

```
                dingdan = dao.findrOrderByOrderId(orderId);
                StringBuffer sb = new StringBuffer();
                int j = 1;
                System.out.println("客户姓名　客户地址　电话号码　建议　　");
                System.out.println(userName + "　　" + address + "　　" + telephone + "　　" +
suggest);
                System.out.println( "菜名　　　　　菜价格　　　　　份数");
                for(int i = 0;i<dingdan.size();i + = 3){
                  System.out.println(dingdan.get(i) + "　　" + dingdan.get(i + 1) + "
 " + dingdan.get(i + 2));
                }
                out.flush();
                out.close();
        }

        public void doPost(HttpServletRequest request, HttpServletResponse response)
                throws ServletException, IOException {
            doGet(request, response);
        }
    }
```

Android 客户端下单界面如图 19 - 9 所示。

显示下单界面的活动是 OrderListActivity，此活动所使用的布局文件是 hyl_order.xml 文件，其源代码如下：

```
    <?xml version = "1.0" encoding = "utf - 8"? >
    <RelativeLayout
    android:id = "@ + id/widget33"
    android:layout_width = "fill_parent"
    android:layout_height = "fill_parent"
    xmlns:android = " http://schemas. android. com/
apk/res/android"
    >
    <Button
    android:id = "@ + id/bcancel"
    android:layout_width = "80px"
    android:layout_height = "wrap_content"
    android:text = "取消"
    android:layout_alignParentBottom = "true"
    android:layout_alignParentRight = "true"
    ></Button>
    <Button
```

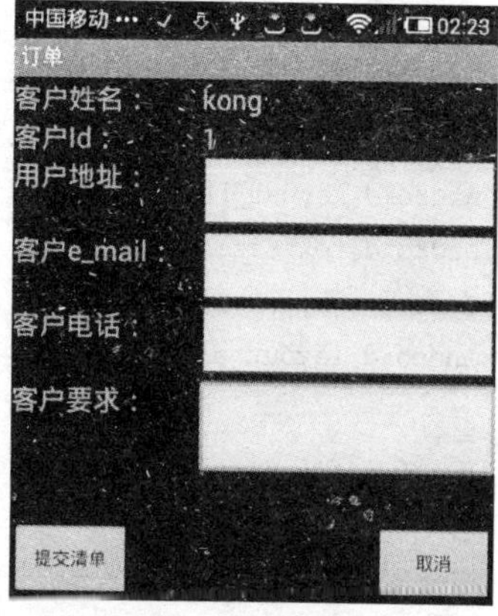

图 19 - 9　下订单界面

```
android:id = "@ + id/bsubmit"
android:layout_width = "80px"
android:layout_height = "wrap_content"
android:text = "提交清单"
android:layout_alignParentBottom = "true"
android:layout_alignParentLeft = "true"
>
</Button>
<EditText
android:id = "@ + id/etsuggest"
android:layout_width = "220px"
android:layout_height = "180px"
android:textSize = "18sp"
android:layout_alignTop = "@ + id/tvsuggest"
android:layout_alignLeft = "@ + id/ettelephone"
>
</EditText>
<TextView
android:id = "@ + id/tvsuggest"
android:layout_width = "wrap_content"
android:layout_height = "wrap_content"
android:textSize = "18sp"
android:text = "客户要求:"
android:layout_below = "@ + id/ettelephone"
android:layout_alignParentLeft = "true"
>
</TextView>
<EditText
android:id = "@ + id/ettelephone"
android:layout_width = "220px"
android:layout_height = "wrap_content"
android:textSize = "18sp"
android:layout_alignTop = "@ + id/tvtelephone"
android:layout_alignLeft = "@ + id/ete_mail"
>
</EditText>
<TextView
android:id = "@ + id/tvtelephone"
android:layout_width = "wrap_content"
android:layout_height = "wrap_content"
android:textSize = "18sp"
android:text = "客户电话:"
```

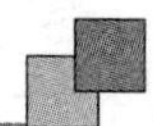

```
android:layout_below = "@ + id/ete_mail"
android:layout_alignParentLeft = "true"
>
</TextView>
<EditText
android:id = "@ + id/ete_mail"
android:layout_width = "220px"
android:layout_height = "wrap_content"
android:textSize = "18sp"
android:layout_alignTop = "@ + id/tve_mail"
android:layout_alignLeft = "@ + id/etaddress"
>
</EditText>
<TextView
android:id = "@ + id/tve_mail"
android:layout_width = "wrap_content"
android:layout_height = "wrap_content"
android:textSize = "18sp"
android:text = "客户 e_mail:"
android:layout_below = "@ + id/etaddress"
android:layout_alignParentLeft = "true"
>
</TextView>
<EditText
android:id = "@ + id/etaddress"
android:layout_width = "220px"
android:layout_height = "wrap_content"
android:textSize = "18sp"
android:layout_alignTop = "@ + id/tvaddress"
android:layout_alignLeft = "@ + id/tvId_show"
>
</EditText>
<TextView
android:id = "@ + id/tvaddress"
android:layout_width = "wrap_content"
android:layout_height = "wrap_content"
android:textSize = "18sp"
android:text = "用户地址:"
android:layout_below = "@ + id/tvId_show"
android:layout_alignParentLeft = "true"
>
</TextView>
```

```
<TextView
android:id="@+id/tvId_show"
android:layout_width="220px"
android:layout_height="wrap_content"
android:textSize="18sp"
android:text="1"
android:layout_alignTop="@+id/tvId"
android:layout_alignLeft="@+id/tvname_show"
>
</TextView>
<TextView
android:id="@+id/tvId"
android:layout_width="wrap_content"
android:layout_height="wrap_content"
android:textSize="18sp"
android:text="客户 Id:"
android:layout_below="@+id/tvname_show"
android:layout_alignParentLeft="true"
>
</TextView>
<TextView
android:id="@+id/tvname_show"
android:layout_width="220px"
android:layout_height="wrap_content"
android:textSize="18sp"
android:text="hello"
android:layout_alignParentTop="true"
android:layout_toRightOf="@+id/tvname"
>
</TextView>
<TextView
android:id="@+id/tvname"
android:layout_width="wrap_content"
android:layout_height="wrap_content"
android:textSize="18sp"
android:text="客户姓名:        "
android:layout_alignParentTop="true"
android:layout_alignParentLeft="true"
>
</TextView>
</RelativeLayout>
```

OrderListActivity 活动的源代码如下：

```
public class OrderListActivity extends Activity {
    private DButil db;
    private TextView tvname_show;      //客户的名字显示
    private TextView tvid_show;        //客户的 Id 显示
    private EditText etaddress;        //客户的地址
    private EditText ete_mail;         //客户的 Email
    private EditText ettelephone;      //客户的电话号码
    private EditText etsuggest;        //客户的要求(要不要加配料等)
    private Button bsubmit;
    private Button bcancel;
    public  String address;
    public  String email ;
    public  String telephone ;
    public  String userId;//客户的 Id
    public  String suggest ;
    public  String userName;//客户的 name
    private ArrayList<ShopCart> orderFoods = new ArrayList<ShopCart>();
                                                        //购物车中的菜的信息
    protected void onCreate(Bundle savedInstanceState) {
        super.onCreate(savedInstanceState);
        Log.e("1111111111", "tttttt");
        setContentView(R.layout.hyl_order);
        SharedPreferences share = this.getSharedPreferences("android",MODE_WORLD_
READABLE);
        userId = share.getString("userId", "");
        userName = share.getString("userName", "");;
        db = new DButil(this);
        orderFoods = db.getFoodFromCart();//从客户端数据库读出购物车中的信息
        final Object foods[][] = new Object[orderFoods.size()][5];
        tvname_show = (TextView) findViewById(R.id.tvname_show);
        tvid_show = (TextView) findViewById(R.id.tvId_show);
        etaddress = (EditText) findViewById(R.id.etaddress);//客户地址的编辑框
        ete_mail = (EditText) findViewById(R.id.ete_mail);//客户邮箱编辑框
        ettelephone = (EditText) findViewById(R.id.ettelephone);//客户电话
        etsuggest = (EditText) findViewById(R.id.etsuggest);//客户的建议要求
        bsubmit = (Button) findViewById(R.id.bsubmit);
        bcancel = (Button) findViewById(R.id.bcancel);
        tvname_show.setText(userName);//显示客户的名字
        tvid_show.setText(userId);//显示客户的 Id
        bsubmit.setOnClickListener(new OnClickListener() {//提交清单按钮被按下
            public void onClick(View v) {
```

```
                    address = etaddress.getText().toString();
                    email = ete_mail.getText().toString();
                    telephone = ettelephone.getText().toString();
                    suggest = etsuggest.getText().toString();
                    if(!"".equals(userId)&&!"".equals(address)&&!"".equals(telephone)
&&orderFoods.size()>0){
                        callPost(foods, address, email, telephone, suggest, userId, user-
Name);//向服务器传数据
                     db.deleteAllShopcart();
                     db.close();
                     finish();
                    }
                    else{
                        showDialog(2);
                    }
                }
                private void callPost(Object[][] foods, String address,
                            String email, String telephone, String suggest,String userId,
String userName) {
                    HttpPost req = new HttpPost("http://211.155.227.204:8080/Destine-
FoodServer/AddOrderServlet");
                    List<NameValuePair> param = new ArrayList<NameValuePair>();
                    param.add(new BasicNameValuePair("userName",userName));
                    param.add(new BasicNameValuePair("userId",userId));
                    param.add(new BasicNameValuePair("address",address));
                    param.add(new BasicNameValuePair("email",email));
                    param.add(new BasicNameValuePair("telephone",telephone));
                    param.add(new BasicNameValuePair("suggest",suggest));
                    param.add(new BasicNameValuePair("foodsNum",orderFoods.size() + ""));
                  for(int i = 0;i<orderFoods.size();i++){//此循环是将购物车中的菜全
部加入到 param 中去。
                     param.add(new BasicNameValuePair("foodId" + i,orderFoods.get(i).
getFoodId() + ""));
                    param.add(new BasicNameValuePair("foodName" + i,orderFoods.get(i).
getFoodName()));
                     param.add(new BasicNameValuePair("foodNum" + i,orderFoods.get(i).
getFoodNum() + ""));
                    param.add(new BasicNameValuePair("foodPrice" + i,orderFoods.get(i).
getFoodPrice() + ""));
                    param.add(new BasicNameValuePair("sumPrices" + i,orderFoods.get(i).
getSumPrices() + ""));
                    }
                  try {
```

```
                req.setEntity(new UrlEncodedFormEntity(param,HTTP.UTF_8 ));
                try {
                    HttpResponse res = new DefaultHttpClient().execute(req);
                    if(res.getStatusLine().getStatusCode() == 200){
                        showDialog(1);
                    }
                } catch (ClientProtocolException e) {
                    e.printStackTrace();
                } catch (IOException e) {
                    e.printStackTrace();
                }
            } catch (UnsupportedEncodingException e) {
                e.printStackTrace();
            }
        }
    });
    bcancel.setOnClickListener(new OnClickListener() {//取消按钮被按下
        public void onClick(View v) {
            db.close();
            Intent intent = new Intent(OrderListActivity.this,ShopCartShowAll.class);
            startActivity(intent);
            finish();

        }
    });
}
protected Dialog onCreateDialog(int id) {
    super.onCreateDialog(id);
    AlertDialog.Builder builder = new AlertDialog.Builder(this);
    switch (id) {
    case 1:
        builder.setTitle("已提交至服务器");
        builder.setPositiveButton("确定", new DialogInterface.OnClickListener() {

          public void onClick(DialogInterface dialog, int which) {
                Intent intent = new Intent(OrderListActivity.this,ShopCart-
ShowAll.class);
              startActivity(intent);
              finish();
          }
      });
      break;
  case 2:
```

```
            builder.setTitle("请检查购物车,输入的地址,电话是否为空!");
            builder.setPositiveButton("确定", new DialogInterface.OnClickListener() {
                public void onClick(DialogInterface dialog, int which) {
                    Intent intent = new Intent(OrderListActivity.this,ShopCartShowAll.class);
                    startActivity(intent);
                }
                });
                break;
            }
            return builder.create();
        }
        protected void onResume() {
          SharedPreferences share = this.getSharedPreferences("android",MODE_WORLD_
READABLE);
            userId = share.getString("userId", "");
            userName = share.getString("userName", "");;
             super.onResume();
        }
    }
```

当用户单击下单界面上的“提交清单”请求的时候,会将订单信息全部提交给 AddOrderServlet 控制器去处理。

除了这些功能以外,此电子订餐系统当中还有一些其他功能:例如清空购物车,继续购物这类功能,笔者在这里就不一一描述了。

参考文献

[1] (印)克曼特内尼，(美)麦克莱恩，(美)哈希米. 精通 Android 3. 杨越，译. 北京：人民邮电出版社，2011.

[2] (法)Hervé Guihot. Android 应用性能优化. 白龙，译. 北京：人民邮电出版社，2012.

[3] (美)李伟梦　著. Android 4 编程入门经典. 何晨光，李洪刚，译. 北京：清华大学出版社，2012.

[4] (美)史密斯(Smith，D.)，(加)弗里森(Friesen，J.). Android 攻略. 陈钢，译. 北京：人民邮电出版社，2012.

[5] (美)梅德尼克斯. Android 程序设计(影印版). 南京：东南大学出版社，2011.

[6] 李兴华. 名师讲坛—Android 开发实战经典. 北京：清华大学出版社，2012.

[7] 吴亚峰，苏亚光. Android 应用案例开发大全. 北京：人民邮电出版社，2011.

[8] 孙更新. Android 从入门到精通. 北京：电子工业出版社，2011.

[9] 李宁. Android 应用开发实战. 北京：机械工业出版社，2012.

[10] (英)梅尔. Android2 高级编程(第 2 版). 王超，译. 北京：清华大学出版社，2010.

[11] 韩超. Android 经典应用程序开发. 北京：电子工业出版社，2012.